STATISTICAL INFERENCE
FOR MANAGEMENT
AND ECONOMICS

David V. Huntsberger, D. James Croft, Patrick Billingsley

Iowa State University, University of Utah, The University of Chicago

2nd Edition

STATISTICAL INFERENCE FOR MANAGEMENT AND ECONOMICS

ALLYN AND BACON, INC.

Boston, London, Sydney, Toronto

To four unique and uncommonly talented teachers:

J. Hazel Whitcomb, Myrtle Austin, Harold Bacon, and Robert Machol

Production editor: Barbara Willette

Preparation buyer: Linda Card

Series editor: Richard Carle

Portions of this book first appeared in ELEMENTS OF STATISTICAL INFERENCE, Third Edition, by David V. Huntsberger and Patrick Billingsley, Copyright © 1973, 1967, 1961 by Allyn and Bacon, Inc.

Library of Congress Cataloging in Publication Data

Huntsberger, David V
  Statistical inference for management and economics.

  Includes index.
  1. Statistics. I. Croft, David James,
1942–   joint author. II. Billingsley,
Patrick, joint author. III. Title.
HA29.H855 1980   519.5'4   79-20995
ISBN 0-205-06803-0

Printed in the United States of America.
10 9 8 7 6 5 4   85

CONTENTS

■The material in sections marked with a square is optional.

■The material in sections marked with a square is optional.

SAMPLE DATA SET APPENDIX A-1

ANSWERS TO ODD-NUMBERED PROBLEMS A-27
INDEX A-39

PREFACE

THE INTENDED AUDIENCE

Statistical Inference for Management and Economics, Second Edition is intended to be used in a brief (one or two quarters or one semester) course designed to meet the needs of business and economics students. There are, however, sufficient materials and problems for a longer course.

Major goals in this revision have been to keep the presentation at an introductory level and to illustrate clearly the applications of statistics in business and economic decision making.

SPECIAL FEATURES

This book contains a number of special features that are found, individually, in only a few statistics books.

Most of the chapters have a summary section in which the important formulas and concepts are presented in tabular form. In these summaries, important assumptions and cautionary notes are listed together with the formulas so that readers will know the appropriate conditions under which the formulas can be applied.

Eighteen cases have been added to the book. Most of these cases are short—some might call them "extended problems." They vary in difficulty, and most of them can be used in connection with more than one chapter. Many cases contain data obtained from actual business situations by one of the authors. The cases emphasize the use of a numerical solution within the context of a dynamic environment and help illustrate the fact that a numerical solution is only a small part of a realistic statistical problem.

A data set has been included as an appendix to the book. It contains information concerning 113 credit applicants. Nine variables are recorded for each of the applicants, and the applicants are divided into two groups: those who were granted credit and those who were denied credit. The data are real and were taken from an anonymous organization. Each chapter contains questions that refer to the Sample Data Set Appendix. These questions serve to tie the chapters together and to demonstrate how numerous statistical techniques can be used to extract information from one data set.

CHANGES FROM THE FIRST EDITION

Many changes have been made in this second edition which strengthen the book:

1. The level of the problems at the end of each chapter has been made more consistent with the level of the material presented in the chapter.
2. Topic coverage has been widened. The hypergeometric and Poisson probability distributions have been introduced formally in the probability chapters rather than in the problem sets. An entirely new chapter on multiple regression has been added. This chapter contains a unique section explaining in detail the meaning of each number that standard computerized multiple regression packages print out. The use of multiple regression in time series has been introduced. The chapter on index numbers presents a discussion of the two new consumer price indices published by the Bureau of Labor Statistics since 1978, the CPI-U and the CPI-W.
3. The special features discussed in the previous section of this preface have all been added since the first edition.
4. The analysis-of-variance chapter has been placed prior to the

regression material in this edition. The regression chapters now have sections that discuss the relationship between regression and analysis-of-variance results.

5. Clarifying material has been added to each chapter in places where readers of the first edition have suggested that further explanation would be helpful. Two specific examples of clarifications are found in the hypothesis-testing chapters, where the statements of null and alternative hypotheses have been made more explicit, and in the simple regression chapter, which has been expanded with new example material illustrating the concepts more clearly.

6. Roughly eighty percent of the problems in this edition are new or are problems from the first edition with new data inserted.

ORGANIZATION

This book has four major sections. The first section (Chapters 1–3) comprises an introduction to the nature of statistical problems and descriptive statistical procedures. The only required mathematical background that might be new to readers is the summation notation, which is handled in the first section of Chapter 3.

The second section of the book covers the subject of probability (Chapters 4–7). Chapter 6 is a unique chapter demonstrating the relationship between the binomial, hypergeometric, Poisson, and normal probability distributions. Chapter 7 presents the central limit theorem and its extensions.

The third section is the heart of the book. Statistical inference is covered in Chapters 8–11 and Chapter 18. Methods for estimating population parameters are discussed in Chapter 8, and various methods of testing hypotheses are presented in Chapters 9–11. Some instructors may choose to omit the more advanced topics of analysis of variance and uses of the chi-square distribution in Chapters 10 and 11. Nonparametric methods are covered in Chapter 18. While these methods are traditionally classified as hypothesis testing methods, they are placed at the end of the book owing to the optional nature of this material.

Chapters 12–17 form the fourth section of the book. These chapters present a collection of highly useful statistical techniques: simple and multiple regression, time series analysis and index numbers, and decision theory.

Instructors will ordinarily take up Chapters 1–9 in sequence, but some may wish to omit Chapter 6, which shows the relationships between several probability distributions. Material from Chapters 10–18 can be selected in a variety of ways depending on the length of the course and interests of the class. Titles of some sections are preceded by a square symbol indicating that these sections are more difficult or are not required in subsequent chapters.

ACKNOWLEDGMENTS

The authors are indebted to the Literary Executor of the late Sir Ronald A. Fisher, F.R.S., to Dr. Frank Yates, F.R.S., and to Longman Group Ltd., London, for permission to reprint Table VI from their book *Statistical Tables for Biological, Agricultural and Medical Research;* Professor George W. Snedecor and the Iowa State University Press for permission to reproduce Table VII; and Professor E. S. Pearson and the Biometrika Trustees for permission to use the material presented in Tables VIII and IX.

The authors have benefited from comments made by reviewers. Especially helpful were: Thomas Yancey, University of Illinois at Urbana; Edward Mansfield, University of Alabama; Rebecca Klemm, Temple University; Gordon H. Otto, the University of Houston; Belva Cooley, Virginia Polytechnic Institute and State University; Neil Polhemus, the University of North Carolina at Chapel Hill; James Sullivan, Bowling Green State University; Carol Pfrommer, the University of Alabama; Chipei Tseng, Northern Illinois University; and John E. Hanke, Eastern Washington University.

In addition, the authors are jointly and separately indebted to many other people in diverse places for their contributions to this book.

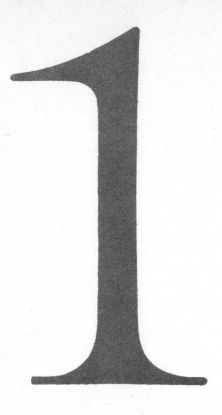

INTRODUCTION

1.1 WHAT IS STATISTICS?

The word *statistics* often conjures up images of numbers piled upon numbers in vast and forbidding arrays and tables, of volume after volume of figures pertaining to births, deaths, taxes, populations, incomes, debits, credits, and the like, ad infinitum. This is so because in common usage the word *statistics* is synonymous with *data,* as, for example, when we speak of the statistics of a football game, the statistics of an election, or the statistics of highway accidents. This conception of the word does not correspond to the discipline that carries the name statistics, nor does it give a clue to the activities of present-day statisticians, who are no longer to be defined as collectors and tabulators of numerical facts.

In the modern sense of the word, statistics is concerned with the development and application of methods and techniques for collecting, analyzing, and interpreting quantitative data in such a way that the reliability of conclusions based on the data may be evaluated objectively by means of probability statements. It is this use of mathematical probability to evaluate the reliability of conclusions and inferences based on data that is the unique and major contribution of the statistical approach to inductive decision making. For this reason, the mathematical theory of probability plays a fundamental role in the theory and application of statistics.

While statistical methods can be used in many fields of work and research, the following chapters will concentrate on their use in decision making by managers in profit and nonprofit organizations. Most managers are very interested in avoiding disastrous decisions which might cost them their jobs, lose customers or clients, or deny the public a needed product or service. Decision makers are interested in statistical methods because these methods can help them avoid making those inept decisions and can help them make intelligent, reliable decisions. Thus they are most interested in the *applications* of statistics to their particular problems, but they must also have a good foundation in the *theory* of probability and statistics in order to use these applications wisely.

Statistical activities are usually classified into the two major areas mentioned previously: theoretical or mathematical statistics, and applied statistics. Theoretical statistics is concerned with the mathematical elements of the subject—with lemmas, theorems, and proofs—and in general with the mathematical foundations of statistical methodology. We look to the mathematical statistician for the development of new theories that will provide new methods with which to attack practical problems. But to the applied statistician and the decision maker, statistics is a means to an end. They face problems and select from the available statistical methods those best fitted to the job at hand. The applied statistician may be asked by the decision maker to participate in designing a sample survey or experiment, or may be consulted about sampling inspection schemes for statistical quality control. Statistical work directly involves the statistician in the subject area whence the problem arose—accounting and auditing, production control, financial management, or whatever it may be. Frequently the statistician is presented with a set of data and asked to provide an analysis and interpretation. This is often a difficult task, since the methods of analysis must depend on the way the data were collected. If the statistician was not consulted on the methods used in obtaining the data, he or she is too often obliged to say that, owing to the way in which the figures were obtained, they are inadequate, or of the wrong kind, to give the desired information.

In the main, statisticians are consultants to administrators, executives, and research workers in business, in government, in academic institutions, and in industry. Although we have made a distinction between theoretical and applied statistics, the majority of statisticians engage in both.

1.2 TYPES OF PROBLEMS IN STATISTICS

Statistical methods can be divided into four areas, depending on the type of problems they are used to solve. The first area is *descriptive statistics*. Methods of organizing, summarizing, and presenting numerical data fall into this area of statistics. Chapters 2 and 3 cover the topics of descriptive statistics. Chapter 2 deals with pictorial and graphic summary of data, and Chapter 3 deals with numerical summaries.

The second area of statistical study is *probability*. Probability problems arise when a statistician takes a sample from a large body of items, called a *population,* and wishes to make statements about the likelihood of the sample's having certain characteristics. A typical probability problem is stated next:

A company knows that 40 percent of its customers are men and 60 percent are women. What is the probability that in a sample of ten customers, half will be men and half will be women?

Note that in this problem statement the population consists of all the company's customers. Also note that the male/female breakdown in the population is given. The sample is the ten customers selected at random, and the question asked concerns what the sample will look like. Chapters 4 through 7 deal with statistical methods that can be used to solve problems of this type—probability problems.

The third area of statistics is called *statistical inference*. Problems involving statistical inference arise when a statistician takes a sample from a population and wishes to make statements about the population's characteristics based on the information contained in the sample. A typical inference problem follows:

The manager of employee benefits in the personnel department of a large corporation questioned 100 people selected from a work force of 7000 concerning their opinions of a proposed change in the company's medical insurance plan. The medical insurance salesperson claimed, "At least 80 percent of your work force will favor this change." However, the manager's sample of 100 showed that only 70 people in the sample favored the change. Is it possible that the insurance salesperson's claim is true? If not, what proportion of the total work force favors the change?

In this problem the company's work force is the population, and the

100 people whose opinions were recorded form the sample. Here, however, we are given the sample results and ask questions about what the population looks like. This is directly opposite to the situation presented in probability problems. In a probability problem situation, the problem solver knows what the population looks like and raises questions about sample results that are likely to occur. In statistical inference situations, the problem solver knows what the sample result looks like and raises questions about the population from which the sample came. Statistical methods which can be used to solve inference problems are presented in Chapters 7 through 11.

The fourth area of statistics covered in this text is difficult to label. We might call it *miscellaneous*. Methods covered under this topic area are used to solve a wide range of statistical problems—from economic forecasting to deciding how large a production run a manager should order. Some statistical experts may not agree entirely, but the methods covered in Chapters 12 through 18 can be classified into this miscellaneous category.

The four areas of statistics described previously build on one another. For instance, *probability* methods would be difficult to understand for someone who did not know *descriptive statistics*. Also, problems in *statistical inference* cannot be solved by someone who does not understand probability. And the tools employed to solve *miscellaneous* types of statistical problems often use the methods of inference.

It is not the purpose of this book to make statisticians of its readers. The objective is to give them an understanding of the fundamental principles of the subject and enough of the language of statistics that they may meet with statisticians on common ground. They will learn enough about statistical techniques to be able to handle many common types of problems themselves. They will also learn to recognize those more complex problems that require consultation with a professional statistician.

1.3 MISUSES OF STATISTICS

Another of the objectives of this book is worthy of special mention. When readers finish studying the statistical methods covered in this book, they should have a good feeling for the difference between the appropriate and inappropriate uses of statistics. This book concentrates, naturally, on the appropriate uses. However, some of the examples, problems, and cases will point out potential statistical abuses of which the reader should be aware.

The misuses of statistics are so varied and broad that entire books have been written on them. One excellent example is Darrell Huff's delightful book entitled *How to Lie with Statistics,* which is referenced at the end of this chapter. A few of the more common problems that lead to misused statistics and incorrect conclusions are discussed next. For the most part, these problems involve the way in which data are gathered and used in some sort of statistical analysis. Even the most careful applications of the statistical methods presented in this book are useless if some of the problems cited here are not avoided.

One of the simplest data problems to overcome in a statistical analysis is the *failure to adjust gross data to a per-item basis.* For instance, a company president might report that the company made $30 million last year and $40 million this year. On the surface this looks like a 33 percent improvement in the company's performance. However, if the company had 15 million shares of stock outstanding last year and 20 million shares outstanding this year, then the per-share performance is unchanged since:

$$\text{Earnings per share last year} = \$30 \text{ million}/15 \text{ million shares}$$
$$= \$2.00/\text{share}$$

$$\text{Earnings per share this year} = \$40 \text{ million}/20 \text{ million shares}$$
$$= \$2.00/\text{share}$$

From this example we can see that a failure to adjust data to per-item figures can result in a misleading impression. For this reason, much of the financial and economic data reported in private and government publications are presented on a per-capita, per-share, per-household, or per-transaction basis. When one sees gross figures reported, one should always ask: Are these figures relevant as they stand or should they be presented on a per-something basis?

Another misleading impression can be produced if the data used in a statistical analysis are *not adjusted for inflation.* The high levels of inflation that economies all over the world have experienced in recent years make adjustments of this type very necessary if correct conclusions are to be drawn from financial and economic figures. For instance, a plant manager may proudly report that the average wages of workers in the plant have risen 60 percent in the past ten years. However, if the general price level of goods purchased by these workers has risen 80 percent during that same period, they are worse off in terms of what their wages will purchase than they were ten years ago.

Consider another example. Let us suppose that a company had a before-tax profit this year of $200 thousand. Further, let us assume that the company has $2 million worth of equipment it purchased 20 years ago and which it is depreciating using the straight-line method over a period of 20 years. The following calculations show that the total of before-tax profits and depreciation is $300 thousand:

Yearly depreciation charge = \$2,000,000/20 years = \$100,000
This year's profits before taxes   = $\underline{\$200,000}$

$\hspace{8cm}$ *Total*   \$300,000

The problem we have here is that the \$100 thousand in depreciation charges is based on the actual cost of assets purchased 20 years ago. Since the depreciation period used was 20 years, we can reasonably assume that the assets are nearing the end of their useful life and will have to be replaced. The high inflation rates of the past 20 years will have raised the cost of replacing these assets. Let us assume, for instance, that assets equivalent to those that must be replaced would cost \$6 million in today's market. If these assets are also depreciated over 20 years on a straight-line basis, the depreciation charge is:

$$\$6,000,000/20 \text{ years} = \$300,000/\text{year}$$

which is equal to this year's depreciation charge plus the *entire profits* for this year. That is, if the firm were to report its depreciation charges, not on the basis of historical costs, but on current replacement costs, it would show no profit at all for this year.* Thus depreciation figures based on historical costs, which are not adjusted for inflation, create an impression that a company is more profitable than it really is in light of today's costs. This problem is so serious that the Securities and Exchange Commission now requires many large companies to report the impact that depreciation of assets costed at replacement values would have on their earnings figures. Methods for making these adjustments are discussed in Chapter 15, which deals with index numbers.

Another problem that can creep into statistical analysis is that of *induced bias*. When the data used in a statistical study are gathered through interviews with people, the people doing the study must take care that the interviewers do not in any way convey the responses they wish or expect to obtain. The average person is really rather nice. If this person is being interviewed, he or she may "read" the interviewer in an attempt to find out what the interviewer wants or likes to hear and adjust his/her responses accordingly. This, of course, produces a bias in the responses toward what the interviewer wanted to hear in the first place. Trained interviewers are very good at hiding their own opinions and expectations during an interview. Sometimes, however, no amount of training can eliminate an induced bias. For instance, if a black person interviews people about their attitude toward

---

*This example is greatly oversimplified. Also, it does not take into account the fact that new equipment might be much more efficient and allow the company to produce at lower cost.

minority groups, that interviewer will likely get somewhat different responses than would a white interviewer.

Often statistical methods are applied to data that have been gathered from questionnaires. Such data obviously have no interviewer-induced bias. But the way in which questions are asked can induce bias in the results. Consider the example of a union official who wishes to know how members of his union feel about a wage settlement recently offered to a group of employees. He might ask the following question in a survey of their opinion:

> Should we accept or reject management's latest offer of a $1.05 per hour increase over the next three years?

His responses to that question might be rather different from those he would receive if he asked his question in this manner:

> Should we accept or turn down flat our tight-fisted management's skimpy offer of a $1.05 per hour increase over the next three years in light of the fact that the guys over at Universal Industries held out and got $1.30 per hour over the same period?

Methods of avoiding induced bias during interviews and in question-naires will not be discussed in this book. But a little common sense and judgment will go a long way in eliminating this problem for those who are aware of it in advance.

A very serious problem which is, unfortunately, rather common involves *inappropriate comparisons of groups*. Statistical studies often involve the comparison of two or more groups with one another. Banks versus savings and loans, purchasers versus nonpurchasers of your produce, married versus nonmarried credit applicants, and western versus midwestern versus eastern versus southern markets are examples of comparisons that are common. There are several reasons why it may be inappropriate to compare two or more groups. We will discuss only two of these: *self-selection* and *hidden differences*.

Let us first deal with the problem of self-selection as it applies to the comparison of two different groups. Consider the situation in which a large, national corporation offered a course in "Selling to Nonprofit Organizations" to any of its salespeople who are interested in enrolling. Approximately 35 of the company's 160 salespeople signed up and took the course. In an effort to show there had been an impact on the sales effort, the training director of the corporation reported to the vice-president of personnel the following comparison:

> In the six months following the training course offered by this office, the people who took the course showed an average sales to nonprofit organizations of $1298 per salesperson per month. In the group that did not take the course, this average was only $706 per salesperson per

month. More customer-oriented sales training programs of this type would be of obvious benefit.

The training director correctly adjusted the sales figures to a per-salesperson and per-month basis. However, the problem here is the self-selection of the training program's participants. Perhaps only those salespeople who had large nonprofit organizations in their territories signed up in the first place. Or perhaps the people who took the course already had substantial nonprofit organizations as accounts and merely wanted the training to see if they could increase the already-high sales level they had to these accounts. A better way to measure the impact of the training course would be to measure the sales level of the participants to nonprofit organizations before the training and then measure it after the training. Care would be needed to make sure that comparable sales periods were measured both before and after the training and that any effects of price changes between the measured periods (inflation) were taken into account.

The problems involved with comparing groups with hidden differences are similar to those involved with comparing groups that have self-selection. In fact, some of the self-selection situations, such as the one described previously, produce hidden differences in the groups being compared—such as the suggested differences in the sales potential to nonprofit organizations in the groups mentioned. However, consider a less obvious problem of comparing two groups that have hidden differences. Not long ago a state official bragged that the climate in his state was obviously healthier than that found in most states since the death rate for the population (expressed in deaths per 100,000 residents) was one of the lowest in the nation. On the surface, again, this sounds like a convincing argument for this state's climate. However, a closer examination of the population living in this state revealed that a higher-than-average proportion of the people in the state were young people, and a much lower-than-average proportion were old. Thus, one would expect a group of people with this type of age distribution to have a lower-than-average death rate, regardless of the climate!

The examples of misusing statistical data cited previously are only a very few that could be mentioned. Other examples of failure to convert data to a per-something basis, failure to adjust for inflation, induced bias, self-selection, and hidden differences will be presented for the readers to discover in the problems following the chapters and in the cases at the end of the text.

1.4 SOURCES OF STATISTICAL DATA

In the preceding example, a state official said that his state had one of the lowest death rates found in the nation. Where, one might ask, would he find information like this? Where would someone who wanted to refute his argument find information on the age distribution of people living in the various states? Someone interested in the answers to these questions would likely check with the United States Bureau of the Census, which keeps records on these matters.

In fact, federal, state, and local government agencies are the largest gatherers and publishers of data in the country. Business and trade organizations also gather and publish a great deal of data concerning their members' activities. And some private data gathering firms are useful sources of information in specialized areas. The difficult problem for the businessperson or economist, however, is trying to find out where the information that he or she is interested in has been published.

One of the most helpful sources of statistical information is the *Statistical Abstract of the United States,* which is published annually by the Bureau of the Census. This publication is subtitled "National Data Book and Guide to Sources." It contains close to 1500 tables of statistics describing subjects including population, education, banking, energy, manufacturing, transportation, and many others. Although the emphasis in these tables is on national statistics, there are tables containing data for regions, states, and some metropolitan areas. The information presented in the *Statistical Abstract* is usually summarized from other detailed sources and publications. The publications in which the more detailed information can be found are usually cited in the tables.

One of the most useful features of the *Statistical Abstract* is its Appendix IV, "Guide to Sources of Statistics." This appendix lists the names of publications that print statistics on a wide variety of subjects. The list is arranged alphabetically by the subject matter for which the information is sought. A researcher who is interested in finding out what data have been published on industrial accidents, say, can look under that subject category in the "Guide to Sources of Statistics" section and find that several government agencies and the National Safety Council publish reports on these accidents. Many states publish their own statistical abstracts, and the names of these publications and the agencies that distribute them are listed in the "Guide to Sources of Statistics" also.

Another government publication that is a very useful source of business and economic data is the *Survey of Current Business,* which is published monthly by the Bureau of Economic Analysis in the U.S. Department of Commerce. Each issue of the *Survey* contains articles on the current economic situation, the future economic outlook, and related topics. In addition, the *Survey* includes a statistical supplement entitled "Current Business Statistics." This supplement, which runs more than forty pages in length, contains monthly and quarterly data on thousands of business measures. For each major industry in the U.S. economy sales, cost, and inventory figures are given for the past year.

Also, indices of activities in these industries are presented. The amount of data available from this supplement is almost too vast to describe. Each issue also carries roughly thirty tables of national economic statistics not broken down by industry.

A number of trade organizations publish summaries of statistics covering activities in their industries. The *Encyclopedia of Associations* is available in most libraries, and it lists organizations and associations in business, health, cultural, athletic, and many other areas of interest. It also shows the publications that are distributed by these organizations, how often they are printed, and their general contents. Someone interested in accounting organizations, for instance, would find that there are dozens of them listed in the *Encyclopedia of Associations*. A person desiring information about a particular industry would be wise to begin by finding out from this source what associations are primarily interested in this industry and what information they publish. This valuable reference book can save a researcher hours of telephoning and mountains of correspondence in finding out who has data for the industry.

PROBLEMS: *Answers to odd-numbered problems are found in the back of the text.*

1. If you were assigned by your boss to examine each of last week's sales transactions, find their average, the difference between the highest and lowest sales figure, and construct a chart showing the differences between charge account and cash customers, what type of statistical problem would this be: descriptive statistics, probability, or inference?

2. Look at the problem statement that follows and determine if it is a probability problem or a statistical inference problem.

   > A quality control engineer took a sample of ten items from a production line and found that 20% of them were defective. What is the proportion of defectives in the population of all items coming off this production line?

3. Look at the next problem statement and determine if it is a probability problem or a statistical inference problem.

   > The manager of a hospital laboratory knows that 5% of the lab's tests have to be rerun. If her subordinates perform ten tests in one morning, what is the chance she will have to run none of them again?

4. When we are dealing with problems in probability and statistical inference, which is usually the larger value in the problem, the population size or the sample size? Why?

5. Which type of statistical problem:

    a. Assumes you know what the population characteristics are?

    b. Assumes you know what the sample characteristics are?

6. When a firm does a test market for a product, are they involved in a probability problem or a statistical inference problem? Why?

7. In the last few years there has been an effort in many corporations to cut energy costs. The vice-president of operations for a regional trucking company was recently shocked to learn that, despite the company's efforts to increase their fleet's fuel economy through improved maintenance and purchase of more fuel-efficient trucks the fuel costs for this year are running 12% ahead of those for last year.

    a. Can you suggest *two* possible explanations for the increase?

    b. Can you suggest a way to measure fuel usage that would more accurately reflect the company's success or failure in reducing fuel usage?

8. Enrollments in many large universities have remained stable in terms of numbers of students taking classes each quarter. However, many students are taking fewer classes due to their having to work to support themselves and to cover the high cost of their educations. Rather than expressing their costs in dollars per student enrolled each term, how should university administrators measure the cost of providing education so that the lighter student loads do not distort their figures?

9. During the past ten years, a manager's income has doubled. Give two reasons why the purchasing power of his income has not doubled.

10. The manufacturers of a certain health food supplement showed in their advertising that a study of people who used their product had a lower incidence of heart attack than that found in the population as a whole. Why might such advertising not be justified?

11. Two islands are located off the East Coast of the United States. Both islands have 1000 residents. The first island, however, has 800 people whose ages are in the twenties and 200 people in their sixties. The second island has the opposite situation—800 people in their sixties and 200 in their twenties.

    a. Which island would likely have the higher birthrate?

    b. Which island would likely have the higher death rate?

    c. If you were the proprietor of the general store located on the second island, what kinds of products might you stock differently than if you were the proprietor of the store on the first island?

12. Consult the *Statistical Abstract of the United States* to determine which three states have the lowest death rates. Also use this source to determine any likely explanation of these states' low death rates from the distribution of people's ages in these states. (*Hint:* What proportion of each of the states' populations is over 65 years old? Compare this with the national average.)

13. Consult the *Statistical Abstract of the United States* and find the table that lists numbers of physicians, dentists, and nurses in the population of each state.

    a. Which area listed has the highest number of physicians per capita?

    b. Why would this location have such a high rate as compared to the others listed?

14. No one knows for sure how many heroin addicts there are in this country. However, the National Institute of Drug Abuse recently released a study in which they guessed that there were well over 500 thousand. Researchers who conducted the study used 24 metropolitan areas and estimated the following number of addicts in each of these five cities: New York, 69,000; Los Angeles, 60,000; Chicago, 47,700; Detroit, 33,200; and San Francisco, 28,600. If the head of the narcotics division of the New York City police force used these figures to argue that a higher proportion of the police department budget should go to this division, since New York has the worst problem in heroin addiction, would there be justification for this? (*Hint:* Use the *Statistical Abstract of the United States* population figures for these cities to adjust them to a per-capita basis.)

15. The National Anti-Union Committee recently sent a survey to people across the nation asking their opinions on proposals to allow unionization of the armed forces. One of the questions was:

    > Do you want to have union bosses exercising tyrannical control over our troops when we are under attack by hordes of ruthless soldiers from Russia or Mainland China? Yes ____ No ____

    a. Name the specific statistical problem that this question would cause.
    b. Change the statement of the question so that the essence of the question remains the same but the emotion is removed.

16. Read the case at the end of the text entitled "The New England Tire Dealers Association." Discuss the problems involved in comparing the two groups mentioned in this case. Are the groups comparable? Is self-selection a problem here? Might there be hidden differences in the groups? If so, which ones? If not, why not?

17. Read the case entitled "Statistics 101" and comment on problems that are associated with comparing the two groups mentioned there.

18. Read the case entitled "Mountain States Feed Exchange" and discuss the hidden differences that could exist in the groups of cattle that are compared in that case.

19. Consult the most recent issue of the *Survey of Current Business* to determine:
    a. The gross national product in current dollars for the fourth quarter of last year.
    b. Final sales of automobiles in current dollars for the latest quarter reported.

20. Consult the "Current Business Statistics" supplement in the most recent *Survey of Current Business* to find the following:
    a. U.S. exports of goods and services (excluding military grant transfers) in millions of dollars.
    b. Estimated total civilian labor force for the latest month on which data are available.

21. Consult the *Encyclopedia of Associations* to determine the location and founding date of the American Women's Society of CPA's.

22. Consult the *Encyclopedia of Associations* and determine the following information about the National Retail Merchants Association:

    a. When was it founded and where are its central offices?

    b. How many members does it currently have?

    c. How many publications does it distribute?

SAMPLE DATA SET QUESTIONS:
Refer to the 113 applicants for credit listed in the Sample Data Set Appendix of this book.

    a. What types of applicants seem to have the value 99 listed for the variable JOBYRS?

    b. Does it appear logical to you that a blank entry (indicated by 9999) and a 0 entry for the variable SPINC mean the same thing?

REFERENCES

1.1 Campbell, Stephen K. *Flaws and Fallacies in Statistical Thinking.* Prentice-Hall, Englewood Cliffs, 1974.

1.2 Fisk, Margaret, editor. *Encyclopedia of Associations,* 11th edition. Gale Research Company, Detroit, 1977.

1.3 Huff, Darrell. *How to Lie with Statistics.* Norton, New York, 1965.

1.4 U.S. Bureau of the Census. *Statistical Abstract of the United States,* 97th edition. Washington, D.C., 1976.

EMPIRICAL FREQUENCY DISTRIBUTIONS

2.1 FREQUENCY DISTRIBUTIONS

Statistical data obtained by means of census, sample survey, or experiment usually consist of raw, unorganized sets of numerical values. Before these data can be used as a basis for inferences about the phenomenon under investigation or as a basis for decision, they must be summarized and the pertinent information must be extracted. The kind of information required naturally depends on the application. If we were investigating the reliability of electronic components, it might suffice for our purposes simply to observe what percentage of the components tested failed to operate satisfactorily for longer than 1000 hours. If, however, we were interested in improving these components, we would require information concerning the reasons for the observed failures.

One type of information that may be desired of a set of data relates to the pattern or grouping into which the data fall. As a first step in developing this type of information, we could construct an *array*; that is, we could order the numerical values from low to high (or high to low). An array may help make the overall pattern of the data apparent. However, if the number of values is large, construction of the array may have to be done on a computer, and even then the array may turn out to be so large that it is difficult to comprehend.

A more useful way to summarize a set of data is to construct a *frequency table* or *frequency distribution*. That is, we divide the overall range of the values in our set of data into a number of *classes*. A class is thus an interval of values within the overall range the data span. Then we count the number of observations that fall into each of these classes. By looking at such a table or distribution we can readily see the overall pattern of the data.

To illustrate the construction of a frequency distribution, we can use the 106 numbers shown in Table 2.1. These values represent the number of employees absent from work at a large manufacturing plant over the last 106 working days.

Before we can construct our frequency distribution, we must decide how many classes to use. This is to a certain extent arbitrary, but it is common to use from five to twenty classes, with more classes for larger sets of data. Suppose we decide to use ten classes in this example. Note that this was just an arbitrary selection. There is nothing magic about using ten classes. (In fact, after we present this example, we will present a rule for selecting the number of classes to use, and this rule will take some of the guesswork out of this decision.)

Having decided to use ten classes, we now specify the *class intervals*

see p. 18

TABLE 2.1 Absences from Work Last 106 Days

146	141	139	140	145	141	142	131	142	140
144	140	138	139	147	139	141	137	141	132
140	140	141	143	134	146	134	142	133	149
140	143	143	149	136	141	143	143	141	140
138	136	138	144	136	145	143	137	142	146
140	148	140	140	139	139	144	138	146	153
148	142	133	140	141	145	148	139	136	141
140	139	158	135	132	148	142	145	145	121
129	143	148	138	149	146	141	142	144	137
153	148	144	138	150	148	138	145	145	142
143	143	148	141	145	141				

(sizes or widths of the classes) and the *class boundaries.* Referring to Table 2.1, we note that the smallest observation is 121 and the greatest is 158. This means that our class intervals must cover the range from 121 to 158; that is, the combined length of the classes must be at least 158 − 121, or 37. Dividing 37 by 10, the number of classes, gives a class interval of 3.7. It would be very inconvenient to use a class interval that is not a whole-number multiple of the basic unit (an absence in this case), so we use an interval of 4. Thus we can use exactly ten classes, each of which is 4 units wide. Notice that all ten of the classes will be 4 units wide. In general, frequency distributions have classes that are all of the same width. Of course, use of equal class intervals is not absolutely necessary, and in the case of very lopsided distributions unequal intervals sometimes give a clearer picture of the situation. But equal intervals result in simpler calculations, and it is therefore best to use them whenever possible.

Since we are using ten classes of width 4 absences, the classes of our frequency distribution will cover $10 \cdot 4 = 40$ absences. But the data range from 121 to 158 absences, so the data cover only 37 absences. This raises the following question: What do we do with the 3 absences where the frequency distribution's classes overlap the data? The answer to this question is that we try to split the overlap approximately in half. This means that we will start the first class in the distribution *below* the first data value. Since our data are integers, we will split the overlap by integers and arbitrarily start the first class 1 unit (1 is *about* half of 3) below the first data value. This will automatically cause the last class to end with a boundary *above* the top data value.

This problem of having the classes overlap the actual data demonstrates why we rounded the class width up from 3.7 to 4. If we had rounded down to the next lower integer, 3, the ten classes would have covered only $10 \cdot 3 = 30$ absences and the data would have covered 37 absences. Thus, the data would not have fit into the ten classes. As a rule, then, once we determine the approximate width of our classes, we *always round up to get our convenient class width.* Unfortunately, there is no rule that tells us how far we should round up. In some cases it is convenient to round up to the next highest integer, but in others we may want to round up to the next multiple of .5 or multiple of 2. The precision with which the data are measured helps determine this decision.

To this point in our example we have determined that we will use ten classes of width 4 absences each and that, due to the 3 absences overlap, we will begin our first class 1 unit below the smallest piece of data. Thus our first class will begin at the value of 120, which is 1 unit below the smallest data value of 121. To prevent ambiguity, we select our class boundaries in such a way that there can be no question about which class a given observation belongs to. Suppose we make the first class' boundaries 120 and 124 (which gives an interval of width 4) and the next class' boundaries 124 and 128. Then an observation of 124 is impossible to classify; we do not know which of the two adjacent classes to put it in. The simplest solution is to select *impossible values* for the boundaries. In the present instance, if a class has boundaries 119.5 and 123.5 (which gives an interval of width 4) and the next has boundaries 123.5 and 127.5, there can be no ambiguity, since all the values are *counts* and are thus recorded as integers and no values such as 119.5, 123.5, or 127.5 will appear in the data.

With these preliminary considerations taken care of, we can now construct the frequency distribution. The result is presented in Table 2.2. The tally shown in the central portion of the table is for convenience only and would not be displayed in the completed distribution. Notice that each class in this frequency distribution is specified by the greatest and least values that can be attained by a member of the class; these two numbers are called the *class limits*. Thus, a class boundary is midway between the upper limit of the class preceding it and the lower limit of the class following it (see Table 2.3). Note that in Table 2.2 the class limit of the last class is 159. This is 1 unit above the highest data value of 158. The lower limit of the first class is 120, which is 1 unit below the lowest data value, 121. In Table 2.3 the class boundaries of these two classes are each 1.5 units away from the extreme values in the data. Thus in this example, the overlap of 3 absences ended up symmetrically distributed at each end of the frequency distribution. This happy circumstance is not always the case. Usually the overlap is split so that *approximately* half end up in the first and half in the last classes.

TABLE 2.2 Distribution of Absences in 106 Days

NUMBER OF ABSENCES	TALLY	NUMBER OF OBSERVATIONS
120–123	/	1
124–127		0
128–131	//	2
132–135	//// //	7
136–139	//// //// //// //// /	21
140–143	//// //// //// //// //// //// //// //// /	41
144–147	//// //// //// ////	19
148–151	//// //// //	12
152–155	//	2
156–159	/	1
	Total	106

TABLE 2.3   Distribution of Absences in 106 Days

CLASS LIMITS	CLASS BOUNDARIES	CLASS MARK	CLASS FREQUENCY	RELATIVE FREQUENCY	PERCENTAGE OF OBSERVATIONS
120–123	119.5–123.5	121.5	1	.009	0.9
124–127	123.5–127.5	125.5	0	.000	0.0
128–131	127.5–131.5	129.5	2	.019	1.9
132–135	131.5–135.5	133.5	7	.067	6.7
136–139	135.5–139.5	137.5	21	.198	19.8
140–143	139.5–143.5	141.5	41	.387	38.7
144–147	143.5–147.5	145.5	19	.179	17.9
148–151	147.5–151.5	149.5	12	.113	11.3
152–155	151.5–155.5	153.5	2	.019	1.9
156–159	155.5–159.5	157.5	1	.009	0.9
		Totals	106	1.00	100.0

It is convenient to select one value from each class to serve as a representative of the class. The value commonly used is the midpoint of the class, the *class mark*. This is the average of the two class limits, or, what amounts to the same thing, the average of the two class boundaries. For the distribution of employee absences, the class mark for the first class is $(120 + 123)/2 = 121.5$ if computed from the class limits, or $(119.5 + 123.5)/2 = 121.5$ if computed from the class boundaries.

The number of observations in any class is the *class frequency* of that class. It is sometimes convenient to present the data in a *relative frequency distribution*. The relative class frequencies are found by dividing the class frequencies by the total number of observations. A *percentage distribution* is obtained by multiplying the relative class frequencies by 100 to convert them to percentages. For example, for the fifth class in Table 2.2, the class frequency is 21, so the relative frequency is $21/106$, or .198, and the percentage is $.198 \times 100$, or 19.8%.

All these concepts are illustrated in Table 2.3 for the employee absence data. In practice, not all the columns of Table 2.3 are displayed.

Earlier in this section we indicated that our choice of using ten classes in this example was somewhat arbitrary. Ten seemed like a reasonable number, and it made some of our calculations rather simple. However, we also indicated that we would present a rule that can take the guesswork out of determining how many classes to use in constructing a frequency distribution. One rule that has been advanced (see Reference 2.4) is this:

If n is the number of pieces of data to be summarized in your frequency distribution, then find the first power of 2 which just equals or exceeds n. This power equals the number of classes you should use.

This rule needs some illustration to make it clear. The powers of 2 up to the ninth power are:

$$2^1 = 2 \qquad 2^4 = 16 \qquad 2^7 = 128$$
$$2^2 = 4 \qquad 2^5 = 32 \qquad 2^8 = 256$$
$$2^3 = 8 \qquad 2^6 = 64 \qquad 2^9 = 512$$

Let us assume that a data set contains 25 items. According to the rule stated previously, they should be grouped into five classes. This is because $2^4 = 16$ is less than 25, but $2^5 = 32$ is the first power of 2 which just exceeds the value of $n = 25$.

As another example let us assume that a data set is made up of $n = 300$ values. This rule indicates that nine classes should be used since $2^9 = 512$ is the first value among the powers of 2 which just exceeds $n = 300$. Again, let us consider a data set with 64 values. Since $2^6 = 64$ and since the rule states the "power of 2 which just *equals* or exceeds n" is to be used, we would use six classes.

In order to obtain the Chapter 2 answers to odd-numbered problems that appear at the end of the text, readers will have to use this rule to determine the number of classes in the frequency distributions they construct. We will not go back and rework the example involving employee absences using the number of classes suggested by this rule. If we were to do that, we would use seven classes since $2^7 = 128$ is the first power of 2 which just exceeds that example's data set size of $n = 106$.

We also indicated earlier in this section that all the classes in a frequency distribution are usually of the same size. But sometimes it happens that our data contain a few observations whose numerical values are much smaller or much larger than the rest. If we include these values in the ordinary way, we may find that a number of our class frequencies are equal to zero. This is the case with the frequency distribution for the waiting times between failures of a certain type of airborne radio equipment given in Table 2.4. Because of the one extreme value in our data, five of the class frequencies in Table 2.4 are equal to zero. We can obtain a clearer picture of the distribution of the time to failure of these equipments without increasing the number of classes if we make use of an *open-ended interval,* one that has no boundary on one end, for the last class. The resulting distribution is given in Table 2.5.

Open-ended intervals permit the inclusion of a wide range of extreme values, but, unhappily, the actual numerical value is lost and we do not know how much larger or smaller the extremes actually were unless some indication is given in a footnote or elsewhere. Open-ended intervals also present difficulties when we want to calculate the descriptive measures of later chapters.

TABLE 2.4   Waiting Times between Failures of Equipments

TIMES BETWEEN FAILURES (HOURS)	*NO. OF FAILURES*
At least 0 but less than 50	3
At least 50 but less than 100	7
At least 100 but less than 150	13
At least 150 but less than 200	18
At least 200 but less than 250	22
At least 250 but less than 300	21
At least 300 but less than 350	12
At least 350 but less than 400	8
At least 400 but less than 450	0
At least 450 but less than 500	0
At least 500 but less than 550	0
At least 550 but less than 600	0
At least 600 but less than 650	0
At least 650 but less than 700	1
Total	105

The class limits in Table 2.4 are defined somewhat differently than they were in the example of employee absences. That is because the employee absences were counts and were thus recorded as whole numbers. But the time between equipment failures is a continuous measure. That is, it may be a fractional value such as 54.667 hours (54 hours and 40 minutes). Thus the seven observations in the second class of Table 2.4 could have values ranging from 50.0 to 99.99999 hours, i.e., from 50.0 up to but not including 100.0.

In Table 2.4, the second class technically has limits 50 and 99.99999,

TABLE 2.5   Waiting Times between Failures of Equipments

TIMES BETWEEN FAILURES (HOURS)	*NO. OF FAILURES*
At least 0 but less than 50	3
At least 50 but less than 100	7
At least 100 but less than 150	13
At least 150 but less than 200	18
At least 200 but less than 250	22
At least 250 but less than 300	21
At least 300 but less than 350	12
At least 350 but less than 400	8
400 or more	1
Total	105

out to as many 9's as we care to express this upper limit. Thus the class mark technically is $(50 + 99.99999 \ldots)/2$, or $74.9999 \ldots$. But the class mark is more conveniently taken to be 75, with no loss in accuracy.

2.2 CUMULATIVE FREQUENCY DISTRIBUTIONS

Frequency distributions as described in Section 2.1 are valuable aids for organizing and summarizing sets of data and for presenting data in such a way that the outstanding features are readily apparent. Sometimes, however, we require information on the number of observations whose numerical value is less than a given value. This information is contained in the *cumulative frequency distribution*. For instance, a stockbroker might be interested in the number of stocks on the New York Stock Exchange which have price-earnings ratios of less than certain values. A salesperson might be interested in the number of companies in a particular industry which have sales under a certain figure. A quality control engineer might be interested in the number of days for which the rejection rate on a production line was less than certain values. The cumulative frequency distribution would be useful in all three of these cases.

If we return to Table 2.2 we see that of the days observed, none had fewer than 120 absences, one had fewer than 124, one $(1 + 0)$ had fewer than 128, three $(1 + 0 + 2)$ had fewer than 132, and so on. To obtain the cumulative frequency of the number of absences less than the lower limit or boundary of a specified class, we add the class frequencies of all the preceding classes. The completed distribution is presented in Table 2.6. Note that Table 2.6 uses the *lower boundary* of each class in order to avoid confusion over where values such as 120 and 124 should be counted.

This table contains further information. The number of days on which there were 140 or more absences is, of course, $106 - 31$, or 75. And the number of days on which there were at least 140 but fewer than 148 absences is $91 - 31$, or 60.

Cumulative distributions may be constructed for relative frequencies and percentages as well as for the absolute frequencies. The procedures are identical except that we add the relative frequencies or percentages, as the case may be, instead of the absolute frequencies. Cumulative distributions which show the number of observations whose value exceeds a certain amount can also be constructed. Such distributions contain exactly the same information as the "less than" cumulative distributions discussed previously.

TABLE 2.6   Cumulative Distribution of Absences

NUMBER OF ABSENCES	NUMBER OF OBSERVATIONS
Less than 119.5	0
Less than 123.5	1
Less than 127.5	1
Less than 131.5	3
Less than 135.5	10
Less than 139.5	31
Less than 143.5	72
Less than 147.5	91
Less than 151.5	103
Less than 155.5	105
Less than 159.5	106

2.3   GRAPHIC PRESENTATION—THE HISTOGRAM

Though frequency distributions are effective in presenting the salient features of a set of data and are indispensable for computations, pictorial representation of the same information often makes the important characteristics more immediately apparent. Here we shall consider only the most basic pictorial forms: histograms and ogives. Horizontal bar charts, compound bar charts, pictographs, and pie diagrams are, in general, adaptations of these basic forms (see Reference 2.3).

A *histogram* is a graphic presentation of a frequency distribution and is constructed by erecting bars or rectangles on the class intervals. Along the horizontal scale we record the values of the variable concerned, marking off the class boundaries. Along the vertical scale we mark off frequencies. If we have equal class intervals, we erect over each class a rectangle whose height is proportional to the frequency of that class. For the distribution of employee absences that we divided into ten classes, we obtain the histogram of Figure 2.1. It shows very plainly that the number of employees absent each day tends to be in the vicinity of 140 and is generally within ten absences of this central value.

The class interval used in constructing the frequency table has a marked effect on the appearance of the histogram. The employee absence data of Table 2.1 are, in Table 2.2, sorted into classes having a class interval of 4. If a class interval of 2 or of 8 is used, the distributions of Table 2.7 result.

The top and bottom histograms in Figure 2.2 come from the two distributions in Table 2.7, and the middle histogram is a repeat of Figure 2.1. The horizontal scale is the same in each case. When raw data are grouped into classes, a certain amount of information is lost,

FIGURE 2.1 Histogram of employee absences

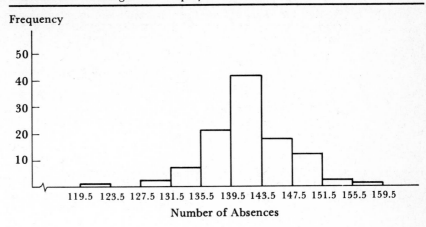

Number of Absences

TABLE 2.7 Distribution for Employee Absences

CLASS INTERVAL = 2		CLASS INTERVAL = 8	
Number of Absences	Number of Observations	Number of Absences	Number of Observations
120–121	1	116–123	1
122–123	0	124–131	2
124–125	0	132–139	28
126–127	0	140–147	60
128–129	1	148–155	14
130–131	1	156–163	1
132–133	4	Total	106
134–135	3		
136–137	7		
138–139	14		
140–141	24		
142–143	17		
144–145	13		
146–147	6		
148–149	11		
150–151	1		
152–153	2		
154–155	0		
156–157	0		
158–159	1		
Total	106		

FIGURE 2.2   Histograms of absent employees, showing varying class intervals

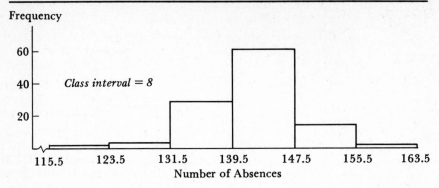

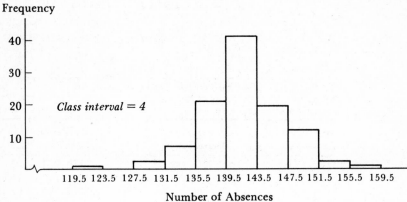

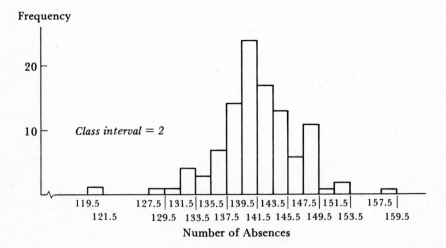

since no distinction is made between observations falling in the same class. The larger the class interval is, the greater is the amount of information lost. For the employee absence data, a class interval of 8 is so large that the corresponding histogram gives very little idea of the shape of the distribution. A class interval of 2, on the other hand, gives a ragged histogram. Little information has been lost in the bottom histogram of Figure 2.2, but the presentation of information is somewhat misleading because the small irregularities in the histogram merely reflect the accidents of sampling. If we had used the rule involving the powers of 2 to determine the number of classes used in presenting this data set, we would have used seven classes of width 6, and this would have produced a histogram between those for the intervals of 4 and 8. Thus, it appears that the powers of 2 rule will likely produce histograms between the extremes of giving too much detail or of giving too little.

Figure 2.3 shows the histogram for the times between failures given in Table 2.5. Observe that although the open class in this distribution cannot be represented by a bar in the histogram, we indicate by a note that there was an extreme value.

In the first section of this chapter we mentioned that in constructing frequency distributions we usually select all the classes to be of the same width. However, this is not always the case. But one of the advantages of equal class intervals is that the areas of the bars will be proportional to the heights of the bars. We therefore can draw

FIGURE 2.3   Histogram for times between failures of airborne radio equipment

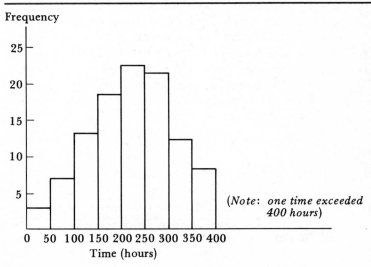

TABLE 2.8   Distribution of
Employee Absences
with the Last Four
Classes Combined

NUMBER OF ABSENCES	NUMBER OF OBSERVATIONS
120–123	1
124–127	0
128–131	2
132–135	7
136–139	21
140–143	41
144–159	34

the bars so that height is proportional to frequency. If the class intervals are unequal, we must modify our approach somewhat. Suppose that in the distribution of employee absences we were to combine the last four classes, as shown in Table 2.8. The rule for constructing a histogram is that the *area* of the bar over a class interval must be proportional to the frequency of the class. In the case of equal class intervals, this is the same as requiring that the height of the bar be proportional to the class frequency, but what if the class intervals are unequal? If the histogram for the new distribution with four combined classes in Table 2.8 were drawn with height proportional to frequency, we would obtain the result shown in Figure 2.4. Here, because we have not followed the area rule, the 34 values in the range 144–159 are given an undue emphasis, and the histogram gives the impression that on a very large proportion of the days more than 143 employees were absent. The correct representation is shown in Figure 2.5, where the areas of the bars are proportional to the frequencies. The height of the rightmost bar is one-quarter what it was in Figure 2.4, since the width of the class is sixteen days—four times the width of the other classes in the histogram. This is correct because the eye naturally interprets size by area.

2.4   GRAPHIC PRESENTATION—THE OGIVE

Just as a frequency distribution can be represented graphically by a histogram, a cumulative frequency distribution is represented graphically by an *ogive*. To construct an ogive, we first lay out the class boundaries on the horizontal scale, just as for a histogram. Above each class boundary

FIGURE 2.4 Incorrectly drawn histogram for Table 2.8

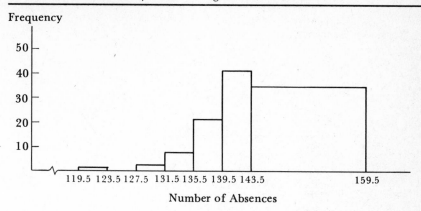

we plot a point at a vertical distance proportional to the cumulative frequency—proportional, in other words, to the number of observations whose numerical value is less than that class boundary. These points are then connected by straight lines. Using Table 2.6 in this way leads to the ogive shown in Figure 2.6.

It is proper to interpolate graphically from an ogive. From the ogive for employee absences we may, for example, get an approximation for the number of observations whose numerical value is less than 145.5 by finding the height of the curve over that point. With the help of the upper dashed line in Figure 2.6, we *estimate* that on 82 days fewer than 145.5 employees were absent (the actual number is 85).

FIGURE 2.5 Correctly drawn histogram for Table 2.8

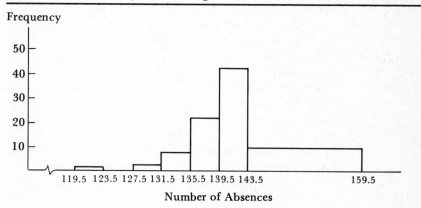

FIGURE 2.6   Ogive for the cumulative distribution of employee absences

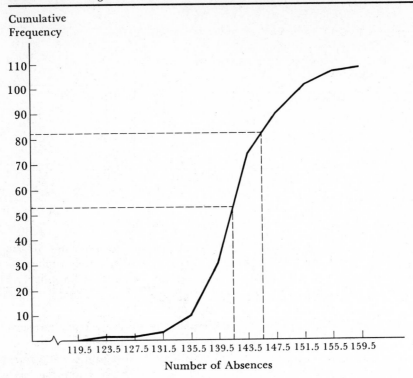

2.5   DATA AND POPULATIONS

It must be stressed that the material of this chapter and the next concerns empirical data—data obtained by selecting a sample or performing an experiment. The actual data in hand are usually only a very small part of some much larger whole. The 105 failure times recorded in Table 2.4, for example, constitute only a fraction of the totality of failure times for all radio equipment of that kind.

Conceptually, we could construct a frequency distribution for the totality of possible values which might have been obtained. Such conceptual distributions we refer to as *theoretical* or *population* distributions; they are to be distinguished from the *empirical* distributions of this chapter and the next. (Theoretical distributions are discussed in detail in Chapter 5.) An empirical distribution *estimates* the corresponding theoretical distribution.

PROBLEMS: *Answers to odd-numbered problems are given in the back of the text.*

1. If the powers-of-2 rule had been used to determine the number of classes in a frequency distribution for the data in Table 2.4, how many classes would have been used?

2. If the powers-of-2 rule were used, how many classes would frequency distributions have if they contained the following numbers of data items:
    a. 50?   **b.** 70?   **c.** 250?   **d.** 1200?   **e.** 4000?

3. Assume that you are given 250 numbers. The smallest is 8.8 and the largest is 79.9.
    a. How many classes would the data's frequency distribution contain?
    b. What would be the width of each class, assuming you want an integer as the class width?
    c. How much overlap is there beyond the frequency distribution's classes and the actual data?
    d. Give the class limits of the first class, assuming that you split the overlap approximately in half.

4. If the class limits in a frequency distribution are 32–36, 37–41, 42–46, and 47–51,
    a. What is the class width?
    b. What are the class boundaries?
    c. What are the class marks?

5. Given next are the amounts paid to 48 part-time clerks in a clothing store.
    a. How many classes should be used in grouping these data?
    b. How wide should each class be, assuming you want the class width rounded up to the nearest dollar?
    c. Construct the frequency distribution.
    d. Construct the cumulative distribution.
    e. Draw the histogram.
    f. Draw the ogive.

181	182	184	193	125	70
168	172	161	149	77	123
115	114	135	136	115	97
112	109	104	108	184	117
80	64	128	115	132	92
123	120	75	131	118	161
84	100	106	71	56	148
111	128	83	114	112	135

6. The following data are the numbers of parts produced on an assembly line during 20 randomly selected hours.
    a. How wide should each class be if you want the class width expressed to the nearest integer?
    b. Construct the frequency distribution.
    c. Find the cumulative frequency distribution.

2786	3217	2618	2845
2832	2524	2427	2838
2312	2697	2220	3010
3218	2855	3137	2581
2426	3346	2978	2714

7. Given the following distribution of the ages (in days) of 100 accounts receivable,
   a. Was the powers-of-2 rule used to construct this distribution? How do you know?
   b. What is the width of each class?
   c. List the class boundaries.
   d. List the class marks.
   e. Draw the histogram.

CLASS	FREQUENCY
25– 49	15
50– 74	25
75– 99	30
100–124	20
125–149	10

8. As an experiment, toss four coins 50 times and construct the frequency distribution for the number of heads which appear on each toss. The values will be 0, 1, 2, 3, and 4. Draw the histogram and the ogive.

9. The following relative frequency distribution shows the distribution of 50 Civil Service examination scores achieved by recent job applicants with the United States Postal Service. Find the original class frequencies.

CLASS	RELATIVE FREQUENCY
20–39	.12
40–59	.28
60–79	.36
80–99	.24

10. The following data are the amounts, in parts per million, of pollutant materials found in 80 water samples taken from a river in the Midwest.
    a. Construct a frequency distribution.
    b. Construct the cumulative distribution.
    c. Draw the histogram.

3.64	5.08	3.87	3.52	3.05	4.98	3.30	3.64
3.20	2.55	4.40	4.61	3.74	4.42	2.76	3.20
3.30	3.63	5.05	2.87	4.50	4.44	4.40	4.74
3.64	3.81	4.61	4.04	3.40	4.74	3.52	5.06
2.76	3.87	3.39	4.39	5.50	3.52	3.86	4.74
3.40	3.05	2.63	3.08	4.48	5.74	3.64	4.08
4.54	3.72	4.50	3.98	3.96	2.74	3.74	2.76
3.30	5.51	3.07	4.42	4.62	4.10	4.30	2.98
2.76	3.22	2.53	2.98	6.06	3.40	4.50	3.42
5.38	3.73	3.06	5.42	3.98	3.66	4.18	6.08

11. The following is the distribution for the number of defective items found in 100 lots of manufactured items.
    a. What percentage of the lots contained more than five defective items?
    b. What is the relative frequency of the lots that contained two or fewer defectives?
    c. What are the class limits? the class boundaries?
    d. Draw the histogram and ogive.

NUMBER DEFECTIVE	FREQUENCY
0	23
1	25
2	19
3	14
4	11
5	5
6	2
7	1

12. What is the width of each class in Problem **11**?

13. All the classes in a certain frequency distribution were 3 units wide, except the last class which was 9 units wide and contained 21 data items. How high should the bar for the last class be drawn in this distribution's histogram?

14. Given the following distribution,
    a. Draw the histogram. Note that the class intervals are not equal.
    b. Draw the ogive to scale.
    c. From the ogive, estimate the age such that 50% are older; that 75% are younger.

Estimated Number of Married Women in the Labor Force of Eastern U.S., 1979

AGE	NUMBER OF MARRIED WOMEN (THOUSANDS)
15–19	2212
20–24	3503
25–29	1313
30–34	939
35–39	3490
40–44	3005
45–54	5096
55–64	2904
65–80	885

15. The following data are the percentages of employees at 96 firms who have had computer programming experience.
    a. Construct a frequency distribution using eight classes of width 1.0 each.
    b. Construct the cumulative distribution.

   c. Sketch the histogram.

   d. How many classes would the powers-of-2 rule suggest be used in this problem?

7.78	6.02	9.68	6.14	7.28	6.23	4.64	7.41
6.75	4.51	8.42	7.92	6.84	5.81	4.67	7.38
5.43	11.28	7.96	6.38	5.92	4.80	10.72	6.17
6.17	9.54	7.82	5.33	5.79	4.76	9.78	5.84
6.00	5.23	6.48	6.53	5.46	4.83	9.51	6.82
7.26	10.90	10.82	9.81	10.73	10.46	10.63	7.79
6.54	9.72	9.74	10.62	4.87	5.23	4.68	7.19
5.28	10.62	7.24	8.38	5.05	4.98	8.62	6.76
5.03	8.41	8.65	8.37	8.46	7.93	7.19	6.53
4.29	8.62	8.50	8.83	7.78	7.64	5.39	7.21
5.31	7.76	6.42	5.13	6.49	6.31	7.72	4.15
4.97	6.13	5.29	6.34	7.82	6.68	6.44	4.68

SAMPLE DATA SET QUESTIONS: *Refer to the 113 applicants for credit listed in the Sample Data Set Appendix of this book.*

   a. If the powers-of-2 rule is used to construct a frequency distribution of JOBINC, how many classes would be used in the distributions of:

      1. All the applicants.

      2. All applicants who were granted credit.

      3. All applicants who were denied credit.

   b. How many classes should be used in a frequency distribution for the SEX variable?

   c. Construct a frequency distribution and draw the histogram for JOBINC of the applicants who were granted credit. Use the powers-of-2 rule to determine the number of classes in the distribution, and use a class width rounded to the nearest $100. Ignore the six applicants who did not list an income and simply add a note to your histogram concerning them.

REFERENCES

2.1 Croxton, Frederick E., Dudley J. Cowden, and Ben W. Bolch. *Practical Business Statistics,* 4th edition. Prentice-Hall, Englewood Cliffs, 1969. Chapter 2.

2.2 Huff, Darrell. *How to Lie with Statistics.* Norton, New York, 1965. Chapters 5 and 6.

2.3 Neter, John, William Wasserman, and G. A. Whitmore. *Fundamental Statistics for Business and Economics,* Abridged 4th Edition. Allyn and Bacon, Boston, 1974. Chapters 3 and 4.

2.4 Sturges, H. A. "The Choice of a Class Interval," *J. Am. Stat. Assoc.,* March 1926.

DESCRIPTIVE
MEASURES

In the last chapter we saw how tabular and graphical forms of presentation may be used to summarize and describe quantitative data. Though these techniques help us to sort out important features of the distribution of the data, statistical methods for the most part require concise *numerical* descriptions. These are arrived at through arithmetic operations on the data which yield *descriptive measures* or *descriptive statistics*. The basic descriptive statistics are the measures of central tendency or location and the measures of dispersion or scatter.

Discussion of these descriptive measures requires some familiarity with the mathematical shorthand used to express them. Since we are concerned with masses of data and since the operation of addition plays a large role in our calculations, we need a way to express sums in compact and simple form. The summation notation meets this requirement.

3.1 SYMBOLS AND SUMMATION NOTATION

If our data consist of measurements of some characteristic of a number of individuals or items such as the annual incomes of some group of persons, the weights of a number of pigs, or the weekly sales of several stores, we designate the characteristic of interest by some letter or symbol, say, X. If we have measured two or more characteristics, we use different letters or symbols for each. If, in addition to obtaining annual income, we also record the age of each person interviewed we could represent income by X and age by Y. A third characteristic, such as educational level, we could represent by the letter Z, and so on.

In order to differentiate between the same kind of measurements made on different items or individuals, or between similar repeated measurements made on the same element, we add a subscript to the corresponding symbol; thus X_1 stands for the income of the first person interviewed, X_2 for that of the second, X_{23} for that of the twenty-third, and so on. In general, any arbitrary observed value would be represented by X_i, where the subscript i is variable in the sense that it represents any one of the observed items and need only be replaced by the proper number in order to specify a particular observation. The income, age, and educational level of the ith, or general, individual would be represented by X_i, Y_i, and Z_i, respectively.

Given a set of n observations which we represent by X_1, X_2, X_3, $\ldots$, X_n, we can express their sum as

$$\sum_{i=1}^{n} X_i = X_1 + X_2 + X_3 + \cdots + X_n$$

where Σ (uppercase Greek sigma) means "the sum of," the subscript i is the index of summation, and the 1 and n that appear respectively below and above the operator Σ designate the range of the summation. The combined expression says, "add all X's whose subscripts are between 1 and n, inclusive." In place of i we sometimes use j as the summation index; any letter will do.

If we want the sum of the squares of the n observations, we write

$$\sum_{i=1}^{n} X_i^2 = X_1^2 + X_2^2 + X_3^2 + \cdots + X_n^2$$

which says, "add the squares of all observations whose subscripts are between 1 and n, inclusive." The sum of the products of two variables X and Y would be written as:

$$\sum_{i=1}^{n} X_i Y_i = X_1 Y_1 + X_2 Y_2 + X_3 Y_3 + \cdots + X_n Y_n$$

If we want a partial sum, the sum of part but not all the quantities involved, the range of the summation is adjusted accordingly. For example:

$$\sum_{i=4}^{8} Y_i^2 = Y_4^2 + Y_5^2 + Y_6^2 + Y_7^2 + Y_8^2$$

$$\sum_{i=2}^{5} Y_i^2 f_i = Y_2^2 f_2 + Y_3^2 f_3 + Y_4^2 f_4 + Y_5^2 f_5$$

and

$$\sum_{i=2}^{3} (Y_i - X_i) = (Y_2 - X_2) + (Y_3 - X_3)$$

The first of the preceding sums includes only the fourth through the eighth items in the Y list. The second sum includes the squares of the second through the fifth items in the Y list multiplied by the corresponding item in the f list. The third sum is merely the sum of the differences between corresponding items in the X and Y lists, taken only over the second and third items. In those cases where the context makes it clear that *the sum is to be taken over all the data* we sometimes simplify the summation notation by omitting the range of summation. Thus, ΣX_i is the sum of all of the numbers and ΣX_i^2 is the sum of the squares of all of the numbers.

We now look at some of the algebraic rules that apply to summations:

Rule 1: **The summation of a sum (or difference) is the sum (or difference) of the summations:**

$$\sum_{i=1}^{n} (X_i + Y_i - Z_i) = \sum_{i=1}^{n} X_i + \sum_{i=1}^{n} Y_i - \sum_{i=1}^{n} Z_i \qquad (3.1)$$

Rule 2: **The summation of the product of a variable and a constant is the product of the constant and the summation of the variable:**

$$\sum_{i=1}^{n} c Y_i = c \sum_{i=1}^{n} Y_i \qquad (3.2)$$

Rule 3: **The summation of a constant is the constant multiplied by the number of terms in the summation:**

$$\sum_{i=1}^{n} c = nc \qquad \textbf{(there are } n \textbf{ terms)} \qquad\qquad (3.3)$$

For example,

$$\sum_{i=3}^{8} c = 6c \qquad \text{(there are 6 terms)}$$

Working through some numerical examples makes it clear that these rules are correct. The examples that follow demonstrate this. (See also Problem 3 at the end of the chapter.)

EXAMPLE 1

Problem: Assume the following values are assigned to each of three variables X, Y, and Z.

$$
\begin{array}{lll}
X_1 = 2 & Y_1 = -1 & Z_1 = 2 \\
X_2 = 0 & Y_2 = 7 & Z_2 = 3 \\
X_3 = 5 & Y_3 = 2 & Z_3 = 6
\end{array}
$$

Referring to Equation (3.1) show that Rule 1 holds for these nine numbers.

Solution: The left side of Equation (3.1) is

$$\sum_{i=1}^{3} (X_i + Y_i - Z_i) = (2 + (-1) - 2) + (0 + 7 - 3) + (5 + 2 - 6)$$

$$= (-1) + (4) + (1)$$

$$= 4$$

The right side of Equation (3.1) is found by summing up each of the three columns (the X column, the Y column, and the Z column) and then adding the X and Y sums and subtracting the Z sum.

$$\sum_{i=1}^{3} X_i + \sum_{i=1}^{3} Y_i - \sum_{i=1}^{3} Z_i = (2 + 0 + 5) + (-1 + 7 + 2) - (2 + 3 + 6)$$

$$= (7) + (8) - (11)$$

$$= 4$$

Since both the left and right sides of Equation (3.1) equal 4, Rule 1 holds for the nine numbers.

EXAMPLE 2

Problem: Referring to Equation (3.2) use the Y values in Example 1 to show that Rule 2 holds if $c = 8$.

Solution: The left side of Equation (3.2) is

$$\sum_{i=1}^{3} cY_i = 8(-1) + 8(7) + 8(2) = -8 + 56 + 16 = 64$$

The right side of Equation (3.2) is

$$c\sum_{i=1}^{3} Y_i = 8(-1 + 7 + 2) = 8(8) = 64$$

Since both the left and right sides of Equation (3.2) equal 64, Rule 2 holds.

EXAMPLE 3

Problem: Referring to Equation (3.3) show that Rule 3 holds if $c = 5$ and $n = 4$.

Solution: The left side of Equation (3.3) is simply

$$\sum_{i=1}^{n} c = \sum_{i=1}^{4} 5 = (5 + 5 + 5 + 5) = 20$$

The right side of Equation (3.3) is $nc = 4 \cdot 5 = 20$, and the two sides of the equation are equal.

Each summation must be read with care, so that its meaning is exactly understood. If one reads summations from the inside out, so to speak, many common errors can be avoided. Consider the summation

$$\sum_{i=1}^{n} (X_i - c)^2$$

Read from the inside out, it tells us to

1.   Subtract the constant c from each X,
2.   Square each of the differences obtained in Step 1, and
3.   Sum the squares obtained in Step 2.

The placement of parentheses in summation formulas has a very profound effect on the result. Note that if the parentheses are removed from the preceding expression, we have

$$\sum_{i=1}^{n} X_i - c^2$$

In this second case each X value is added, and then c^2 is subtracted once from the total at the end. Parts **c** and **d** of Problem 5 at the end of this chapter demonstrate the importance of noting parentheses.

Two important but sometimes misread sums are the **sum of squares**

$$\sum_{i=1}^{n} X_i^2 = X_1^2 + X_2^2 + \cdots + X_n^2$$

and the **square of the sum**

$$\left(\sum_{i=1}^{n} X_i \right)^2 = (X_1 + X_2 + \cdots + X_n)^2$$

The former is found by squaring each X and then adding the squares, the latter by adding up the X's and then squaring the sum. These two expressions are usually unequal and often occur together in the same mathematical statement.

EXAMPLE 4

Problem: Use the values of X given in Example 1 to find both the *sum of the squares* and the *square of the sum*.

Solution: First, the sum of the squares is

$$\sum_{i=1}^{3} X_i^2 = (2)^2 + (0)^2 + (5)^2 = 4 + 0 + 25 = 29$$

Second, the square of the sum is

$$\left(\sum_{i=1}^{3} X_i \right)^2 = (2 + 0 + 5)^2 = (7)^2 = 49$$

It is obvious that in this case the sum of squares and the square of the sum are not equal.

Now that we have gotten the preliminaries of summation notation out of the way, we are prepared to discuss measures of location and measures of dispersion.

3.2   MEASURES OF LOCATION

When we work with numerical data and their frequency distributions, it soon becomes apparent that in most sets of data there is a tendency for the observed values to group themselves about some interior value; some central value seems to be characteristic of the data. This phenomenon, referred to as *central tendency*, may be used to describe the data in the sense that the central value locates the "middle" of the distribution. The statistics we calculate for this purpose are *measures of location,* also called measures of central tendency. For a given set of data, the measure of location we use depends upon what we mean by *middle,* different definitions giving rise to different measures. We consider here four such measures and their interpretations: the mean, the weighted mean, the median, and the mode.

3.3 THE MEAN

All of us are familiar with the concept of the mean or average value. We read and speak of batting averages, grade-point averages, mean annual rainfall, the average weight of a boxcar load of coal, and the like. In most cases, the term *average* used in connection with a set of numbers refers to their *arithmetic mean.* For the sake of simplicity we call it the **mean.**

For a given set of n values, X_1, X_2, X_3, ..., X_n, the mean is their sum divided by n, the number of values in the set. It is denoted by μ (lower case Greek mu) and may be expressed as

$$\mu = \frac{1}{n} \sum_{i=1}^{n} X_i \tag{3.4}$$

If, for example, we receive three carloads of coal and their weights are 10, 13, and 16 tons, their mean weight is

$$\mu = \frac{10 + 13 + 16}{3} = \frac{39}{3} = 13 \text{ tons}$$

The record of employee absences for 106 days in Table 2.1 has a sum equal to 15,005, and the mean number of employees absent per workday is

$$\mu = \frac{15,005}{106} = 141.6 \text{ employees}$$

Notice that the unit for the mean is the same as the unit for the observations themselves (tons, employees, dollars, etc.). Notice also that even though the observations may be whole numbers (like employees), the mean may have fractional or decimal parts (like 141.6 employees).

When we wish to find the mean of a set of numbers, and when those numbers form the population in which we are interested, then we use Formula (3.4) and call our mean μ. However, there are times when the numbers we have at our disposal are only sample values which we are using to estimate the mean of a larger population. For instance, the three carloads of coal whose mean weight was 13 tons might be a sample of many carloads on a particular railroad siding. If we are interested in using the three weights to estimate the mean weights of all the carloads, then we perform the same calculations as those in Formula (3.4), but we call our resulting mean $\overline{X}$. Thus,

$$\overline{X} = \frac{1}{n} \sum_{i=1}^{n} X_i \tag{3.4a}$$

is the way we represent the mean of a sample. The only difference between this formula and Formula (3.4) is the way we denote the result of our calculations. When someone reviewing our work sees that our mean is denoted by $\overline{X}$, they know that the numbers we used to calculate the mean were a sample. However, if they see that the mean is denoted by μ, they know that our mean was figured from numbers that form the entire population whose mean we seek. It is only the context of the situation or the problem statement which will allow the reader to determine if he or she is dealing with a sample and should call the mean $\overline{X}$ (or $\overline{Y}$ if we are calling our variable Y_i) or with a population and should call the mean μ.

The mean of a set of numbers is their "middle" in the sense that it is their center of gravity. Suppose we have the four observations 1, 8, 7, and 1, with a mean of $(1 + 8 + 7 + 1)/4$, or 4.25. Imagine a seesaw with a scale marked off along its edge, and imagine, for each of the four observations, a one-pound weight positioned according to this scale, as shown in Figure 3.1. The mean 4.25 is the point at which the fulcrum of the seesaw must be placed in order to make it balance.

3.4 THE WEIGHTED MEAN

When we compute the simple arithmetic mean of a set of numbers we assume that all the observed values are of equal importance and we give them equal weight in our calculations. In situations where

FIGURE 3.1

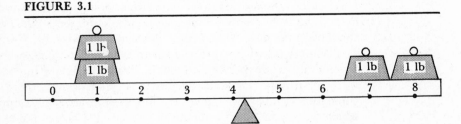

the numbers are not equally important we can assign to each a weight which is proportional to its relative importance and calculate the *weighted mean*.

Let X_1, X_2, X_3, ..., X_n be a set of n values, and let w_1, w_2, w_3, ..., w_n be the weights assigned to them. The weighted mean is found by dividing the sum of the products of the values and their weights by the sum of the weights; that is,

$$\mu = \frac{w_1 X_1 + w_2 X_2 + w_3 X_3 + \cdots + w_n X_n}{w_1 + w_2 + w_3 + \cdots + w_n} \tag{3.5}$$

or, in summation notation,

$$\mu = \frac{\Sigma w_i X_i}{\Sigma w_i} \tag{3.5a}$$

Naturally, if the X values come from a sample, we would denote the left-hand values in the preceding formulas as $\overline{X}$.

Most students are familiar with the concept of the weighted mean, for the grade-point average is such a measure. It is the mean of the numerical values of the letter grades weighted by the numbers of credit hours in which the various grades are earned. If a student makes A's in two 3-credit courses, a B in a 5-credit course, a C in a 4-credit course, and a D in a 2-credit course, and if the numerical values of the letter grades are A = 4, B = 3, C = 2, and D = 1, the grade-point average for the term is

$$\mu = \frac{3 \times 4 + 3 \times 4 + 5 \times 3 + 4 \times 2 + 2 \times 1}{3 + 3 + 5 + 4 + 2} = \frac{49}{17} = 2.882$$

Failure to weight the values when combining data is a common error. Suppose that a chemical compound is made up of two ingredients, chemical A and chemical B. If chemical A costs $5 per gallon and chemical B costs $10 per gallon, one might be tempted to say that the average cost of ingredients for the compound is ($5 + $10)/2 = $7.50 per gallon. This would not be correct unless the compound were made up of equal parts of A and B. For example, if one gallon of the compound consists of .4 gallon of A and .6 gallon of B, then the true mean cost of ingredients per gallon is

$$\mu = \frac{.4 \times \$5 + .6 \times \$10}{.4 + .6} = \$8$$

in accordance with Equation (3.5) for two (rather than n) weighted values.

3.5   THE MEDIAN

Roughly speaking, the **median** of a set of data is a number such that half the observations are less than that number and half the observations are greater than that number. To be specific, suppose there are n observed values and that n is an odd number. The array is formed (the values are lined up in either increasing or decreasing order), and the median is by definition the observation in the middle position. For example, the number of people being supervised by each of 11 department heads in a VA hospital is arrayed next.

$$1 \quad 1 \quad 2 \quad 3 \quad 3 \quad \mathbf{8} \quad 11 \quad 14 \quad 19 \quad 19 \quad 20$$

The median is 8 people (there are five observations to its left and five to its right). If n is even, there is no one observation in the middle position. In this case, we take the median to be the average of the *pair* of observations occupying the two central positions. In the next array we see the number of people supervised by 12 department heads in a private hospital.

$$2 \quad 5 \quad 5 \quad 6 \quad 7 \quad \mathbf{10} \quad \mathbf{15} \quad 21 \quad 21 \quad 23 \quad 23 \quad 25$$

The observations 10 and 15 occupy the two center positions (there are five observations to the left of 10 and five to the right of 15), and so the median is $(10 + 15)/2$, or 12.5 people.

3.6   THE MODE

The **mode** of a set of numbers is the most frequently occurring value in the data set. The mode can be useful as a measure of location in cases where we desire to know which number comes up most often. For instance, a contractor looked at the winning bids for jobs on which he had submitted bids. He found that the winning bids expressed as a percentage of his estimated costs were:

140, 125, 130, 125, 125, 110, 105, 125, 135, 125, 105

The winning bid most often seems to be 125 percent of his estimated costs (it occurs five times). Thus, he might wish to enter a bid at 124 percent of his estimated costs on the next job he wants to get. Since the 125 percent bid seems to be the winning bid most often, he will increase his chances of winning the contract with this knowledge of the mode.

The mode is not as useful a measure of location as the mean and the median. This is due to the fact that when the data consist of only a few numbers, it is often the case that each value occurs only once, and thus all values are modal values. If, however, one value does occur more than once, there is no guarantee that this value will in any way show the central tendency of the data if there are only a few values in the data set. For the numbers

$$7, 12, 18, 22, 31, 31$$

the mode is 31 since it appears twice and all other numbers appear only once. But 31 could hardly be considered a measure of central tendency for these data since it is at the extreme high end of the values.

3.7 MEASURES OF VARIATION

The measures of location discussed in the preceding section describe one characteristic of a set of numbers, the typical or central value. However, we often wish to know about another characteristic of a number set, the variation or scatter among the values. For instance, a student about to graduate from college has accepted a job with a company. She has her choice of working at the company's offices in city A or city B. Since she likes outdoor activities, she decides to check on the climate in each of the cities. Her investigations tell her that both cities have a mean daily high temperature of 70° F. This figure alone, however, may be misleading. Further investigation reveals that city A's high temperature varies from 65° F in the winter to 75° F in the summer. But city B's high varies from 20° F in the winter to 120° F in the summer. This information may have a rather significant effect on the student's location selection.

The statistics we calculate to measure the variation or scatter in a set of numbers are referred to as *measures of variation* or *measures of dispersion.* Since several sets of data could have the same—or nearly the same—mean, median, and mode but vary considerably in the extent to which the individual observations differ from one another, a more complete description of the data results when we evaluate one of the measures of variation in addition to one or more of the measures of location. Four such measures are considered in the following sections: the range, mean deviation, variance, and standard deviation.

3.8 THE RANGE

The **range** is the difference between the largest and smallest values in the data. In Table 2.1 we presented data concerning the number of employees absent from their work at a large manufacturing plant over a 106-day period. The smallest number of absences on any of the 106 days was 121 and the largest was 158; hence the range is 158 − 121, or 37, employees.

Although the range is easy to calculate and is commonly used as a rough-and-ready measure of variability, it is *not* generally a satisfactory measure of variation for several reasons. In the first place, its calculation involves only two of the observed values regardless of the number of observations available; therefore, it utilizes only a fraction of the available information concerning variation in the data, and reveals nothing with respect to the way in which the bulk of the observations are dispersed within the interval bounded by the smallest and largest values. Secondly, as the number of observations is increased the range generally tends to become larger; therefore, it is not proper to use the ranges to compare the variation in two sets of data unless they contain the same numbers of values. Finally, the range is the least stable of our measures of variation for all but the smallest sample sizes; that is, in repeated samples taken from the same source the ranges will exhibit more variation from sample to sample than will the other measures.

The range differs from most of our statistical measures in that it is a relatively good measure of variation for small numbers of observations, but becomes less and less reliable as the sample size increases. Because it is easy to calculate and is reasonably stable in small samples, it is commonly used in statistical quality control where samples of four or five observations are often sufficient.

In recent years the range and statistics based on the range have found favor with people who are interested in the development of rapid statistical methods for arriving at answers with a minimum of lengthy or involved computations. It also is the basis for other recently developed techniques that are beyond the scope of this book (see Reference 3.5).

3.9 DEVIATIONS FROM THE MEAN

Since the disadvantages of the range limit its usefulness, we need to consider other measures of variation. Suppose we have a population of n numbers, X_1, X_2, X_3, $\ldots$, X_n, whose mean is μ. If we were to plot these values and μ on the X-axis as in Figure 3.2 (only five values are shown) and measure the distance of each of the X's from the mean, μ, it seems reasonable that the average, or mean, of these distances should provide a measure of variation. These distances, or *deviations from the mean*, are equal to $(X_1 - \mu)$, $(X_2 - \mu)$, $\ldots$, $(X_n - \mu)$. The general term $X_i - \mu$ is the amount by which X_i, the ith

FIGURE 3.2 Deviations from the sample mean

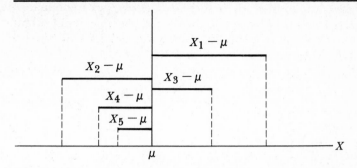

value of X, differs from the population mean. If $X_i - \mu$ is 5, then the value of X_i is five higher than the mean. If, on the other hand, $X_i - \mu$ is -8, then the value of X_i is eight below the mean. Thus, the deviations from the mean will be either positive or negative depending on whether X_i is above or below the mean.

The mean of these deviations is found by taking their sum and dividing it by the number of deviations there are. This value is

$$\frac{1}{n} \sum_{i=1}^{n} (X_i - \mu)$$

But this quantity is always equal to zero, since the algebraic sum of the deviations of a set of numbers from its mean is always equal to zero.* That is,

$$\sum_{i=1}^{n} (X_i - \mu) = 0 \tag{3.6}$$

The same result holds true for deviations around the sample mean $\overline{X}$. That is,

$$\sum_{i=1}^{n} (X_i - \overline{X}) = 0$$

---

*This can be shown as follows: By Rules 1 and 3 in Section 3.1,

$$\Sigma(X_i - \mu) = \Sigma X_i - \Sigma\mu = \Sigma X_i - n\mu$$

since μ is a constant. But

$$\mu = \frac{1}{n} \Sigma X_i \quad \text{or} \quad n\mu = \Sigma X_i$$

Therefore,

$$\Sigma(X_i - \mu) = \Sigma X_i - \Sigma X_i = 0$$

The trouble is that we have used *signed* distances rather than unsigned distances of the X_i to the mean μ. We can eliminate the algebraic signs on these values in two ways: we may simply ignore them, or we may square them. If we ignore the signs on these distances, we then have the *mean deviation*

$$\frac{1}{n} \sum_{i=1}^{n} |X_i - \mu| \tag{3.7}$$

where $|X_i - \mu|$, the *absolute value* of $X_i - \mu$, is just $X_i - \mu$ with the sign converted to $+$ if it happens to be $-$.

The mean deviation, though it appears to be relatively simple, is not of great interest. It is not particularly easy to calculate, and the absolute values make the mean deviation difficult to use in the mathematical operations we wish to perform in later chapters. Since the mean deviation does not have the mathematical properties we desire, it will be best if we eliminate the signs on the deviations around the mean by squaring them.

3.10   THE VARIANCE AND STANDARD DEVIATION

Suppose we do use the square instead of the absolute value as a measure of deviation. In the squaring process the negative signs will disappear; hence, the sum of the squares of the deviations from the mean will always be a positive number greater than zero unless all the observations have the same value. In symbols, the sum of squares of deviations from the mean is

$$\sum_{i=1}^{n} (X_i - \mu)^2 \tag{3.8}$$

This quantity, for simplicity, will be referred to as the **sum of squares.** Hereafter, whenever the phrase *sum of squares* appears it should be taken to mean **the sum of squares of deviations from the mean.** To avoid ambiguity, if the sum of squares of the observations, or the sum of squares for any other quantities, is meant, the phrase will be qualified accordingly.

It is clear that the sum of squares provides a measure of dispersion. If all the observed values are identical, the sum of squares is equal to zero. If the values tend to be close together the sum of squares will be small, but if they scatter over a wide range the sum of squares will be correspondingly large. To illustrate this characteristic and, incidentally, to show how the sum of squares is calculated according to the definition given as Equation (3.8), we use the numbers that follow. They show the sales of ten salespeople (in thousands) for last month.

$$12, 6, 15, 3, 12, 6, 21, 15, 18, 12$$

Their mean is

$$\mu = \frac{1}{n} \Sigma X_i = \frac{120}{10} = 12$$

Notice that we denoted the mean as μ rather than $\overline{X}$. This indicates that we are interested in *this* set of values and do not consider it to be a sample. Thus, these salespeople are all the salespeople in the company and not just a sample. If they were a sample, we would have denoted their mean as $\overline{X}$. To find the deviations from the mean we subtract $\mu = 12$ from each of the X values. We get

$$0, -6, 3, -9, 0, -6, 9, 3, 6, 0$$

We can use the fact that the sum of the deviations from the mean is equal to zero to check our arithmetic:

$$(3 + 9 + 3 + 6) - (6 + 9 + 6) = 0$$

To complete our calculations we square each of the deviations and sum them:

$$0 + 36 + 9 + 81 + 0 + 36 + 81 + 9 + 36 + 0 = 288$$

The sum of squares for these ten numbers is

$$\sum_{i=1}^{10} (X_i - \mu)^2 = 288$$

The calculations are summarized in the following table.

X_i	$X_i - \mu$	$(X_i - \mu)^2$
12	0	0
6	−6	36
15	3	9
3	−9	81
12	0	0
6	−6	36
21	9	81
15	3	9
18	6	36
12	0	0
120	0	288

$$n = 10 \qquad \mu = 12$$

In contrast to the preceding example, which displays a fair amount of variation, suppose we have the numbers

12, 10, 12, 14, 10, 13, 12, 11, 14, 12

We find the sum of squares as before (see next table). As could be expected, since the variation in the second set of numbers is less than in the first set, the sum of squares, 18, is considerably smaller than that calculated for the first set. Notice that the mean is the same as

X_i	$X_i - \mu$	$(X_i - \mu)^2$
12	0	0
10	−2	4
12	0	0
14	2	4
10	−2	4
13	1	1
12	0	0
11	−1	1
14	2	4
12	0	0
120	0	18

$$n = 10 \qquad \mu = 12$$

in the preceding example; the sum of squares measures spread *from* the mean, whatever the mean may be.

It is obvious that if we have a large number of values, or if the mean is not a whole number, this procedure for calculating the sum of squares could become rather tedious. Fortunately, there is another method for computing the sum of squares that does not necessitate finding the individual deviations from the mean. It can be shown*

---

*Algebraically we know that $(a - b)^2 = a^2 - 2ab + b^2$. Therefore, we can rewrite the sum of squares as

$$\Sigma(X_i - \mu)^2 = \Sigma(X_i^2 - 2\mu X_i + \mu^2)$$

If we now apply Rules 1, 2, and 3 of Section 3.1, we get

$$\Sigma(X_i - \mu)^2 = \Sigma X_i^2 - 2\mu\Sigma X_i + n\mu^2$$

But

$$\mu = \frac{1}{n}\Sigma X_i \qquad \text{and} \qquad \mu^2 = \frac{1}{n^2}(\Sigma X_i)^2$$

Therefore

$$\Sigma(X_i - \mu)^2 = \Sigma X_i^2 - 2\frac{1}{n}(\Sigma X_i)^2 + \frac{1}{n}(\Sigma X_i)^2$$

so that, combining the two right-hand terms,

$$\Sigma(X_i - \mu)^2 = \Sigma X_i^2 - \frac{1}{n}(\Sigma X_i)^2$$

that the sum of the squared deviations from the mean can be written as

$$\Sigma (X_i - \mu)^2 = \Sigma X_i^2 - \frac{1}{n} (\Sigma X_i)^2 \qquad (3.9)$$

The expression on the right-hand side of Equation (3.9) provides an alternative procedure for computing a sum of squares that is especially useful when a calculator is available. For this reason it is often referred to as the *machine formula* for sums of squares, and it should be used for all but the most simple sets of data. To apply the formula, we obtain the sum of the squares of the observed values as well as their sum, square the sum, divide it by n (the number of observations), and subtract the result from the sum of squares of the observed values.

Although the sum of squares is a measure of variation, it is usually more convenient to use the mean of the squared deviations as a dispersion measure. This mean of squared deviations is called the **variance, σ^2** (lowercase Greek sigma squared), and it is found by dividing the sum of squares by n, the number of items in the data set:

$$\sigma^2 = \frac{1}{n} \sum_{i=1}^{n} (X_i - \mu)^2 \qquad (3.10)$$

Thus, the variance is the mean value of the squares of the deviations. When the machine formula is used to obtain the sum of squares, the variance is

$$\sigma^2 = \frac{1}{n} \left[\Sigma X_i^2 - \frac{1}{n} (\Sigma X_i)^2 \right] \qquad (3.11)$$

The **standard deviation,** or root mean squared deviation, of the numbers in a sample of size n is the square root of the variance; that is,

$$\sigma = \sqrt{\sigma^2} \qquad (3.12)$$

The standard deviation σ is used partly because its units are those of the original data. If the X_i are measurements in pounds, then μ has pounds as its unit. The variance σ^2, however, being a sum of squares of deviations in pounds, has square pounds as its unit; but its square root σ has the pound as its unit again, which is what we want: The spread should be measured in terms of the units involved.

For the moment, we will concentrate on how the standard deviation is computed. Then we will turn to the problem of what it really measures and why it is useful. Let us work out a sample calculation of the variance and standard deviation, using the following data on the number of defectives found by a quality control inspector during nine successive inspection hours:

$$3, 6, 2, 5, 3, 8, 6, 7, 5$$

STEP 1.   Find the sum of squares.

$$\Sigma X_i = 45$$

$$\mu = \frac{1}{n} \Sigma X_i = \frac{45}{9} = 5$$

Thus the mean value is 5 defects per hour.

$X_i - \mu$	-2	1	-3	0	-2	3	1	2	0	0
$(X_i - \mu)^2$	4	1	9	0	4	9	1	4	0	32

$$\Sigma (X_i - \mu)^2 = 32$$

Thus the sum of squares is 32.

STEP 2.   To get the variance, divide the sum of squares by n.

$$\sigma^2 = \frac{1}{n} \Sigma (X_i - \mu)^2 = \frac{32}{9} = 3.56$$

Thus the variance has a value of 3.56 measured in units of (defects)^2.

STEP 3.   To obtain the standard deviation take the square root of the variance.

$$\sigma = \sqrt{\sigma^2} = \sqrt{3.56 \, (\text{defects})^2} = 1.89 \text{ defects}$$

Thus, the standard deviation is 1.89 defects and has the same units of measure as do the individual data values and the mean.

If we were to use Machine Formula (3.9) for the sum of squares in Step 1, we would find:

$$\Sigma X_i = 45$$

$$\Sigma X_i^2 = 9 + 36 + 4 + 25 + 9 + 64 + 36 + 49 + 25 = 257$$

$$\Sigma X_i^2 - \frac{(\Sigma X_i)^2}{n} = 257 - \frac{45^2}{9} = 257 - 225 = 32$$

Since finding the standard deviation involves taking a square root, dividing by n, and summing the squared deviations, the standard deviation is sometimes called the *root mean squared deviation*, as was indicated earlier.

3.11   THE SAMPLE STANDARD DEVIATION

The reader may recall that in Section 3.3 we made a distinction between μ, the mean of a population, and $\overline{X}$, the mean of a sample. While the methods for computing these means were shown to be the same, we used the different notations to indicate whether our data set was a population, or a sample chosen to represent the population.

The same type of distinction is made between the population standard deviation σ, which we have been discussing, and the standard deviation of a sample, which we denote by s. If the numbers in our data set form the population of interest, then we use

$$\sigma = \sqrt{\frac{1}{n} \sum (X_i - \mu)^2} \tag{3.13}$$

to find the standard deviation. [Formula (3.13) is just Formulas (3.10) and (3.12) combined. Also, Formula (3.11) could be used to obtain the expression under the square root sign.]

However, if the numbers in our data set form a sample and we wish to use them to estimate the population standard deviation, then we compute the sample mean $\overline{X} = (1/n) \sum X$ and then find the sample standard deviation as

$$s = \sqrt{\frac{1}{n-1} \sum (X_i - \overline{X})^2} \tag{3.14}$$

The only differences between the formulas for the population and sample standard deviations are:

1. In the population standard deviation formula we call the mean μ and in the sample standard deviation formula we call it $\overline{X}$, and
2. In the population formula for σ we divide the sum of squared deviations by n, but in the sample formula we divide that sum by $n - 1$.

The first difference is only a notational one. We use μ and $\overline{X}$ to denote whether we are working with population or sample data, but the mathematical computations are the same. The second difference, however, is more substantial. Dividing by $n - 1$ when we calculate the standard deviation from sample data produces a mathematically different result from what we obtain using population data and dividing by n.

An intuitive reason why we divide the sample sum of squares by $n - 1$ rather than by n is shown in Figure 3.3. We must remember that the standard deviation is a measure of dispersion, spread, or scatter of a set of values around the mean. Figure 3.3 shows a picture of a population and the range over which its values are spread. A typical set of randomly selected values from the population is denoted by the X's. Note that the range over which the sample values are spread is smaller than that over which the population is spread. It would

FIGURE 3.3

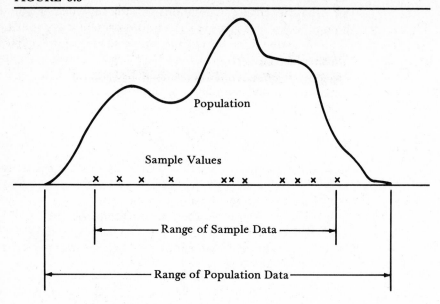

be a rather unusual sample which included both the largest and smallest values in the population. Thus, we can see intuitively that the spread of values in a sample will typically be less than the spread in the population. If we are using sample data to estimate the population standard deviation, then we must adjust our calculations somehow to make up for the smaller spread in the sample. Dividing the sum of squares by $n - 1$ rather than by n adjusts the sample standard deviation up to where it better estimates the population standard deviation.*

One might logically ask, "Why not divide by $n - 2$ or by $n - 3$?" This is because it can be shown mathematically that dividing the sample sum of squares by $n - 1$ gives a better estimate of the population variance, and thus standard deviation, than if any other divisor is used. We will not present that mathematical proof. However, it does involve the concept of *degrees of freedom.*

There is associated with any sum of squared quantities an expression or number known as its degrees of freedom, which may be defined as the number of squares minus the number of independent restrictions

---

*Actually the division of the sample sum of squares by $n - 1$ makes the sample variance estimate, s^2, the best estimator of the population variance, σ^2, in the sense that it is not biased downward. Since the standard deviation is the positive square root of the variance, we will proceed using the simplifying, though somewhat erroneous, notion that s is the best estimator of σ.

imposed upon the quantities involved. For n numbers there are n squares of deviations from the mean, of which only $n - 1$ are independent; that is, when $n - 1$ of them are specified the nth one is also determined. This follows from the fact that the sum of the deviations from the mean must be zero, a restriction. The sum of squares of deviations of the n values from their mean has, therefore, $n - 1$ degrees of freedom. We shall make it a practice always to give the degrees of freedom associated with any new sum of squares whenever it first arises.

EXAMPLE 1

The real estate multiple listing service in a certain city would like to know how long it takes homes which it lists to sell. Thus, the director of the service took a sample of $n = 10$ homes listed last year and found they sold in the following number of weeks (rounded to the nearest whole week): 4, 8, 22, 1, 12, 7, 9, 5, 20, 6. Find the best estimates of the mean and standard deviation of the time to sale for the population of all homes listed by the service last year.

Since these data form a sample and we are interested in the population, we will find $\overline{X}$, the sample mean, and s, the sample data's estimate of the population standard deviation.

X_i	$X_i - \overline{X}$	$(X_i - \overline{X})^2$
4	−5.4	29.16
8	−1.4	1.96
22	12.6	158.76
1	−8.4	70.56
12	2.6	6.76
7	−2.4	5.76
9	−0.4	0.16
5	−4.4	19.36
20	10.6	112.36
6	−3.4	11.56
94		416.40

$$\overline{X} = \frac{1}{n} \sum_{i=1}^{10} X_i = \frac{1}{10} (94) = 9.4 \text{ weeks}$$

$$s = \sqrt{\frac{1}{n-1} \sum_{i=1}^{10} (X_i - \overline{X})^2} = \sqrt{\frac{1}{9} 416.40} = 6.8 \text{ weeks}$$

EXAMPLE 2

In the previous section we worked an example in three steps. We calculated the mean and standard deviation of the number of defects found by a quality control inspector during nine successive inspection hours. We found the mean and standard deviation of those nine figures were $\mu = 5$ defects and $\sigma = $

1.89 defects, respectively. Those calculations were made under the assumption that the inspector wanted the mean and standard deviation of *those nine values,* i.e., of that population of nine numbers. In a real situation, however, a quality control inspector typically takes a sample of roughly this size and tries to estimate the population mean and standard deviation from the sample. If we view the nine values in that example as a sample, then we find that the sample mean, $\overline{X}$ = 5 defects, remains unchanged except that the notation of $\overline{X}$ rather than μ is used. The sample standard deviation, s, is found by dividing the sum of squared deviations around the mean by $n - 1 = 8$ rather than by 9. This is because the population of all hours of operation of this process will likely have greater variation or spread than do these nine values. The sum of squared deviations in that example was 32. Thus

$$s = \sqrt{32/8} = \sqrt{4} = 2 \text{ defects}$$

Note that $s = 2$ defects is larger than the 1.89 defects we found in the example where we assumed the given data formed the population of interest.

The reader may wonder how he or she is to determine whether to use the formula of σ, the population standard deviation, or s, the sample standard deviation. The answer is that most statistical problems involve sampling. Thus the formula for s, using $n - 1$ as the divisor, is the most commonly used formula. But in situations where the reader is unsure, he or she should ask this question: "Am I interested in these particular values or are they just representative of some larger group which I would examine if I have the time, resources, and patience?" If the numbers you have are what you are interested in, then calculate μ and σ, using n as the divisor, to describe them. If you are interested in some larger group, then calculate $\overline{X}$ and s, using $n - 1$ as the divisor, as your descriptive statistics.*

---

*In those special cases where we are sampling from a finite population of N values, then we can adjust our calculation of the sample standard deviation to account for the fact that a sample making up a large percentage of the population will give a more accurate estimate of σ. To do this we multiply the sample standard deviation by the *finite population correction factor* as shown next:

$$s = \sqrt{\frac{1}{n-1} \sum (X_i - \overline{X})^2} \times \sqrt{\frac{N-n}{N-1}}$$

When the sample is small relative to the total population size, we usually ignore this correction since $(N - n)/(N - 1)$ is very close to 1.0. Some statisticians suggest that when the sample is less than 10% of the finite population, the correction factor should not be used. This is most often the case in business and economics statistical problems.

3.12 WHAT DOES THE STANDARD DEVIATION TELL US?

So far we have concentrated on how to *calculate* the standard deviation of a set of numbers. We have said very little about what the standard deviation *means,* other than to say that it is a measure of spread or dispersion in the data set. A data set with a large standard deviation has much dispersion with values widely scattered around its mean, and a data set with a small standard deviation has little dispersion with the values tightly clustered about its mean.

The range of a data set, the largest minus the smallest value, is a rough measure of this spread and it is very understandable. The standard deviation, on the other hand, has little intuitive appeal. It is the mathematical properties of the standard deviation, which we will use in subsequent chapters, which justify its wide use in statistics. Since we will be using this measure often, we will spend some time at this point to relate more fully what the standard deviation tells us about a set of numbers.

The following simple rule has been developed both from theory and from empirical observations. It may be useful in helping the reader obtain a more intuitive feel for the meaning of the standard deviation.

> For any set of numbers, at least 90 percent of the values lie within plus or minus 3 standard deviations from the mean.

Let us examine what this rule tells us. Assume that a set of numbers has a mean of 50 and a standard deviation of 5. Then the rule tells us that at least 90 percent of the numbers in the data set lie in the interval $50 \pm (3)(5) = 50 \pm 15$, or from 35 to 65.

In the previous section we presented an example in which the estimated mean and standard deviation of time to sale of homes listed with a multiple listing service were $\overline{X} = 9.4$ weeks and $s = 6.8$ weeks, respectively. If we form an interval that is 3 standard deviations on either side of 9.4 weeks, we have $9.4 \pm (3)(6.8) = 9.4 \pm 20.4 = -11.0$ to 29.8 weeks. Thus, the director of this service can say that at least 90 percent of the homes listed with his service sell within 0 to 29.8 weeks of their listing. (It would be silly to apply any meaning to the lower end of the interval since -11.0 weeks is meaningless.)

The quality control inspector in Example 2 of the previous section estimated the population mean and standard deviation for the number of hourly defectives to be $\overline{X} = 5$ defects and $s = 2$ defects. If she were to form an interval of 3 standard deviations on either side of the mean, she would obtain $5 \pm (3)(2) = 5 \pm 6 = -1$ to 11. Then she could say that at least 90 percent of the hours of operation for her process produce between 0 and 11 defects.

A more exact statement about the proportion of a data set that lies in an interval can be made if we have data with special properties. That is, if we have a data set whose histogram shows that the numbers tend to follow a bell-shaped curve, we can make the following statement:

> About two-thirds, or 67 percent, of the values lie within the interval $\mu \pm \sigma$. About 95 percent of the values lie within the interval $\mu \pm 2\sigma$. And virtually all of the values lie within the interval $\mu \pm 3\sigma$.

Three things should be noted about this rule for bell-shaped data. The first is the restriction that it can be used only when there is reason to believe that the data are symmetrical about their mean and have a histogram which looks something like the one shown in Figure 3.4. Typically, physical measurements such as weights, lengths, and pressures follow a bell-shaped curve. This more powerful rule can then be applied to this type of data.

The second thing to note is that this rule is consistent with the simpler rule stated earlier. The simpler rule holds for *any* set of data, while this rule holds only for bell-shaped data. But bell-shaped data qualify as *any* data set and the "virtually all" statement in the preceding rule qualifies as "at least 90 percent" we get from the simpler rule for the interval 3 standard deviations on either side of the mean.

The third thing we should note about the rule explaining the meaning of the standard deviation in bell-shaped data is that the range of the values, the highest minus the lowest, is about 6 standard deviations (plus 3 and minus 3 standard deviations gives an interval that is 6 standard deviations wide).

As an example of applying this rule, assume you are given the information that trucks traveling on a particular highway in your state

FIGURE 3.4   Data with a bell-shaped histogram

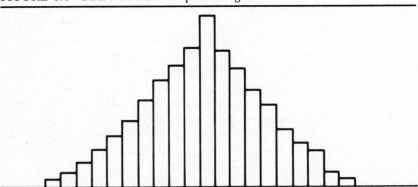

weigh an average of 12.5 tons and have a standard deviation of their weights that is 2.2 tons. What does this standard deviation tell you? Since weights often follow a bell-shaped histogram when plotted, you could estimate that about 67 percent of the trucks on that highway weigh between $12.5 - 2.2 = 10.3$ tons and $12.5 + 2.2 = 14.7$ tons. In addition, you could estimate that about 95 percent of the trucks weigh between 8.1 and 16.9 tons (plus 2 and minus 2 standard deviations). Virtually all of the trucks would weigh between 5.9 and 19.1 tons.

It would not be wise for us to apply this bell-shaped data rule to the previous examples involving the time to sell homes and the number of defects each hour on a production line. This is because we have no evidence that those data follow bell-shaped histograms when they are plotted, and they are discrete counts, not continuous physical measurements like the weights of trucks.

3.13   DESCRIPTIVE STATISTICS FOR GROUPED DATA

There are times when we want to find the descriptive statistics, mean, median, mode, and standard deviation of a data set which has been grouped into a frequency distribution. It may be that someone has already grouped the data for us and that we do not have access to the raw data to make these calculations as they have been presented in the previous sections of this chapter. Or sometimes if we are faced with finding the mean for a very large number of observed values and do not have powerful electronic calculators available, we can materially decrease the labor involved by grouping the data into a frequency distribution and then finding the statistics for the grouped data. It should be borne in mind that by grouping the data we have lost information, and the statistics obtained from the grouped data will therefore only approximate those of the ungrouped data. If the number of observations is large and the class intervals are small, the approximation will be very good.

To illustrate the calculation of the mean for grouped data, we apply the procedures to the data summarized in Table 3.1. To find the mean of this distribution, we operate as if each observed value in a given class were equal to the class mark for that class. Now the observations are $X_1, X_2, \ldots, X_n$, where $n = 300$ for this set of data, and these are grouped into k classes, where $k = 10$ in this case. Let v_j denote the class mark for the jth class, and let f_j denote the class frequency for that class. Now if each observation X_i falling into the jth class had a value of exactly v_j instead of approximately v_j, the sum of all the observations in the jth class would be exactly $v_j f_j$. In Table 3.1, the class mark of the fourth class (1100 but less than

TABLE 3.1   Annual Earnings of 300 Part-time Employees

ANNUAL EARNINGS	CLASS MARK v_j	NUMBER OF EMPLOYEES f_j
$ 500 but less than 700	$ 600	12
700 but less than 900	800	21
900 but less than 1100	1000	52
1100 but less than 1300	1200	70
1300 but less than 1500	1400	68
1500 but less than 1700	1600	36
1700 but less than 1900	1800	16
1900 but less than 2100	2000	11
2100 but less than 2300	2200	9
2300 but less than 2500	2400	5
	Total	300

1300) is $v_4 = 1200$ and the class frequency is $f_4 = 70$. If each employee in the fourth class made exactly $1200, the sum of the earnings for these 70 employees would be $v_4 f_4 = 1200 \times 70$, or $84,000. This will *approximate* their total earnings even if their individual earnings are spread out over the range $1100–1300. The total earnings for all 300 employees is approximated by summing the individual class totals:

$$\sum_{j=1}^{k} v_j f_j = v_1 f_1 + v_2 f_2 + \ldots + v_k f_k$$

Notice that, since each of the n observations goes into exactly one of the k classes, the class frequencies f_j must add to the total number of observations n:

$$\sum_{j=1}^{k} f_j = f_1 + f_2 + \ldots + f_k = n$$

The mean for the grouped data is thus given by

$$\mu = \frac{\Sigma v_j f_j}{\Sigma f_j} = \frac{1}{n} \sum_{j=1}^{k} v_j f_j \qquad (3.15)$$

If the data with which we are working form a sample, we find the mean just as is shown in Formula (3.15), but we call it $\overline{X}$ rather than μ. For the data of Table 3.1, which we will view to be a sample of 300 earnings figures, we calculate the mean as follows:

CLASS MARK v_j	FREQUENCY f_j	PRODUCT $v_j f_j$
$ 600	12	7,200
800	21	16,800
1000	52	52,000
1200	70	84,000
1400	68	95,200
1600	36	57,600
1800	16	28,800
2000	11	22,000
2200	9	19,800
2400	5	12,000
$n = \Sigma f_j = 300$		$\Sigma v_j f_j = 395,400$

$$\overline{X} = \frac{395,400}{300} = \$1318$$

For a second example, consider the distribution of employee absences of Chapter 2. The calculations necessary to obtain the grouped mean of the 106 days are presented in Table 3.2. The class interval used in these calculations is four employees wide. Note that the class marks have values like 121.5, 125.5, etc. Thus in calculating the mean of the grouped data, we operate as though on all 21 days in the 136–139 class there were 137.5 employees absent. This is, of course, impossible

TABLE 3.2   Calculation of Mean Daily Absences Using Grouped Data

NUMBER OF ABSENCES	CLASS MARK v_j	FREQUENCY f_j	PRODUCT $v_j f_j$
120–123	121.5	1	121.5
124–127	125.5	0	0.0
128–131	129.5	2	259.0
132–135	133.5	7	934.5
136–139	137.5	21	2,887.5
140–143	141.5	41	5,801.5
144–147	145.5	19	2,764.5
148–151	149.5	12	1,794.0
152–155	153.5	2	307.0
156–159	157.5	1	157.5
		$\Sigma f_j = 106$	$\Sigma v_j f_j = 15,027.0$

$$\mu = \frac{15,027.0}{106} = 141.8 \text{ employees}$$

since the original data registered only whole numbers of employees. But this somewhat artificial procedure allows us to calculate a mean for the grouped data that is usually quite close to the mean of the ungrouped data.

The difference between the mean of 141.8 employees obtained from the grouped data and that calculated from the ungrouped data, the 141.6 found in Section 3.3, is due to the loss of information through grouping. The individual daily absence records lost their identities in the process, and we assumed that on every day in a given class the number of employees absent was equal to the class mark. Grouping according to a class interval two employees wide rather than four (see Table 2.7) gives a grouped mean of 141.6, which exactly equals the ungrouped mean; a class interval eight employees wide gives a grouped mean of 142.1, which is farther away. In general, the wider the class interval, the farther the grouped mean tends to be from the actual, ungrouped mean.

The mean for a grouped data set is also its center of gravity. If we draw the histogram on some stiff material of uniform density, such as plywood or metal, cut it out, and balance it on a knife-edge arranged perpendicular to the horizontal scale, as shown in Figure 3.5, the X value corresponding to the point of balance is the mean μ.

Since we often have our data in the form of a grouped frequency distribution, we need a way of estimating the *median for grouped data*. Recall that when raw data are grouped into classes, information is lost because the distinction between the various observations in each individual class is lost. The following method assumes that the observations in each class are more or less evenly spread over that class, so that this loss of information is of little importance.

FIGURE 3.5   The mean as a center of gravity

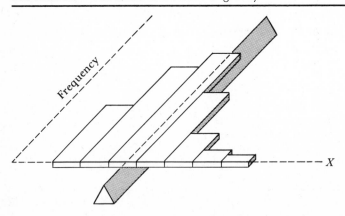

The method is most easily understood and carried out on the ogive. Consider the ogive in Figure 2.6 for the 106 records of employee absences. The height of the curve over a given value on the horizontal scale represents approximately the number of observations whose numerical value is less than that given. Since we want the value for which about 53 (half of 106) observations are less than it, we should look for that value on the horizontal scale for which the height of the curve above it is 53. The way to do this is to locate the cumulative frequency 53 on the vertical scale, trace horizontally over to the curve and then vertically down to the scale at the bottom (see the lower and inside dotted lines in Figure 2.6), and read off the value of the point where the traced line intersects the horizontal scale: approximately 141.5. Thus, the median is about 141.5. The actual median according to the preceding definition, obtained by forming the array, is 141.

There is an arithmetic procedure equivalent to this graphical one; it is less convenient if the ogive has been constructed, but it does not require the ogive. From the cumulative frequency distribution for the employee absences as given in Table 2.6, we see that there are 31 observations less than 139.5 and 72 observations less than 143.5; therefore, since 53 lies between 31 and 72, the median must lie in the class bounded on one side by 139.5 and on the other by 143.5. Now there are $72 - 31$, or 41, observations in this class, and of these $53 - 31$, or 22, are less than or equal to the median. We need for the median a value, call it m, such that 53 out of all 106 observations are less than or equal to m, which is the same thing as a value m such that 22 of the observations in the range 139.5–143.5 are less than or equal to m. If the 41 observations in this range are essentially evenly spread out, then m should be about 22/41 of the distance from the lower class boundary to the upper class boundary, that is,

$$m = 139.5 + \frac{22}{41} \times 4 = 141.6$$

This value should be the same as the value arrived at with the graphical procedure; the difference in the answers, 141.6 as opposed to 141.5, is due merely to the difficulty of reading the scale in Figure 2.6 with great accuracy. Since the median according to our original definition is 141, the discrepancy is immaterial anyway.

The formula for this arithmetic procedure is

$$m = b + \frac{d}{f} \times c \tag{3.16}$$

where m = median,

b = lower boundary of the class containing the median,

c = width of the class containing the median,

f = frequency of the class containing the median,

d = $n/2$ minus the cumulative frequency at the lower boundary b of the class containing the median, n being the total number of observations.

As a second example of the arithmetic procedure, consider the cumulative frequency distribution for the earnings of part-time employees in Table 3.1.

ANNUAL EARNINGS	*NUMBER OF EMPLOYEES*	*CUMULATIVE FREQUENCY*
$ 500 but less than 700	12	12
700 but less than 900	21	33
900 but less than 1100	52	85
1100 but less than 1300	70	155
1300 but less than 1500	68	223
1500 but less than 1700	36	259
1700 but less than 1900	16	275
1900 but less than 2100	11	286
2100 but less than 2300	9	295
2300 but less than 2500	5	300
Total	300	

Since 150 (half of 300, the number of employees) is between 85 and 155, the median must fall in the fourth class. The lower boundary for that class is 1100, the width of the class is 200, its frequency is 70, $n/2$ is 150, and the cumulative frequency at the lower boundary of the fourth class is 85. Thus,

$$b = 1100 \qquad c = 200 \qquad f = 70 \qquad d = 150 - 85 = 65$$

The formula gives

$$m = 1100 + \frac{65}{70} \times 200 = 1100 + 185.7 = 1285.7$$

so the median earnings is about $1286.

When data are grouped, we often find that there is one class which has maximum frequency and that the frequencies of the other classes tend to fall away continually as we move away from the maximum class in either direction. This class is called the *modal class*, and we define the **mode** as the class mark of that class.

For example, for the distribution of employee absences given in Table 2.2, the class with the greatest frequency is 140–143, the frequency being 41. Moreover, as a glance at the histogram in Figure 2.1 shows, the frequencies taper off on either side of this one with the trivial exception that the frequency of the first class exceeds that of the second. This is therefore the modal class, and its midpoint or class mark 141.5 is the mode M:

$$M = 141.5$$

If there is such a class with maximum frequency, then the distribution is said to be *unimodal*. If there is no such class, the distribution is said to be *multimodal*, and the mode is undefined. In a histogram like that in Figure 3.6, the distribution is multimodal; A is the major mode, and B and C are minor modes. The concentration of values is greatest around A, and there are secondary concentrations around B and C.

So far in this section we have identified methods for finding the mean, median, and mode of data that have been grouped into a frequency distribution. These three measures, however, are all measures of location—measures of the typical value in the data set. We often want to measure the variation in a number set too. Since the variance and standard deviation are the most commonly used measures of variation in statistics, we will only consider methods of finding these two measures in grouped data.

When we want to calculate the variance and standard deviation for data which have been grouped or classified to form a frequency table we follow the three-step procedure outlined previously, but the calculation of the sum of squares requires a slight modification. As when computing the mean μ, we proceed as though each value in a given class were equal to the class mark. To find the sum of squares, we calculate the mean for the distribution by the procedure of Equation (3.15) and then find the deviations of the class marks v_j from the mean. The square of the deviation of a class mark from the mean, $(v_j - \mu)^2$, is the amount each value in the class contributes to the sum of squares. If there are f_j items in the jth class, the total contribution of the class is $(v_j - \mu)^2 f_j$. Therefore, the sum of squares is the sum of the contributions of all of the classes,

FIGURE 3.6 A histogram showing major and minor modes

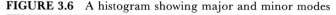

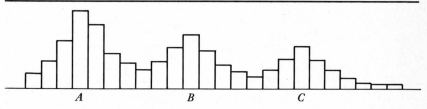

$$\Sigma \, (v_j - \mu)^2 f_j$$

and the variance of grouped data is

$$\sigma^2 = \frac{1}{n} \sum_{j=1}^{n} (v_j - \mu)^2 f_j \tag{3.17}$$

Of course, if the grouped data we are using form a sample, then we use $\overline{X}$ to represent the mean of the sample and the sample variance is calculated as

$$s^2 = \frac{1}{n-1} \Sigma \, (v_j - \overline{X})^2 f_j \tag{3.18}$$

where $n = \Sigma f_j$ is the total number of observations in the data.

Let us use the numbers in the following frequency distribution to go through a sample calculation. Suppose the following table shows the age distribution in a *sample* of 101 children who regularly watch a particular television program (and are thus exposed to its advertising).

AGE	CLASS MARK v_j	FREQUENCY f_j
0– 4	2	30
5– 9	7	51
10–14	12	10
15–19	17	10
	Total	101

The calculations required to obtain the variance and standard deviation for this distribution are displayed in the next table.

v_j	f_j	$v_j f_j$	$v_j - \overline{X}$	$(v_j - \overline{X})^2$	$(v_j - \overline{X})^2 f_j$
2	30	60	−5	25	750
7	51	357	0	0	0
12	10	120	5	25	250
17	10	170	10	100	1000
	101	707			2000

$$\overline{X} = \frac{1}{n} \Sigma v_j f_j = \frac{707}{101} = 7 \text{ years}$$

$$s^2 = \frac{1}{n-1} \Sigma \, (v_j - \overline{X})^2 f_j = \frac{2000}{100} = 20 \text{ (years)}^2$$

$$s = \sqrt{s^2} = \sqrt{20} = 4.47 \text{ years}$$

Note that in this example the deviations of the class marks from the mean do *not* sum to zero. For grouped data it is the sum of the products of the frequencies times the deviations of the class marks from the mean that is equal to zero.

The machine formula for the sum of squares of grouped data is

$$\Sigma v_j^2 f_j - \frac{1}{n} (\Sigma v_j f_j)^2 \tag{3.19}$$

where v_j is the class mark. If we use the machine formula to find the sum of squares for the preceding table, the calculations are as follows:

v_j	v_j^2	f_j	$v_j f_j$	$v_j^2 f_j$
2	4	30	60	120
7	49	51	357	2499
12	144	10	120	1440
17	289	10	170	2890
		101	707	6949

$$\Sigma v_j^2 f_j - \frac{1}{n} (\Sigma v_j f_j)^2 = 6949 - \frac{(707)^2}{101}$$

$$= 6949 - 4949 = 2000$$

$$s^2 = \frac{2000}{100} = 20 \ (\text{years})^2$$

$$s = \sqrt{20} = 4.47 \ \text{years}$$

Because of the information lost in grouping, the variance and standard deviation are, in general, not exact when calculated from the grouped data. They are good approximations, however, and the smaller the class interval, the better the approximation. Neither the mean nor the variance can be calculated when the distribution contains open-ended classes.

3.14   SELECTING A MEASURE OF LOCATION

We have presented four measures of location in this chapter: the mean, weighted mean, median, and mode. We have also presented four measures of variation: the range, mean deviation, variance, and standard deviation. We have indicated that due to certain mathematical properties possessed by the variance and standard deviation (which are virtually the same measure of variation) we most often use them to describe the variation, dispersion, or scatter in a data set. However, we have not indicated when it is appropriate to use the different measures of location. That is the purpose of this section.

In deciding which measure of location to report for a set of data, a primary consideration is the use to which the results are to be put. In addition, we need to know the advantages and disadvantages of each measure of location as regards its calculation and interpretation.

If the distribution of the data is symmetric and unimodal, then the mean μ, the median m, and the mode M will all coincide; but as the distribution becomes more and more skewed (lopsided), the differences among these measures will become greater. This is illustrated in Figure 3.7.

The mean is sensitive to extreme values. If in a small town the average annual income of the 100 heads of household is reported as $9990, this figure is correct but very misleading if one head of household is a multimillionaire with an income of $900,000 and the remaining 99 are paupers with incomes of $1000. A few extreme values have little or no effect on the median and the mode. For the numbers

$$1, 3, 4, 6, 6, 9, 13$$

we have $\mu = 6$, $m = 6$, and $M = 6$. If we add the number 70 to this set of numbers the mean will be equal to 14, a shift of 8 units, but the median and mode remain unchanged. For this reason, when

FIGURE 3.7   Mean, median, and mode for symmetrical and assymmetrical distributions

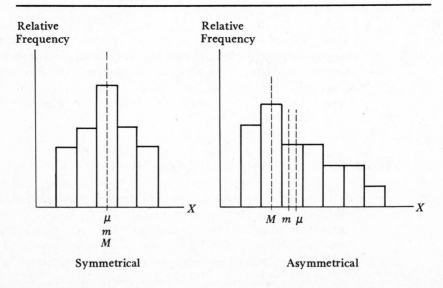

Symmetrical

Asymmetrical

the data are skewed the median or the mode may be more characteristic and therefore provide a better description.

With the increasing availability of electronic calculators and computers, ease of calculation is a less important consideration. It should be noted, however, that ungrouped data must be arrayed to find the median; hence, if an adding machine or calculator is at hand it may be easier to find the mean. Once the data have been grouped, both the median and the mode are calculated more simply than the mean.

The effects of grouping must also be considered. If the frequency distribution includes open-ended classes, the mean cannot be evaluated accurately, if at all. Open-ended classes, in general, will not affect the median and mode. If the classes have unequal widths all three measures may be found, but they will usually be poorer approximations to the values obtained from the ungrouped data. The mode is especially sensitive to grouping in that alternative sets of class intervals may result in modes that differ considerably.

If we want to combine measures for several sets of data, the algebraic properties of the mean give it a distinct advantage. We have seen that we can use the weighted mean for this purpose. The median and the mode are not subject to this type of algebraic treatment.

If the evaluation of a measure of location is a first step toward making inferences about the source of the data, the mathematical and distributional properties of the mean give it a distinct advantage. In the realm of statistical inference a primary consideration is statistical stability. It can be shown that if a large number of sets of data are taken from the same source and all three measures are calculated for each set, there will be less variation among the means than among the medians or modes; hence, the mean is more stable. This, coupled with the fact that it is more amenable to mathematical and theoretical treatment, makes the mean an almost universal choice for all but purely descriptive purposes. The three measures of location are usually ranked according to their overall desirability in the following order: mean, median, and mode.

3.15   THE COEFFICIENT OF VARIATION

One final descriptive statistic combines the standard deviation and the mean. It is called the **coefficient of variation.** The following example will show why the coefficient of variation is a useful measure of relative dispersion.

Consider two stocks, A and B. If we take a random sample of the daily closing prices of these stocks, we might find that the respective standard deviations of these closing prices are:

$$s_A = \$.50 \quad \text{and} \quad s_B = \$5.00$$

Since the standard deviation measures variation in a set of numbers, we might conclude that the second stock's closing prices vary much more than those of the first stock. In fact, we might wish to avoid investment in such a volatile stock and choose to put our funds into stock A.

Before we call our broker, however, it might be wise to note that

$$\overline{X}_A = \$1.00 \quad \text{and} \quad \overline{X}_B = \$100.00$$

Now the situation looks rather different. Stock A is really the one with volatile price movements because its standard deviation of price closings is 50 percent of its mean value. On the other hand, stock B's standard deviation of price closings is only 5 percent of its mean value.

In order to convert the standard deviation to a value that can be compared between two number sets of rather different magnitudes, we compute the coefficient of variation as:

$$CV = (\sigma/\mu) \cdot 100 \quad \text{or} \quad CV = (s/\overline{X}) \cdot 100 \qquad (3.20)$$

Thus, the coefficient of variation is a unitless figure that expresses the standard deviation as a percentage of the mean. We intuitively found the coefficients of variation for the two stocks in our example:

$$CV_A = (\$.50/\$1.00) \cdot 100 = 50\%$$

and

$$CV_B = (\$5.00/\$100.00) \cdot 100 = 5\%$$

The coefficient of variation is sometimes used by financial managers to measure and compare the riskiness of competing portfolios of investments. Those portfolios with higher coefficients of variation go through wider fluctuations in their market value from period to period than do those portfolios with smaller coefficients of variation.

3.16 DESCRIPTIVE MEASURES AND ELECTRONIC COMPUTERS

The reader has, no doubt, recognized that the calculation of descriptive measures for a large set of data can be rather involved and require a vast number of computations. In recent years small electronic calculators have become widely available to the general public. Many university and public libraries will rent out these machines for a small fee. These machines can cut the time required for obtaining descriptive measures in a large data set from hours to minutes. Some calculators even have automatic function keys which allow the user to find the mean, standard deviation, and coefficient of variation of a data set by merely entering the data and pushing one or two keys. Readers are urged to work a good share of the problems at the end of this chapter without using such automatic functions. They will find that they understand the descriptive measures better if they resist these work-savers—at least in the beginning.

If one does use a calculator with automatic function keys, two cautions are needed. First, the mean and standard deviation of grouped data cannot typically be figured using such keys. Second, the user should determine if the automatic function for computing the standard deviation divides the sum of squared deviations by n or by $n - 1$. Most calculators with this feature use $n - 1$ as the divisor.

When extremely large data sets are involved, statisticians usually employ a large electronic computer to do the calculations required to obtain the desired descriptive measures. Many computers come with programs (sets of instructions to the computer) already written which will calculate any of the descriptive measures we have discussed in this chapter. Someone using such a program must simply supply the computer with the code number for the program the user wants run—for example, the program for calculating descriptive measures—and then supply the computer with the data. The code number for the descriptive measures program differs from computer to computer. Users must rely on their local computer systems experts for code information. Most of these programs use the machine formula in calculating the variance and the standard deviation. The formula saves arithmetic computations and thus saves computer time (and money). The data can be supplied to the computer in many ways. The most common are: (1) punching the numerical values on tabulating cards from which the computer can read the information in the punched holes, and (2) typing the numerical values on the keyboard of a teletypewriter terminal wired directly into the computer.

PROBLEMS: *Answers to odd-numbered problems are found in the back of the text.*

1. Was the powers-of-2 rule given in Chapter 2 used to determine the number of classes used in Table 3.1? If not, how many classes would that rule suggest for the data in Table 3.1?

2. Write out the following summations as sums of the terms involved.

   a. $\displaystyle\sum_{i=1}^{5} i^2$

   b. $\displaystyle\sum_{i=1}^{4} a^i Y_i$

   c. $\displaystyle\sum_{i=2}^{5} Y_i^i$

   d. $\displaystyle\sum_{i=3}^{6} (Y_i - a^i)$

   e. $\displaystyle\sum_{i=1}^{3} Y_i^2 - 3$

   f. $\displaystyle\sum_{i=4}^{7} (-1)^{i+1} Y_i$

3. Express the following in summation notation.
   a. $X_1 Y_1 + X_2 Y_2 + X_3 Y_3 + X_4 Y_4$
   b. $Y_1^2 f_1 + Y_2^2 f_2 + Y_3^2 f_3 + Y_4^2 f_4$
   c. $(Y_1 + Y_2 + Y_3)^2$
   d. $X_1 + X_2 + X_3 - Y_1 - Y_2 - Y_3$
   e. $Y_1 - Y_2^2 + Y_3^3 - Y_4^4 + Y_5^5 - Y_6^6$
   f. $Y - \dfrac{Y^2}{2} + \dfrac{Y^3}{3} - \dfrac{Y^4}{4} + \dfrac{Y^5}{5}$

4. Given that

$$X_1 = 3 \qquad X_4 = 6 \qquad X_7 = 3$$
$$X_2 = 2 \qquad X_5 = 1 \qquad X_8 = 5$$
$$X_3 = 4 \qquad X_6 = -2 \qquad X_9 = 1$$

find:

   a. $\displaystyle\sum_{i=1}^{9} X_i$
   b. $\displaystyle\sum_{i=1}^{9} X_i^2$
   c. $\left(\displaystyle\sum_{i=1}^{9} X_i\right)^2$

   d. $\displaystyle\sum_{i=4}^{9} X_i$
   e. $\displaystyle\sum_{i=2}^{5} (X_i - 3)^2$

5. Use the same data given for X_1 through X_9 values in Problem 4 to find:

   a. $\displaystyle\sum_{i=1}^{6} (X_i^2 + 2)$
   b. $\displaystyle\sum_{i=1}^{5} X_i$
   c. $\displaystyle\sum_{i=1}^{9} (X_i - 4)$

   d. $\displaystyle\sum_{i=1}^{9} X_i - 4$
   e. $\displaystyle\sum_{i=6}^{9} i X_i$
   f. $\displaystyle\sum_{i=1}^{4} 8$

6. Show that
   a. $\displaystyle\sum_{i=1}^{n} (Y_i - c)^2 = \sum_{i=1}^{n} Y_i^2 - 2c \sum_{i=1}^{n} Y_i + nc^2$
   (*Hint:* Expand the $(Y_i - c)^2$ term. Then apply the first, second, and third summation rules, in that order.)

   b. $\displaystyle\sum_{i=1}^{n} (Y_i - \bar{Y}) = 0,$ where $\bar{Y} = \dfrac{1}{n} \displaystyle\sum_{i=1}^{n} Y_i$
   (*Hint:* Rearrange the expression on the right to read $n\bar{Y} = \Sigma Y$ and substitute for ΣY in the expression on the left.)

7. Given that $X_1 = 3 \qquad X_2 = 4 \qquad X_3 = 2 \qquad X_4 = -3$
                $Y_1 = 2 \qquad Y_2 = 1 \qquad Y_3 = 5 \qquad Y_4 = 7$
   find:

   a. $\displaystyle\sum_{i=1}^{4} X_i Y_i$
   b. $\left(\displaystyle\sum_{i=1}^{4} X_i\right)\left(\displaystyle\sum_{i=1}^{4} Y_i\right)$

   c. $\displaystyle\sum_{i=1}^{4} (X_i - Y_i)$
   d. $\displaystyle\sum_{i=1}^{4} (X_i - 4)(Y_i - 5)$

e. $\sum_{i=1}^{4} X_i Y_i - \frac{1}{4} \left(\sum_{i=1}^{4} X_i \right) \left(\sum_{i=1}^{4} Y_i \right)$ **f.** $\sum_{i=1}^{4} X_i^2 Y_i$

8. For each of the following *samples* find:
   - **a.** The mean.
   - **b.** The median.
   - **c.** The mode or modes.
   - **d.** The range.
   - **e.** The variance.
   - **f.** The standard deviation.

   (i)   2, 5, 9, 11, 13
   (ii) 1, 3, 3, 5, 6, 6
   (iii) 0, 0, 0, 0
   (iv) 3, 6, 2, 5, 3, 8, 6, 7, 5
   (v)   −4, 2, −6, 0, −4, 6, 2, 4, 0
   (vi) 1, 2, 3, 4, 5
   (vii) 16, 2, 22, 8, 6, 20, 24, 14
   (viii) −1, 2, −2, 1, −1, 4, 2, 3, 1

9. The following numbers represent the weights of 30 cattle (in pounds) after they have been in a feedlot for 60 days. For these *sample* data find:
   - **a.** The mean.
   - **b.** The median.
   - **c.** The range.
   - **d.** The variance.
   - **e.** The standard deviation.
   - **f.** The coefficient of variation.
   - **g.** The standard deviation if you consider the data to be a population.

982	1205	258	927	620	1023
395	1406	1012	762	840	960
1056	793	713	736	1582	895
1384	862	1152	1230	1261	624
862	1650	368	358	956	1425

10. An auditor found that the following billing errors occurred on invoices of a certain type. Positive errors indicate the customer was billed too much, and negative errors indicate the customer was billed too little. For these sample data find:
    - **a.** The mean error.
    - **b.** The median error.
    - **c.** The range.
    - **d.** The mean deviation.
    - **e.** The variance.
    - **f.** The standard deviation.

CUSTOMER	DATE	SIZE OF ERROR
Jensen	Mar. 4	$32
Lease	Jan. 8	− 45
Stewart	Dec. 12	66
Miller	June 7	2
Johnson	May 23	− 8
Blodgett	Sept. 9	− 51
Semenik	Feb. 5	12
Glenn	Oct. 30	18

11. The weights (in tons) of boxcars coming into a company's warehouse were found to be: 30.6, 26.2, 44.3, 21.9, and 27.0. For these figures find:

   a. The mean.   b. The median.

   c. The range.   d. The variance (assuming these values form the population).

   e. The variance (assuming these values form a sample).   f. The standard deviation from parts d and e.

12. Find the mean and standard deviation of the beginning weights for the 28 animals whose weight data are given in the case entitled "Mountain States Feed Exchange" at the end of the text.

13. Use your answer from Problem 12 to determine which set of weights, the beginning weights or the ending weights, of the cattle in "Mountain States Feed Exchange" have more *relative* variation. Discuss why your result makes sense; that is, why does the group you find with more relative variation logically have more variation?

14. Which of the following statements are true?

   a. For symmetrically distributed data, the mean, median, and mode are all about the same numerical value.

   b. The range of a set of numbers is usually about three times the standard deviation.

   c. If a set of numbers is skewed to the right (has a few values quite a bit larger than the rest of the numbers), then the mean will exceed the median.

   d. If all the numbers in a data set have the same value, then the variance and standard deviation are both zero.

15. During a recent maintenance check, the maximum thrust of the engines of 40 DC9's was measured. The differences between the thrusts of the port and starboard engines, measured in pounds, are given next. Find the mean, variance, and standard deviation for this sample.

70	480	−50	−100	10
0	190	−150	220	−240
310	−330	440	30	90
−340	360	−180	210	−80
50	−270	−40	−30	20
−40	60	200	−230	−450
−20	−400	80	90	110
−90	−20	60	130	−200

16. The following is the distribution for the number of defectives found in 404 lots of manufactured items. Find the mean, median, mode, variance, and standard deviation of the number defective per lot. (*Hint:* Use the formulas for grouped data as though the numbers 1, 2, . . . , 12 were class marks, v_j.)

NUMBER OF DEFECTIVE ITEMS	NUMBER OF LOTS
0	53
1	110
2	82
3	58
4	35
5	20
6	18
7	12
8	9
9	3
10	1
11	2
12	1

17. Given the following percentages of defective items in eight samples, find the percentage of defective items when the samples are combined into one large sample.

SAMPLE SIZE	PERCENT DEFECTIVE
50	6.0
20	5.0
35	20.0
150	0.0
100	1.0
75	4.0
40	2.5
200	0.5

18. Three sets of data had means of 15, 20, and 24 based on 30, 35, and 50 observations, respectively. What is the mean if these three sets are combined?

19. The following frequency distribution shows the ages of new-car purchasers at a large midwest automobile dealership. Find the mean, variance, and standard deviation of this distribution. (Assume this is a population.)

CLASS	FREQUENCY
33–37	10
38–42	10
43–47	51
48–52	30

20. The following frequency distribution shows the ages of used-car purchasers at the same large midwest automobile dealership mentioned in Problem **19.** Find the mean, variance, and standard deviation of this distribution. Do the new-car and used-car departments appear to be selling to different

age groups? What implications does this information have for the management of this dealership?

CLASS	FREQUENCY
16–21	15
22–27	16
28–33	5
34–39	5

21. Find the mean and median.

CLASS	RELATIVE FREQUENCY
20 but less than 40	.12
40 but less than 60	.28
60 but less than 80	.36
80 but less than 100	.24

22. Find the mean and median.

CLASS	RELATIVE FREQUENCY
5 but less than 10	.2
10 but less than 15	.5
15 but less than 20	.3

23. The following frequency distribution shows the ages of homebuyers in
a Chicago neighborhood. Find the mean, variance, standard deviation,
and coefficient of variation for this sample distribution.

CLASS	FREQUENCY
18–32	5
33–37	10
38–42	10
43–47	30
48–52	35
53–67	10

24. For the estimated numbers of married women in the labor force of the
Eastern U.S. (1979) as given in Problem **14** of Chapter 2 find:
  a. The mean age.   **b.** The median age.
  c. The variance.   **d.** The standard deviation
  e. The coefficient of variation.

25. In an investigation of electronic equipment reliability, a record of the
time between equipment malfunctions was kept for nearly two years. Forty
pieces of equipment were involved and gave rise to 850 observations
on times between malfunctions.
  a. Find the median and the mode.
  b. Find the mean, variance, and standard deviation of this sample.

TIME BETWEEN MALFUNCTIONS (HOURS)	NO. OBSERVATIONS
$0 \leq X < 100$	20
$100 \leq X < 200$	43
$200 \leq X < 300$	60
$300 \leq X < 400$	75
$400 \leq X < 500$	95
$500 \leq X < 600$	138
$600 \leq X < 700$	240
$700 \leq X < 800$	97
$800 \leq X < 900$	62
$900 \leq X < 1000$	20
Total	850

26. If a set of grouped data has an open-ended class for either the first or the last class, which of the following *cannot* be found from the grouped distribution:
  a. The class marks of all the classes.
  b. The boundaries of all the classes.
  c. The mean.
  d. The range.
  e. The standard deviation.
  f. The median.

27. A set of numbers has a mean of 50 and a standard deviation of 3. What is the interval within which you are sure that at least 90% of the numbers lie?

28. A store manager found that the store's mean daily receipts totaled $550 with a standard deviation of $200. What can this manager say about the proportion of days on which receipts lie between $0 and $1,150?

29. The balances in the escrow account for a savings and loan association's mortgage portfolio have a mean of $305 and a standard deviation of $85.
  a. According to the simple rule defining the meaning of a 3-standard deviation interval, what proportion of the accounts have balances between $50.00 and $560.00?
  b. According to the rule for bell-shaped data, what proportion of the accounts have balances between $220.00 and $390.00?

30. The weights of packages coming off a production line have a mean of 10.20 pounds and a standard deviation of 0.10 pounds.
  a. Which rule can likely be applied to these measurements: the simple rule, the bell-shaped data rule, or both?
  b. What proportion of the packages have weights between 10.10 and 10.30 pounds?
  c. What proportion weigh between 10.00 and 10.40 pounds?
  d. If the packages are labeled as containing 10.0 pounds, what proportion of the packages will be underweight?

31. Use the data given in the statement of Problem **30** to estimate the proportion of packages with weights between 10.10 pounds and 10.40 pounds.

SAMPLE DATA SET QUESTIONS: *Refer to the 113 applicants for credit listed in the Sample Data Set Appendix of this book.*

a. Find the mean and standard deviation of JOBYRS for all applicants who were employed for wages. That is, leave out applicants who appear to have been retired, nonworking students, or housewives. (*Hint: n* = 106)
b. Find the mean and standard deviation of ADDINC for all applicants who were denied credit.
c. Use the frequency distribution found in question **c** of the Sample Data Set Questions for Chapter 2 and the formulas for the mean and standard deviation of grouped data to find the mean and standard deviation of JOBINC for the applicants who were granted credit.

REFERENCES

3.1 Croxton, Frederick E., Dudley J. Cowden, and Ben W. Bolch. *Practical Business Statistics,* 4th edition. Prentice-Hall, Englewood Cliffs, 1969. Chapters 3 and 4.

3.2 Hoel, Paul G. *Elementary Statistics,* 3rd edition. John Wiley & Sons, New York, 1971. Chapter 2.

3.3 Huff, Darrell. *How to Lie with Statistics.* Norton, New York, 1965. Chapters 5 and 6.

3.4 Mendenhall, William, and James E. Reinmuth. *Statistics for Management and Economics,* 3rd edition. Duxbury Press, Belmont, California, 1977. Chapters 2 and 3.

3.5 Snedecor, George W., and William G. Cochran. *Statistical Methods,* 6th edition. Iowa State University Press, Ames, 1967. Chapter 5.

ELEMENTARY
PROBABILITY

In this chapter we begin our study of the second major topic area in the subject of statistics—probability. As pointed out in Chapter 1, the mathematical theory of probability provides a basis for evaluating the reliability of the conclusions we reach and the inferences we make when we apply statistical techniques to the collection, analysis, and interpretation of quantitative data. Since probability plays so important a role in the theory and applications of statistics, we need an acquaintance with the elements of the subject.

In this chapter we consider briefly some of the basic ideas of probability theory. These ideas are often illustrated here by examples involving cards and dice. The connections with decision making in management and economics are made here to a limited extent, but are made much more fully in later chapters.

4.1 THE MEANING OF PROBABILITY

Imagine a *random experiment* or *observation;* that is, imagine an operation which can result in any one of a definite set of possible outcomes, but which is governed by chance so that the actual outcome cannot be predicted with complete certainty. We have in mind the drawing of a card from a well-shuffled deck, for example, the rolling of a pair of dice, the selection of an invoice during an audit, or the inspection of a part coming off a production line.

If the random experiment consists of drawing a card from an ordinary deck, there are 52 possible outcomes—the 52 cards in the deck. This set of all possible outcomes or results of the experiment is called the *sample space.** Since 13 of the 52 possible outcomes in the sample space are spades, almost everyone will say that, if the deck was shuffled well, the chance of drawing a spade (the chance that the outcome is a spade) is 13/52. Similarly, almost everyone will say that the chance of drawing an ace is 4/52. This is because each of the outcomes in the sample space is regarded as being equally likely.

For rolling a pair of dice, the sample space—the set of all 36 possible outcomes or results—is exhibited in full in Table 4.1. Here 3–1 denotes the result that one die shows three dots and the other shows one dot. The outcome 1–3, where the order is reversed, is also listed in the table, because the two dice are to be distinguished from each other. (Perhaps it is best to think of a red die and a green one, even though the dice of an actual pair are made so similar that telling them apart would require very close examination.) If the dice are fair ones (not loaded), everyone expects the 36 outcomes to be equally likely. Since the total number of dots showing is 4 for three of the outcomes (namely, for 3–1, 2–2, and 1–3), we expect the probability of rolling a 4 to be 3/36. Since the total showing is 7 for six of the outcomes (namely, for 6–1, 5–2, 4–3, 3–4, 2–5, and 1–6), we expect the probability of rolling a 7 to be 6/36.

In the auditing example, we may have 100 invoices in which 95 are correctly written and 5 are not. If we select only one invoice, then the sample space consists of two possible outcomes: "correct" which has 95 elements and "error" which has 5 elements. The chance of

---

*Sometimes it is called the *outcome space;* this is a better term, but *sample space* has become standard.

TABLE 4.1 Sample Space for Rolling Dice

1–1	1–2	1–3	1–4	1–5	1–6
2–1	2–2	2–3	2–4	2–5	2–6
3–1	3–2	3–3	3–4	3–5	3–6
4–1	4–2	4–3	4–4	4–5	4–6
5–1	5–2	5–3	5–4	5–5	5–6
6–1	6–2	6–3	6–4	6–5	6–6

selecting a correctly written invoice is 95/100, or .95. The chance of selecting one containing an error is 5/100, or .05.

These examples are instances of the classical conception of probability: There is a definite, finite set of possible outcomes or results or cases (the 52 cards, the 36 combinations for the rolled pair of dice, or the 100 invoices); this is the sample space. Within the sample space, there is some smaller set of outcomes (the spades, the dice-pairs that total 7, or the invoices that are correctly written) whose probability we seek. This smaller set is called an *event,* or, variously, a *subset,* and the outcomes or cases making up the event are called the *favorable cases.* The *probability of the event* is taken to be the ratio of the number of outcomes in the event to the number of outcomes in the sample space, that is, the ratio of the number of favorable cases to the total number of cases:

$$\text{Probability} = \frac{\text{Number of favorable cases}}{\text{Total number of cases}} \tag{4.1}$$

In drawing a card, the probability of the event "ace"—the chance of getting an ace—is 4/52, because the event "ace" comprises 4 of the 52 cases. In rolling dice, the probability of the event "seven"—the chance of rolling a total of 7—is 6/36, because the event "seven" comprises 6 of the 36 cases. In the auditing example, the probability of the event "error"—the chance of selecting an invoice containing an error—is 5/100, because the event "error" comprises 5 of the 100 cases.

Formula (4.1) is to be viewed as a *way of computing* a probability and *not* as a *definition* of the probability. Whether or not the computation leads to a result that corresponds with reality depends on the circumstances. The chance of drawing an ace is 4/52 if the deck was well shuffled—but not if a sleight-of-hand expert did the shuffling. The chance of rolling a 7 is 6/36 if the dice are fair—but not if they are loaded. The chance of selecting an invoice with an error, in our example, is 5/100 if the selection is random—but not if the erroneous invoices are hidden away somewhere. Formula (4.1) leads to the right answer if all the possible cases (all the outcomes in the sample space) are equally likely.

The words "right answer" in the previous sentence presuppose the existence of a right answer. Most of us feel that, if a particular pair of dice is rolled, even a loaded pair, there does exist a certain definite probability that the result will be a 7, even though we may be entirely ignorant of the actual numerical value of that probability. This parallels the fact that most of us feel that the sixth moon of Jupiter has a certain definite mass in tons, even though we may be entirely ignorant of the actual numerical value of that mass. And most of us feel that, if a particular coin (even a bent one) is tossed, there does exist a certain definite probability that it will land heads upward. To give a satisfactory definition of this probability is very difficult, just as it is very difficult to define with precision concepts such as mass and force in physics. As in the case of physical concepts, the fruitful procedure is to worry less about how to define probability and more about how to measure it. To get an idea of the probability that a particular coin will show heads, the obvious procedure is to toss the coin a number of times and check the fraction of times it comes up heads. If 500 tosses yield 255 heads (and 245 tails), the probability of heads must be something like 255/500, or .51.

Experience shows that if a coin is tossed repeatedly, the relative frequency (fraction) of heads tends to stabilize at a definite value, which we take to be the probability of heads. Figure 4.1 shows the results of such an experiment. The horizontal scale in this figure represents the number of tosses of the coin, and the height of the curve over any given point on the horizontal scale represents the relative frequency of heads up to that point. In this experiment the coin was tossed 600 times in all. Since the relative frequency seems to be settling down at a value near 1/2, we can say that the coin is well balanced. Figure 4.2 shows the same effect for a different coin. Here, the relative frequency is converging to something like .3, and the coin is seen to be unbalanced. It is typical for events to show this kind of stability. Suppose we repeat many times an experiment such as inspecting parts coming off a production line, and keep track of an event such as "part rejected." Suppose that at each repetition of the experiment we compute the relative frequency with which the event has occurred in the sequence of trials up to that point. Quite generally, this relative frequency will stabilize at some limit which we take to be the probability of the event. This probability is to be regarded as a natural characteristic of the experiment and the event itself, just as the mass of the sixth moon of Jupiter is a physical characteristic of that body.

By now it should be clear that the probability of an event can be defined in terms of relative frequency. That is, the probability that

FIGURE 4.1   Relative frequencies of heads for a balanced coin

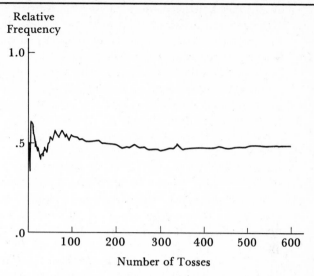

a particular event will be the result of a random experiment is the *proportion of time* that this event is the outcome of the experiment *in the long run.* Here the term *long run* means that the experiment is performed many, many times. It is not necessary, however, to perform an experiment repeatedly and observe the results in order to determine the probabilities of the events associated with the experiment. For

FIGURE 4.2   Relative frequencies of heads for an unbalanced coin

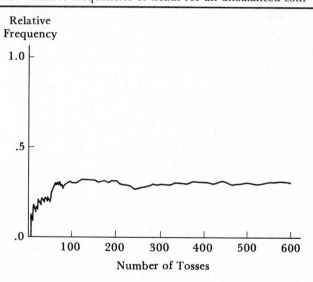

instance, one does not have to toss dice for a long time in order to determine that the probability of rolling two 1's with two dice is 1/36. We were able to show this in Table 4.1 without tossing a single die. This is because it is possible to define probabilities in terms of the physical structure of the experiment as well as in terms of relative frequencies. That is, we know that a die has six sides, and from this knowledge and our understanding of Formula (4.1) we can calculate the *theoretical probability* of various outcomes for rolling dice. But we could also determine the same probabilities using the relative frequency concept. There is a third concept of probability called *subjective probability*. This will be treated briefly in Sections 4.7 and 17.1.

Just as physical quantities like force and mass can often be computed mathematically, so can the probability that belongs to a given event. The formula applicable in many cases of interest to us is Formula (4.1). The probability as given by this formula will often agree closely with the value at which the relative frequency of the event would stabilize in a long sequence of empirical trials of the experiment. This kind of agreement will exist when the outcomes of the experiment (the elements of the sample space) are equally likely or nearly so, and this in turn will be the case in the presence of an appropriate symmetry: when the coin is balanced, for example, or the dice are uniformly made or the invoices are shuffled so that no invoice has precedence over any other.

Computing a probability by Formula (4.1) requires computing the numerator (the number of favorable cases) and the denominator (the total number of cases). In the examples considered thus far, these values were found by actually counting out the cases. This cannot usually be done and is often unnecessary. It has been stated that we can determine the theoretical probabilities associated with tossing dice. But to do so we need some mathematical techniques for computing these numbers. Often they can be found by what we shall call the *addition principles* and the *multiplication principle*.

4.2 THE ADDITION PRINCIPLES

If someone has counted the 11 girls in a class and the 13 boys, he or she knows without counting the whole class over again that there are 24 students altogether. This is because of the principle of addition of disjoint sets. Two sets of objects are said to be *disjoint* or *mutually exclusive* if there is no object that belongs to both sets.

> **An Addition Principle:** **If one set contains *a* objects and a second set contains *b* objects, and if the two sets are mutually exclusive, then the two sets taken together contain *a* + *b* objects.**

For example, a company has 40 people in its manufacturing department and 30 people in its marketing department. Thus, it must have 70 people who are in either manufacturing or marketing. Another simple example of this principle is found in playing cards. There are 4 aces in an ordinary deck of cards and 4 kings, so there are 4 + 4, or 8, cards that are *either* ace *or* king. But how about the case of aces and spades? There are 4 aces and 13 spades, but the number of cards that are either ace or spade is not 4 + 13 = 17, because the set of aces and the set of spades share a card—the ace of spades. This leads us to a more general addition principle.

> **The General Addition Principle:** **If one set contains *a* objects and a second set contains *b* objects, then the two sets taken together contain *a* + *b* − *c* objects, where *c* is the number of objects that are in both sets.**

Let us assume that the company mentioned earlier has a labor force composed in the following manner.

	MANUFACTURING	*MARKETING*	*TOTAL*
Male	30	25	55
Female	10	5	15
Total	40	30	70

The number of people who are either female or in manufacturing is 15 (total females) plus 40 (total manufacturing) minus 10 (those who are female *and* in manufacturing), or 45.

The general addition principle is also applicable to the cases of the more limited rule of addition of mutually exclusive sets. If two sets are mutually exclusive, then c, the number of objects contained in both sets, is zero. Thus the two disjoint, or mutually exclusive, sets taken together contain $a + b - 0 = a + b$ objects, as stated in the first addition principle. Thus, the first principle is a special case of the general principle.

The addition principle extends in an obvious way to three or more sets: Since there are 4 jacks, 4 queens, and 4 kings, there are 4 + 4 + 4, or 12, face cards all told.

4.3 THE MULTIPLICATION PRINCIPLE

We tend to use the addition principle correctly and without reflection. The multiplication principle is more complicated. A good starting point in coming to a definition of the multiplication principle is the old nursery rhyme,

As I was going to St. Ives,
I met a man with seven wives.
Every wife had seven sacks,
Every sack had seven cats,
Every cat had seven kits.
Kits, cats, sacks, and wives,
How many were going to St. Ives?

The usual answer is "one": Only "I" was going to St. Ives. Let us compute the answer that is unacceptable in the nursery, the number of kits, cats, sacks, and wives coming *from* St. Ives. The number of wives is 7. Since each wife has 7 sacks, the total number of sacks is 7 · 7, or 49. Since each of these 49 sacks has 7 cats, the total number of cats is 49 · 7, or 343. Since each of these 343 cats has 7 kits, the total number of kits is 343 · 7, or 2401.

Wives	7
Sacks	49
Cats	343
Kits	2401
Total	2800

The number of kits, cats, sacks, and wives coming from St. Ives is, therefore, 2800. If we include the man as well, the total increases to 2801, a number less round than 2800 and hence more arresting.

The successive multiplications here are instances of what we shall call the multiplication principle—it could well be called the St. Ives principle. It can be used to solve a surprising variety of counting problems.

> **The Multiplication Principle:** We are to make two choices, a first-stage choice followed by a second-stage choice:
>
> **(i)   The number of alternatives open to us at the first stage is *a*.**
> **(ii) No matter which alternative we select at the first stage, the number of alternatives open to us at the second stage is *b*.**
>
> In these circumstances, the number of two-stage alternatives, or simply results, is *a* · *b*.

If at the first stage we are to choose one of the St. Ives man's wives, the number of alternatives, *a*, is 7. If at the second stage we are to choose one of the sacks in the keeping of this wife, then, no matter which wife we happen to have selected at the first stage, the number of alternatives now open to us, *b*, is 7. According to the principle, the number of sacks is *a* · *b* = 7 · 7, or 49.

Starting afresh with another task, that of choosing a sack at the first stage and then choosing one of this sacks's cats at the second stage, we know from the first computation that the number of alternatives, a, at the new first stage is 49. And no matter which sack we select at the first stage, the number of second-stage alternatives b—the number of cats belonging in this sack—is 7. According to the principle, the number of cats is $a \cdot b = 49 \cdot 7$, or 343.

Finally, for choosing a cat and then one of the cat's kits, a is 343 (the number of cats, just derived) and b is 7 (the number of kits per cat), so that there are $a \cdot b = 343 \cdot 7$, or 2401, kits in all. To show that there are 2800 kits, cats, sacks, and wives *in toto,* we need only apply the addition principle.

EXAMPLE 1

Problem: A distributor of small electrical appliances sells hand mixers, toasters, and blenders. She can obtain hand mixers made by Westinghouse, General Electric, Hamilton Beach, and Sunbeam. But her toasters and blenders come from Westinghouse, General Electric, Proctor-Silex, and West Bend. How many appliance-manufacturer combinations can a customer purchase?

Solution: Assume that the customer first chooses an appliance and then the manufacturer. The alternatives at the first stage are hand mixer, toaster, and blender. Their number, a, is 3. The set of alternatives open to the customer at the second stage varies with the selection made at the first stage; the following table lists them. Although the *set* of possible manufacturers varies with the appliance chosen, the number of possible manufacturers for each appliance does not: the number, b, is 4.

CHOICE OF APPLIANCE	ALTERNATIVES AT THE SECOND STAGE			
Hand mixer	Westinghouse	GE	Hamilton Beach	Sunbeam
Toaster	Westinghouse	GE	Proctor-Silex	West Bend
Blender	Westinghouse	GE	Proctor-Silex	West Bend

By the multiplication principle, the number of appliance-manufacturer pairs is $a \cdot b = 3 \cdot 4$, or 12. An alteration of the table preceding leads to an exhibit of all twelve pairs. With the second-stage alternatives, we simply couple the appliance with each of the manufacturers in its row:

Mixer–Westinghouse	Mixer–Hamilton Beach
Toaster–Westinghouse	Toaster–Proctor Silex
Blender–Westinghouse	Blender–Proctor Silex
Mixer–GE	Mixer–Sunbeam
Toaster–GE	Toaster–West Bend
Blender–GE	Blender–West Bend

This exhibit shows why the multiplication principle works: We all know that to find the number of boxes in a rectangular array we can multiply the number of rows times the number of columns in each row; there is no need to count the boxes individually.

For a case in which the multiplication principle does not apply, suppose that the distributor finds that she cannot obtain Proctor-Silex and West Bend toasters but that she can sell Hoover blenders:

CHOICE OF APPLIANCE	ALTERNATIVES AT THE SECOND STAGE				
Hand mixer	Westinghouse	GE	Hamilton Beach	Sunbeam	
Toaster	Westinghouse	GE			
Blender	Westinghouse	GE	Proctor-Silex	West Bend	Hoover

Again *a* is 3, but the number of alternatives at the second stage is no longer constant. There *is* no *b*, because in the definition of the *b* value for the multiplication principle we stated that *b* is the number of alternatives open to us at the second stage "no matter which alternative we select at the first stage." In our example here, the number of second-stage alternatives varies with what we select at the first stage. Thus, the multiplication principle does not apply. While the number of appliance-manufacturer pairs can be *counted up* (there are eleven of them), it cannot be *computed*. Of course, the answer for the original set of manufacturers also could have been obtained by counting up all the possibilities, but the idea is to derive the answer from a general principle that works even when an exhaustive count would be prohibitive.

EXAMPLE 2

Problem: An executive in New York may take a plane to Chicago or to St. Louis. From Chicago he may fly on to Seattle, San Francisco, or Los Angeles. From St. Louis he may fly on to San Francisco, Los Angeles, or San Diego. To how many West Coast cities may he go?

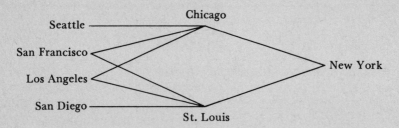

Solution: At the first stage he has *a* = 2 alternatives, and at the second he has *b* = 3. By the multiplication principle, the number of destinations is *a* · *b* = 2 · 3, or 6. This answer is wrong—the number of destinations is only 4. The trouble is not far to seek: The various first-stage selections coupled with the various second-stage selections do not lead to distinct results (destinations).

Thus, to the statement of the multiplication principle we must add this proviso:

Each first-stage alternative coupled with each one of its second-stage successors leads to a distinct result.

This condition holds in our St. Ives example and in the first part of Example 1. It also holds in Example 2 if we ask not for the number of *destinations,* but for the number of *routes* to the West Coast.

The multiplication principle also works if we have a sequence of three choices instead of just two. Here the first-stage and second-stage choices are to satisfy conditions (i) and (ii) of the multiplication principle, and the third-stage choice must satisfy the analogous condition:

(iii)   No matter which alternative we select at the first stage, and no matter which alternative we select at the second stage, the number of alternatives open to us at the third stage is c.

In these circumstances, the number of different results (we assume again that different choices lead to different results) is $a \cdot b \cdot c$.

EXAMPLE 3

Problem: Suppose the executive in Example 2 may from Seattle or San Francisco fly on to Tokyo or to Hong Kong and may from Los Angeles or San Diego fly on to Hong Kong or to Manila. How many *routes* to the Orient are available to him?

Solution: As before, a equals 2 and b equals 3. No matter which West Coast city our executive is in after the first two legs of the journey (and, in the cases of San Francisco and Los Angeles, no matter whether he got there via Chicago or via St. Louis), he has $c = 2$ third-stage alternatives open to him. By the three-stage multiplication principle, there are $a \cdot b \cdot c = 2 \cdot 3 \cdot 2$, or 12, routes all told.

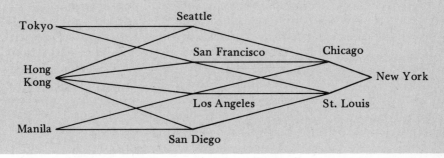

The multiplication principle extends in the obvious way to a sequence of four or more choices. It is sometimes useful to think of the multiplication principle in terms of *representation* rather than stages. For instance, in determining the number of appliance-manufacturer combinations available in Example 1, we could think of the problem as one in which we had to choose an appliance "representative" and then a manufacturer "representative." We found the total number of pairs by multiplying the ways to get the appliance representative (3) times the number of ways we could get the manufacturer representative (4). In Example 3, the executive had to choose a representative from among the 2 available routes for the first leg of the trip, a representative from the 3 available routes on the second leg of the trip, and a representative from the 2 available routes on the third leg of the trip. Thus, he had $2 \cdot 3 \cdot 2 = 12$ routes available. See Problems 13 through 18 at the end of the chapter for more examples where representation is involved.

4.4 PERMUTATIONS AND COMBINATIONS

At this point, it would be well for the reader to recall Formula (4.1):

$$\text{Probability} = \frac{\text{Number of favorable cases}}{\text{Total number of cases}} \qquad \textbf{(4.1)}$$

We remind our readers that the only reason for learning the addition and multiplication principles is that they are useful in computing the numbers of cases that go in the numerator and denominator of the fraction in Formula (4.1). Before we turn to the actual computations involving probabilities, we will examine two more principles that are useful in counting the number of cases that have certain characteristics. These are the permutation and combination principles.

We begin with a simple example, that of counting the number of three-letter "words" that can be made up from the five letters A, B, C, D, and E. The "word" need not make sense (CDA is counted as well as CAD), a letter may be repeated (BDB is counted), and order matters (AED and ADE are counted as different). Situations like this often arise in computer programming. A programming problem may require several three-letter codes made up from a group of five particular letters.

We must successively choose letters to fill the three blanks __ __ __. Since in each blank we can use any of the five letters, the three-stage multiplication principle with a, b, and c equal to 5 shows that there are $5 \cdot 5 \cdot 5$, or 125, possible words or codes. In general, from an "alphabet" of a letters one can make up a^n "words" of length n.

Let us now examine a variation of this situation.

EXAMPLE 1

Problem: Count the three-letter computer code words that can be made from A, B, C, D, and E when repeats are prohibited (BDB is not counted).

Solution: As before, we must successively choose letters to fill the three blanks ___ ___ ___. Since we may fill the first blank with any of the five letters, the number of first-stage alternatives is $a = 5$:

$$
\begin{array}{l}
A \ \_\_ \ \_\_ \\
B \ \_\_ \ \_\_ \\
C \ \_\_ \ \_\_ \\
D \ \_\_ \ \_\_ \\
E \ \_\_ \ \_\_
\end{array}
$$

If the first blank was filled with A, the second may be filled with B, C, D, or E; if the first blank was filled with B, the second may be filled with A, C, D, or E; and so on. Thus b equals 4 this time; there are always four alternatives at the second stage. Finally, the third blank may be filled with any of the three unused letters; for example, from BC__ we may go on to BCA, BCD, or BCE. So c equals 3, and the total number of "words" is thus $a \cdot b \cdot c = 5 \cdot 4 \cdot 3$, or 60. Here is a systematic table of them.

TABLE 4.2   The Sixty Permutations of A, B, C, D, E, Taken Three at a Time

ABC	ABD	ABE	ACB	ACD	ACE	ADB	ADC	ADE	AEB	AEC	AED
BAC	BAD	BAE	BCA	BCD	BCE	BDA	BDC	BDE	BEA	BEC	BED
CAB	CAD	CAE	CBA	CBD	CBE	CDA	CDB	CDE	CEA	CEB	CED
DAB	DAC	DAE	DBA	DBC	DBE	DCA	DCB	DCE	DEA	DEB	DEC
EAB	EAC	EAD	EBA	EBC	EBD	ECA	ECB	ECD	EDA	EDB	EDC

Thus the five letters A, B, C, D, and E can be lined up in $5 \cdot 4 \cdot 3$ ways when taken three at a time. A lining up, or ordering, is called a **permutation.** If we take three people and line them up, we may have Tom, Dick, and Harry *in that order.* That is one permutation. If Tom and Dick change places, we have Dick, Tom, and Harry. This is a different permutation or ordering, even though it is the same combination or group of people. Many of the "words" in Table 4.2 have the same letters in them, but they are counted as separate permutations since the order of the letters differs from one "word" to the next.

If we denote the number of permutations of five distinct things taken three at a time by $_5P_3$, what we have shown is that $_5P_3 = 5 \cdot 4 \cdot 3$.

Answers like this are most conveniently expressed by using the

factorial notation. The symbol $n!$, read *n-factorial,** is the product of
the integers from 1 to n:

$$n! = n(n-1)(n-2) \cdots 3 \cdot 2 \cdot 1 \tag{4.2}$$

For example,

$$5! = 5 \cdot 4 \cdot 3 \cdot 2 \cdot 1 = 120$$

One can also consider *n*-factorial to be the number of permutations
of n things taken n at a time, $_nP_n = n!$. That is, *n*-factorial is the
number of orderings possible for n distinguishable items. Thus, five
distinguishable items can be arranged into $5! = 120$ different orderings;
one could make up 120 different codes, or words, with five *different*
letters. If the same letter could be used more than once, then $5 \cdot
5 \cdot 5 \cdot 5 \cdot 5$, or 3125, possible codes could be formed.

From the definition of $n!$ it is clear that if r is a number between
1 and n, then

$$n! = n(n-1)! = n(n-1)(n-2)!$$

$$= n(n-1)(n-2) \cdots (n-r+1)(n-r)! \tag{4.3}$$

and, if we divide both sides of Equation (4.3) by $(n-r)!$, we see
that

$$\frac{n!}{(n-r)!} = n(n-1)(n-2) \cdots (n-r+1) \tag{4.4}$$

Our case of five things taken three at a time, where $n = 5$ and $r
= 3$, helps us test this result:

$$_5P_3 = \frac{5!}{2!} = 5 \cdot 4 \cdot 3$$

It is convenient to define $0!$ as 1:

$$0! = 1 \tag{4.5}$$

We can now state the general rule governing permutations, or
orderings.

Permutations: **The number of permutations of *n* distinct things taken
r at a time $(1 \le r \le n)$ is:**

$$_nP_r = n(n-1)(n-2) \cdots (n-r+1) = \frac{n!}{(n-r)!} \tag{4.6}$$

---

*This use of the exclamation mark is to be distinguished from the ordinary one.
The gravity of mathematical discourse makes confusion unlikely!

The argument for Relation (4.6) is the same as for Example 1. We can think of the n objects as distinct symbols or letters with which we are to fill in a succession of r blanks, using a different letter in each blank. The set of alternatives at any stage is the set of letters as yet unused, and the numbers of these for the first stage, second stage, third stage, etc., are n, $n - 1$, $n - 2$, etc. According to the multiplication principle, the number of permutations is the product of these numbers, of which there are r (one for each blank). This gives Relation (4.6).

If r equals n, we have a special case described earlier. *All n items are to be arranged in order, and we suppress the phrase "taken r at a time."*

Permutations: **The number of permutations of n distinct things is***

$$_nP_n = n! = n(n - 1) \cdots 3 \cdot 2 \cdot 1 \tag{4.7}$$

The number of permutations of the digits 0 through 9 is

$$10! = 3,628,800$$

and the number of permutations of the 26 letters of the alphabet is

$$26! = 403,291,461,126,605,635,584,000,000$$

The number of different ways 7 distinct products can be lined up on a display shelf is

$$7! = 5040$$

In these examples the answer certainly must be arrived at *not* by counting out all the cases, but by the application of general principles.

To arrive at a rule for permutations, we found in Example 1 the number of three-letter computer code words that can be made from A, B, C, D, and E with repetitions disallowed. What was involved was essentially the order in which letters lined up. Since ABC and ACB represented different orderings, they represented different permutations as well. Suppose we change the terms of our problem in such a way that ABC is no longer recognized as different from ACB. This will lead us to another general principle of statistics.

---

*This rule is a special case of Rule (4.6). That is,

$$_nP_n = \frac{n!}{(n - n)!} = \frac{n!}{0!} = \frac{n!}{1} = n!$$

This operation shows the necessity for defining 0! as 1.

EXAMPLE 2

Problem: A smaller grouping of objects chosen from a larger one is called a *subset* of the larger *set*. How many subsets of three people can be formed from a group consisting of office workers named Tom, Dick, Harry, Sue, and Mary? The subset (Tom,Dick,Harry) is distinguishable from (Tom,Dick,Sue), since they contain different people. But (Tom,Dick,Harry) and (Harry,Dick,Tom) differ merely in the order in which the names appear. They are the same subset. In a subset, committee, or group the objects are not considered as being lined up at all; we could write them in a jumble:

Harry
Dick
Tom   or   Harry
Dick
Tom

The usual mathematical notation is {Tom,Dick,Harry}.

Solution: Let x be the number of subsets. We can arrive at a solution in two stages. At the first stage we choose any one of the possible subsets x. At the second stage we take the three names chosen and line them up in some order. For example, we may choose the subset {Tom,Dick,Sue} and then line them up in the order Sue,Tom,Dick. The number of alternatives at the first stage is x. The number of alternatives at the second stage is simply the number of permutations of the three names chosen at the first stage, and this number we know to be 3!. So $a = x$ and $b = 3!$. But after making these two choices, what we arrive at is some permutation of three of the five names. Each permutation can arise in exactly one way from such a pair of choices. By the two-stage multiplication principle, the product $a \cdot b$, or $x \cdot 3!$, is the same as the number $_5P_3$ of permutations of five things taken three at a time. Since we already know that $_5P_3$ is equal to $5!/2!$, we can conclude that

$$x \cdot 3! = 5!/2!$$

Solving for x gives our answer,

$$x = \frac{5!}{3!\,2!} = 10$$

Here is a list of the ten subsets

{Tom,Dick,Harry}   {Tom,Sue,Mary}
{Tom,Dick,Sue}   {Dick,Harry,Sue}
{Tom,Dick,Mary}   {Dick,Harry,Mary}
{Tom,Harry,Sue}   {Dick,Sue,Mary}
{Tom,Harry,Mary}   {Harry,Sue,Mary}

In the argument in the preceding example, we may replace 5 by a general number n and 3 by r, where we suppose $r \leq n$. If from a set of n distinct objects we are to choose a subset of size r, the number of ways this can be done is called the number of **combinations** of n things taken r at a time. This is denoted by $_nC_r$ or, more commonly, by $\begin{pmatrix} n \\ r \end{pmatrix}$. We can choose a combination in $_nC_r$ ways, and then permute the r objects in the combination in $r!$ ways, arriving at one of the $_nP_r$ permutations. By the two-stage multiplication principle: $_nC_r \times r! = {_nP_r} = n!/(n-r)!$; and we can divide through by $r!$ to get our answer:

Combinations: **The number of combinations of n distinct objects taken r at a time (the number of subsets of size r) is**

$$\begin{pmatrix} n \\ r \end{pmatrix} = {_nC_r} = \frac{n!}{r!(n-r)!} \qquad \textbf{(4.8)}$$

Notice that for the case where r is equal to n, the number of combinations is $n!/(n!\ 0!)$, which is 1 by the convention shown in Relation (4.5). This is the right answer: There is but one subset that contains all the n objects. The answer is also 1 for the case where r is equal to zero. This is itself just a convention: There exists exactly one set with nothing in it.

It is sometimes hard to tell whether a problem requires combinations or whether it requires permutations. To answer this question, we ask ourselves whether or not the arrangement or order of the objects is relevant. If we need to take order into account, permutations are called for; if order is irrelevant, combinations are called for. As an aid to distinguishing between situations that involve permutation and those that involve combinations, the reader may wish to determine if the following key words apply to the problem at hand: "arrangements," "sequences," "orderings," or "permutations." If these words apply, then the situation is one involving permutations. On the other hand, the following key words apply to situations involving combinations: "groups," "committees," "sets," "subsets," "teams" (unless changing the members' positions on the team gives a new "team"), and "delegations."

EXAMPLE 3

Problem: Assume there are five boxcars that need to be unloaded at a dock. But there is only enough time left in the day to unload three of them. The goods in the car unloaded first will be delivered today. Those unloaded second will be delivered tomorrow morning, and those unloaded last will be delivered tomorrow afternoon. In how many ways can three from among the five cars be unloaded in first, second, and third order?

Analysis: Here the dock superintendent is definitely interested in order since goods in each of the cars are needed by customers. Since the order of unloading is important, we use permutations; the answer is

$$_5P_3 = \frac{5!}{2!} = 60$$

EXAMPLE 4

Problem: How many ways can an executive *committee* of 5 be chosen from a board of directors consisting of 15 members?

Analysis: Here order is irrelevant; Brown, Jones, Smith, Black, and Williams form the same executive committee regardless of the order in which we list their names. Since order is irrelevant, we use combinations. The answer is

$$\binom{15}{5} = \frac{15!}{5! \; 10!} = 3003$$

Table I in the Appendix at the back of this text will be useful in evaluating combinatorial expressions like the one in Example 4 preceding. Table I gives the combinations of n things taken r at a time for selected values of n and r. For instance, the answer to Example 4 is found in the $n = 15$ row under the $r = 5$ column of Table I. This table will prove helpful in later chapters too.

SUMMARY. In this section and in the preceding section, we have looked into the beginnings of *combinatorial analysis*, the branch of mathematics that deals with counting. The facts used subsequently are these:

1. If we have a ways to select the first stage or representative, b ways to select the second, and c ways to select the third, then there are $a \cdot b \cdot c$ different possible outcomes. Key words: stages or representatives.

2. The number of permutations or orderings of n distinct things is:

$$_nP_n = n! = n(n-1) \cdots 3 \cdot 2 \cdot 1$$

Key words: arrangements, sequences, and orderings.

3. The number of permutations of n distinct things taken r at a time is:

$$_nP_r = n(n-1) \cdots (n-r+1) = \frac{n!}{(n-r)!}$$

Key words: same as for Item 2.

4. The number of combinations of n distinct things taken r at a time (that is, the number of subsets of size r) is:

$$\binom{n}{r} = \frac{n!}{r!(n-r)!}$$

Key words: groups, committees, sets, subsets, teams, and delegations. See Appendix Table I for selected values of n and r.

4.5 COMPUTING PROBABILITIES

Recall that probabilities are to be computed by the formula

$$\text{Probability} = \frac{\text{Number of favorable cases}}{\text{Total number of cases}} \tag{4.9}$$

We can often compute the numerator and denominator by the rules of the previous sections.

EXAMPLE 1

Problem: If the board of directors mentioned in Example 4 of Section 4.4 consists of 4 women and 11 men, what is the probability that a randomly selected executive committee of 5 will consist of 2 women and 3 men?

Analysis: In Example 4 of Section 4.4, we saw that the total number of possible executive committees is $\binom{15}{5}$. If the selection of the executive committee is truly random, then each one of these committees is equally likely to be selected. Thus the denominator of Formula (4.9) is $\binom{15}{5}$. If the executive committee is to have 2 women on it, then there are $\binom{4}{2} = 6$ ways (see Appendix Table I) in which a subgroup of 2 women can be selected from the 4 on the board. Thus there are 6 combinations of 2 women for the executive committee. But for each of these combinations, there are $\binom{11}{3} = 165$ ways (see Appendix Table I) in which a subgroup of 3 men can be selected from the 11 on the board. Thus there are 165 combinations of 3 men for the executive committee for each of the 6 combinations of women. Hence the numerator of Formula (4.9) is $\binom{4}{2} \cdot \binom{11}{3}$, and so the probability that the executive committee will consist of 2 women and 3 men is

$$\frac{\binom{4}{2} \cdot \binom{11}{3}}{\binom{15}{5}} = \frac{6 \cdot 165}{3003} = \frac{990}{3003} = .32967$$

EXAMPLE 2

Problem: In Example 1 of the previous section, we saw that there are 60 three-letter codes or words (without repeated letters) that can be made of the letters A, B, C, D, and E. If one of the 60 words is chosen at random (all 60 outcomes in the sample space assumed equally likely), what is the probability it starts with either A or E?

Solution: The denominator is $5 \cdot 4 \cdot 3$. The number of words starting with A is the number of ways of filling the blanks in A __ __ with letters from B, C, D, and E, and this number is $4 \cdot 3$, or 12. This is also the number of words starting with E, so the numerator is $2 \cdot 4 \cdot 3$, and the probability is:

$$\frac{2 \cdot 4 \cdot 3}{5 \cdot 4 \cdot 3} = \frac{2}{5}$$

Many examples can be carried through in the same way; see the problems at the end of the chapter.

We turn now to some general properties of probabilities. We illustrate each one by examples in which the outcomes are equally likely, so that probabilities are given by Formula (4.9). Each property also holds, however, even if the outcomes are not equally likely. We denote events by A and B, etc., and their probabilities by $P(A)$ and $P(B)$, etc.

The first property is obvious:

$$0 \le P(A) \le 1 \qquad\qquad (4.10)$$

The second property has to do with adding probabilities. An example will make clear its meaning and truth.

EXAMPLE 3

For rolling a pair of dice, the sample space consists of the 36 outcomes in Table 4.1. If A is the event that the total (sum of the dice) is 4, we know that A contains three outcomes:

$$A = \{3\text{-}1, 2\text{-}2, 1\text{-}3\}$$

The event B that the total is 3 contains two outcomes:

$$B = \{2\text{-}1,\ 1\text{-}2\}$$

Now "A or B" stands for the event that the total is *either* 4 *or* 3. Clearly, this event contains five outcomes: the three of A and the two of B.

$$A \text{ or } B = \{3\text{-}1,\ 2\text{-}2,\ 1\text{-}3,\ 2\text{-}1,\ 1\text{-}2\}$$

By three applications of Formula (4.9) for computing probabilities,

$$P(A \text{ or } B) = \frac{5}{36} = \frac{3}{36} + \frac{2}{36} = P(A) + P(B)$$

The general rule is this:

> ***Addition of Probabilities:***   **If the events A and B are mutually exclusive, then**
>
> $$P(A \text{ or } B) = P(A) + P(B) \qquad (4.11)$$

As explained in Section 4.2, A and B are mutually exclusive, or disjoint, if, as in the preceding example, they have no outcomes in common. In this case, the number of outcomes in "A or B" is the number in A plus the number in B, so Equation (4.11) is a consequence of Formula (4.9).

Events A and B may be said to be mutually exclusive *if they cannot both happen at the same time.* Understanding this point will remove a common source of confusion. The sample space represents *a single trial* of the experiment, and A and B are mutually exclusive if they cannot both occur in one trial. The dice in Example 3 cannot total both 4 and 3 at once; but none of this prevents them from totaling 4 on one trial and 3 on another trial.

$$P(A \text{ or } B \text{ or } C) = P(A) + P(B) + P(C) \qquad (4.12)$$

The same thing holds for four or more events, as in the next example.

EXAMPLE 4

A group of people consists of 5 production workers, 7 clerks, 4 secretaries, 6 staff trainees, 3 managers, and 2 executives. This is a total of 27 people. The probability that a randomly selected person is a secretary, staff trainee, manager, or executive is:

$$\frac{4}{27} + \frac{6}{27} + \frac{3}{27} + \frac{2}{27} = \frac{15}{27} = \frac{5}{9}$$

If A and B are not disjoint, Equation (4.11) does not apply, but there is another formula that does:

$$P(A \text{ or } B) = P(A) + P(B) - P(A \text{ and } B) \qquad (4.13)$$

EXAMPLE 5

A card is drawn at random from an ordinary deck; A is the event of drawing a spade, and B is the event of drawing a face card. Here "A or B" is the event of drawing a card that is a spade *or* a face card *or both*—the word "or" does not exclude the possibility that *both* events occur. Of course, "A and B" is the event that the card is both a spade *and* a face card. The number of outcomes in A is 13, the number in B is 12, and the number in "A and B" is 3. By Equation (4.13), the probability of drawing a spade or a face card is:

$$P(A \text{ or } B) = \frac{13}{52} + \frac{12}{52} - \frac{3}{52} = \frac{22}{52}$$

Each of the two terms $P(A)$ and $P(B)$ on the right in Equation (4.13) includes the outcomes that are both in A and in B; subtracting $P(A \text{ and } B)$ compensates for this double counting. As a special case Formula (4.13) contains Rule (4.11), which applies when A and B are disjoint events: If A and B are disjoint, then the event "A and B" cannot happen and hence has a probability of zero.

EXAMPLE 6

Problem: The accompanying table gives the breakdown of a company's 245 customers by the frequency with which they place an order and their payment terms. We would like to find the probability that a randomly selected customer is either a regular or a credit customer.

	CASH	*CREDIT*	*TOTAL*
REGULAR ORDERING	10	15	25
IRREGULAR ORDERING	20	200	220
Total	30	215	245

Solution: The selection of a regular customer can be called event A, and the selection of a credit customer can be called event B. Thus we desire $P(A$ or $B)$. If the selection of the customer is random, each of the 245 customers has an equal opportunity of being selected. The number of customers in event A is 25, and the number in event B is 215. The number in "A and B" is the number that are both regular and credit, and this number is 15. Thus, by Equation (4.13) the probability of randomly selecting a customer that is either a regular or a credit customer is:

$$P(A \text{ or } B) = \frac{25}{245} + \frac{215}{245} - \frac{15}{245} = \frac{225}{245}$$

Note that the 25 regular customers in the first term of the preceding equation include 15 credit customers. But these same 15 credit customers are part of the 215 credit customers in the second term. Since these 15 are double counted, they are subtracted from the sum in the third term.

It should be noted that when a situation can be phrased in terms of "either" event A "or" event B "or" event C, etc., then Addition Formulas (4.11) through (4.13) should be considered. That is, the *either-or-or* sequence implies that probabilities should be added (with the possible need to subtract out the probabilities that the events occur together).

The *complement* of an event A is the "opposite" event, the one that happens exactly when A does not. We denote it by $\overline{A}$.

Complementary Events: The probabilities of A and $\overline{A}$ are related by:

$$P(\overline{A}) = 1 - P(A) \tag{4.14}$$

EXAMPLE 7

In a stack of invoices, 25 are for single items, 30 are for two items, and 20 are for three or more items. The probability that a randomly selected invoice is for two or more items can be found as

$$P(\text{two or more}) = P(\text{either two or three-or-more})$$

$$= 30/75 + 20/75 = 50/75 = 2/3$$

However, it may be easier to consider the complement event:

$$P(\text{two or more}) = 1.0 - P(\text{single item}) = 1.0 - 25/75$$

$$= 50/75 = 2/3$$

The answer is the same regardless of the way the problem is worked, but in many problems, finding $1.0 - P(\text{complement event})$ is much easier than attacking the problem head-on.

Conditional Probability: The conditional probability of an event B given another event A, which we denote $P(B|A)$, is

$$P(B|A) = \frac{P(A \text{ and } B)}{P(A)} \tag{4.15}$$

The probability in the numerator of the previous fraction is called the *joint probability* of A and B. The ordinary probability $P(B)$ we sometimes call an *unconditional probability* to distinguish it from the conditional probability $P(B|A)$. When any two of the three terms in Formula (4.15) are known, we can solve for the third. For instance, multiplying the formula by $P(A)$ gives

$$P(A \text{ and } B) = P(A) \cdot P(B|A) \qquad\qquad (4.16)$$

Sometimes we know $P(A)$ and $P(A \text{ and } B)$ in advance and we put them into Formula (4.15) to find the value of $P(B|A)$. On the other hand, sometimes it is $P(A)$ and $P(B|A)$ that we know in advance, and we put them into Formula (4.16) to find the value of $P(A \text{ and } B)$. The next several examples demonstrate how Formulas (4.15) and (4.16) can be used to find either conditional or joint probabilities.

EXAMPLE 8

Consider the following table which classifies the equipment used in a particular company's plant.

	HIGH-USE	MODERATE-USE	LOW-USE	TOTAL
IN-WORKING-ORDER	12	18	10	40
UNDER-REPAIR	8	6	2	16
Total	20	24	12	56

Find the probability that a randomly selected piece of equipment is a High-Use item given that it is In-Working-Order. If we use Equation (4.15), then we find:

$$P(\text{High-Use}|\text{In-Working-Order}) = \frac{P(\text{High-Use and In-Working-Order})}{P(\text{In-Working-Order})}$$

$$= \frac{\dfrac{12}{56}}{\dfrac{40}{56}} = \frac{12}{40} = \frac{3}{10}$$

In the preceding double fraction of Example 8, we note that the value 56 is in the denominator of both fractions. When this value is divided out, we are left with 12/40. This leads us to the general conclusion that when we have a table such as the one in Example 8, the "given" part of the conditional probability restricts our attention to one single row or column of the table. For example, we were "given" that the piece of equipment selected was In-Working-Order. That restricted our attention to the first row of the table, where there were 40 pieces of equipment. Given these 40, there were only 12 that were High-Use, and thus $P(\text{High-Use}|\text{In-Working-Order}) = 12/40$. Consider this same approach in the following example.

EXAMPLE 9

Use the table in Example 8 to find the probability that a randomly selected piece of equipment is Under-Repair given that it is Moderate-Use. We can use Equation (4.15) as follows:

$$P(\text{Under-Repair}|\text{Moderate-Use}) = \frac{P(\text{Under-Repair and Moderate-Use})}{P(\text{Moderate-Use})}$$

$$= \frac{\dfrac{6}{56}}{\dfrac{24}{56}} = \frac{6}{24} = \frac{1}{4}$$

However, we can consider the fact that the "given Moderate-Use" statement restricts our attention to the second column of the table, where there are 24 pieces of equipment. Of these 24, only 6 are Under-Repair. Thus $P(\text{Under-Repair}|\text{Moderate-Use}) = 6/24$, or $1/4$.

EXAMPLE 10

For the experiment of drawing a single card from a deck of 52 cards, let A be the event "spade," and let B be the event "face card." Then "A and B" consists of the three face cards that are also spades, so $P(A \text{ and } B)$ is $3/52$. Since $P(A)$ equals $13/52$, using Equation (4.15) to determine the probability of a face card, given that the card in question is known to be a spade, we obtain:

$$P(B|A) = \frac{\dfrac{3}{52}}{\dfrac{13}{52}} = \frac{3}{13}$$

As before, this makes sense: If the referee draws a card and tells you it is a spade, as far as you are concerned the thirteen spades are now the possible outcomes and, of these, three favor the event "face card."

Notice in Example 10 that since $P(B) = 12/52 = 3/13 = P(B|A)$, the conditional and unconditional probabilities coincide, which is not true in Examples 8 and 9. This is because knowledge of a card's suit does not influence the probability that it is a face card—all suits have the same proportion of face cards. Whenever A gives no information about the probability of B, then $P(B|A) = P(B)$. In the following example, this is not the case.

EXAMPLE 11

Consider again the 245 customers in Example 6 of this section. Suppose you are told that a particular customer pays cash. What is the probability that this customer orders irregularly as well?

Let A be the event that a cash customer is selected and B be the event that a customer who orders at irregular intervals is selected. The event "A and B" is the event that an irregular cash customer is selected. The table that accompanies Example 6 tells us that it has the probability 20/245. $P(A)$ equals 30/245, so by Equation (4.15)

$$P(B|A) = \frac{\dfrac{20}{245}}{\dfrac{30}{245}} = \frac{20}{30} = \frac{2}{3}$$

This again is the correct answer, since if a referee tells you that the selected customer pays cash, you limit your consideration to the 30 cash customers, 20 of which order irregularly.

The following examples demonstrate how Formula (4.16) can be used.

EXAMPLE 12

The experiment consists of drawing two cards in succession from a deck, without replacing the card after the first drawing. Let A be the event that the first card drawn is a spade, and let B be the event that the second card drawn is a diamond. The chance of A, $P(A)$, is 13/52. Now, if the first card drawn is indeed a spade, then, just prior to the second drawing, the deck consists of 51 cards, of which 13 are diamonds, so the conditional chance of a diamond on the second draw must be $13/51:P(B|A) = 13/51$. By Formula (4.16), the chance of a spade and then a diamond is

$$P(A \text{ and } B) = \frac{13}{52} \cdot \frac{13}{51}$$

(In this example, interpreting $P(B|A)$ requires no conceptual referee; there is no difficulty in imagining yourself between the two draws, knowing the first card was a spade but ignorant of what the second will be.)

EXAMPLE 13

Again consider the 245 customers in Examples 6 and 11. We can use Equation (4.16) to find the probability that a randomly selected customer has *both* of any two characteristics. Suppose we desire the probability that a randomly selected customer both is a credit customer (call this event A) and orders irregularly (call this event B). From the table in Example 6 it can be seen that $P(A)$ equals 215/245 and that $P(B|A)$ equals 200/215. Thus by Equation (4.16) the probability that a randomly selected customer both is a credit customer and orders irregularly is

$$P(A \text{ and } B) = \frac{215}{245} \cdot \frac{200}{215} = \frac{200}{245}$$

Of course, if one had the table in Example 6, there would not be any need for calculating these probabilities. The number of favorable events is immediately apparent from the table and need only be divided by 245 to give the appropriate probabilities. The calculations have been made here only to show that the relationships in various equations are true.

It should be noted that when the situation can be phrased in terms of the words "both" event A "and" event B, then Formula (4.16) applies.

Independence: **The events A and B are called independent if**

$$P(A \text{ and } B) = P(A) \cdot P(B) \tag{4.17}$$

To see the idea behind this definition, divide both sides of Formula (4.17) by $P(A)$, which gives $P(A \text{ and } B)/P(A) = P(B)$. By Definition (4.15), then, Formula (4.17) is the same thing as

$$P(B|A) = P(B) \tag{4.18}$$

In other words, A and B are independent if the conditional probability of B, given A, is the same as the unconditional probability of B. Another way of stating the concept of independence is the following: if knowing that one event has already occurred does not change the probability that the other event has also occurred, then the two events are independent.

Assume for a moment that a coin is to be tossed twice and that you have a bet that a head will occur on the second toss. Your chance of winning is 1/2. Now assume that a friend tells you: "A tail came up on the first toss." This information does not change your chance of winning since the probability of a head on the second toss remains 1/2. This is because the tosses of a coin produce independent results— knowing the outcome of one toss does not change the probabilities of what will happen on subsequent tosses.

On the other hand, consider Example 8 again. The table from that example is reproduced next.

	HIGH-USE	MODERATE-USE	LOW-USE	TOTAL
IN-WORKING-ORDER	12	18	10	40
UNDER-REPAIR	8	6	2	16
Total	20	24	12	56

Let us assume that a piece of equipment has been selected randomly and that you are betting it is In-Working-Order (as you might well be if you have been boasting to a customer about the company's ability to handle any work that is given to it). Your unconditional probability that the equipment is In-Working-Order is 40/56. However, assume that a friend tells you, "The equipment selected is a Low-Use item." Now your probability of the equipment's being In-Working-Order shifts to 10/12. This is because the usage of the equipment and its repair status are not independent—knowing the usage level tells us something about whether a piece of equipment will be In-Working-Order or Under-Repair.

In the next example we show two events that are independent.

EXAMPLE 14

The experiment consists of rolling a die twice. Let A be the event that the first roll produces a 5. Let the event B be that the second roll produces an even number. We know that the probability of A is 1/6 and the probability of B is 1/2. Since the rolls of a die are independent.

$$P(A \text{ and } B) = \frac{1}{6} \cdot \frac{1}{2} = \frac{1}{12}$$

In Example 10 $P(B|A)$ is equal to $P(B)$. The probability of a face card is 3/13, and this is so even if you know the card is a spade. In this example, A does not influence B: The occurrence of A leaves unaltered the chance that B occurs. In this sense, A and B are independent.

Equation (4.17) defines independence. Dividing it through by $P(A)$ gives Equation (4.18), which is perhaps more intuitive as a condition for independence. We could just as well divide through by $P(B)$, which leads from Equation (4.17) to:

$$P(A|B) = P(A) \tag{4.19}$$

Here the occurrence of B neither increases nor decreases the probability of A. Thus, Equations (4.17), (4.18), and (4.19) all mean the same thing, and A has no influence on B if and only if B has no influence on A.

The concepts of dependence and independence of two events can be demonstrated one last time using the Example 6 data on our well-documented 245 customers. The frequency of ordering and the payment terms of customers are not independent. Let us see why.

	CASH	CREDIT	TOTAL
REGULAR ORDERING	10	15	25
IRREGULAR ORDERING	20	200	220
Total	30	215	245

EXAMPLE 15

Let B be the event that a cash customer is selected. From the table it can be seen that $P(B)$ is 30/245. Let A be the event that a regular-ordering customer is selected. If we are told in advance that a regular customer has been selected, then we limit our attention to the first row of the table. *Now* the probability that a cash customer has been selected is $P(B|A) = 10/25$, which is very much different from $P(B) = 30/245$. Thus, A and B are *dependent*. Further examination would show that knowledge of either the payment terms or the frequency of ordering for a selected customer will change the probability of any event from what it would be prior to obtaining that knowledge. Thus, the payment terms and frequency of ordering are dependent characteristics in this data set.

The concepts *independent* and *disjoint* (or mutually exclusive) are often confused with each other. They are, in fact, diametrically opposed: Disjoint events cannot, in general, be independent. Suppose that A and B are disjoint—that they cannot both happen. If you know that A has occurred, then you automatically know that B cannot have occurred, so $P(B|A)$ must be zero. Thus $P(B|A)$ and $P(B)$ must differ (except in the uninteresting case where $P(B)$ is zero, in which B is impossible in the first place), so A and B are not independent.

SUMMARY. The main probability facts and definitions needed in what follows are these:

1. If A and B are disjoint (if they cannot both occur at the same time), then

$$P(A \text{ or } B) = P(A) + P(B)$$

But if A and B can occur together, then

$$P(A \text{ or } B) = P(A) + P(B) - P(A \text{ and } B)$$

2. If A and $\overline{A}$ are complementary (if the occurrence of $\overline{A}$ is the same thing as the nonoccurrence of A), then

$$P(\overline{A}) = 1 - P(A)$$

3. The conditional probability of B, given A, is defined by:

$$P(B|A) = \frac{P(A \text{ and } B)}{P(A)}$$

This equation is sometimes used in the form

$$P(A \text{ and } B) = P(A) \cdot P(B|A)$$

4. The events A and B are defined to be independent (to reflect the idea that the occurrence of one does not alter the probability of occurrence of the other) if

$$P(A \text{ and } B) = P(A) \cdot P(B)$$

This condition is the same as

$$P(B|A) = P(B)$$

and the same as

$$P(A|B) = P(A)$$

The reader may more easily determine probabilities involving two or more events by remembering that when events are related by the word *or*, the probabilities of the individual events are *added* together and the probability that the events occur together is subtracted from this sum. When events are related by the word *and*, the probabilities of the events are *multiplied* together as in Items 3 and 4 of the preceding summary.

4.6   PROBABILITY TREE DIAGRAMS

Many of the concepts presented in the previous section can be illustrated using probability tree diagrams. These diagrams are useful in presenting a visual picture of a probability situation.

A probability tree diagram is a sequence of lines, which represent possible events, that branch into all the possible sequences of events that could occur in that situation. Each branch of the tree is labeled to indicate which event it represents. Along each branch is given the probability of the event's occurrence, given that the sequence of events has reached that point in the tree. Thus, all the branches that emanate from one point must be: (1) mutually exclusive and (2) collectively exhaustive. That is, they must represent distinct events and they must account for all possible events that can occur at that point.

As an example consider the breakdown of 245 customers of Examples 6, 11, 13, and 15 of the previous section. The table associated with those examples is presented again as follows.

	CASH	CREDIT	TOTAL
REGULAR ORDERING	10	15	25
IRREGULAR ORDERING	20	200	220
Total	30	215	245

The probability tree diagram that describes the random selection of a customer is shown next. This diagram illustrates several features

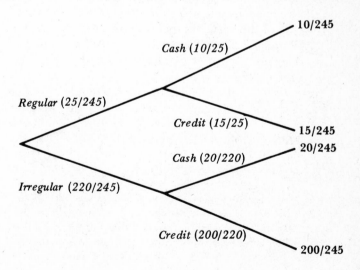

of probability trees. First, all the branches emanating from one point have probabilities that add up to unity. This is because the events represented by the branches are mutually exclusive and collectively exhaustive. Second, the probabilities associated with all the branches after the first set are conditional probabilities. That is, the 10/25 probability that someone is a cash customer on the top branch is the conditional probability that someone is a cash customer *given* that he or she is a regular customer, i.e., given that we are on the branch labeled "Regular." Third, the probabilities at the ends of the branches are joint probabilities and can be found easily by multiplying together all the probabilities along the branches leading to the end point. That is, the 200/245 probability found at the end of the bottom set of branches is the joint probability that a customer is both Irregular and Credit. This value can be found by multiplying the two probabilities along these branches: $(220/245) \cdot (200/220) = 200/245$.

There is nothing about the construction of probability tree diagrams that says we had to draw the tree in this example by using the Regular/Irregular classification first. We could just as easily have drawn a Cash/Credit branch first and then followed with Regular/Irregular branches at the ends. The way in which we are given the probability information, however, often dictates which branches are first. See Example 1 following.

Tree diagrams can be used to find joint probabilities rather easily, since, as was illustrated previously, one merely has to multiply all the conditional probabilities along the branches leading to the joint event to obtain the correct value. Unconditional probabilities are also easy to calculate. For instance, the unconditional probability of selecting a regular customer is found by adding up all the probabilities at the ends of the branches that were reached by going down a branch labeled "Regular." Thus $P(\text{Regular}) = (10/245) + (15/245) = 25/245$. This does not seem too profound since the "Regular" branch itself has a probability of 25/245.

However, consider the more complex problem of finding $P(\text{Cash})$ from this diagram. We merely add up the probabilities at the ends of the branches that were reached by going down any "Cash" branch. Thus $P(\text{Cash}) = (10/245) + (20/245) = 30/245$.

EXAMPLE 1

The personnel manager of a company classifies job applicants as qualified or unqualified for the jobs they seek. The manager says that only 25% of the job applicants are qualified and of those that are qualified, 20% list high school as their highest level of education. But 30% of the qualified applicants list trade school and 50% list college. The situation is different among the

unqualified applicants in that 40% of them list high school as their highest
level of education, another 40% list trade school, and only 20% list college.
Draw a probability tree diagram of this situation, find the joint probability
that an applicant is both qualified and a college graduate, and find the
unconditional probability that an applicant comes from a trade school background.

Solution: Since the problem states "of those that are qualified . . . ," the
probabilities that follow this statement are conditional—they assume the condition that the person is qualified to begin with. Thus, we are constrained to
drawing out a probability tree with the Qualified/Not-qualified branches as
our first branching. The other branches and associated probabilities can be
listed by noting carefully the description of the situation in the problem statement.
The tree is presented next with the probabilities that are given in the problem
statement. All of the joint probabilities can be found by multiplying the

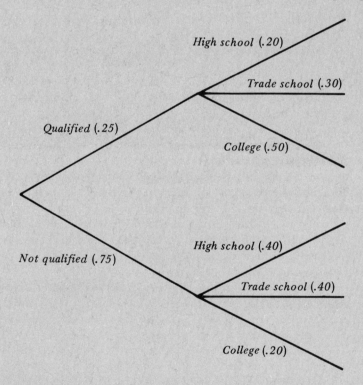

probabilities along the branches together. Thus the problem asked us to find
the joint probability, P(Qualified and College). This probability is found by
going up the first branch and down the third branch that starts there. Thus
P(Qualified and College) = (.25) · (.50) = .125. This same result could have
been obtained using Formula (4.16), which adapted to this situation would

state: P(Qualified and College) $= P$(Qualified) $\cdot P$(College|Qualified) $= (.25)$ $\cdot (.50) = .125$. The problem also asks us to find the unconditional probability that an applicant comes from a trade school. To do this we merely compute and add up all the joint probabilities found at the ends of any branches that are reached by going through "Trade School." There are two such branches, and the resulting computations are $(.25) \cdot (.30) + (.75) \cdot (.40) = .075 +$ $.300 = .375$. If we had used Formulas (4.11) and (4.16), we could have obtained the same result, but in a much more cumbersome fashion:

Let T represent "Trade School," Q represent "Qualified," and NQ represent "Not qualified."

$P(T) = P$ [either $(T$ and $Q)$ or $(T$ and $NQ)$]

$\quad = P(Q) \cdot P(T|Q) + P(NQ) \cdot P(T|NQ)$

$\quad = (.25) \cdot (.30) + (.75) \cdot (.40)$

$\quad = .075 + .300$

$\quad = .375$, which is the same answer obtained using the tree diagram.

Since tracing through the semantics of the probability formulas can be rather difficult, probability tree diagrams such as the one presented in this example can be very useful in simplifying the computations of joint and unconditional probabilities.

Tree diagrams can also be used to demonstrate the concept of statistical independence. Consider the following figure. The tree diagram on the top demonstrates that the events A and B are independent. This is because regardless of whether A or Not-A occurs first, the chance that event B will occur remains at .7 and the chance of Not-B remains at .3. That is, the fact that the probabilities assigned to events B and Not-B at the ends of the A and Not-A branches are the same, indicates that knowing about event A has no influence on the probabilities we assign to event B.

The tree diagram on the bottom demonstrates that events C and D are dependent. Note that the probabilities for events D and Not-D are .2 and .8, respectively, if event C has already occurred. However, if C has not occurred, the chances of D and Not-D are .6 and .4. Thus, knowing about event C's outcome changes the probabilities assigned to event D.

In general then, two events A and B are independent if the pair of probabilities assigned to B and Not-B at the end of the A branch are the same as those assigned to B and Not-B at the end of the Not-A branch. The two events are dependent if the pair of probabilities assigned to B and Not-B at the end of the A branch differ from the pair of probabilities for B and Not-B at the end of the Not-A branch.

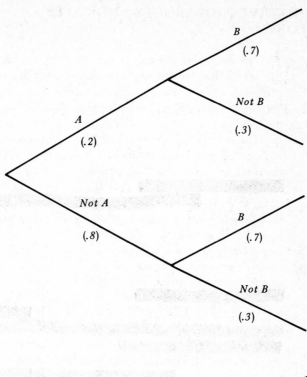

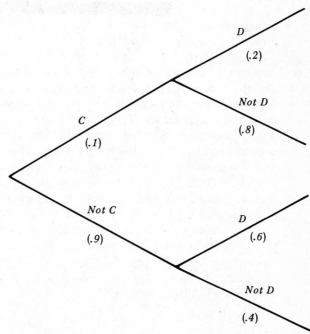

4.7   SUBJECTIVE PROBABILITY AND BAYES' THEOREM

The probabilities discussed in this chapter have all been *objective* in the sense that we have taken the probability of an event as being a property of the event itself. Statistics is sometimes placed in the framework of a different conception of probability: *subjective,* or *personal,* probability.

Consider a company which has a contract that requires it to build a piece of equipment which it has never built before. Suppose the contract requires that the project be completed within 90 days. Let H denote the hypothesis that the project will be completed within 90 days, and let $\overline{H}$ denote the opposite hypothesis that the project will take longer than 90 days to complete. Certainly people in the company have varying degrees of belief in H and say things such as, "It is fairly likely that we will complete the project in 90 days or less," or "It is improbable that we can finish the project in 90 days." Adherents of the theory of subjective probability hold that it is possible to assign to H a probability $P(H)$ which represents numerically the degree of a person's belief in H. The idea is that $P(H)$ will be different for different people, because they have different information about H and assess it in different ways. This conception of probability does not require ideas of repeated trials and the stabilizing of relative frequencies.

Setting aside the problem of how subjective probabilities are to be arrived at in the first place, we shall consider here only a method for modifying them in the light of new information or data.

Suppose that our contracted company hires a work methods engineer to determine how long it will take to complete the project and that she says that it can be completed within the specified time. This new information, call it D (for data), will certainly alter our attitude toward the hypothesis H. Still, we know that the methods engineer is not always correct. Suppose that after asking some questions we determine that 80% of the projects that are completed on time have been correctly forecast as being completed on time by this engineer. That is, $P(D|H) = .8$. Suppose that we also know that for projects that were not completed on time, the engineer forecasted completion by the deadline 10 percent of the time; thus $P(D|\overline{H}) = .10$.

We are to compute a *new* probability $P(H|D)$ of the hypothesis, given the data from the engineer. Since personal probabilities are assumed to obey the rules of Section 4.5, we can proceed as follows. By Definition (4.15) for conditional probability, $P(H|D)$ is equal to $P(H \text{ and } D)/P(D)$, and applying Formula (4.16) to the numerator gives

$$P(H|D) = \frac{P(H) \cdot P(D|H)}{P(D)}$$

Now D happens if "H and D" happens *or* if "$\overline{H}$ and D" happens. Therefore, by Addition Rule (4.11), $P(D) = P(H \text{ and } D) + P(\overline{H} \text{ and } D)$. Using Formula (4.16) on each of these last two terms now gives $P(D) = P(H) \cdot P(D|H) + P(\overline{H}) \cdot P(D|\overline{H})$. Substituting this for the denominator in the formula displayed above gives the answer:

$$P(H|D) = \frac{P(H) \cdot P(D|H)}{P(H) \cdot P(D|H) + P(\overline{H}) \cdot P(D|\overline{H})} \qquad \textbf{(4.20)}$$

This formula constitutes *Bayes' rule* or *Bayes' theorem*. Given the *prior* probabilities $P(H)$ and $P(\overline{H}) = 1 - P(H)$, and the respective probabilities $P(D|H)$ and $P(D|\overline{H})$ of observing the data D if H and $\overline{H}$ hold, we can use Bayes' rule to compute the *posterior* probability $P(H|D)$, the personal probability for H after the information in D has been taken into account. In our example, $P(D|H) = .8$ and $P(D|\overline{H}) = .1$, so the formula gives

$$P(H|D) = \frac{.8 \times P(H)}{.8 \times P(H) + .1 \times P(\overline{H})}$$

If $P(H) = .3$, say, so that $P(\overline{H}) = .7$, then $P(H|D) = .77$. Notice that $P(H|D)$ exceeds $P(H)$ here; observing D increases our personal probability of H because H "explains" D better than $\overline{H}$ does. In statistical notation, $P(D|H) > P(D|\overline{H})$.

Equation (4.20) is somewhat difficult to comprehend at first (or even second) glance. However, it is much easier to understand if we use the probability tree diagrams discussed in the previous section. The example of trying to complete the project in 90 days or less is diagrammed next. The joint probabilities at the ends of the branches were found by multiplying the probabilities along the branches. For instance, the .06 is the probability that the project can be done in 90 days or less and that the engineer will not predict that it can be done in 90 days or less.

In the statement of this situation, however, we desire to know $P(H|D) = P(\text{Project complete in 90 days or less}|\text{Engineer says it can be done})$. Since the engineer says the project can be done by the deadline, that means that we are at the end of the first or third branches in our diagram. The chance of our being at either one of these points is .24 + .07 = .31. The fact remains, however, that we *are* at the end of one of those branches, that is, it has been *given* that the engineer says "yes" the project can be done by the deadline. Thus, if we want

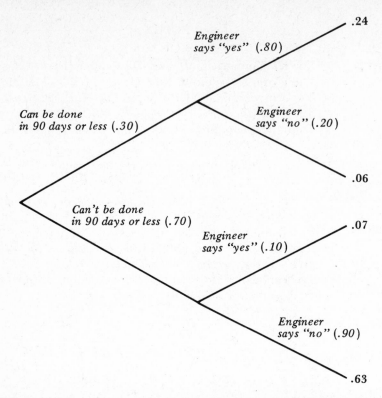

to know if the project really can be done by that time in the face of our original prior estimate of P(Project can be done in 90 days or less) $= .30$ and in light of the engineer's prediction, then we must have come down the "Can be done in 90 or less" branch with a probability that is proportional to the .24 contribution which this branch makes to the two situations where we have "Engineer says 'yes'." Thus

$$P(\text{Can be done in 90 or less}|\text{Engineer says "yes"}) = \frac{.24}{.24 + .07} = .77$$

The preceding calculation is a direct application of Formula (4.20).

EXAMPLE 1

The chance that someone has a particular disease is .02. A test that identifies this disease gives a positive reaction in 97% of the people who have the disease, but it also gives a positive reaction in 5% of the people who do not have the disease. Given that you just received a positive reaction from this test, what is the probability that you have the disease?

Solution: The situation is diagrammed next. Since it was *given* that you have

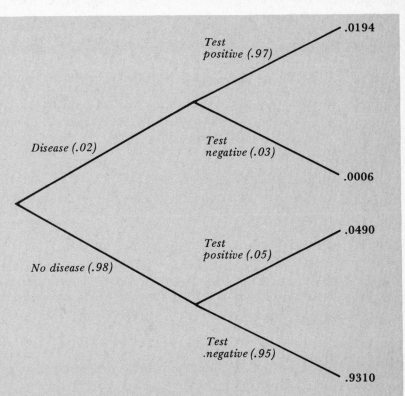

a positive test, you are out at the end of the top branch or the third one down. The chance of a positive test is just the sum of the probabilities at the ends of these two branches. Thus $P(\text{Positive}) = .0194 + .0490 = .0684$. But the chance that you came down the "Disease" branch to get the positive test is just that branch's share of this sum. Thus

$$P(\text{Disease}|\text{Positive}) = \frac{.0194}{.0194 + .0490} = \frac{.0194}{.0684} = .2836$$

A direct application of the cumbersome Formula (4.20) yields the same numerical result if we let H denote the presence of the disease and D denote the positive test.

It is sometimes said that Bayes' theorem is controversial. This is only partly true. If probabilities satisfy the rules of Section 4.5, then they must also satisfy Equation (4.20). The question that is disputed is whether probabilities can sensibly be assigned to hypotheses in the first place. In this book we shall formulate most statistical concepts within the framework of objective, or frequency, theory of probability. We shall postpone further treatment of subjective probabilities and extensive use of Bayes' theorem until Chapter 17.

PROBLEMS

1. List the different possible outcomes when three coins are tossed. How many are there? How many result in two heads? two or more heads?

2. List the different possible outcomes when a die is tossed. How many are there? How many outcomes are even? How many greater than four?

3. In an Air Force unit there are 50 people who have had experience flying fighters and 8 people who have had experience flying transport planes. Ten percent of the fighter pilots have also flown transports. Use the addition principle to determine how many total pilots there are in this unit.

4. A book publisher has 3000 books that are obsolete due to their being superseded by another edition. He also has 1600 books that have been returned to him by bookstores because they are damaged in some way. His records show that 800 books are both obsolete and damaged. He is going to hold a special sale to get rid of the books that are obsolete or damaged. How many books does he have that he can sell?

5. There are six employees who work for one supervisor. The supervisor must rank these people. How many different *rankings* are possible?

6. In a contest with five finalists, in how many ways can a winner, a first runner-up, and a second runner-up be selected?

7. If there are nine starters in a race, in how many different ways can first, second, and third prizes be awarded?

8. A display shelf has room to exhibit 5 items of merchandise. The store manager has 8 items she would like to display. How many different displays can she construct if:

    a. She counts it as a new display every time she changes the *order* of the 5 items—even though the same 5 items may be on the shelf?

    b. She counts it as a new display only if she takes one of the 5 items off the shelf and puts a new one on (orderings within the five don't get counted separately)?

9. A company sells a product called Make-Your-Doll. It is a toy which consists of three doll bodies, five doll heads, four sets of legs, and two sets of arms. A child assembles the doll he or she wishes to play with. How many complete dolls should the marketing manager advertise that the kit can make if a doll consists of one representative from each of the following groups: body, head, legs, and arms?

10. In how many ways can a committee of three be selected from ten individuals? (*Hint:* You may wish to use Table I in the Appendix.)

11. How many 13-card bridge hands can be dealt from a deck of 52 cards?

12. A company has 12 geographical areas in which it can market its new product. In how many ways can the marketing vice president plan a marketing campaign with:

    a. One major and one minor target area?

    b. One major and first and second minor target areas?

    c. One major and two minor target areas if it is not necessary to order the minor areas as first and second?

13. The manager of a new hotel is getting ready to determine which shops he will rent space to in the lobby and mezzanine of his hotel. He has

determined that he will allow two clothing shops, one flower shop, three airline offices, and two gift shops in the hotel. He has requests to locate from five airlines, seven clothiers, six florists, and four gift stores. How many different ways could he select the shops he will accept? (*Hint:* Table I in the Appendix will give you a helpful *start* on this problem.)

14. A company offers a holiday gift package in which the buyer can order any three of the five different jams the company makes, any two of the four nut rolls, and any three of seven dried fruit packets. How many different assortments of the gift package can be constructed?

15. If a group consists of five men and five women, in how many ways can a committee of four be selected if:
   a. The committee is to consist of two men and two women?
   b. There are no restrictions on the numbers of men and women on the committee?
   c. There must be at least one man?
   d. There must be at least one of each sex?

16. A construction firm builds tract homes for moderate-income families. Its homes have three basic floor plans. Also, each home can be built with a single or double garage or a carport. Each can be outfitted with one of three different types of kitchen cabinets, each can have one of four different fireplaces, and each can have one of seven different carpets laid. How many variations of this firm's tract homes can be built?

17. A menu offers a choice of six appetizers, five salads, eight entrees, four kinds of potatoes, seven vegetables, and ten desserts. If a complete meal consists of one of each, in how many ways can one select a dinner?

18. The board of directors of a certain corporation has decided to form an Environmental Impact Committee made up of three management representatives, two union representatives, and one major consumer-group representative. The company's management has designated five executives it feels are qualified to serve on the committee, the union has designated three of its people as possible members, and the consumer group has submitted the names of seven of its members who would serve. How many different Environmental Impact Committees could be formed?

19. From the six numbers 1, 2, 3, 4, 5, and 6, how many three-digit numbers can be formed if:
   a. A given integer can be used only once?
   b. A given integer can be used any number of times?
   c. The last digit must be even and a given integer may be used only once?

20. A company is going to send a delegation of three people to a convention being held in Hawaii. Naturally, many people would like to go. The president of the company has indicated that only one person may go from each of the major departments: production, marketing, and finance. There are four production, six marketing and three finance people who are eligible to attend the convention. How many different delegations could this company send to the convention?

21. Give an example of two events that are mutually exclusive (can't occur together).

22. Give an example of two events that are independent of each other

(knowing that one has occurred doesn't change the probability that the other has occurred).

23. The payments for hospital bills at the Midstate Valley Hospital are either private payments (made by the patients) or third-party payments (made by insurance companies or government agencies). The payments made at Midstate Valley are divided between private and third-party in such a way that the probability of a third-party payment is .75. Two-thirds of those third-party payments are for surgical bills and one-third are nonsurgical in nature. The private payments, however, are split 50-50 between payments for surgery and nonsurgery items.

   a. Find the probability that a randomly selected payment is both private and for a surgical item.
   b. Find the probability that a randomly selected payment is both third-party in nature and does not involve a surgical item.
   c. Find the probability that a randomly selected payment is for a surgical item.

24. A large western medical clinic serves 4000 regular patients. Clinic records show that 2500 of these patients are covered by private medical insurance, 1000 are on Medicare or welfare programs which cover their medical costs, and 500 do not have any medical coverage. If a patient is selected at random, find the probability that:

   a. The patient is covered by private insurance.
   b. The patient is covered by some third-party program.
   c. The payment for services will not involve filing governmental forms.

25. A box contains ten red tags numbered from 1 through 10 and ten white tags numbered from 1 through 10. If one tag is selected at random, what is the probability that it is:

   a. Red?   b. An even number?
   c. Red and even?   d. Red or even?

26. A parts bin contains 100 parts, 5 of which are defective and 7 of which are used. All the used parts are good. Find the probability that a randomly selected part is:

   a. Neither defective nor used.   b. Neither used nor good.
   c. Both defective and new.   d. Both defective and used.

27. A major oil company is drilling test wells both in Alaska and off the coast of Texas. They feel that the chance of oil in Alaska is .6 and the chance off the coast of Texas is .8. Find the probability that:

   a. Both wells find oil.
   b. Exactly one well finds oil.
   c. Neither well finds oil.

28. A hospital has two emergency sources of power. Historical records show that there is a .97 chance that the one source will operate during a total power failure and that there is a .92 chance that the other source will operate. Find the probability that in a total power failure neither emergency power source will operate.

29. Two balls are drawn from an urn containing three white and seven black balls. What is the probability of obtaining two white balls if:

   a. The first is replaced before drawing the second?

b. The first is *not* replaced before drawing the second?

30. There are 30 employees in one department, and only 10 of them are enrolled in the company's savings bond program. If 2 employees from this department are selected at random, what is the probability that both of them are enrolled in the program?

31. A national automobile rental firm claims that 95% of its customers are satisfied with the service they receive. If you were to interview two randomly selected customers, find the probability that both of them would be dissatisfied with the service.

32. About 8% of people in the population have an allergic reaction to materials used in the production process at Southwest Chemical Co. If two people are hired by Southwest Chemical tomorrow, what is the probability that:

   a. Neither is allergic to the materials?
   b. Both are allergic to the materials?
   c. Exactly one is allergic to the materials?

33. There are three automatic control boxes on a production line. Two were replaced last year. The third one needs to be replaced this year but no one can remember which box needs replacement and which ones don't. If two boxes are randomly selected for replacement, what is the probability that the oldest box (one not replaced last year) will get replaced?

34. A pipeline has three safety shutoff valves in case the line springs a leak. These valves are designed to operate independently of one another. There is only a 7% chance that valve #1 will fail at any one time, a 10% failure chance for valve #2, and a 5% failure chance for valve #3. If there is a leak, what is the probability that:

   a. All three valves operate correctly?
   b. All three valves fail?
   c. Only one valve operates correctly? (*Hint:* You might want to draw a tree diagram).
   d. At least one valve operates correctly?

35. A company has 800 employees. Twenty percent of the employees have college degrees, but half of those people are in nonmanagement positions. Thirty percent of the nondegree people are in management positions.

   a. How many managers does the company have?
   b. What is the conditional probability of being a college graduate given a person is a manager?
   c. What is the conditional probability of being a manager given a person has a college degree?

36. Seventy-five percent of the graduates of a certain business education program end up in management. It has been estimated that 25% of the graduates have both management jobs and good public speaking skills that such a job usually requires. Find the conditional probability that graduates of this program will have good public speaking skills given that they have management jobs.

37. In Problem **35**, find the probability of randomly selecting someone who is either a college graduate or a manager.

38. Company records show that 80 employees have been reprimanded more than once, and 300 have never been reprimanded. The company has

400 employees. The people who have zero or one reprimand on their records are split evenly between men and women. The people who have two or more reprimands are 75% men. Draw a probability tree to help you determine the probability of randomly selecting a person who is:

a. A woman with one reprimand.

b. A male.

c. One with no reprimand given that he is male.

39. A furniture store has found in its historical records that 30% of its customers are urban and 70% are suburban. It also classifies customers as "good" (60%), "marginal" (30%), and "poor" (10%). Of the good customers 20% are urban and 80% suburban. But marginal customers are split 40% to 60% in the urban-suburban breakdown. Find:

a. P(Both Good and Suburban).

b. P(Both Poor and Urban).

c. P(Either Suburban or Good).

d. P(Marginal).

e. P(Suburban|Poor).

40. A type of electron tube, A, and its ruggedized version, B, are installed at random in single-tube units. Thirty percent of the tubes are of type B. The probability that type B will fail in the first hundred hours of continuous operation is .1; the probability that type A will fail is .3. If a particular unit fails, what is the probability it had a type A tube installed? a type B? Use Bayes' rule (4.20).

41. A company has two suppliers which sell it batches of a certain chemical. The only way to tell if the chemical is good is to use it in the company's production process. Vendor X supplies 40% of the batches and vendor Y supplies 60%. Eighty percent of the batches from vendor X are good, and 85% of the batches from vendor Y are good. If one of the company's production runs is spoiled due to a bad chemical batch, find the probabilities that the batch came from vendor X and vendor Y, respectively.

SAMPLE DATA SET QUESTIONS: *Refer to the 113 applicants for credit listed in the Sample Data Set Appendix of this book.*

a. Construct a classification table in which the rows are: Married and Unmarried; and the columns are: Credit Granted and Credit Denied. Find the number of people who fall into each of this table's four cells.

b. Use the table constructed above to find the following probabilities:

1. P(Credit Granted|Married)

2. P(Credit Granted|Unmarried)

c. Construct a probability tree in which the first branches deal with marital status and the second set of branches deals with whether credit was granted or denied.

REFERENCES

4.1 Feller, William. *An Introduction to Probability Theory and Its Applications,* 3rd edition, Volume I. John Wiley & Sons, New York, 1968.

4.2 Mosteller, Frederick, Robert E. K. Rourke, and George B. Thomas, Jr. *Probability with Statistical Applications,* 2nd edition. Addison-Wesley, Reading, Massachusetts, 1970.

4.3 Parzen, Emanuel. *Modern Probability Theory and Its Applications.* John Wiley & Sons, New York, 1960.

POPULATIONS, SAMPLES, AND DISTRIBUTIONS

5.1 POPULATIONS

In the broadest meaning of the term, a statistical **population,** or **universe,** is simply a set or collection; the things of which the population is composed are called its *elements*. Sometimes, the population is specified by a complete list of its elements, as when the voting population of a town is explicitly listed on the voting rolls. More commonly, a population is specified by a definition of some kind, or by the singling out of a characteristic common to its elements, as when we speak of the population of transistors produced by a given manufacturer during a given year, the population of patients receiving a certain medical treatment, or the population of infected trees of a species under attack by a particular disease. Such examples extend the original notion of the population of a geographic area.

The populations in the preceding examples are all *finite*. Statistics also involves certain hypothetical *infinite populations*. Consider an experiment to measure a physical quantity, such as the speed of light or the percent copper in a sample of brass. Different executions of the experiment will give different answers, because of the combined effect of small errors that creep in. This is why experiments are customarily carried through a number of times. Then variations in the results are due to random causes, as long as the experiment has none of the problems discussed in Section 1.3. Those problems produce systematic error in measurements which can be eliminated only through modification of the experimental design. If we have a well-designed experiment, then the experimenter analyzes the data by considering not only the measurements obtained, but the hypothetical, infinite population of all measures obtained if the experiment were to be repeated infinitely often under identical conditions.

The analysis of a manufacturing process often involves an infinite population, the hypothetical population of all the items (all the transistors, say) that the process would produce if it were to run indefinitely under constant conditions. Sometimes, random experiments like those in the examples in Chapter 4 are viewed as infinite populations: There is an infinite population of tosses of a coin or rolls of a pair of dice, and to toss the coin or roll the dice is to observe an element of the population.

Often a finite population is so large as to be effectively infinite for the purposes of statistical analysis. An effective way to acquire an initial understanding of what it means for a population to be infinite is to imagine the population to be finite but so very large that removing a number of elements from it has no discernible effect on the composition of the population. That infinite populations are constructs of the human mind does not make them less important in practical affairs.

Ordinarily, a statistical analysis involves only certain aspects of a population—only certain attributes or characteristics of the elements of the population. In a population of voters, we are perhaps concerned only with the voter's party affiliation, and not with his height, blood type, etc. In a population of trees, we may be concerned with height but not age. In a population of transistors, we may be concerned only with whether or not one is defective according to certain standards.

Since the primary objective of inductive statistics is to make inferences about populations, it is important that the population concerned be carefully specified. To know how the population of property owners in a town feels about a new shopping mall is not necessarily to know how the population as a whole feels. And the population appropriate to one method of measuring a physical quantity may be inappropriate to another method.

5.2 SAMPLES

For inquiring into the nature of a population it would be ideal if we could with ease and economy examine each one of its elements, but this is usually out of the question. Sometimes it is impossible because some of the elements are physically inaccessible. In other cases it is uneconomic. Obviously, we will not test every item produced if the test destroys the item. Even when all the elements are available, they may be so numerous that a complete census is not justified; often, for most practical purposes, sufficiently accurate results may be more quickly and inexpensively obtained by examining only a small part of the population. And of course it is in the nature of an infinite population that its elements cannot all be examined.

In most situations, then, we must be content with investigating only a part of the whole population. The part investigated is a **sample.** Samples may be collected or selected in a variety of ways. A *systematic sample* is one selected according to some fixed system—taking every hundredth name from the phone book, for example, or selecting machined parts from a tray according to some definite pattern. Most statistical techniques presuppose an element of randomness in the sampling, and we shall be concerned almost exclusively with **random samples.**

Consider first a finite population, say of size 50, from which we take a sample, say of size 5. There are $\binom{50}{5}$ such samples, a sample being a subset or combination of size 5 (see Section 4.4). The sample is random if it is selected in such a way that all $\binom{50}{5}$ samples have the same probability of being chosen. (Notice that randomness is really a property of the selection procedure rather than of the particular sample that happens to result.) In general, there are $\binom{N}{n}$ samples of size n in a finite population of size N, and random sampling means choosing one in such a way that all are equally likely to be chosen.

Although the concept of a random sample is fairly easy to grasp, there are situations in which it is not clear how to obtain one. If we can list all the elements of a population and number them, we can get a random sample by drawing numbered tags or tickets from a bowl, or we can use a table of random numbers to decide which elements to include in the sample. Table II in the Appendix is a table of random numbers. As its name implies, this table consists of numbers which have no pattern or scheme. The numbers are grouped into columns that are five digits wide. Each column of digits can be thought of as a column of one-digit random numbers.

EXAMPLE 1

Suppose that the marketing manager of a firm with 1000 customers desires to take a random sample of five customers to determine their attitudes concerning a new product. The marketing manager could obtain a list of the customers and number them from 1 to 1000. Then he could satisfy himself that he obtained a truly random sample by going to the random numbers of Table II in the Appendix and picking any three-digit random numbers. These numbers would then represent the numbers of the customers whose opinions would be sought. If, for simplicity, we assume that the marketing manager began at the upper left-hand corner of the random number table and selected the first three digits in the left column for his first number, he would obtain the number 395. He might then move down to the second row and select the number 463. Continuing in this manner he would obtain a sample consisting of customers numbered 395, 463, 995, 67, and 695 respectively. It might at first appear that in using three-digit random numbers the marketing manager would never select customer number 1000. However, the random number 000 could be assigned to this customer.

But if we cannot enumerate the population, these techniques are useless. How, for instance, would we obtain a random sample of oranges from a tree, fish from a river, or trees from a forest? How can a public opinion poller obtain a random sample of people in a city? We shall not go into such questions of sampling technique here (see References 5.1 and 5.3). We assume we have successfully obtained a random sample from the population in which we are interested, and investigate the inferences that can be based on such a sample.

Random sampling as defined previously is sometimes called sampling *without replacement*, because an element sampled is not returned to the population before the next element of the sample is drawn. Sampling that is done *with replacement*, of course, involves putting each element back into the population before the next one is chosen. The distinction between sampling with replacement and without replacement will be very important in determining the answers to probability questions we will approach in subsequent sections.

5.3 RANDOM VARIABLES

Many of the random experiments or observations discussed in Chapter 4 have outcomes that can be characterized in a natural way by numbers.

In rolling a pair of dice, one can observe the total number of dots; in tossing a coin three times in a row, one can count the number of times heads turns up; in a poker hand, one can count the number of aces. Such a number, determined by a chance mechanism, is called a **random variable.** We denote random variables by X and Y, etc.

Before looking further into this idea, consider a physically determined variable that has no element of randomness. To find the area of a square, you measure the length l of one of its sides and then multiply that length by itself, computing l^2. Now l is a variable; it may be 1.5 feet or 8.3 cubits, depending on the square in question. To keep in mind this element of variability, it helps to regard l as a name for the length of the side *before the side is measured.* Measuring the side of a specific square converts l into a specific number. The variable l is useful for making general statements such as, "The area is l^2."

And now, let X be the total on a pair of dice. The dice may show a total of 7 when actually rolled, in which case X is 7; or they may show 5, in which case X is 5; and so on. Now X is a number whose value depends on the outcome of a random experiment, and the outcome cannot be predicted. To keep in mind this variability and unpredictability, it helps to regard the random variable as a name for the number connected with the outcome of the experiment *before the experiment is performed.* Carrying out the experiment converts the random variable into a specific number.

Selecting a random sample from a population, finite or infinite, is a random experiment, and most random variables of concern to us will arise from sampling. The number of defectives in a sample of transistors is a random variable. In a sample of people, their individual heights are all random variables, and so is the average of these heights.

There is a classification of random variables that is based on the set of values the variable can take on. The random variable is **discrete** if there is a definite distance from any possible value of the random variable to the next possible value. The number of defectives in a sample is discrete, the distance between successive possible values being 1. A height measured to the nearest tenth of an inch takes as value one of the numbers . . . , 60.0, 60.1, 60.2, . . . ; it is discrete because the distance from one possibility to the next is 0.1. A **continuous** random variable, on the other hand, is assumed able to take any value in an interval. A height measured with complete accuracy can, in principle, take on any positive value, and so is a continuous variable. So can measurements such as weight, temperature, pressure, time, and the like, provided they are measured with complete accuracy. In reality, there is a limit to the precision with which measurements can be made, and continuous random variables, like infinite populations, are only a useful idealization. Throughout the remainder of this text, we will consider physical measurements such as lengths, weights, times, pressures, and forces to be continuous variables. Discrete variables are usually *countable* measures that are referred to as "the number of . . ." that have a particular characteristic.

5.4   GENERAL PROBABILITY DISTRIBUTIONS FOR DISCRETE RANDOM VARIABLES

The behavior of a discrete random variable can be described by giving the probability with which it takes on each of its distinct, discrete values when the experiment is carried out. Thus the behavior of a discrete random variable can be described using a table of values or, sometimes, a formula.

EXAMPLE 1

We know from Chapter 4 that the probability of rolling a 7 with a pair of fair dice is $1/6$. Letting X be the discrete random variable that represents the total for the two dice (X is a name for the total "before the dice are rolled"), we can express this fact by writing: $P(X = 7) = 1/6$. Similarly, the probability of rolling a 4 is $3/36$: $P(X = 4) = 3/36$. Here is a table of all the probabilities (they can be determined by counting cases in Table 4.1, Section 4.1):

$$P(X = 2) = \frac{1}{36} = .028 \qquad P(X = 3) = \frac{2}{36} = .056$$

$$P(X = 4) = \frac{3}{36} = .083 \qquad P(X = 5) = \frac{4}{36} = .111$$

$$P(X = 6) = \frac{5}{36} = .139 \qquad P(X = 7) = \frac{6}{36} = .167$$

$$P(X = 8) = \frac{5}{36} = .139 \qquad P(X = 9) = \frac{4}{36} = .111$$

$$P(X = 10) = \frac{3}{36} = .083 \qquad P(X = 11) = \frac{2}{36} = .056$$

$$P(X = 12) = \frac{1}{36} = .028$$

This collection of probabilities is called the *distribution* of the discrete random variable X. The distribution can be used to answer any question about X. For instance, the chance of "X is 7 or 11" is:

$$P(X = 7 \text{ or } X = 11) = P(X = 7) + P(X = 11)$$

$$= \frac{6}{36} + \frac{2}{36} = \frac{2}{9} = .222$$

To continue with our example. A pair of dice was rolled 100 times and a record kept of the number of times each possible value for the sum of the faces occurred. Figure 5.1(a) shows the observed relative frequencies; Figure 5.1(b) shows the corresponding probabilities. The two figures would agree more closely if the dice had been rolled 1000 times, say. The probabilities represent a theoretical limit toward which the observed relative frequencies should tend as the number of rolls increases beyond bound.

EXAMPLE 2

A warehouse receives goods from its manufacturing suppliers. Historical records show that during the past 50 days, the warehouse manager indicated 6 trucks arrived on 8 of those days, 7 trucks arrived on 25 of those days, and 8 trucks arrived on 17 of those days. Thus, the historical probability distribution for the discrete random variable called "number of truck arrivals" can be shown in the following table:

NUMBER OF TRUCK ARRIVALS		PROBABILITY
6	$\frac{8}{50} =$	.16
7	$\frac{25}{50} =$	.50
8	$\frac{17}{50} =$	.34
		1.00

We should note two things here. First, the sum of the discrete probabilities is 1.0. This indicates that the values of the random variable listed in the table are the only ones that can occur and that they totally describe the possible outcomes. This leads us to our second point. Since the values less than 6 and greater than 8 do not appear in the table, this distribution indicates that those values cannot occur, regardless of our subjective feelings to the contrary. However, for the sake of simplicity, we will assume that since the historical records show no days with fewer than 6 or more than 8 truck arrivals, that these events cannot occur.

FIGURE 5.1   (a) Distribution of the outcomes when a pair of dice was rolled 100 times.
(b) Theoretical distribution of the outcomes when a pair of dice is rolled

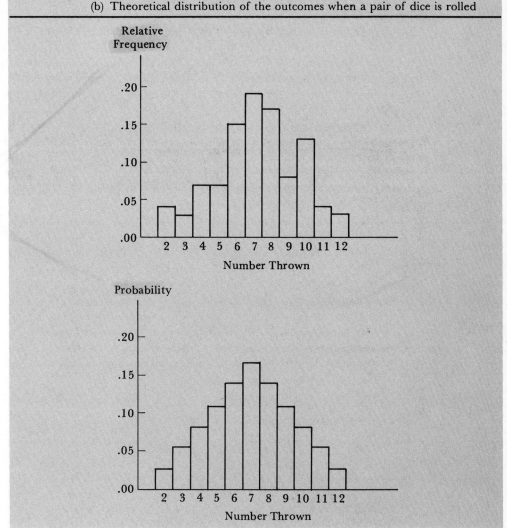

5.5   GENERAL PROBABILITY DISTRIBUTIONS FOR CONTINUOUS RANDOM VARIABLES

A *discrete* random variable is described by its distribution—the list of probabilities for its various possible values. This sort of distribution does not work for a continuous random variable, because such a variable takes on any given value with probability *zero*: If X is continuous, then $P(X = 3.1)$ is 0, $P(X = 2.854)$ is 0, etc. This seems paradoxical at first, since X must assume *some* value when the experiment is carried out. But consider a line segment; each point in it has length zero, whereas all the points together give a segment with positive length. In the same way, X has probability 0 of taking on any one given number but has probability 1 of taking on *some* number (some number not specified in advance of the experiment). Consider an example involving weights. Weight is a continuous measure. What is the probability of finding a bag of sugar in your grocery store that weighs exactly 10 pounds? This chance is zero since, if you found a bag that weighed 10 pounds on your scale, then someone else could theoretically find a more accurate scale that would show your 10-pound bag to be slightly different from 10 pounds. Then if you found a bag that weighed 10 pounds on that finer scale, someone else could theoretically find an even more accurate scale, and show you that your bag did not weigh exactly 10 pounds out to, say, the fifty-first decimal place. In the extreme, we have to concede that it would be virtually impossible to find a bag of sugar that weighed 10 pounds exactly (say, out to a billion decimal places). Thus $P(X = 10.0) = 0$.

Since there are infinitely many possible values for a continuous variable, we cannot possibly list those values in a table as we did with probability distributions for discrete random variables. In consequence, the distribution of a continuous X must be given not by a list or table of probabilities, but by a *continuous curve*.

EXAMPLE 1

A construction company is involved in blasting a path for a new highway through a mountain pass. The weight X in tons of rock that is hauled off after each small blast is measured. The behavior of X is described by a curve such as that in Figure 5.2. The curve lies entirely above the horizontal axis, and the area under the curve is 1—that is, the area between the curve and the horizontal axis is 1. The probability that, when the experiment is carried out, X will assume a value in a given interval equals the area under the curve and over that interval. The chance that X lies between 1 and 3 is the area of the shaded region R in Figure 5.2; the chance that it lies between 4.5 and 5.8 is the area of the shaded region R'. Notice that, since an individual point has probability 0, $P(4.5 < X < 5.8)$ is the same as $P(4.5 \leq X \leq 5.8)$; it makes no difference whether or not the endpoints of the interval are included. This feature does not hold for discrete probability distributions.

We have all along interpreted probabilities in terms of limiting relative

FIGURE 5.2   Theoretical distribution of blast yields

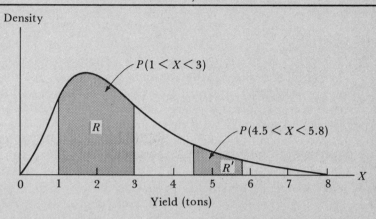

frequencies, and we can give a similar interpretation to the curve in Figure 5.2 by using the histograms of Chapter 2. If we were to set off 100 blasts and measure the rock yield to the nearest ton, using a class interval of one ton, then the histogram would resemble that in Figure 5.3(a). If we were to set off 1000 blasts instead of 100, we could use shorter class intervals (say, quarter-ton intervals) and more of them, and the histogram would resemble that in Figure 5.3(b).

If we continue to set off blasts, using more and more intervals of smaller and smaller size, the histogram will approach the curve in Figure 5.2. In each histogram the area of a bar is proportional to the observed relative frequency

FIGURE 5.3   (a) Distribution for 100 blasts. (b) Distribution for 1000 blasts

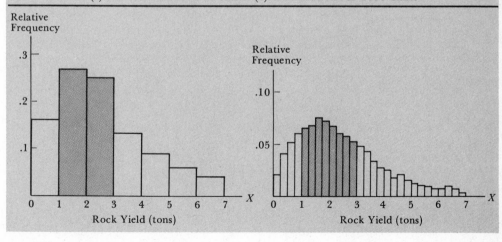

in that class, and so an area under the histogram (like the areas of the shaded regions in Figure 5.3) represents an observed relative frequency which converges to a probability such as that represented by the area of the shaded region R in Figure 5.2.

The distribution of any continuous random variable is specified by a curve like that in Figure 5.2. The curve is called a *frequency curve*, and the area under the curve between two limits on the horizontal scale is the probability that the random variable will take on a value lying between those two limits. The height of the curve over a point on the horizontal scale is called a **probability density.** Unlike the areas under the frequency curve, it has no direct probability meaning.

It should be noted that probability is associated with area even in the histograms that describe discrete probability distributions. In Figure 5.1(b) we tend to think of the probabilities as being represented by the *heights* of the bars. But since the bars are all 1-unit wide, the height and the area are the same. Thus, in Figure 5.1(b) the areas in the bars represent probabilities. In the following material it will be useful to think of areas in probability histograms *and* under frequency curves as representing probabilities.

If the distribution of a continuous variable is specified by a frequency curve, the question arises as to how the frequency curve is itself specified. This is only rarely done by an actual curve carefully drawn on graph paper. Sometimes, as in the following example, the distribution is specified by a geometric description.

EXAMPLE 2

Suppose X always has a value between 0 and 2, and suppose its distribution is given by the straight line in Figure 5.4. The equation of the line is: Density = $\frac{1}{2}X$. By the rule for the area of a triangle, the area under the line is 1. To get the probability that X lies between $1/2$ and $3/2$, we compute the area of the shaded trapezoid in the figure. The base of the trapezoid has length 1 and the sides have lengths $1/4$ and $3/4$ (found by substituting $1/2$ and $3/2$ in the density equation). Then the area rule for trapezoids gives

$$1 \times \frac{\dfrac{1}{4} + \dfrac{3}{4}}{2} = \frac{1}{2}$$

and this is the probability sought:

$$P\left(\frac{1}{2} \leq X \leq \frac{3}{2}\right) = \frac{1}{2}$$

FIGURE 5.4   A triangular distribution

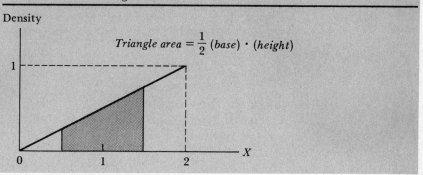

Density

$$Triangle\ area = \frac{1}{2}\ (base) \cdot (height)$$

Usually, distribution curves for continuous random variables are specified by mathematical formulas of greater or lesser complexity [for example, Formula (5.21) in Section 5.12], and the relevant probabilities—that is, the relevant areas—are determined by the methods of integral calculus. Such determinations lie outside the scope of this book. For various important frequency curves often encountered in statistical practice, tables of these areas have been constructed. Several such tables are given in the Appendix; they represent the frequency curves we shall discuss and use in succeeding chapters.

5.6   EXPECTED VALUES OF RANDOM VARIABLES

For a discrete random variable X, the expected value of X, denoted $E(X)$, is the weighted mean of the possible values of X, the weights being the probabilities of these values [see Formula (3.5) in Section 3.4]. Thus

$$E(X) = \sum_r rP(X = r) \tag{5.1}$$

where the sum extends over all possible values r of X. For example, suppose X can assume the values 1, 2, and 3 and has the distribution in Table 5.1. This might be the distribution for the number of cars arriving at a toll booth during a given minute. Here

$$E(X) = \sum_{r=1}^{3} rP(X = r) = 1 \times .25 + 2 \times .40 + 3 \times .35 = 2.10$$

The words "expected value" are misleading in that, in this example, 2.10 is not to be expected at all as a value of X, since 1, 2, and 3 are the only possible values. That is, we do not "expect" 2.10 cars

TABLE 5.1

r	1	2	3
$P(X = r)$	.25	.40	.35

to arrive during any single minute. But $E(X)$ is what is to be expected in an *average* sense. The expected value of X is also called the *mean* of X. This mean is similar to the mean for a set of numerical data, treated in Section 3.3.

Just as the mean for numerical data can be viewed as a center of gravity (see Figure 3.2 in Section 3.9), so can $E(X)$. For the distribution in Table 5.1, imagine weights positioned on a seesaw at each of the possible values of X, each weight being proportional to the probability of the value it represents, as in Figure 5.5; the expected value 2.10 is where a fulcrum will balance the seesaw.

For Examples 1 and 2 of Section 5.4, the expected values are, respectively,

$$2 \times \frac{1}{36} + 3 \times \frac{2}{36} + 4 \times \frac{3}{36} + 5 \times \frac{4}{36} + 6 \times \frac{5}{36} + 7 \times \frac{6}{36}$$

$$+ 8 \times \frac{5}{36} + 9 \times \frac{4}{36} + 10 \times \frac{3}{36} + 11 \times \frac{2}{36} + 12 \times \frac{1}{36} = 7$$

and

$$6(.16) + 7(.50) + 8(.34) = 7.18$$

For a continuous X, as for a discrete one, the mean $E(X)$ represents an average. It is the value that X can be expected to have on the average, the actual value on a trial being greater or less than $E(X)$ and distributed in such a way as to balance out at $E(X)$. Just as in the case of the discrete variable, $E(X)$ can be viewed as a center of gravity in the continuous case as well. If we were to draw the frequency curve on some stiff material of uniform density, such as plywood or metal, cut it out, and balance it on a knife-edge arranged perpendicular to

FIGURE 5.5   The expected value as a center of gravity: discrete case

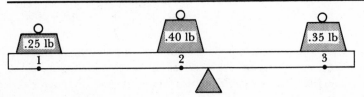

the horizontal scale, the value corresponding to the point of balance would be the expected value. Figure 5.6 illustrates this for the curve of Figure 5.2; compare it with the analogous Figure 3.5 in Section 3.13, which relates to the mean for grouped numerical data.

For the frequency curve in Example 2 of the preceding section, the expected value works out to 4/3. The computation of expected values in the continuous case, requiring as it does the methods of integral calculus, lies outside the scope of this book. It will be possible nonetheless to use the results of such computations for statistical purposes.

Suppose a random variable has expected value $E(X)$ equal to μ (Greek mu). The *variance* of the random variable is defined as

$$\text{Variance} (X) = E(X - \mu)^2 \qquad (5.2)$$

and this definition applies in both the discrete and continuous cases. Now $(X - \mu)^2$, which is itself a random variable, is always positive, and the more distant X is from μ, the greater is $(X - \mu)^2$. The expected value of $(X - \mu)^2$ measures the amount by which the distribution is spread around the mean μ, much as does the variance for a set of numerical data as defined in Section 3.10.

In all cases, variance (X) is greater than or equal to zero. In the extreme case in which it is equal to zero, the value of X equals μ all the time, and X is not really random at all.

The *standard deviation* of X is:

$$\text{Standard deviation} (X) = \sqrt{\text{Variance}(X)} = \sqrt{E(X - \mu)^2} \qquad (5.3)$$

For the distribution in Table 5.1, the mean μ was 2.10, and so the variance is

$$.25 \times (1 - 2.1)^2 + .40 \times (2 - 2.1)^2 + .35 \times (3 - 2.1)^2 = .59$$

and the standard deviation is $\sqrt{.59}$, or .768.

FIGURE 5.6   The expected value as center of gravity: continuous case

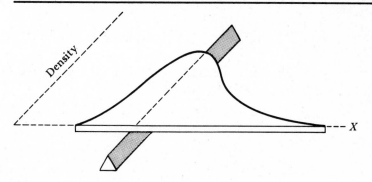

The mean is often denoted by μ, the variance by σ^2 (Greek sigma, squared), and the standard deviation by σ; sometimes μ_X, σ_X^2 and σ_X are used instead, to indicate that the terms refer to computations involving X. Whatever the units of X (feet, trucks, degrees, etc.), its mean and standard deviation have these same units, too.

The mean and standard deviation for a numerical data set were discussed in Chapter 3. The mean and standard deviation for random variables are related to this earlier discussion in a natural way. Since a random variable is a numerical measure which results from a random experiment or observation, an experimenter or observer making a series of observations will obtain a numerical data set. The mean and standard deviation for the random variable are the same as the mean and standard deviation the experimenter or observer would obtain for his numerical data set were he theoretically to take infinitely many observations.

If X is a random variable, discrete or continuous, and if A and B are fixed numbers, then

$$U = AX + B$$

is another random variable. The expected values of these two random variables are related by the equation

$$E(U) = AE(X) + B \qquad (5.4)$$

If X is a random temperature in celsius degrees, the same temperature in Fahrenheit degrees, U, is $\frac{9}{5}X + 32$; it is only natural that the expected values of U and X should be related by the formula $E(U) = \frac{9}{5}E(X) + 32$.

The variance and standard deviation are given by the following rules:

$$\text{Variance}(U) = A^2 \cdot \text{Variance}(X) \qquad (5.5)$$

and

$$\text{Standard deviation}(U) = A \cdot \text{Standard deviation}(X) \qquad (5.6)$$

for positive A.

Suppose X has mean μ and variance σ^2 (standard deviation σ), and consider the related variable

$$U = \frac{X - \mu}{\sigma} = \frac{1}{\sigma}X - \frac{\mu}{\sigma}$$

According to Equation (5.4),

$$E(U) = \frac{1}{\sigma} E(X) - \frac{\mu}{\sigma} = \frac{1}{\sigma}\mu - \frac{\mu}{\sigma} = 0$$

and according to Equation (5.5),

$$\text{Variance}\,(U) = \left(\frac{1}{\sigma}\right)^2 \cdot \text{Variance}\,(X) = \left(\frac{1}{\sigma}\right)^2 \cdot \sigma^2 = 1$$

Thus

$$E\left(\frac{X - \mu}{\sigma}\right) = 0 \qquad\qquad (5.7)$$

and

$$\text{Variance}\left(\frac{X - \mu}{\sigma}\right) = 1 \qquad\qquad (5.8)$$

if X has mean μ and standard deviation σ. Subtracting μ from X centers it at 0, and dividing by σ standardizes the variability; the term $(X - \mu)/\sigma$ is the random variable X *standardized* to have mean 0 and standard deviation 1.

5.7 SETS OF RANDOM VARIABLES

Random variables often come in pairs. This happens whenever there is a pair of numbers associated with each of the various outcomes of an experiment. If a man is drawn at random from some population, his height X in inches is one random variable and his weight Y in pounds is another. These random variables are associated with one another because they are associated with one experiment (the drawing of the person from the population); X and Y attach to the same man.
Expected values of sums satisfy the formula

$$E(X + Y) = E(X) + E(Y) \qquad\qquad (5.9)$$

For example, if X is the income of the husband in a family and Y is the income of the wife, $X + Y$ is the family income (other sources excluded). If in a population of families the average income for husbands is \$9,000 and the average income for wives is \$5,000, certainly the average income for families must be \$14,000. Formula (5.9) expresses the general form of this fact.

In Chapter 4 we gave a definition of independence of events [Equation (4.17)], a definition embodying the idea that knowing whether or not one of two events occurred does not in any way help us to guess whether or not the other occurred. There is a similar notion of independence of random variables X and Y; it embodies the idea that knowing the value of X does not in any way help us to guess the value of Y, and vice versa.

For instance, suppose two dice are rolled, one die colored red and one green to keep them straight; let X be the number showing on the red die and let Y be the number showing on the green die. Since there is no interaction between the dice, the value that X takes on when the dice are rolled has no influence on the value Y takes on; X and Y are independent. On the other hand, if X and Y are the height and weight of a man, then X and Y are not independent: If you know that X is very large, then you know that Y is likely to be large also.

The fact that will be of use to us is that, *if X and Y are independent, then**

$$\text{Variance}(X + Y) = \text{Variance}(X) + \text{Variance}(Y) \qquad (5.10)$$

We shall often be concerned with whole sets of random variables. Suppose we have at hand a random sample of size n from some population (say, a sample of corporate stocks), and suppose we have measurements $X_1, X_2, \ldots, X_n$ (say, the prices of the stocks), one measurement for each element of the sample. Choosing a random sample is a random experiment, and each X_i is a number associated with the outcome; thus, $X_1, X_2, \ldots, X_n$ are random variables. The procedures of Chapter 3—the computations of means and variances for sets of data—are usually performed on samples.

Rule (5.9) extends to sets of random variables:

$$E\left(\sum_{i=1}^{n} X_i \right) = \sum_{i=1}^{n} E(X_i) \qquad (5.11)$$

If $X_1, X_2, \ldots, X_n$ come from a random sample of size n from an infinite population, then they are independent. As explained in Section

---

*Without giving a full derivation of this fact, we can give a partial argument for it by considering a case of dependence. Suppose that X is some random variable (with positive variance), and suppose that Y is its negative: $Y = -X$. Certainly, X and Y are dependent: To know X is to know Y exactly. Now, for every outcome of the experiment, $X + Y$ is equal to 0, so $X + Y$ has no variability at all and variance$(X + Y)$ is 0. Thus in Equation (5.10), the right side, certainly positive, exceeds the left side, which vanishes.

Here, Equation (5.10) fails because X and Y vary in such a way that in $X + Y$ their variability cancels out. In other cases, their variability reinforces in such a way that the left side of Equation (5.10) exceeds the right. But if X and Y are independent, they can interact in no way at all, so this sort of canceling and reinforcing is impossible and the variances exactly add up in accordance with Equation (5.10).

5.2, this is the essential feature of samples from infinite populations. Formula (5.10) can be extended to more than two random variables *if they are independent* of one another:

$$\text{Variance}\left(\sum_{i=1}^{n} X_i\right) = \sum_{i=1}^{n} \text{Variance}\,(X_i) \qquad (5.12)$$

5.8   SPECIFIC DISCRETE PROBABILITY DISTRIBUTIONS

In Section 5.4, we discussed discrete probability distributions in general terms. In that section, we pointed out that a discrete probability distribution can be represented by a table or formula. Example 2 of that section discussed the probability distribution for the number of trucks arriving at a warehouse. That example's probability distribution is presented again as follows.

NUMBER OF TRUCKS ARRIVING	PROBABILITY
6	.16
7	.50
8	.34
	1.00

The probabilities in the right-hand column were developed from historical records at the warehouse and they are thus empirical probabilities. Since the values in the left-hand column are integers, this distribution is discrete. This table completely specifies the probability distribution.

There are many situations, however, where we do not have any historical records from which we can calculate probabilities. But sometimes we are fortunate in that the conditions surrounding our probability problem are such that we can apply certain probability formulas to develop our probability distribution. One of the skills of a good probability problem solver is that he or she is adept at recognizing the situations in which a simple probability formula can be applied to develop a probability distribution.

The next three sections of this chapter are devoted to helping the reader develop this skill. Three commonly used probability formulas are presented. The conditions under which these formulas can be used to calculate probability distributions are discussed, along with the formulas themselves and some work-saving probability tables. The probability formulas to be presented are for the following discrete probability distributions: the binomial, the hypergeometric, and the Poisson.

5.9   THE BINOMIAL PROBABILITY DISTRIBUTION

Consider the following example, which illustrates the conditions under which we can apply the **binomial probability formula.**

EXAMPLE 1

A coin is to be flipped three times. We would like to know the probability that there will be two heads and one tail in these three flips.

The conditions which must exist in order for us to use the binomial probability formula are met in this example. Those conditions are the following:

1. There is a series of trials with only two possible outcomes for each trial.
2. The trials are independent of one another. That is, the chance of getting either of the two outcomes on each trial does not change from trial to trial.
3. We desire to find the probability of a certain mixture of the outcomes *given* a fixed number of trials.

Let us examine how Example 1 meets these conditions. First, the three flips of the coin can be considered a "series of trials with only two possible outcomes for each trial." That is, the three flips result in "heads" or "tails" on each flip. In general, we label the outcomes of trials like these "success" and "failure." These terms have no moral connotations; we could easily have used more neutral terms like "blue" and "green," but tradition and some logic dictate that we use "success" and "failure" to denote the two possible outcomes. For simplicity in this example we will call heads a success and tails a failure.

Second, the trials are independent of one another in the coin-flipping example. That is, the chance of success remains unchanged from flip to flip. The chance of a head is .5 on the first flip, .5 on the second flip, and .5 on the third flip.

Finally, the problem asks us to find the probability of getting two heads and one tail in the three flips of the coin. Thus we want to know the probability of getting the mixture of two successes and one failure in three trials. Note that the number of trials is fixed in advance. That is, we are not asking how many trials it will take to get a certain outcome. We know that the trials are fixed to begin with.

Let us consider another example for which the binomial formula's conditions are met.

EXAMPLE 2

A die is to be rolled three times. Find the probability of getting one 4 in the three rolls. Also find the probability of getting two 4's, three 4's, and no 4's in three rolls.

Note that the rolls of a die result in six possible outcomes, but in this problem we have reduced those six outcomes to two: 4's and "anything else." The chance of a 4 on any of the three rolls is $1/6$; thus the three trials are independent of one another. Finally, the number of trials is fixed at three, and we desire to know the probability of getting a mixture of one 4 and two non-4's in three rolls. We are also asked to find three other mixtures: (two 4's and one non-4), (three 4's and no non-4's), and (no 4's and three non-4's).

Now that we have determined this example meets the conditions required for the application of the binomial formula, we will continue the discussion of it in developing the general binomial probability formula. Let us denote the occurrence of a 4 by S, for success, and the occurrence of anything else by F, for failure. Table 5.2 lists all the possible results for the three rolls (ignore the list of probabilities for the moment). The sequence SSS indicates that all three rolls resulted in success (a 4); SSF indicates that the first two rolls resulted in success and the last roll resulted in failure (non-4); and so on.

If the die is rolled three times in a row, what is the probability of obtaining some particular sequence, say SSF? If the die is fair, the probability of success on any one trial is $1/6$, and so the probability of failure is $5/6$. We know, as in Example 1 of this section, that there is independence from trial to trial—that the result on one trial can in no way influence the result on a different trial. Hence, we multiply probabilities. The chance of getting S on the first roll and then S on the second roll and then F on the third roll is $(1/6)(1/6)(5/6)$. The other probabilities in Table 5.2 are computed the same way.

What is the probability of getting exactly one success (and two failures)? There are three ways this can happen: Table 5.2 shows them to be SFF, FSF, and FFS. Each of these sequences has probability $(1/6)(5/6)^2$—one factor of $1/6$ for the one S and two factors of $5/6$ for the two F's. Thus the probability of getting *exactly one* success is:

$$P(1) = 3 \left(\frac{1}{6}\right)\left(\frac{5}{6}\right)^2 = \frac{75}{216} = .347$$

The sequences containing *two* S's and one F are SSF, SFS, and

TABLE 5.2   Independent Trials
with $p = 1/6$

SEQUENCE	PROBABILITY
SSS	$\left(\dfrac{1}{6}\right)^3$
SSF	$\left(\dfrac{1}{6}\right)^2\left(\dfrac{5}{6}\right)$
SFS	$\left(\dfrac{1}{6}\right)^2\left(\dfrac{5}{6}\right)$
SFF	$\left(\dfrac{1}{6}\right)\left(\dfrac{5}{6}\right)^2$
FSS	$\left(\dfrac{1}{6}\right)^2\left(\dfrac{5}{6}\right)$
FSF	$\left(\dfrac{1}{6}\right)\left(\dfrac{5}{6}\right)^2$
FFS	$\left(\dfrac{1}{6}\right)\left(\dfrac{5}{6}\right)^2$
FFF	$\left(\dfrac{5}{6}\right)^3$

FSS, and each has probability $(1/6)^2(5/6)$, so the probability of getting exactly two successes is:

$$P(2) = 3 \left(\frac{1}{6}\right)^2 \left(\frac{5}{6}\right) = \frac{15}{216} = .069$$

Finally, the chance of *three* successes is the chance of SSS, namely,

$$P(3) = \left(\frac{1}{6}\right)^3 = \frac{1}{216} = .005$$

and the chance of *no* successes is the chance of FFF, namely,

$$P(0) = \left(\frac{5}{6}\right)^3 = \frac{125}{216} = .579$$

The general principle is this: We repeat an experiment, singling out an event we call success. We let

p = the probability of success on a single trial
$1 - p$ = the probability of failure on a single trial
n = the number of trials
r = the number of successes

The probability that S occurs exactly r times (so that F must occur $n - r$ times) in n trials where the chance of an S on each trial is p is:

$$P(r|n,p) = \binom{n}{r} p^r (1 - p)^{n-r} \tag{5.13}$$

The symbol $P(r|n,p)$ is usually read "the binomial probability of r successes *given* n trials, where the chance of success on each trial is p." For this formula to be valid, the three conditions listed at the beginning of this section must hold.

Rule (5.13) may be derived in this way: One sequence which contains r S's and $n - r$ F's is:

$$\underbrace{SS \ldots S}_{\substack{r \\ \text{times}}} \underbrace{FF \ldots F}_{\substack{n - r \\ \text{times}}}$$

By independence, the probability of r S's in a row followed by $n - r$ F's in a row is $p^r (1 - p)^{n-r}$. We do not insist that the r S's occur in the *first* r trials; we insist only that exactly r of the trials produce S and $n - r$ of them produce F. The probability of each such sequence is $p^r (1 - p)^{n-r}$. Equation (5.13) follows because the number of such sequences is $\binom{n}{r}$. That is, from the n trials there are $\binom{n}{r}$ ways to choose a combination of r trials in which to put S's, and the other trials must take F's.

To demonstrate that this formula gives correct answers that might otherwise have to be derived through detailed study of a problem, we will demonstrate how it could have been used to solve the first part of Example 2 of this section. There we sought to find the probability of getting one 4 in three rolls of a die. In that case, we called getting a 4 a success. Thus $n = 3$ trials, $p = 1/6$, and $r = 1$. Thus Formula (5.13) gives the following:

$$P\left(r = 1 \middle| n = 3, p = \frac{1}{6}\right) = \binom{3}{1}\left(\frac{1}{6}\right)^1 \left(\frac{5}{6}\right)^2$$

$$= \frac{3!}{1!2!} \left(\frac{1}{6}\right)^1 \left(\frac{5}{6}\right)^2 = .347$$

as was shown previously using Table 5.2 and some additional reasoning. The answer to Example 1 of this section can also be found using Formula (5.13) since that example met the conditions for application of the binomial formula. There we sought to find the probability that there will be two heads and one tail in three flips of a coin. Thus if we call a head a success, we have $r = 2$, $n = 3$ again, and $p = 1/2$. Formula (5.13) gives the probability we desire as:

$$P\left(r = 2 \middle| n = 3, p = \frac{1}{2}\right) = \binom{3}{2}\left(\frac{1}{2}\right)^2\left(\frac{1}{2}\right)^1$$

$$= \frac{3!}{2!1!}\left(\frac{1}{2}\right)^2\left(\frac{1}{2}\right)^1 = \frac{3}{8} = .375$$

According to the binomial theorem of algebra,

$$(x + y)^n = \sum_{r=0}^{n} \binom{n}{r} x^r y^{n-r}$$

The quantities $\binom{n}{r}$ are therefore called *binomial coefficients*, and thus the probabilities in Equation (5.13) are called *binomial probabilities*. We will return to discuss the binomial coefficients again in a later section. The values of selected binomial coefficients are given in Table I of the Appendix and were used previously to evaluate the number of ways r things can be taken from n things.

The situations in which the binomial formula can be used to calculate probabilities are not limited to those involving coins and dice. Consider the following example.

EXAMPLE 3

A production process produces items, and the defective rate is 15 percent. Assume a random sample of 10 items is drawn from the process. Find the probability that two of them are defective.

In this example we have a series of trials ($n = 10$) where there are only two possible outcomes for each trial (defective and nondefective). The chance of getting a success (defective) remains at 15 percent ($p = .15$) from item to item selected for the sample; thus the trials are independent. The conditions for use of the binomial formula are satisfied, and we seek:

$$P(r = 2 | n = 10, p = .15) = \binom{10}{2}(.15)^2(.85)^8 = \frac{10!}{2!8!}(.15)^2(.85)^8$$

$$= .2759$$

EXAMPLE 4

For another example of binomial probabilities, consider the experiment of performing a market survey. Suppose that one-fourth of the consumers in a particular market purchase your product. Let success S be the random selection of a consumer who purchases your product. In this case, $p = 1/4$, and $1 - p = 3/4$. If we randomly select $n = 5$ people to talk to about your product, the probability that exactly 3 people in the group will be purchasers of your product is the probability of 3 successes in 5 trials where the probability of success on each trial is $1/4$:

$$P\left(r = 3 \middle| n = 5, p = \frac{1}{4}\right) = \binom{5}{3}\left(\frac{1}{4}\right)^3\left(\frac{3}{4}\right)^2 = \frac{90}{1024} = .0879$$

So far we have discussed using the binomial formula to calculate the probability that there will be exactly a certain number of success in a fixed number of trials. In practical problems, however, we are usually interested not in the probability that r assumes some individual value, but instead in the probability that r will exceed a specific value or that r will lie in a given interval. These probabilities can be found more easily if we have the complete table of binomial probabilities that describe all the outcomes of some binomial experiment. This table can be developed using Formula (5.13). If we expand Example 1 of this section, we can fully demonstrate the calculation and use of such a binomial table.

Suppose we toss a coin five times. The probability p of a head (success) is $1/2$, and the probability $1 - p$ of a tail (failure) is $1/2$. Hence, the probability of getting exactly r heads in the five tosses is:

$$P\left(r \middle| n = 5, p = \frac{1}{2}\right) = \binom{5}{r}\left(\frac{1}{2}\right)^r\left(\frac{1}{2}\right)^{5-r} \qquad r = 0, 1, 2, 3, 4, 5$$

The probability of exactly three heads is

$$P\left(r = 3 \middle| n = 5, p = \frac{1}{2}\right) = \binom{5}{3}\left(\frac{1}{2}\right)^3\left(\frac{1}{2}\right)^2 = 10\left(\frac{1}{2}\right)^5 = \frac{10}{32}$$

and the probability of no heads is

TABLE 5.3   Binomial Probability Distribution, $p = 1/2$, $n = 5$

r	0	1	2	3	4	5
$P(r)$	$\dfrac{1}{32}$	$\dfrac{5}{32}$	$\dfrac{10}{32}$	$\dfrac{10}{32}$	$\dfrac{5}{32}$	$\dfrac{1}{32}$

$$P\left(r = 0 \middle| n = 5, p = \frac{1}{2}\right) = \binom{5}{0}\left(\frac{1}{2}\right)^0 \left(\frac{1}{2}\right)^5 = 1 \cdot 1 \cdot \left(\frac{1}{2}\right)^5 = \frac{1}{32}$$

Calculating the remaining probabilities the same way gives the distribution in Table 5.3. Note that these probabilities add to 1. Table 5.3 is an example of a specific *binomial probability distribution*. This is because the probabilities for this discrete distribution were derived from Equation (5.13).

To find the probabilities that the number of successes, r, lie in a given interval, we add the terms in the distribution. For example, the probability of two or more heads in five tosses is:

$$P\left(r \geq 2 \middle| n = 5, p = \frac{1}{2}\right) = \sum_{r=2}^{5} P(r) = P(2) + P(3) + P(4) + P(5) = \frac{26}{32}$$

The probability that the number of heads will be greater than two but less than or equal to four is:

$$P\left(2 < r \leq 4 \middle| n = 5, p = \frac{1}{2}\right) = \sum_{r=3}^{4} P(r) = P(3) + P(4) = \frac{15}{32}$$

The probability that there will be at least one head is:

$$P\left(r \geq 1 \middle| n = 5, p = \frac{1}{2}\right) = \sum_{r=1}^{5} P(r) = \frac{31}{32}$$

Note that this last probability may more easily be found by using the rule for complementary events [Formula (4.14)]; that is, the probability of at least one head is:

$$P(r = 1) = 1 - P(r = 0) = 1 - \frac{1}{32} = \frac{31}{32}$$

Notice that for writing probabilities associated with discrete variables it is important to distinguish between *greater than or equal to* ($\geq$) and *greater than* ($>$), and also between *less than or equal to* ($\leq$) and *less than* ($<$). This distinction is unnecessary in the case of a continuous variable, since there the probability of a single specified value is 0.

Evaluating Expression (5.13) every time we want to find a binomial probability or build a binomial probability distribution can be rather tedious. It becomes even more difficult if we desire to know a cumulative probability like $P(r \leq 10|n = 25, p = .6)$. That is, if we want to know the probability of ten or fewer successes in twenty-five trials where the chance of success on each trial is .6, we would have to evaluate Equation (5.13) for the eleven values of r from 0 to 10. Fortunately, a set of binomial tables has been included in the Appendix of this book to aid in finding some complicated probabilities like the preceding one. These tables were produced by a computer programmed to evaluate Expression (5.13) for various values of n, p, and r and then sum the resulting probabilities from 0 up to r. Thus Table III in the Appendix is a cumulative table and gives

$$P(r \leq k|n, p) = \sum_{r=0}^{k} \binom{n}{r} p^r (1 - p)^{n-r}$$

Suppose that we want to know the probability of obtaining three or fewer successes in five trials in a situation where the probability of success on each trial is .4. This probability can be found directly from the subtable for $n = 5$ in Table III. It is found in the $p = .40$ column and the $r = 3$ row. The value is .9130. The probability of three or fewer successes in $n = 10$ trials can be found in the subtable for $n = 10$. If the probability of success on each trial is again $p = .4$, then this value is found in the $p = .40$ column and the $r = 3$ row and is .3823.

Sometimes we would like to know the probability of *exactly r* successes in n trials, not r or fewer. This probability can be found using Table III also. In order to do this, we must use the relationship shown in Expression (5.14).

$$P(r = X) = P(r \leq X) - P(r \leq X - 1) \tag{5.14}$$

This expression means that to find the probability of, for example, *exactly three*, we find the probability of three or fewer and subtract the probability of two or fewer. Thus the probability of, say, three successes in fifteen trials where the probability of success on each trial is .2 can be found in the subtable for $n = 15$ in Table III. Expression (5.14) is then used in the following way:

$$P(r \leq 3) = .6482$$
$$\underline{P(r \leq 2) = .3980}$$
$$P(r = 3) = .2502$$

Table III can also be used to find the probability of r *successes or more.* This requires the use of the complement of the probabilities given in Table III. The probability of r successes or more is found in Expression (5.15).

$$P(r \geq X) = 1.0 - P(r \leq X - 1) \tag{5.15}$$

This expression means that to find the probability of, say, three or more successes we find the probability of two or fewer and subtract that figure from 1.0. Thus the probability of three successes or more in $n = 20$ trials where the probability of success on each trial is .1 can be found in the subtable for $n = 20$ in Table III. Expression (5.15) is then used in the following way:

$$
\begin{array}{rr}
 & 1.0000 \\
P(r \leq 2) = & .6769 \\
\hline
P(r \geq 3) = & .3231
\end{array}
$$

EXAMPLE 5

Problem: Use Table III to find the following binomial probabilities:

a) $P(r \leq 8 | n = 10, p = .70)$
b) $P(r = 7 | n = 15, p = .40)$
c) $P(3 \leq r \leq 9 | n = 20, p = .30)$
d) $P(r \geq 6 | n = 10, p = .40)$
e) $P(r = 12 | n = 15, p = .85)$
f) $P(r = 25 | n = 25, p = .90)$

Solution:

a)   This value can be read directly from Table III: .8507.
b)   Formula (5.14) must be used here: $.7869 - .6098 = .1771$.
c)   To find this value we must take $P(r \leq 9) - P(r \leq 2) = .9520 - .0355 = .9165$. Note that we subtracted $P(r \leq 2)$, which left us with the $P(r = 3)$ still in the answer as we desired since we wanted the chance that the number of successes ranged from 3 to 9, *inclusive.*
d)   We must employ Formula (5.15) here to get $1.0 - .8338 = .1662$.
e)   This answer cannot be found using Table III since there is no column for $p = .85$. Thus we will use Formula (5.13) to obtain the answer:

$$\frac{15!}{12! \ 3!} (.85)^{12} (.15)^3 = .2184$$

Note that we did not interpolate between the columns for $p = .80$ and $p = .90$ to obtain this answer. If we did that, we would have obtained an answer of .1894, which is rather different from the correct answer of .2184 given by Formula (5.13). In general, it is not wise to interpolate between columns of the binomial table, and we must never try to interpolate between sections of the table for different n values.

f)   We must use Formula (5.14) to solve this problem, but when we go to Table III to obtain $P(r \leq 25 | n = 25, p = .90)$, we find that the 25th row in the table is missing. This is not a mistake, but involves the assumption that the reader realizes that $P(r \leq n | n, p) = 1.0000$ since this is the probability of n or fewer successes in n trials and takes into account all possibilities. The reader should note that none of Table III's subsections contain a final row. Thus $P(r = 25 | n = 25, p = .90) = 1.0000 - .9282 = .0718$.

EXAMPLE 6

Problem: Seventy percent of the people arriving at a toll plaza for a bridge have the correct change. If twenty-five cars pass through the toll plaza in the next five minutes, what is the probability that between ten and twenty cars, inclusive, have the correct change?

Solution: This is a binomial sampling situation since we have $n = 25$ trials, with success (correct change) or failure (no change) at each trial. Knowing that the last driver had correct or incorrect change tells nothing about the status of the next driver. Thus, $p = .70$ from trial to trial. We desire to know

$$P(10 \leq r \leq 20 | n = 25, p = .70) = P(r \leq 20) - P(r \leq 9) = .9095 - .0005$$

$$= .9090$$

Formulas (5.1) and (5.2) can be used to find the mean and variance of any randomly distributed variable. However, when the variable follows the binomial probability distribution, finding the mean and variance of the variable can be accomplished using the two formulas that follow:

$$E(r) = np \tag{5.16}$$

and

$$\text{Variance}(r) = np(1 - p)$$

or

$$\text{Standard Deviation}(r) = \sqrt{np(1 - p)} \tag{5.17}$$

For example, if we took many samples (say fifty thousand samples of size $n = 5$ with $p = 1/2$), we would get fifty thousand numbers— the number of successes in each of the fifty thousand samples. We could find the mean of those fifty thousand numbers, and it would be approximately $np = 5(1/2) = 2.5$. The variance of those fifty thousand numbers would be about $np(1 - p) = 5(1/2)(1/2) = 1.25$. The standard deviation, of course, would be $\sqrt{1.25} = 1.12$. We will use our knowledge of Formulas (5.16) and (5.17) in the next chapter.

5.10   THE HYPERGEOMETRIC PROBABILITY DISTRIBUTION

Consider Example 4 of the previous section once again. In that example, we looked at a situation where a market survey was being performed. We assumed that the probability of a success ($p = 1/4$) remained constant from trial to trial. That is, we assumed the trials were independent of one another. This would be true only if the market in which surveying was being done were large (say, several thousand people) or if selection were being done *with replacement*, allowing the possibility that once a person was interviewed he or she might be randomly selected and interviewed again later. A different computational technique must be used if the surveying takes place in a small, finite population or if the selection is done without replacement. That is, we cannot use the binomial probability distribution to compute the probability of various survey outcomes since the second condition required for using the binomial distribution (that of independent trials) is not met. Let us change Example 4 to the one that follows and demonstrate how it should be handled.

EXAMPLE 1

As in Example 4 of the preceding section, let us suppose we perform a market survey. This time, suppose that the survey takes place in an industrial market where one-fourth of the companies use your product. Thus $p = 1/4$. If there are only 40 companies in this market and if once a company is selected it is not selected again, then the outcomes of each trial are not independent and the binomial formula of the preceding section cannot be used. If there are 40 companies and $p = 1/4$, then there are 10 companies that use your product. But once the first company is selected for interviewing, the value of p changes. If the first company is a user of your product, then $p = 9/39$ on the next selection. If the first company is not a user of your product, then $p = 10/39$ on the next trial. That is, the outcome of one trial changes the probabilities of the outcomes on the next trial.

Thus a new computational technique is called for. If we desire the probability of getting exactly 3 successes (companies who use your product) in a sample of $n = 5$, the total number of ways in which 5 customers can be selected from 40 is $\binom{40}{5}$, the combination of 40 companies taken 5 at a time. How many favorable cases are there? To obtain exactly three successes, we must make two choices. First, we must choose 3 of the 10 product users, and there are $\binom{10}{3}$ ways to do this. Then we must choose 2 of the 30 nonusers, and there are $\binom{30}{2}$ ways to do this. By the multiplication principle there are

$\dbinom{10}{3} \cdot \dbinom{30}{2}$ favorable cases, so the probability of exactly 3 product users

in a survey of 5 companies is

$$\frac{\dbinom{10}{3}\dbinom{30}{2}}{\dbinom{40}{5}} = \frac{725}{9139} = .0793$$

Notice that this probability differs from the probability .0879 in Example 4 of the last section. This difference reflects the effect of selection from a finite population and the resulting variation in the value of p.

The general rule is this: Suppose we have $a + b$ distinct objects divided into two classes, say a class of successes and a class of failures. Suppose there are a successes and b failures.

And suppose we take at random a sample of size n and ask for the probability that exactly r of the objects in it are successes. Thus, we let

a = the number of successes in the population of interest,
b = the number of failures in the population of interest,
$a + b$ = the total number of objects in the population,
n = the number of objects drawn in the sample,
r = the number of successes in the sample.

The probability of getting exactly r successes (and $n - r$ failures) is:

$$P(r|n) = \frac{\dbinom{a}{r}\dbinom{b}{n-r}}{\dbinom{a+b}{n}} \tag{5.18}$$

These quantities have the forbidding name of **hypergeometric probabilities**. Sometimes it is easier to visualize the three combinations involved in Formula (5.18) if they are presented in word form as they are here:

$$P(r|n) = \frac{\dbinom{\text{Number of Successes in the Population Taken } r \text{ at a Time}}{}\dbinom{\text{Number of Failures in the Population Taken } n - r \text{ at a Time}}{}}{\dbinom{\text{Total Population Taken } n \text{ at a Time}}{}} \tag{5.19}$$

The notation for the probabilities in Formulas (5.18) and (5.19) is $P(r|n)$, which is similar to the binomial probability notation of $P(r|n,p)$ except that the probability of a success changes from trial to trial in the hypergeometric situation due to sampling from a finite population without replacement. Thus there is no p value that can be indicated in the notation, and $P(r|n)$ is read as "the hypergeometric probability of r successes in n trials."

In Example 1, there were 10 successes (product users) in the population and 30 failures (nonusers). We wanted the probability of getting 3 successes in a sample of 5 from the total population of 40 people. Thus by Formula (5.18) we arrive at the same answer we reasoned to in Example 1:

$$P(r = 3|n = 5) = \frac{\binom{10}{3}\binom{30}{2}}{\binom{40}{5}} = .0793$$

EXAMPLE 2

Suppose nine families live on a street and six of them own their homes while three of them rent. If we interview four families at random without replacement (that is, if we take a random sample of size 4), and if we let r be the number of homeowners in the sample, then r is a random variable which can take the values 1, 2, 3, and 4 (there must be at least one homeowner since we select four families and there are only three renters on the street). The probability of getting r homeowners in a sample of four is the same as the probability of getting r successes and $4 - r$ failures. This is given by Formula (5.18) or (5.19):

$$P(r|n = 4) = \frac{\binom{6}{r}\binom{3}{4-r}}{\binom{9}{4}} \qquad r = 1, 2, 3, 4$$

Table I in the Appendix could be used to assist in evaluating the three preceding combinations. The individual probabilities (shown in Figure 5.7) are:

$$P(r = 1) = \frac{6}{126} = .048 \qquad P(r = 2) = \frac{45}{126} = .357$$

$$P(r = 3) = \frac{60}{126} = .476 \qquad P(r = 4) = \frac{15}{126} = .119$$

This set of probabilities for Example 2 is the distribution of r. They sum to 1, and they suffice to answer any question about the probability

FIGURE 5.7   Distribution of number of homeowners

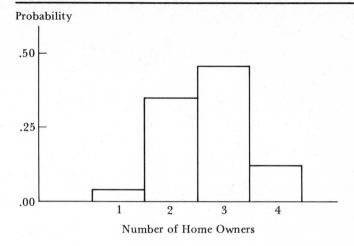

Number of Home Owners

that r will have a given property. For instance, the probability that r is even is:

$$P(r \text{ is even}) = P(r = 2) + P(r = 4) = \frac{60}{126}$$

The probability that r is 3 or more is:

$$P(r \geq 3) = P(r = 3) + P(r = 4) = \frac{75}{126}$$

EXAMPLE 3

Problem: A group consists of eight management and four union people. A committee of three is to be selected. Find the probability that a randomly selected committee will consist of one management person and two union people.

Solution: In this problem, we have a series of $n = 3$ trials where there are two possible outcomes (management or union person) on each trial. But the trials are dependent due to sampling from a population of 12 people without replacement. Thus, Formulas (5.18) and (5.19) hold, and, if we call selection of a management person a "success," we have:

$$P(r = 1 | n = 3) = \frac{\binom{8}{1}\binom{4}{2}}{\binom{12}{3}}$$

This expression can be evaluated by consulting Table I in the Appendix to find the three combination values.

$$P(r = 1 | n = 3) = \frac{8 \cdot 6}{220} = .2182$$

There are many sampling situations in which we sample without replacement from a finite population and thus have a hypergeometric probability situation. However, when the population is large, say several thousand, and the sample is small, the probability of a success does not change very much from trial to trial. Thus the Binomial Probability Formula (5.13) will give a good approximation to the Hypergeometric Formula (5.18). Situations like this and means of handling them are discussed in Chapter 6.

5.11   THE POISSON PROBABILITY DISTRIBUTION

The third discrete probability distribution that is useful to managers and administrators is the **Poisson distribution,** which was named for the man who developed it, Simeon Poisson (1781–1840). This distribution is useful in quality control situations, waiting line problems, and numerous other applications of probability.

There are few obvious conditions under which the Poisson distribution is directly applicable to a probability problem. However, the sampling medium is usually not a discrete trial as it is for the binomial and hypergeometric distributions. The sampling is often done on some continuous medium such as time, distance, or area. For instance, the Poisson distribution might be used to answer a question such as: what is the probability of finding seven paint blisters in a paint job that covers 100 square feet of area? The number of paint blisters is discrete and takes on the values $r = 0, 1, 2, \ldots, 7, 8$, etc. But the sampling medium is the painted area, which has no obvious "trials."

The formula for the Poisson distribution is

$$P(r | \theta) = \frac{e^{-\theta} \theta^r}{r!} \qquad r = 0, 1, 2, \ldots \qquad \textbf{(5.20)}$$

where θ is the expected number of Poisson-distributed events over the sampling medium that is being examined. The number e is a mathematical constant, like the Greek pi $\pi = 3.1416$, which occurs at many places in mathematics and statistics. Its value is $e = 2.718$.

EXAMPLE 1

Problem: The number of bubbles found in plate glass windows produced by a certain process is Poisson distributed with a rate of .004 bubbles per square foot. A 20-by-5-foot plate glass window is about to be installed. Find the probability that it will have no bubbles in it. Also, what is the probability that it will have two bubbles in it?

Solution: The continuous sampling medium in this problem is the area of the plate glass. That area is (20 ft.)(5 ft.) = 100 ft.^{2} Thus the expected number of bubbles in this piece of plate glass is θ = (100 ft.^{2})(.004 bubbles/ft.^{2}) = .4 bubbles. Now we can find the two probabilities asked for using Formula (5.20):

$$P(r = 0 | \theta = .4) = \frac{(2.718)^{-.4}(.4)^0}{0!} = 2.718^{-.4} = .6703$$

and

$$P(r = 2 | \theta = .4) = \frac{(2.718)^{-.4}(.4)^2}{2!} = .0536$$

Evaluating expressions like those in the preceding example requires the use of a moderately sophisticated calculator or the use of logarithms. To avoid this problem and obtain the Poisson probabilities of Formula (5.20), the reader can use Table IV in the Appendix. Table IV gives Poisson probabilities that have been calculated from Formula (5.20). The answers to Example 1 can be found under the θ = .4 column and in the r = 0 and r = 2 rows.

Table IV has two features which should be noted. The first is that only selected values of θ are listed. A probability like $P(r = 5 | \theta = 3.25)$ cannot be read from this table since there is no θ = 3.25 column. However, interpolation between the columns is valid. For instance, we can find the value midway between $P(r = 5 | \theta = 3.2)$ = .1140 and $P(r = 5 | \theta = 3.3)$ = .1203; this value is .11715, which can be used as an approximation to $P(r = 5 | \theta = 3.25)$. The exact value probability found by using Expression (5.20) is .11720, which is virtually the same as the interpolated value.

The second feature which should be noted about Table IV is that it is a table of individual probabilities. The binomial table, Table III, is cumulative. There is no inherent advantage of one type of table over the other. Tables III and IV were presented as cumulative and individual values, respectively, to demonstrate to the reader the use of the two types of tables.

EXAMPLE 2

Problem: The number of people arriving at a bank teller's window is Poisson distributed with a mean rate of .75 people per minute. What is the probability that two or fewer people will arrive in the next 6 minutes?

Solution: In this problem, $\theta = (.75 \text{ people/min})(6 \text{ min}) = 4.5$ people. We desire $P(r \le 2|\theta = 4.5)$ so we must add up three probabilities in the $\theta = 4.5$ column of Table IV corresponding to $r = 0$, 1, and 2: .0111 + .0500 + .1125 = .1736.

Each column in Table IV represents a different Poisson probability distribution. A unique feature of the Poisson distributions is that $E(r) = \theta$ and Variance$(r) = \theta$. That is, the mean and variance of a Poisson distribution *both* equal θ. This could easily be verified using Formulas (5.1) and (5.2) on any column in Table IV.

In the examples of this section, the problem statements informed us that the variables involved were Poisson distributed. For the time being, this is the way we will recognize a Poisson probability problem—we will have to be told that the variable involved is Poisson distributed. In Section 6.2, we will discuss some conditions under which the Poisson distribution can be used and how the reader can recognize those conditions without being told directly to use Poisson probabilities.

5.12   NORMAL PROBABILITY DISTRIBUTIONS

The probability distributions discussed in the previous three sections (the binomial, hypergeometric, and Poisson distributions) are discrete probability distributions. That is, they can be used to determine the probability that r, the *number* of successes in a sampling situation or the *number* of Poisson-distributed events, will take on some value or set of values. But r is a variable that is discrete and occurs in countable, integer form.

Section 5.5 discussed probability distributions for *continuous* variables in general terms. In this section we will discuss one specific continuous probability distribution—the normal distribution. This distribution can be used to compute the probability that many continuous measures like weights, lengths, and times will be within a particular range of values. The normal distribution is the most commonly used of all probability distributions, so the reader should pay particularly close attention to this section.

A *normal distribution*, or normal frequency curve, is given by the formula

$$f(X) = \frac{e^{-(X-\mu)^2/2\sigma^2}}{\sigma\sqrt{2\pi}} \quad -\infty < X < +\infty \tag{5.21}$$

Here π is the mathematical constant 3.1416 . . ., and e is another constant which often occurs in mathematics; its value is 2.718. . . . The values of μ and σ are also constant, but their particular values depend on the probability problem at hand. For any value of X substituted into Formula (5.21), $f(X)$ is the height of the normal frequency curve at that value of X.

Formula (5.21) is not important to us. It is far too complex for the beginning statistician to use. What is important to us is the fact that the normal frequency curve it describes is a special continuous probability distribution. It is the symmetrical, bell-shaped distribution shown in Figure 5.8. Any variable whose values follow the normal frequency distribution occurs with probabilities which can be developed from Formula (5.21). In our computations, we shall sidestep the formula and complex mathematics associated with it and find the probabilities for normally distributed variables by using a table in the Appendix.

There is a normal distribution for each pair μ and σ^2, where μ is any number and σ^2 is positive and μ and σ^2 are the mean and variance of the distribution, respectively. If we fix μ and let σ^2 vary, we get a family of curves with the same mean but different variances, as shown in Figure 5.9. The lower and more spread out the curve, the greater the variance σ^2. If we fix σ^2 and let μ vary, we get a family of curves with the same shape but different locations along the axis, as shown in Figure 5.10. The further to the right the curve, the greater the mean μ.

Experience has shown that many continuous random variables in diverse fields of application have distributions for which a normal distribution may serve as a mathematical or theoretical model or a good approximation. For instance, boxcars arriving at a certain ware-house are filled with insulation material. The weights of material shipped

FIGURE 5.8 A normal curve

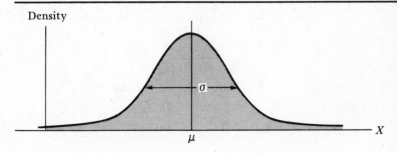

FIGURE 5.9　Normal distributions with the same mean, different variances

Density

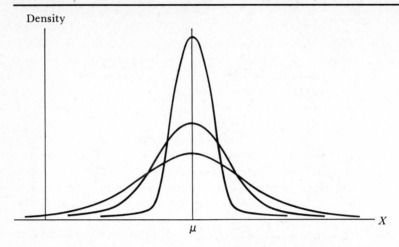

μ

X

in these cars are approximately normally distributed with a mean weight of 25 tons per car and a standard deviation of .5 tons per car. The warehouse operator may be interested in knowing what proportion of the boxcars arriving at his warehouse have a load of less than 24 tons. That is, he would like to know the probability of randomly selecting a boxcar of insulation material and finding that it contains less than 24 tons. Formula (5.21) could be used to find the height of the normal curve at any weight X by substituting into the formula the value for X, $\mu = 25$, and $\sigma = .5$. But it is the areas under the normal curve that give probabilities. In particular, the warehouse operator would be interested in the area under the normal curve to the left of 24 tons. Before he could use the table in the Appendix to find that area, however, he would have to know about the **standard normal distribution.**

As we have seen, there is a normal distribution for each pair μ and σ^2. Although it would be impossible to construct a table for each of them, we can select one, tabulate its areas, and use this table with appropriate conversion formulas to find probabilities for any normally

FIGURE 5.10　Normal distributions with the same variance, different means

Density

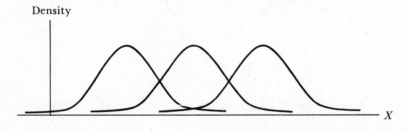

X

distributed variable. The distribution tabulated in Table V of the Appendix is the *standard normal distribution*, and it is defined as that normal distribution which has *mean 0 and variance 1*.

Let Z be a standard normal variable. That is, Z has a mean value $\mu_Z = 0$ and a variance $\sigma_Z^2 = 1$. The height of the normal curve which describes the variation of Z could be found for any value of Z using Formula (5.21), with $\mu = 0$, $\sigma^2 = 1$, and $X = Z$. But for continuous frequency curves we are interested not in the height of the curve but in the area under the curve. In the table of areas for the normal curve, Table V, values of Z from .00 to 3.89 are given in the body of the table. The table gives the area under the curve *between zero and the given value of Z*. For example, if we want the area between zero and 1.53, we find the $Z = 1.5$ row and the .03 column (for $Z = 1.5 + .03 = 1.53$) and read .4370 at the intersection. Since the standard normal curve is symmetric about zero, the area between $-Z$ and zero is equal to the area between zero and Z; therefore, only positive values of Z are given in the table. Since area is proportional to probability, the total area under the curve is equal to 1.

We can illustrate the use of the table of standard normal areas by several examples. A rough sketch of the curve together with the area desired is a great help in finding probabilities by means of the table.

EXAMPLE 1

Problem: Find the probability that a single random value of Z will be between .53 and 2.42.

Solution: The shaded area in Figure 5.11 is equal to the desired probability. The area given in the table for $Z = 2.42$ is .4922, but this is the area from zero to 2.42. The tabled area for .53 is .2019. Referring to the sketch, we see that the area between .53 and 2.42 is the difference between their tabular values:

$$P(.53 < Z < 2.42) = .4922 - .2019 = .2903$$

FIGURE 5.11

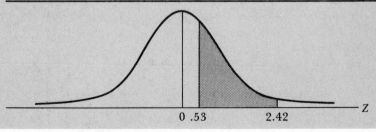

0 .53   2.42 Z

FIGURE 5.12

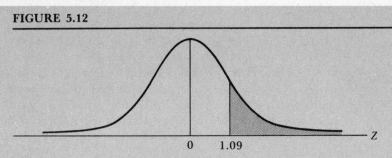

EXAMPLE 2

Problem: What is the probability that Z will be greater than 1.09?

Solution: The shaded area to the right of $Z = 1.09$ is the probability that Z will be greater than 1.09. In the table the area between zero and 1.09 is .3621. The total area under the curve is equal to 1 so that the area to the right of zero must be .5000, therefore (Figure 5.12),

$$P(Z > 1.09) = .5000 - .3621 = .1379$$

EXAMPLE 3

The probability that Z will be greater than $-.36$ is the total area to the right of $Z = -.36$ and consists of the area between $-.36$ and zero plus the area under the right-hand half of the curve (Figure 5.13):

$$P(Z > -.36) = .1406 + .5000 = .6406$$

EXAMPLE 4

The probability that a random value of Z is between -1.00 and 1.96 is found by adding the corresponding areas as indicated by Figure 5.14:

$$P(-1.00 < Z < 1.96) = .3413 + .4750 = .8163$$

As a general rule, if we desire to know the area under the normal curve between two values a and b and if a and b are on the *same*

FIGURE 5.13

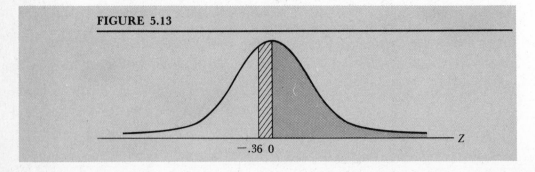

FIGURE 5.14

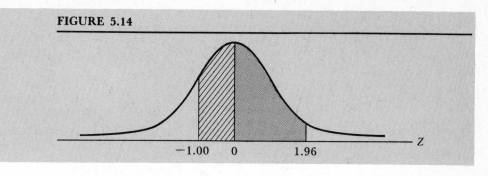

$$-1.00 \quad 0 \qquad 1.96 \qquad Z$$

side of the origin, we subtract the corresponding areas. If they are on *opposite sides, we add.*

A slightly different use of the table is required in the following examples.

EXAMPLE 5

Problem: Find a value for Z such that the probability of a larger value is equal to .0250.

Solution: The probability of a larger value is the area to the right of Z. The area between zero and Z must be .5000 − .0250, or .4750. We look in the body of the table until we find .4750, and take the desired value, $Z =$ 1.96 since .4750 appears in the 1.9 row and the .06 column (Figure 5.15).

EXAMPLE 6

Problem: Find the value of Z such that the probability of a larger value is .7881.

Solution: Since the area to the right of Z is greater than .5000, we know that Z must be negative and that the area between Z and zero must be .7881 − .5000, or .2881. Reading in the body of the table, we locate .2881 and find that the corresponding Z is −.80 (Figure 5.16).

FIGURE 5.15

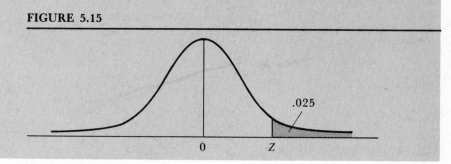

.025

$$0 \qquad\qquad Z$$

FIGURE 5.16

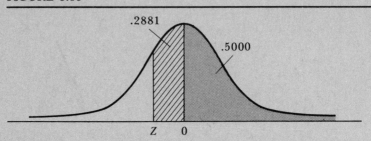

.2881

.5000

Z   0

EXAMPLE 7

Problem: Find the value for b such that

$$P(-b < Z < b) = .9010.$$

Solution: The endpoints $-b$ and b are symmetrically placed; therefore, half the area between them must be between zero and b. In the table we find the area value .4505 in the 1.6 row and .05 column, so $Z = 1.65$ (Figure 5.17).

FIGURE 5.17

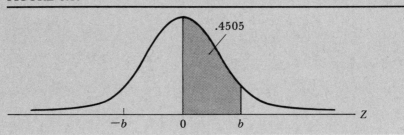

.4505

$-b$ 0 b Z

Now that we have seen how the table of normal areas is used to find probabilities for the standard normal variable, we turn our attention to the general case of the normal variable with mean μ, usually some value other than 0, and variance σ^2, usually some value other than 1. The key to using the standard normal areas for finding probabilities for the general normal variable is to pass to the *standardized* variable. If X is a random variable with mean μ and variance σ^2, then, by Equations (5.7) and (5.8), the standardized variable

$$Z = \frac{X - \mu}{\sigma} \qquad (5.22)$$

has mean 0 and variance 1. If X is normally distributed, Z is a *standard normal variable*.

Let X be normal with mean μ and variance σ^2, and suppose we want to find the probability that a randomly selected value for X will be between a and b. We use the standardization Formula (5.22) and find that when $X = a$,

$$Z = \frac{a - \mu}{\sigma}$$

and that when $X = b$,

$$Z = \frac{b - \mu}{\sigma}$$

Therefore, whenever X is between a and b, the standard variable Z will be between $(a - \mu)/\sigma$ and $(b - \mu)/\sigma$. We have, then,

$$P(a < X < b) = P\left(\frac{a - \mu}{\sigma} < Z < \frac{b - \mu}{\sigma}\right)$$

Thus, we can turn to the tables and find the probability that Z is between $(a - \mu)/\sigma$ and $(b - \mu)/\sigma$.

EXAMPLE 8

Problem: We are now ready to return to the example of the warehouse operator who receives boxcars loaded with insulation materials. The weights of the materials are normally distributed with a mean of 25 tons and a standard deviation of .5 tons. The operator desires to know the proportion of the cars that have less than 24 tons in them.

Solution: The standardization formula becomes

$$Z = \frac{X - 25}{.5}$$

and

$$P(X < 24) = P\left(Z < \frac{24 - 25}{.5}\right)$$

$$= P(Z < -2.0)$$

$$= .5000 - .4772 = .0228$$

The warehouse operator may conclude that about two and one-quarter percent of the cars arrive with less than 24 tons in them.

EXAMPLE 9

Problem: Suppose that a certain type of wooden beam has a mean breaking strength of 1500 pounds and a standard deviation of 100 pounds, and that we want to know the relative frequency of all such beams whose breaking strength is between 1450 and 1600 pounds.

Solution: The standardization formula becomes

$$Z = \frac{X - 1500}{100}$$

and

$$P(1450 < X < 1600) = P\left(\frac{1450 - 1500}{100} < Z < \frac{1600 - 1500}{100}\right)$$

$$= P(-.50 < Z < 1.00)$$

$$= .1915 + .3413 = .5328$$

We may conclude that about 53% of the beams have breaking strengths between 1450 and 1600 pounds.

5.13   SUMMARY

In the last few sections of this chapter, we have discussed several probability distributions and formulas. The reader cannot be faulted if he or she is somewhat confused about when to use the different distributions and formulas that have been presented.

We must answer a series of questions before we can determine which formula, table, or relationship applies to a probability problem. Figure 5.18 presents a diagram designed to lead us through the questions which must be answered to arrive at a workable solution to a problem.

To solve a probability problem, we must first determine if we are dealing with a discrete or continuous random variable. If the random variable is discrete, there are three possible probability distributions which might apply: binomial, hypergeometric, or Poisson. If none of these apply, the problem might still be solvable using Formula (4.13) or (4.16) or some combination of the two. Note that with our current level of knowledge in probability, we must be told a discrete random variable is Poisson distributed before we use the Poisson distribution and Table IV. If we are not informed that the discrete random variable follows the Poisson distribution, then we must determine if the probability situation involves a series of "trials" for which there are only two possible outcomes at each trial: success or failure. If this is the case, then the hypergeometric distribution may apply (if sampling is without replacement from a finite population) or the binomial distribution may apply

FIGURE 5.18

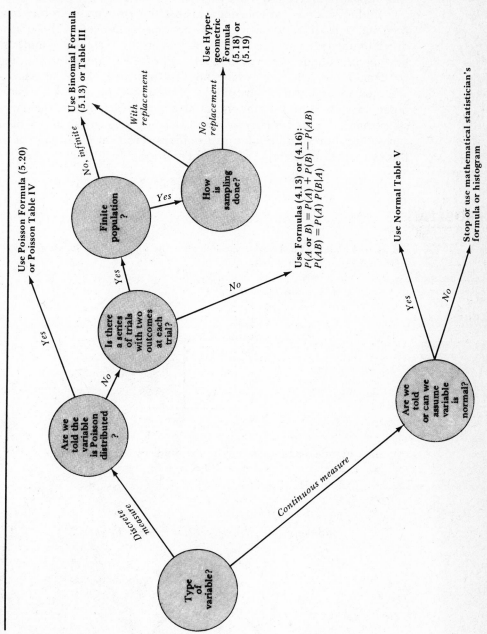

(if the trials are independent due to an infinite population or sampling with replacement from a finite population).

If the random variable in a probability problem is continuous, then the only distribution we have that can be applied is the normal distribution. However, we need good reason to be able to assume the normal distribution describes the behavior of the random variable, and we cannot use it blindly. Many physical measures, however, can be assumed to be normally distributed without too much risk or error. Sometimes a continuous random variable is described by a mathematical statistician's density function or histogram and we can use areas under such a function to determine probabilities. Problem **60** at the end of this chapter is an example of this situation.

PROBLEMS

1. Given that X is a random variable with the distribution shown as follows, find the probability that:

    a. $.5 \le X \le 1$ **b.** $X \ge .75$
    c. $.20 \le X \le .80$ **d.** $X \le .15$

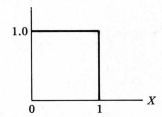

2. If X has the distribution shown next, find:

    a. $P(X \ge 2)$ **b.** $P(0 \le X \le 3)$
    c. $P(X \le -1)$ **d.** $P(-2 \le X \le 2)$
    e. $P(X \ge -2)$

$$\text{(Remember: Triangle area} = \frac{1}{2}\text{ base} \cdot \text{height)}$$

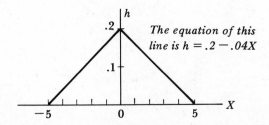

The equation of this line is $h = .2 - .04X$

3. If X has the distribution shown next find:
   a. a
   b. $P(X \le 2)$
   c. $P(X \ge 3)$
   d. $P(1 \le X \le 4)$
   e. C such that $P(X \le C) = .6$

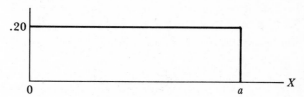

4. Given that X has the distribution shown next, find:
   a. θ
   b. $P(-1 \le X \le .5)$
   c. $P(X \ge .6)$
   d. C such that $P(-C \le X \le C) = .6$
   e. C such that $P(X \ge C) = .8$

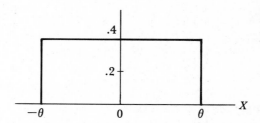

5. The following table gives the amount of time X in seconds by which an automated manufacturing process misses the design completion time when operating on certain types of jobs. Negative values indicate early completion and positive values indicate late completion.

r	-1	0	1	2
$P(X = r)$	.1	.2	.3	.4

   a. Find the mean and variance of X.
   b. On the average, how do the completion times for this type of job compare with the designed completion times?

6. Given that X has the probability distribution given in the following table, find the mean μ and the variance σ^2.

r	0	1	2	3	4
$P(X = r)$	$\dfrac{1}{16}$	$\dfrac{4}{16}$	$\dfrac{6}{16}$	$\dfrac{4}{16}$	$\dfrac{1}{16}$

7. The following distribution shows the number of boxcars that are unloaded each day by a warehouse crew over the past year.

X	P(X)
6	.10
7	.25
8	.50
9	.10
10	.05
	1.00

a. Explain the meaning of the .25 associated with $X = 7$.
b. Find the mean number of boxcars unloaded per day.
c. Find the standard deviation of this distribution.
d. How often during the past year has the crew unloaded five or fewer boxcars? Will the same be true for next year?

8. Given that the mean for a continuous probability distribution is located at its center of gravity, find the mean values for the continuous distributions shown in:
a. Problem 1
b. Problem 2
c. Problem 3

9. Use Table III in the Appendix to find the following binomial probabilities:
a. $P(r \leq 5 | n = 10, p = .6)$
b. $P(r = 12 | n = 20, p = .4)$
c. $P(r \geq 16 | n = 25, p = .8)$
d. $P(3 \leq r \leq 8 | n = 10, p = .7)$

10. Use Table III in the Appendix to find the following binomial probabilities:
a. $P(r = 3 | n = 5, p = .5)$
b. $P(r \leq 1 | n = 25, p = .05)$
c. $P(r \geq 7 | n = 15, p = .3)$
d. $P(18 \leq r \leq 22 | n = 25, p = .9)$

11. Use Formula (5.13) to compute the following binomial probabilities:
a. $P(r = 1 | n = 4, p = .5)$
b. $P(r = 3 | n = 5, p = .15)$
c. $P(r \leq 2 | n = 6, p = .2)$
d. $P(r = 0 | n = 7, p = 1.0)$

12. Find the mean and standard deviation of the number of successes in binomial distributions characterized by:
a. $n = 20, p = .5$ b. $n = 100, p = .9$
c. $n = 30, p = .7$ d. $n = 50, p = .4$

13. Find the mean and variance of the number of successes in the binomial distribution where $n = 5$ and $p = .4$:
a. Using Formulas (5.16) and (5.17).
b. Using Formulas (5.1) and (5.2) and the probabilities for $r = 0,1,2,3,4,5$ that can be obtained from Table III in the Appendix.

14. A shop foreman wants the supervisor to look at one of the defectives which a new automatic drilling machine is producing. The machine has been producing defects randomly at the rate of one every three parts. What is the probability that the shop foreman will find exactly one defective

in a randomly selected sample of $n = 4$ parts? (*Hint:* Use Formula (5.13) with $p = 1/3$.)

15. Large sheets of plate glass are inspected for structural defects at each of five inspection stations. The probability of successfully identifying a defect at any one station is $1/2$. What is the probability that a defective piece of glass is identified at (*Hint:* Use Table III in the Appendix):
    a. All five inspection stations?
    b. At least one inspection station?
    c. Two of the five inspection stations?

16. In a triangle taste test, the taster is presented with three samples, two of which are alike, and is asked to pick the odd one by tasting. If a taster has no well-developed sense and can pick the odd one only by chance, what is the probability that in six trials he will make five or more correct decisions? no correct decisions? at least one correct decision? (*Hint:* Formula (5.13) will be useful here.) Discuss the notion of independence for these trials.

17. A process produces 10% defective items. If we take a sample of 20 items, what is the probability we will find:
    a. No defectives?   b. Not more than one defective?

18. A process is considered to be in control if it produces no more than 10% defectives. The process is stopped and checked if a sample of 15 contains more than one defective. What is the probability it will be stopped needlessly when it is producing 5% defectives?

19. A door-to-door salesperson has a 20% sales success rate. Find the probability that during the next 20 calls (assume these calls' results are independent of one another) this person will:
    a. Make 5 or fewer sales.
    b. Make 3 sales.
    c. Make 4 or more sales.
    d. Fail to make 16 sales.

20. A tree surgeon loses about 5% of his "patients." Tomorrow he will operate on five trees. What is the probability that:
    a. He will lose one or fewer patients from the five?
    b. He will lose your tree assuming it is to be operated on and the operation outcomes are independent of one another.

21. In a large orchard 10% of the apples are wormy. If four apples are selected at random, what is the probability that:
    a. Exactly one will be wormy?
    b. None will be wormy?
    c. At least one will be wormy?

22. An automatic lathe is said to be out of control and is checked if a sample of five turnings contains any that are defective. If 1% of its output is defective, what is the probability that the lathe will be declared out of control after a given sample is checked? Assume the probability of a defective is the same for every item selected.

23. If a fair die is rolled 20 times,
    a. What is the probability that an even number will occur between 3 and 5 times, inclusive?

b.   What is the probability an odd number will occur fewer than 2 times?

24.   Solution of this problem requires a computer or powerful calculator. If a process produces 5% defective items, and if we take a sample of 100, what is the probability of fewer than five defectives?

25.   Solution of this problem requires a computer or powerful calculator. An operator for a telephone answering service answers calls for two insurance agents. The two agents are charged the same monthly fee for this service under the assumption that they will each receive approximately half the calls coming in. (The operator receives calls only for these two agents.)

    a.   If this assumption is correct, find the probability that in the next 225 calls 125 or more will be for agent A.

    b.   Find the probability that one of the agents will get 130 or more of the next 225 calls.

    c.   During the past week the operator has answered 400 calls. Of these, 300 were for agent A. Is this sufficient evidence to indicate that agent A gets more calls than agent B and thus should be charged more? (*Hint:* Find the probability of 300 or more calls for A in 400 trials where the probability of a call for A is $p = .5$.)

26.   A population of 20 items contains 12 successes and 8 failures. The hypergeometric probability that a sample of 7 items from this population will have 5 successes in it is given by the expression:

$$P(r = 5 | n = 7) = \frac{\binom{12}{5}\binom{8}{2}}{\binom{20}{7}}$$

Use Table I in the Appendix to help evaluate this probability expression.

27.   There are seven people who work in an office. Of the seven, four would like to be transferred. If three people from this office are randomly selected for transfer, what is the probability that:

    a.   All three will be people who wanted the transfer?

    b.   Two of the three will be those who wanted a transfer?

28.   A dishonest accountant has "adjusted" 6 of the entries in a particular account. That account has 18 total entries, and an auditor is about to randomly select 4 of the entries to be examined in detail. Find the probability that:

    a.   Two of the "adjusted" entries will be examined.

    b.   One or more of the "adjusted" entries will be examined.

29.   A salesperson has 15 major accounts. Eight of those accounts were sent the wrong billing last month (they were billed too much by the accountant in Problem 28). The salesperson wants to call her accounts to explain what happened, but she only has time to call ten of them before quitting time (assume a call to a wrongly billed account takes just as long as a call to one with correct billing). If she selects accounts and calls them randomly, what is the probability that:

    a.   She will reach six of the eight she wants to contact?

   b. She will reach only five?

   c. She will reach none of those incorrectly billed?

30. In order to test the skills of its auditors, an accounting firm asks its auditors to examine 25 accounting transactions, 5 of which are in error. Assume the auditors are told in advance that there are 5 incorrect transactions. If one of the auditors is simply guessing, what is the probability that he will correctly select the 5 incorrect transactions? What is the probability that he will select 2 of the incorrect transactions?

31. Production inspection records show that 8 of the 35 items in a production lot are defective, but the defective items were not marked. A rush order for 5 of these items has just arrived. If 6 items are selected from the lot and sent, what is the probability that 5 or more of them will be good?

32. In Problem **31**, what is the probability that 2 or fewer of the items are good?

33. Use Table IV in the Appendix to find the following Poisson probabilities:

   a. $P(r = 3|\theta = 4.5)$ **d.** $P(r = 12|\theta = 9.2)$

   b. $P(r = 8|\theta = 2.8)$ **e.** $P(r \le 4|\theta = .8)$

   c. $P(r \ge 9|\theta = 7.7)$ **f.** $P(r = 3|\theta = 1.65)$

34. Use Table IV in the Appendix to find the following Poisson probabilities:

   a. $P(r = 7|\theta = 6.2)$ **d.** $P(5 \le r \le 7|\theta = 3.8)$

   b. $P(r \le 2|\theta = 3.4)$ **e.** $P(3 \le r \le 9|\theta = 5.5)$

   c. $P(r = 6|\theta = 1.3)$ **f.** $P(r \ge 2|\theta = 8.8)$

35. The number of paint blisters produced by an automated painting process is Poisson distributed with a rate of .06 blisters per square foot. The process is about to be used to paint an item that measures 9 by 15 feet. What is the probability that the finished surface will have:

   a. No blisters in it?

   b. Between five and eight blisters, inclusive?

   c. More than two blisters?

36. During the off hours, people arrive at a toll booth at a rate of .5 people per minute. If the arrivals are Poisson distributed, what is the probability that during the next ten minutes:

   a. Seven people will arrive?

   b. None will arrive?

   c. Three or more people will arrive?

   d. Between six and eight people inclusive will arrive?

37. The defects in an automated weaving process are Poisson distributed at a rate of .00025 per square foot. The process is set up to run 1000 square yards of weaving.

   a. What is θ for this problem?

   b. What is the probability that this process will produce five defects on this run?

   c. What is the probability that it will produce between one and three defects inclusive on this run?

38. The process described in Problem **37** is to be used to weave a piece of material that is 5 by 16 yards.

   a. What is θ for this problem?

   b. What is the probability this piece will have no defects?

    c. What is the probability it will have one defect?

39. If the number of telephone calls passing through a switchboard has a Poisson distribution with mean θ equal to $3t$ where t is the time in minutes, find the probability of:

    a. Two calls in any one minute.

    b. Five calls in two minutes.

    c. At least one call in one minute.

40. At a particular location on a river, the number of fish caught per man-hour of fishing effort has a Poisson distribution with θ equal to 1.3 fish per man-hour. Find the probability that:

    a. Four fish will be caught by one man fishing two hours.

    b. Four fish will be caught by two men fishing one hour.

    c. Six fish will be caught by three men fishing twenty minutes.

    d. Eight fish will be caught by three women fishing for two hours.

41. A central computer system has several input stations located in an organization's various departments. Typically, the central computer can respond to eight input jobs per minute without causing any noticeable delay at the input stations. That is, when the jobs submitted in one minute are eight or less, each input station behaves as though it had complete control of the central computer. Internal computer accounting records show that jobs are usually submitted at a Poisson distributed rate of .08333 jobs per second. What is the probability that, during a randomly selected minute of operation, the input stations will experience some processing delays?

42. Given that Z is the standard normal variable, use the table of normal areas to find:

    a. $P(0 \leq Z \leq 1)$ **b.** $P(Z \geq 1)$

    c. $P(-1 \leq Z \leq 1)$ **d.** $P(Z \geq -1)$

43. If Z is the standard normal variable, find:

    a. $P(Z \leq .75)$ **b.** $P(.25 \leq Z \leq .75)$

    c. $P(-.25 \leq Z \leq .75)$ **d.** $P(Z \leq -.25)$

44. Given Z is the standard normal variable, find:

    a. $P(-1.96 \leq Z)$ **b.** $P(-1.96 \leq Z \leq -1.5)$

    c. $P(.38 \leq Z \leq 1.42)$ **d.** $P(-.49 \leq Z \leq 1.05)$

45. Given Z is the standard normal variable, find:

    a. $P(Z \leq 1.23)$ **b.** $P(Z \leq -2.12)$

    c. $P(Z \geq 1.17)$ **d.** $P(Z \geq -.62)$

    e. $P(-1.56 \leq Z \leq -.64)$ **f.** $P(-.72 \leq Z \leq 1.89)$

46. Given Z is the standard normal variable, find C such that:

    a. $P(Z \geq C) = .025$ **b.** $P(Z \leq C) = .0287$

    c. $P(-C \leq Z \leq C) = .95$

47. Given that Z is the standard normal variable, find C such that:

    a. $P(Z \leq C) = .9554$ **b.** $P(Z \leq C) = .3085$

    c. $P(Z \geq C) = .9998$ **d.** $P(Z \geq C) = .0322$

    e. $P(-C \leq Z \leq C) = .4515$ **f.** $P(1 \leq Z \leq C) = .1219$

48. Given that X is a normal variable with $\mu = 50$ and $\sigma = 10$, find:

    a. $P(X \leq 65)$ **b.** $P(X \leq 25)$

    c. $P(42 \leq X \leq 62)$ **d.** $P(38 \leq X \leq 47)$

49. Given that X is normal with $\mu = 15$ and $\sigma^2 = 25$, find:

   a. $P(X \leq 20)$ **b.** $P(X \leq 13)$

   c. $P(10 \leq X \leq 18)$ **d.** $P(19 \leq X \leq 40)$

50. Given that X is normal with $\mu = .05$ and $\sigma = .012$, find:

   a. $P(X \geq .074)$ **b.** $P(.071 \leq X \leq .077)$

51. Given that X is normal with $\mu = .130$ and $\sigma^2 = .000625$, find:

   a. $P(X \geq .126)$ **b.** $P(.110 \leq X \leq .165)$

52. If the resistances of carbon resistors of 1300 ohms nominal value are normally distributed with $\mu = 1300$ ohms and $\sigma = 150$ ohms:

   a. What proportion of these resistors would have resistances greater than 1000 ohms?

   b. What proportion would have resistances that do not differ from the mean by more than 1% of the mean?

   c. What proportion do not differ from the mean by more than 5% of the mean?

53. If the service lives of electron tubes in a particular application are normally distributed, and if 92.5% of the tubes have lives greater than 2160 hours while 3.92% have lives greater than 17,040 hours, what are the mean and standard deviation of the service lives?

54. The mean monthly carbon monoxide count for June in a certain western city is approximately normally distributed with a mean of 7.5 parts per million and a standard deviation of .8 parts per million. The local air pollution control agency has asked the major employers in the area to allow their employees to stay at home when the carbon monoxide count reaches or exceeds 9.5 parts per million. If the companies agree to this proposal, on what proportion of June days will their employees receive "air pollution emergency vacations"?

55. A chemical company sells its major product HBC-50K in gallon containers. The company's records show that the amount of caustic materials in the product averages 4.8 ounces with a standard deviation of .5 ounces. What proportion of the gallon containers will have:

   a. More than 5 ounces of caustic material?

   b. Less than 2 ounces of caustic material?

   c. Between 3.2 and 5.2 ounces of caustic material?

56. A new brand of disposable flashlight is guaranteed to last for one year of normal use. Tests indicate that the length of life of these lights under normal usage is approximately normally distributed with a mean of 1.5 years and a standard deviation of .4 years.

   a. What proportion of the flashlights will fail to meet the guarantee?

   b. What proportion of the flashlights will last longer than one year and nine months?

   c. Would you expect that the proportion of flashlights returned under the guarantee would be higher or lower than the proportion determined in Part **a**?

57. Historical records of the Boonville Redistribution Center show that weights of the trucks arriving at their docks are normally distributed with a mean of 7 tons and a standard deviation of 2 tons.

   a. What proportion of the trucks weigh more than 9 tons?

    b. What proportion of the trucks weigh less than 5.24 tons?

    c. What proportion of the trucks weigh between 4.6 and 10.6 tons?

The following problems are somewhat more challenging than the others in this chapter.

58. The World Series terminates when one team wins its fourth game. If the two teams are evenly matched, i.e., if the probability that either team will win any one game is $1/2$, what is the probability that the series will terminate at the end of the fourth game? the fifth game? the sixth game? the seventh game?

59. If teams A and B are competing in the World Series and the probability that A wins any one game is .6, find the probability that:

    a. A wins in four games.

    b. B wins in five games.

    c. Six games are required to complete the series.

60. The following probability distribution represents the distribution of times until failure of those transistors of a certain type which do not pass inspection during a rather short test of their electrical conductivity. The distribution applies only to those transistors which fail during the two-second test. The density function for the time to failure during the test (for those transistors which fail) is:

$$f(X) = .25(1 + X) \quad \text{for} \quad 0 \le X \le 2 \text{ sec.}$$

    a. Draw a graph of this density function. It should show that a small proportion of the transistors fail almost immediately after electrical current is applied to them and that those that fail tend to fail at a steadily increasing rate the longer the test goes on.

    b. Find the probability that a transistor that fails will fail in less than .5 seconds after the test starts.

    c. Find the probability that a transistor that fails will fail in between .2 and 1.0 seconds from the start of the test.

    d. What is the expected value of the time to failure for the transistors which fail? (This answer requires integral calculus. Calculus can be used to find the answers to Parts **b** and **c** above, or the formula Area of a Trapezoid = (base)(average height of the sides) can also be used.

SAMPLE DATA SET QUESTION: *Refer to the 113 applicants for credit listed in the Sample Data Set Appendix of this book.*

    a. A national survey recently showed that 20% of the applicants for credit at department stores are women who apply for credit in their own names. Find the binomial probability that exactly 19 women in a sample of 113 applicants would apply for credit in their own names as they did in this sample, if the 20% figure is correct.

b. Discuss why the probability found in Question **a** is so small.

c. Find the same probability found in Question **a** using the Poisson formula with $\theta = (.20)(113) = 22.6$.

d. Which probability is the correct figure—the binomial or the Poisson probability?

REFERENCES

5.1 Cochran, W. G. *Sampling Techniques,* 2nd edition. John Wiley & Sons, New York, 1963.

5.2 Hogg, Robert V., and Allen T. Craig. *Introduction to Mathematical Statistics,* 3rd edition. Macmillan, New York, 1970.

5.3 Kish, L. *Survey Sampling.* John Wiley & Sons, New York, 1965.

5.4 Parzen, Emanuel. *Modern Probability Theory and Its Applications.* John Wiley & Sons, New York, 1960.

PROBABILITY
APPROXIMATIONS

In the previous chapter, we discussed the binomial, hypergeometric, Poisson, and normal probability distributions. These four distributions are related to one another, and in some situations two or more of the distributions can be used to calculate the probability of the same event. This characteristic of the four distributions is useful since we may desire a hypergeometric probability, which is cumbersome to calculate using Formula (5.18). But if we can get a good approximation using the convenient binomial or Poisson tables, we may be willing to use the approximation to save some tedious calculations.

In this chapter, we will discuss the conditions under which some of the distributions can be used to approximate probabilities calculated from the other distributions. The examples in this chapter will give

comparisons of the accuracy of these approximations. The approximations to be discussed are: the binomial approximation of the hypergeometric, the Poisson approximation of the binomial, and the normal
approximation of the binomial.

6.1   THE BINOMIAL APPROXIMATION OF THE HYPERGEOMETRIC

The binomial and hypergeometric distributions are similar in that they
are both used when we have a series of trials that result in "success"
or "failure" on each trial. Both distributions tell us the probability of
r successes in n trials. But they differ in that we use the binomial
distribution when the chance of success remains constant from trial
to trial at a value p; we use the hypergeometric distribution when
the chance of a success changes from trial to trial due to our sampling
from a finite population without replacement. However, when the
population is large relative to the sample size, the chance of a success
does not change much from trial to trial, and thus the binomial probability
distribution can be used to give approximately the same answer that
the exact hypergeometric distribution would give.

   To understand the notion that the chance of a success changes
very little from trial to trial when the population from which the sample
is drawn is large relative to the sample, consider a finite population
of 100 transistors, of which 10 are defective and 90 are nondefective.
In a random sample of size 1, the chance of getting any one particular
transistor is .01, but the chance of getting some one of the 10 defective
transistors is .1. In a random sample of size 2, the chance that both
elements are defective is:

$$\frac{\binom{90}{0}\binom{10}{2}}{\binom{100}{2}} = \frac{10 \cdot \dfrac{9}{2}}{100 \cdot \dfrac{99}{2}} = \frac{10}{100} \cdot \frac{9}{99}$$

This follows from Formula (5.18) since from the 90 nondefectives none
are chosen, and from the 10 defectives 2 are chosen. The factor 10/100
on the right here is .1—which is the chance of a defective in a sample
of just 1—and the other factor, 9/99, is .091. The second factor is
slightly less than .1 because if one element of the sample is defective
then one defective has been "used up," lessening by a small amount
the chance that the other element of the sample is also defective.

   Now consider a population of 1000 transistors of which 100 are defective and 900 are nondefective, so that the proportions of defective and nondefective elements are 10% and 90%, just as before. Again we take a sample of size 2. The chance that both are defective is:

$$\frac{\binom{900}{0}\binom{100}{2}}{\binom{1000}{2}} = \frac{100 \cdot \dfrac{99}{2}}{1000 \cdot \dfrac{999}{2}} = \frac{100}{1000} \cdot \frac{99}{999}$$

Again the first factor is .1. The other factor, 99/999, or .099, is closer to .1 than 9/99, or .091, was. For the population of 1000, one defective element in the sample affects less than it did for the population of 100 the chance that the other element is also defective. This is to be expected, because one element "uses up" a smaller proportion of the population of 1000 than of the population of 100. Thus, for a finite population of transistors of which 10% are defective, the chance of two defectives in a sample of size 2 is very close to .1 × .1, *if the population is very large.* For random sampling from an infinite population there is exact equality. In this ideal case, drawing one element from the population and putting it in the sample "uses up" a completely negligible proportion of the population.

In general, the chance of a success changes very little from trial to trial if sampling is done without replacement in a finite population which is large relative to the sample. The implication of this statement is that the Binomial Probability Formula (5.13) and the associated binomial tables can be used to give approximate answers to hypergeometric probability problems *if the population is large relative to the sample.* But how large is "large"? A rough rule of thumb is that if the sample is less than 10 percent of the population, the binomial probability answer, which assumes constant probability of success from trial to trial, will be approximately the same as the exact hypergeometric answer.

EXAMPLE 1

Problem: A set of 25 accounts is classified so that 15 are current and 10 are past due. If a sample of $n = 2$ accounts is chosen at random, what is the probability that there will be one current and one past due account in the sample?

Solution: The exact correct answer to this problem is given by Formula (5.18). If we call drawing a current account a success,

$$P(r = 1 | n = 2) = \frac{\binom{15}{1}\binom{10}{1}}{\binom{25}{2}} = \frac{15 \cdot 10}{300} = .5000$$

Since the sample size of $n = 2$ is less than 10% of the population (10% of 25 is 2.5), the binomial probability of one current account in two trials with $p = (15/25) = .60$ should be approximately the same as the probability just calculated. Using Formula (5.13) (we cannot use Table III, which has no $n = 2$ section), we find:

$$P(r = 1 | n = 2, p = .6) = \frac{2!}{1!1!} (.6)^1 (.4)^1 = .48$$

Thus the answers are quite close, although .50 is the exact, correct answer, and .48 is only the approximation.

EXAMPLE 2

Problem: Only some of the 500 workers at the Benton Plant are members of the retirement program—300 belong and 200 do not. If the union selects 20 people to be on a bargaining committee, what is the chance that 12 of the people will be members of the retirement fund program?

Solution: Here there is a series of $n = 20$ trials where we have a success (member of the program) or failure (nonmember) on each trial. The population is finite, so the chance of success changes from trial to trial since we would sample without replacement here. The exact chance of 12 members in a sample of 20 people at the plant is:

$$P(r = 12 | n = 20) = \frac{\binom{300}{12}\binom{200}{8}}{\binom{500}{20}}$$

This expression is rather difficult to evaluate without a computer or sophisticated calculator. Thus, one can see why the binomial approximation might be more desirable in this case since binomial probabilities can be evaluated so easily using Table III in the Appendix with $n = 20$ and $p = (300/500) = .6$:

$$P(r = 12 | n = 20, p = .6) = .5841 - .4044 = .1797.$$

The exact hypergeometric answer was found using a computer, and $P(r = 12 | n = 20) = .1834$. The difference between the exact hypergeometric answer and the approximate binomial answer is only .0037. This difference is small enough that we might give up that amount of accuracy in our answer in order to solve the problem using Table III rather than a computer or sophisticated calculator.

6.2 THE POISSON APPROXIMATION OF THE BINOMIAL

Consider the following binomial probability: $P(r = 6|n = 1000, p = .004)$. We cannot find this probability in Table III of the Appendix since Table III has no section for $n = 1000$ and no columns for $p = .004$. Thus we must find the answer by resorting to Formula (5.13).

$$P(r = 6|n = 1000, p = .004) = \frac{1000!}{6!994!} (.004)^6 (.996)^{994} = .1043$$

Once again, a sophisticated calculator was required to evaluate this expression.

However, it can be shown mathematically that when there are many trials (n is large) and when the chance of success on each trial is small (p is close to zero), then the binomial probability $P(r|n, p)$ is approximately equal to the Poisson probability $P(r|\theta)$, when $\theta = np$, the mean of the binomial distribution. In the problem just stated, $np = (1000)(.004) = 4.0$. Table IV in the Appendix can be used to find:

$$P(r = 6|\theta = 4.0) = .1042$$

Thus we can see for this example that the Poisson probability approximates the binomial to within .0001.

EXAMPLE 1

Problem: Roughly eight people in one thousand have an allergic reaction to the materials used in the production process at your plant. If you are about to hire 250 new workers at the plant, what is the probability that four or fewer of them will have the allergic reaction?

Solution: Let us call finding a person with the reaction a "success." Technically, this is a hypergeometric problem where the population from which we are sampling is all the people in the world, .008 of whom have the allergic reaction. However, we will assume that the population is so large that the chance of finding a person with the reaction remains constant from one person to the next. Thus we desire to find:

$$P(r \le 4|n = 250, p = .008) = \sum_{r=0}^{4} \frac{250!}{r!\,(250 - r)!} (.008)^r (.992)^{250-r}$$

$$= .9481$$

This formidable expression was evaluated once again with a powerful calculator. But using the Poisson approximation allows us to save the work of that evaluation. We find that $\theta = (250)(.008) = 2.0$. The $P(r \le 4|\theta = 2.0)$ is the sum of the first five probabilities (for $r = 0,1,2,3,4$) under the $\theta = 2.0$ column of Table IV:

$$.1353$$
$$.2707$$
$$.2707$$
$$.1804$$
$$\underline{.0902}$$

$$P(r \leq 4|\theta = 2.0) = .9473$$

Once again, the approximation is very close to the exact binomial probability we sought.

EXAMPLE 2

Problem: Past records show that 99 out of 100 packages coming off the production line have the correct weight of goods in them. Find the probability that in a shipment of 350 packages there will be 348 or more containing the correct weight.

Solution: If we call a package that contains the correct weight a "success," then we desire to know

$$P(r \geq 348|n = 350, p = .99) = \sum_{r=348}^{350} \frac{350!}{r! \, (350 - r)!} (.99)^r \, (.01)^{350-r}$$

$$= .3195$$

It would appear at first glance that we cannot obtain a Poisson approximation to the exact binomial probability just given. This is because we suggested that the approximation is good when n is large and p is small. In this problem, $p = .99$ is close to unity. However, we can restate the problem somewhat. If we obtain 348 or more successes in 350 trials, we must necessarily have 2 or fewer failures. Thus if we let f stand for the number of failures, and $q = 1 - p$ be the probability of a failure, the probability we seek can also be stated as:

$$P(f \leq 2|n = 350, q = .01) = .3195$$

Now we can see that the Poisson approximation to this value is possible by letting $\theta = nq = (350)(.01) = 3.5$ and finding:

$$P(f \leq 2|\theta = 3.5) = .0302 + .1057 + .1850 = .3209$$

The Poisson approximation is only .0014 from the exact binomial answer to this problem.

The results of Example 2 indicate that we must revise our statement about the conditions under which the Poisson distribution can be used to obtain good approximations to the binomial. Initially we said that the binomial n had to be large and the value of p had to be small. Now we restate this to say that n must be large and p must be *extreme* (close to zero or close to unity). When p is close to unity, we restate the problem in terms of failures and proceed as we did in Example 2.

People often wonder how large n must be and how extreme p must be in order for the Poisson probabilities to be good approximations to the binomial. One suggested rule of thumb is that when $\theta = np$ or $\theta = nq$ are calculated for use in the approximation, the value of θ should be 5.0 or less if the approximation is to be accurate.

Since the binomial distribution can be used to approximate the hypergeometric when the conditions are right and since the Poisson distribution can be used to approximate the binomial, it follows that under certain conditions the Poisson distribution can be used to approximate the hypergeometric. The following example demonstrates this situation.

EXAMPLE 3

Problem: Two hundred people in a certain county were exposed to a dangerous virus. The county health department wants to contact some of these people to run tests on them. Thus, they mailed questionnaires to 600 of the county's 40,000 residents. What is the probability that 5 or fewer of the exposed people will be reached by the questionnaires?

Solution: This is a hypergeometric problem, and the following expression must be evaluated to obtain the exact probability:

$$P(r \leq 5 | n = 600) = \sum_{r=0}^{5} \frac{\binom{200}{r}\binom{39800}{600-r}}{\binom{40000}{600}}$$

This expression looks rather imposing so we may wish to try the binomial approximation to the hypergeometric. This is legitimate since the sample size $n = 600$ is less than 10% of the population size, which is 40,000. Thus p is approximately $(200/40{,}000) = .005$. The binomial approximation would then be:

$$P(r \leq 5 | n = 600, p = .005) = \sum_{r=0}^{5} \frac{600!}{r!(600-r)!}(.005)^r (.995)^{600-r}$$

This expression doesn't look much easier to handle than the one associated with the hypergeometric probability. But it has a large n and small p, so we might try to approximate this expression using the Poisson probability distribution with $\theta = (600)(.005) = 3.0$. The probability is found by adding the first six values ($r = 0,1,2,3,4,5$) in the Poisson table under the $\theta = 3.0$ column.

$$P(r \le 5 | \theta = 3.0) = .9160$$

A computer was used to obtain the exact hypergeometric answer and the binomial approximation. These values were .9151 and .9166, respectively. The closeness of the three answers suggests that either of the two approximation methods can be used when conditions are suitable.

6.3 THE NORMAL APPROXIMATION TO THE BINOMIAL

It can be shown that, when the sample size (or number of trials) n is large, and p is close to .5, the binomial distribution becomes very close to a normal distribution. The normal distribution which can be used to approximate the binomial must have the same mean and standard deviation as the binomial distribution that is to be approximated. In Section 5.9 we found that a binomial distribution has mean np and a standard deviation $\sqrt{np(1-p)}$.

Figure 6.1 shows why the normal curve can be used to give binomial probability approximations. The bars in Figure 6.1 represent the binomial probabilities for the distribution of r successes in $n = 100$ trials where $p = .5$ is the chance of a success on each trial. Earlier we mentioned that it is the *area* in the bars of the histogram that rep-

FIGURE 6.1 Normal approximation to a binomial distribution $P(r | n = 100, p = .5)$

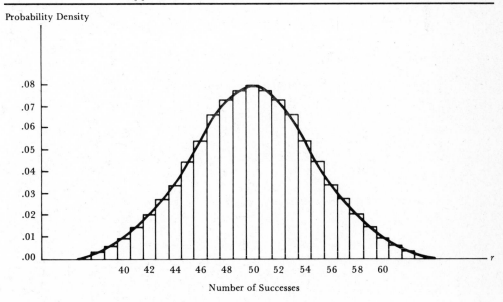

Number of Successes

resents probabilities, and it is the *area* under the normal curve that represents normal distribution probabilities. The normal curve that is superimposed over the binomial histogram has a mean $\mu = np = (100)(.5) = 50$ and a standard deviation $\sigma = \sqrt{np(1-p)} = \sqrt{100(.5)(.5)} = 5$. It is easy to see that the area in any set of bars (the binomial probabilities) is approximately equal to the corresponding area under the curve (the normal probabilities).

Thus, when n is large and p is close to .5, the binomial distribution of r, the number of successes in n trials, is nearly normal around the mean of np with a standard deviation of $\sqrt{np(1-p)}$. In order to use the normal approximation to the binomial, the number of successes, r, must be standardized. The standardized binomial variable

$$Z = \frac{r - np}{\sqrt{np(1-p)}} \qquad (6.1)$$

is approximately a standard normal variable [see Equations (5.7) and (5.8), and also Equation (5.22)]. Therefore, the probability that r will lie between a and b will be approximately the probability that a standard normal variable Z will lie between

$$a' = \frac{a - np}{\sqrt{np(1-p)}} \qquad \text{and} \qquad b' = \frac{b - np}{\sqrt{np(1-p)}}$$

Thus,

$$P(a \le r \le b) \cong P(a' \le Z \le b')$$

where the symbol $\cong$ means *is approximately equal to.*

EXAMPLE 1

Problem: Suppose that the Mountain Manufacturing Company is located in a right-to-work state and that 36% of its 7000 employees belong to unions. If the personnel manager of the company takes a random sample of 100 employees in order to find out union members' attitudes toward management, he would consider the selection of a union member in his sample a "success." We ask for the probability that the number r of successes will be between 24 and 42, inclusive.

Solution: Here the mean, variance, and standard deviation of r are:

$$\mu = np = 100 \times .36 = 36$$

$$\sigma^2 = np(1-p) = 100 \times .36 \times .64 = 23.04$$

$$\sigma = \sqrt{np(1-p)} = \sqrt{23.04} = 4.8$$

The standardization formula is:

$$Z = \frac{r - 36}{4.8}$$

Hence (here Z represents a random variable with exactly a standard normal distribution),

$$P(24 \leq r \leq 42) = P\left(\frac{24 - 36}{4.8} \leq \frac{r - 36}{4.8} \leq \frac{42 - 36}{4.8}\right)$$

$$\cong P\left(\frac{24 - 36}{4.8} \leq Z \leq \frac{42 - 36}{4.8}\right)$$

$$= P\left(-\frac{12}{4.8} \leq Z \leq \frac{6}{4.8}\right) = P(-2.5 \leq Z \leq 1.25)$$

From the table for normal areas we get

$$P(-2.5 \leq Z \leq 1.25) = .4938 + .3944 = .8882$$

The exact probability (taken from binomial tables) is .9074. Our approximation, .8882, is accurate to about 2%. Rounding off .8882 to .89, we see that the normal approximation gives

$$P(24 \leq r \leq 42) \cong .89$$

EXAMPLE 2

Problem: Suppose again that Mountain Manufacturing Company's personnel manager takes a sample of 100 employees. Again ask for the probability that r, the number of union members that appear in the sample, or the number of successes, will be between 24 and 42. This time, however, suppose the proportion of union members working for the firm is .40.

Solution: Here

$$\mu = np = 100 \times .40 = 40$$
$$\sigma^2 = np(1 - p) = 100 \times .40 \times .60 = 24$$
$$\sigma = \sqrt{np(1 - p)} = \sqrt{24} = 4.9$$

The standardization is

$$Z = \frac{r - 40}{4.9}$$

so

$$p(24 \leq r \leq 42) = P\left(\frac{24 - 40}{4.9} \leq \frac{r - 40}{4.9} \leq \frac{42 - 40}{4.9}\right)$$

$$\cong P\left(-\frac{16}{4.9} \leq Z \leq \frac{2}{4.9}\right) = P(-3.26 \leq Z \leq .41)$$

By the normal table,

$$P(-3.26 \leq Z \leq .41) = .4994 + .1591 = .6585$$

Rounding off gives

$$P(24 \leq r \leq 42) \cong .66$$

It is also possible to approximate probabilities defined by a single inequality.

EXAMPLE 3

Suppose we ask for the probability that the number r of successes (union members) in the personnel manager's sample of 100 is 45 or more. If p, the proportion of union members working for the company, is .36, we standardize by the mean and standard deviation of Example 1:

$$P(r \geq 45) = P\left(\frac{r - 36}{4.8} \geq \frac{45 - 36}{4.8}\right)$$

$$\cong P\left(Z \geq \frac{45 - 36}{4.8}\right) = P(Z \geq 1.88)$$

$$= .5000 - .4699 = .0301$$

If p is .40, we use the standardization of Example 2:

$$P(r \geq 45) = P\left(\frac{r - 40}{4.9} \geq \frac{45 - 40}{4.9}\right)$$

$$\cong P\left(Z \geq \frac{45 - 40}{4.9}\right) = P(Z \geq 1.02)$$

$$= .5000 - .3461 = .1539$$

Rounding off to two places in each result, we have

$$P(r \geq 45) \cong .03 \qquad \text{if } p = .36$$

and

$$P(r \geq 45) \cong .15 \qquad \text{if } p = .40$$

Of course, the larger p is, the more likely we are to get 45 or more successes.

It is sometimes convenient to work not with the number r of successes in n trials, but instead with the fraction $f = r/n$ of successes. Dividing both the numerator and denominator in Formula (6.1) by n gives:

$$Z = \frac{f - p}{\sqrt{p(1 - p)/n}} \qquad (6.2)$$

This ratio thus has approximately a standard normal distribution for large n; it is, in fact, f standardized, because f has mean p and variance $p(1 - p)/n$.

EXAMPLE 4

Problem: Suppose n is 100 and p is .36. What is the probability that the fraction f of successes will lie between .24 and .42?

Solution: Here the mean, variance, and standard deviation of f are:

$$E(f) = p = .36$$

$$\text{Variance}(f) = p(1 - p)/n = .36 \times .64/100 = .002304$$

$$\text{Standard deviation}(f) = \sqrt{p(1 - p)/n} = .048$$

Hence, by Formula (5.22), the standardized f is

$$Z = \frac{f - .36}{.048}$$

and

$$P(.24 \le f \le .42) = P\left(\frac{.24 - .36}{.048} \le \frac{f - .36}{.048} \le \frac{.42 - .36}{.048}\right)$$

$$\cong P\left(\frac{.24 - .36}{.048} \le Z \le \frac{.42 - .36}{.048}\right)$$

$$= P(-2.5 \le Z \le 1.25)$$

$$= .8881 \cong .89$$

Note that the answer arrived at in Example 4 is the same as the answer to Example 1, because n and p are the same in the two examples and because $.24 \le f \le .42$ is the same thing as $24 \le r \le 42$. Any given problem can be solved in terms of r or in terms of f; the answer will be the same in either case.

We noted in Example 1 that the normal approximation gave an answer about 2% off from the true answer. This accuracy suffices for many practical purposes. Further accuracy can be achieved by using the *continuity correction*. The exact value of the probability sought in Example 1 is

$$P(24 \le r \le 42) = \sum_{r=24}^{42} \binom{100}{r} (.36)^r (.64)^{100-r}$$

It is represented by the combined areas of the bars in Figure 6.2. In Example 1, we approximated this probability by the area under the normal curve (the one for a μ of 36 and a σ of 4.8) between 24 and 42. Examination of the figure shows that we should obtain a better approximation if we find the area under the normal curve between 23.5 and 42.5, because this will include the areas of the two shaded regions, omitted in the previous approximation. With this procedure we get:

$$P(24 \le r \le 42) = P(23.5 \le r \le 42.5)$$

$$= P\left(\frac{23.5 - 36}{4.8} \le \frac{r - 36}{4.8} \le \frac{42.5 - 36}{4.8} \right)$$

$$\cong P(-2.60 \le Z \le 1.35)$$

$$= .4953 + .4115 = .9068$$

The error in this second approximation is less than 0.1%. All of the problems at the end of this chapter which use the normal approximation to the binomial should be worked using the continuity correction,

FIGURE 6.2   Normal approximation to a binomial distribution

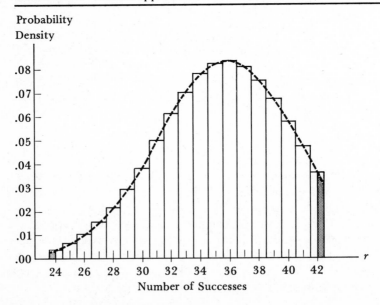

Number of Successes

if the reader desires to obtain the same answers as those listed in the back of the text.

In general, to apply the continuity correction, we subtract ½ from the lower value on the r scale and add ½ to the upper value and then proceed as before. This enables us also to approximate the probability that r will take on a single specified value, as shown in the following example.

EXAMPLE 5

Suppose that a batch of $n = 80$ items is taken from a manufacturing process which produces a fraction $p = .16$ of defectives. We ask for the probability that this batch will contain exactly 20 defectives (call finding a defective a success). In this case

$$\mu = np = 80 \times .16 = 12.8$$
$$\sigma = \sqrt{np(1 - p)} = \sqrt{80 \times .16 \times .84} = 3.279$$

Using the continuity correction gives

$$P(r = 20) = P(19.5 \leq r \leq 20.5)$$
$$= P\left(\frac{19.5 - 12.8}{3.279} \leq \frac{r - 12.8}{3.279} \leq \frac{20.5 - 12.8}{3.279}\right)$$
$$\cong P(2.04 \leq Z \leq 2.35) = .4906 - .4793 = .0113$$

The exact binomial probability is .0122.

Readers sometimes wonder: how large must n be and how far can p depart from .5 before the normal distribution cannot be used to approximate binomial probabilities? A general rule of thumb is that the value of $np(1 - p)$ should equal or exceed 5 if the approximation is to be used. This is the variance of the binomial distribution being approximated. The reader will note that in all the examples of this section $np(1 - p) \geq 5$.

6.4 PROBABILITY SUMMARY

In the last three chapters we have discussed numerous formulas, probability distributions, and relationships. The reader cannot be faulted if he or she is somewhat confused about when to use the probability formulas and relationships that have been presented.

A person must answer a whole series of questions before he or she can determine which formula, table, or relationship applies to a probability problem. Figure 6.3 presents a diagram which is an extension of Figure 5.18. This new figure shows (with dashed lines) the relationships between the various probability distributions. Thus the reader can use the total figure to determine the type of distribution that applies in a *stated* problem and then any convenient approximation that might be used. Example 3 of Section 6.2 is repeated here to demonstrate how the diagram in Figure 6.3 can be used.

EXAMPLE 1

Problem: Two hundred people in a certain county were exposed to a dangerous virus. The county health department wants to contact some of these people to run tests on them. Thus, questionnaires were mailed to 600 of the county's 40,000 residents. What is the probability that five or fewer of the exposed people will be reached by the questionnaires?

Solution: In tracing this problem through the diagram in Figure 6.3 we find that the variable of interest, "number of exposed people contacted," is discrete.

FIGURE 6.3

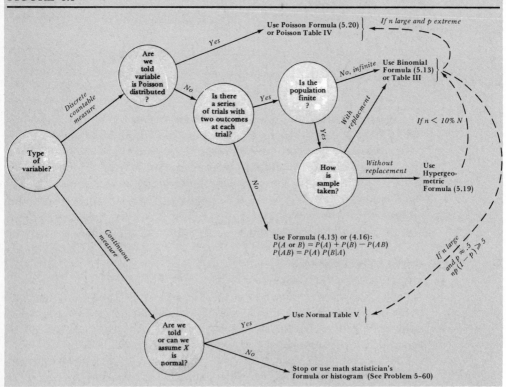

Thus, we proceed from the left side of the diagram along the upper branch to the question about being told the variable is Poisson distributed. Since the problem statement did not mention the Poisson distribution, we proceed to the question about the series of trials. Our response is "yes" since we have a series of $n = 600$ trials where we may or may not reach an exposed person. Next, we respond that the population from which we are sampling is finite (it has $N = 40,000$ members). The sampling is done without replacement since two questionnaires would not purposely be mailed to the same person. Thus, we have determined that the problem is a hypergeometric probability problem. However, our attempts to evaluate the hypergeometric formula in this situation convince us that we might wish to find an approximation to the exact hypergeometric answer. Thus, we conclude that we can use the binomial approximation since $n = 600$ is less than $(.1)N = (.1)(40,000) = 4000$. But even the binomial approximation is difficult to calculate, so we find that we can use the Poisson approximation with $\theta = (600)(200/40,000) = (600)(.005) = 3.0$ due to the fact that $p = .005$ is an extreme value and the calculated θ value is less than 5.0.

PROBLEMS

1. Find the binomial approximation to the following hypergeometric probabilities:

   a. $\dfrac{\dbinom{20}{5}\dbinom{180}{20}}{\dbinom{200}{25}}$

   b. $\dfrac{\dbinom{1500}{12}\dbinom{1500}{8}}{\dbinom{3000}{20}}$

   c. Which binomial approximation will be more accurate and why?

2. Find the Poisson approximation to the following binomial probabilities:
   a. $P(r = 6 | n = 100, p = .04)$
   b. $P(r \le 3 | n = 150, p = .002)$
   c. $P(r = 7 | n = 800, p = .01)$
   d. $P(6 \le r \le 9 | n = 200, p = .012)$
   e. $P(r \ge 18 | n = 450, p = .02)$
   f. $P(r = 6 | n = 100, p = 1.0)$

3. Find the Poisson approximation to the following probabilities:

   a. $\dfrac{\dbinom{40}{2}\dbinom{3960}{98}}{\dbinom{4000}{100}}$

   c. $P(r \le 2 | n = 250, p = .02)$

   b. $\dfrac{\dbinom{100}{5}\dbinom{4900}{20}}{\dbinom{5000}{25}}$

   d. $P(2 \le r \le 8 | n = 300, p = .012)$

4. Use Formula (5.13) to find the binomial approximation to the probability in Problem **3b.**

5. According to Internal Revenue Service records at a particular service center, 40% of the returns examined have mathematical errors on them.

   a. An IRS auditor is about to look over 15 returns for mathematical errors. Find the binomial probability that exactly 5 will contain mathematical errors.

   b. Assume the auditor found 5 returns with mathematical errors but failed to mark them. Upon returning to work the next morning, the auditor randomly selected 4 out of the 15 returns from the previous day. Find the hypergeometric probability that 3 of the returns with mathematical errors were in the sample.

   c. Compute the binomial approximation to your answer in Part **b.**

   d. Explain why the answers in **b** and **c** are not too close to one another.

6. A company has 100 customers, 60 of whom are located out of state. Five orders were randomly selected recently. Find:

   a. The exact hypergeometric probability that 2 of the 5 will be from out-of-state customers.

   b. The binomial approximation to the answer to Part **a.**

   c. Why would the binomial approximation in this problem be better than the binomial approximation in Problem **5c?**

7. In a batch of 200 bottles of their product, the Restin Chemical Company knows that 10 bottles were contaminated. Assume that these 10 bottles were spread randomly through the batch. Restin's best customer was recently sent 6 bottles from this batch and claims that 4 of the bottles were contaminated. Your boss would like to know the probability that this could happen.

   a. *Set up* the hypergeometric solution to this problem. (The numerical answer is .000046.)

   b. *Set up* the binomial solution to this problem. (The numerical answer is .000085.)

   c. Find the Poisson answer to this problem.

   d. Which of the three answers is the exact answer?

8. Assume that your boss sent an inspector to the customer's plant in Problem **7** and that the inspector verified that 4 of the bottles were contaminated. In view of the probabilities computed in Problem **7**, what are some possible explanations as to why the customer received 4 contaminated bottles?

9. There are 5000 customers in the market for your product. Your company has 20% of that market. If you randomly select the names of 25 recent purchasers of this type product:

   a. *Set up* the hypergeometric probability that between 8 and 12 of these purchasers, inclusive, will be your customers.

   b. Use the binomial approximation to the hypergeometric to find the answer to Part **a.**

   c. Use the normal approximation to the binomial to find the approximation to the answer in **b**, which is an approximation to the answer in **a.** (*Hint:* Remember the continuity correction.)

    d. Use the Poisson approximation to the answer in **b** as a possible answer to **a**.

    e. Use Figure 6.3 to determine which approximation to the binomial answer **b** would be closer—the normal approximation or the Poisson approximation to the binomial. Explain which answer is closer and why.

10. A bank receives 42% of its bank card applications from unmarried people. What is the probability that in the next 150 applications 66 or fewer applicants will be unmarried? (*Hint:* remember the continuity correction.)

11. An automotive supply house has found in the past that 4% of the spark plugs it receives from its supplier are misgapped. A shipment of 1000 spark plugs just arrived from this supplier.

    a. What is the expected number of misgapped plugs?

    b. What is the probability that there will be 35 or fewer misgapped plugs in the shipment? Use the normal approximation.

    c. What is the probability of 50 or fewer? Use the normal approximation again. (*Hint:* Remember the continuity correction.)

12. Explain why the Poisson approximation and Table IV could not be used to solve the last two parts of Problem **11** legitimately.

13. An employment agency claims that it finds jobs for 85% of the people who use their services. A newspaper reporter investigating the claims of the agency (reputed to have Mafia connections) did a survey of people who had used the agency. The results showed that, of the 90 people polled, only 70 had found jobs through the agency. Find the probability that 70 or fewer would have found jobs through the agency if the agency's 85% placement success claim is true.

14. The administrators at a state welfare agency have found that 65% of the applicants they process need help in filling out the agency's forms. One hundred twenty people are expected to make applications this week. What is the probability that:

    a. 85 or more will need help with the forms?

    b. Between 70 and 90 inclusive will need help?

    c. Fewer than 60 will need help?

15. Solve Problem **24** in Chapter 5 using the normal approximation to the binomial.

16. Solve Problem **25** in Chapter 5 using the normal approximation to the binomial.

17. A company has 800 female and 1000 male employees. If a group of 50 employees is randomly selected to serve on a grievance review board, what is the probability that 30 or more females will be on the board?

    a. Use Figure 6.3 to trace out the exact probability distribution that applies here. Name that distribution.

    b. Continue tracing through Figure 6.3 to determine the best approximate distribution to apply to this problem. Name that distribution.

    c. Use the distribution found in Part **b** to answer the original question.

18. In a military unit consisting of 1600 people, only 96 have had the special training needed by a secret combat unit that is being formed. If 10 people are randomly selected for this unit, find the probability that 2 or more will have the special training.

    a. Use Figure 6.3 to trace this problem through to the binomial approximation to the answer. Find the binomial approximation.

    b. Use Figure 6.3 to determine if the Poisson distribution would give a reasonable approximation to this probability. If it will not, explain why. If it will, find the Poisson approximation.

SAMPLE DATA SET QUESTIONS:

Refer to the 113 applicants for credit listed in the Sample Data Set Appendix of this book.

    a. A protest committee of five people is to be selected from the people who were denied credit and who are listed in the sample data set. Find the hypergeometric probability that this committee will contain four or more women.

    b. Use the binomial formula to calculate the probability of the event described in Question **a**.

    c. Use the normal approximation to the binomial distribution to find the probability of the event described in Question **a**.

    d. Which of the three probabilities is the exact probability, and which are the approximations?

SAMPLING DISTRIBUTIONS

7.1 SAMPLING AND INFERENCE

It is in order to discover facts about populations that we draw samples from them. Imagine the employees of a large firm split between those who belong to a union and those who do not. The personnel manager of the firm might try to determine what proportion of the employees belong to a union by taking a poll or sample of 100 randomly selected employees rather than by examining the personnel files of all employees. If in the population of employees a proportion of .36 belong to a union (and a proportion .64 do not), what is the chance that the number of employees in the sample who belong to a union will be 45 or more? In Example 3 of Section 6.3 we approximated this probability by using the normal distribution (finding a union member in the sample was called a success). If r is the number of employees in the sample who belong to a union, then

$$P(r \geq 45) = .03 \qquad \text{if } p = .36$$

If a proportion .36 of the employees belong to a union, then the chance that $r \geq 45$ is about .03.

It is natural to challenge the relevance of this sort of calculation to the problem of making inferences about the population of employees. Indeed, if the personnel manager actually knew that the proportion of the population belonging to a union was .36—the assumption underlying the computation—then he would not have gone to the trouble of drawing a sample in the first place. He drew the sample, after all, in order to get some idea what proportion of all employees do in fact belong to a union. This proportion is a number p which is unknown to the personnel manager and the person he might designate to poll the 100 randomly selected employees. Thus why should we assume p is .36?

Now the personnel manager makes guesses about p on the basis of r, the number in the sample of 100 who belong to a union; r for his sample will have some specific value like 72 or 18. Even without benefit of statistical theory, we do know that if r is 72 the personnel manager should guess that p is somewhere around .72; that p is .36 would be a bad guess. If r is 18 the personnel manager should guess that p is something like .18; that p is .95 would be a bad guess. We also know that the personnel manager could make a more accurate guess about p if the sample size were 500 instead of 100. But we want to go beyond these initial ideas to a more detailed understanding of how the personnel manager should guess at p and what kind of precision he can hope for.

In the case where r is 72, we regard .36 as a very bad guess for p because if p really *is* only .36 then it is very strange indeed that r should be as large as 72. And this is the relevance of probability computations based on the assumption that p is .36. Our computation showed that if p were .36 then the chance would be only about .03 that r is even as great as 45. A detailed understanding of how the personnel manager ought to guess requires a detailed knowledge of the distribution of r for the case where p equals .36. But of course there is nothing special about .36, and we must also know the distribution of r in the case where p is .40, say. In Example 3 of Section 6.3 we treated this case as well:

$$P(r \geq 45) \cong .15 \qquad \text{if } p = .40$$

It is a virtue of mathematics that, by letting p be general, we can in principle consider all values of p at the same time; that is, we can give a description of the distribution of r for the general p between 0 and 1; Formula (5.13), Section 5.9, provides the description. This description we require in a general form precisely because the personnel manager does not know the actual value of p in advance.

We draw samples to discover facts about a population, and these facts are usually expressed in terms of numbers called **parameters.** In the preceding discussion, the parameter was p, the proportion of employees who belong to a union. In general, a parameter is a number describing some aspect of a population. In making inferences about a parameter on the basis of a sample, we usually deal with **statistics,** which are numbers that can be computed from the sample. In the preceding discussion, the statistic was r, the number of employees in a sample of 100 who belong to a union. The personnel manager is ignorant of the value of the parameter p; once he has drawn his sample, he knows the value of the statistic r and can use it to make inferences (guesses) about p.

Our example is typical. Generally, we are concerned with one or more parameters that help describe a population. We do not know the values of the parameters (and usually never will). We draw a sample and compute one or more statistics on the basis of the sample. We have actual numerical values for the statistics. And we use the statistics to make inferences about the parameters. In order to know how to make these inferences, we must know about the distribution of the statistics.

For a further example, consider the average weight μ (in pounds) of fertilizer in bags coming off an automatic packaging line. We do not know the value of the parameter μ and never will. To get an idea of what μ is, we take a sample of bags and compute $\overline{X}$, the mean of their weights (as in Chapter 3); the statistic $\overline{X}$ is something we find the actual numerical value for. On the basis of $\overline{X}$ we make inferences about μ. Without statistical theory, we know that $\overline{X}$ somehow estimates μ, but to know just how to make the inferences and how exact they will be, we need to know the distribution of the random variable $\overline{X}$. We need to know, that is, its **sampling distribution.**

7.2   EXPECTED VALUES AND VARIANCES

The mean and variance of a population of numbers are defined as the mean and variance of all possible single random observations from that population, that is, $E(X_1)$ and variance (X_1) for all random samples X_1 of size 1. The population is said to be discrete or continuous according as a single observation X_1 has discrete or continuous distribution. A discrete population may be finite or infinite; a continuous population must be infinite, though it may be approximated by a large but finite population.

Suppose the population has the five elements 2, 4, 6, 8, and 10. In this case the population mean μ is:

$$\mu = \frac{1}{5} (2 + 4 + 6 + 8 + 10) = 6$$

Now suppose we take a random sample of size 2 without replacement. There are $\binom{5}{2}$, or 10, different outcomes—different samples of size 2—and each of them has a sample mean $\bar{X}$:

SAMPLE	2,4	2,6	2,8	2,10	4,6	4,8	4,10	6,8	6,10	8,10
$\bar{X}$	3	4	5	6	5	6	7	7	8	9

The sample being random, each outcome has probability $1/10$. The probability that $\bar{X}$ has the value 5 is the probability of getting the sample 2,8 or the sample 4,6, so $P(\bar{X} = 5) = 2/10$. This and analogous computations give the sampling distribution of $\bar{X}$:

$\bar{X}$	3	4	5	6	7	8	9
$P(\bar{X})$	.1	.1	.2	.2	.2	.1	.1

By the Definition (5.1), Section 5.6, of the expected value for a discrete random variable,

$$E(\bar{X}) = 3 \times .1 + 4 \times .1 + 5 \times .2 + 6 \times .2 + 7 \times .2$$
$$+ 8 \times .1 + 9 \times .1 = 6$$

The point is that $E(\bar{X})$ coincides with the population mean μ.

This is always so. Let $X_1, X_2, \ldots, X_n$ be a random sample of size n from a population, finite or infinite, discrete or continuous. The sample mean is, as in Chapter 3,

$$\bar{X} = \frac{1}{n} \sum_{i=1}^{n} X_i \tag{7.1}$$

and always

$$E(\bar{X}) = \mu \tag{7.2}$$

Equation (7.2) is sometimes expressed as: The mean of the mean is the mean. Each *mean* here has a different meaning, and the statement is short for: The expected value E (first *mean*) of the sample mean $\bar{X}$ (second *mean*) is the population mean μ (third *mean*).

In less elegant terms, the meaning of Equation (7.2) can be demonstrated with an example. Let us suppose that we are interested in the weights of packages coming off a production line. Further, let us assume that the production line fills the packages to a mean weight of 10 pounds. Thus, the population mean is $\mu = 10$. Each day at noon a quality control engineer takes a random sample of nine packages that have come off the line during the previous 24 hours and computes the sample mean weight, $\overline{X}$, of the nine packages. If this procedure were followed every day for 400 working days, the engineer would have collected 400 sample means. These 400 sample means themselves could be added up and divided by 400 to obtain their mean. Equation (7.2) indicates that the 400 sample means would have a mean close to 10 pounds. In the long run, after infinitely many working days, the mean of the infinitely many sample means would be *exactly* 10 pounds.

If the population is infinite, so that the elements $X_1, X_2, \ldots,$ X_n in the random sample are *independent,* then we have for the variance of $\overline{X}$ the formula*

$$\text{Variance } (\overline{X}) = \frac{\sigma^2}{n} \tag{7.3}$$

where σ^2 is the population variance. The **standard deviation of the sample means** is:†

---

*This and the other numbered formulas in this section will be used constantly in the rest of the book. Although we give derivations of them in footnotes, the derivations are not needed in the sequel; an understanding of the *meaning* of the formulas will suffice. The derivation of Formula (7.2) can be shown by using the rules for manipulating expected values. Applying to Formula (7.1) Expression (5.4) and then (5.11) of the fifth chapter, we have:

$$E(\overline{X}) = \frac{1}{n} E\left(\sum_{i=1}^{n} X_i\right) = \frac{1}{n} \sum_{i=1}^{n} E(X_i)$$

Since each individual X_i is a single observation from the population, $E(X_i)$ has the same value as μ. Equation (7.2) follows by Rule 3 for summations in Section 3.1.

†To see why Formula (7.3) holds for an infinite population, use Formula (5.5) and then 5.12; the latter formula is applicable because the X_i are independent:

$$\text{Variance}(\overline{X}) = \text{var}\left(\frac{1}{n} \sum_{i=1}^{n} X_i\right) = \frac{1}{n^2} \text{var}\left(\sum_{i=1}^{n} X_i\right) = \frac{1}{n^2} \sum_{i=1}^{n} \text{var}(X_i)$$

But each variance (X_i) is σ^2, since X_i is a single observation from the population, and so Rule 3 for summations in Section 3.1 yields Formula (7.3).

Formulas (7.3) and (7.4) apply only when *the population is infinite*. If the population is finite, Formula (7.3) must be replaced by the formula

$$\text{Standard deviation} (\bar{X}) = \frac{\sigma}{\sqrt{n}} \qquad (7.4)$$

Equations (7.3) and (7.4) give us a method of computing the variance and standard deviation of the 400 mean values collected by the quality control engineer. If in addition to having a mean package weight of 10 pounds the package filling process fills individual packages with a standard deviation of $\sigma = .6$ pounds, then the standard deviation of the 400 mean values around 10 will be, according to Equation (7.4), $\sigma/\sqrt{n} = .6/\sqrt{9} = .2$. This says that the *means* of many samples of size $n = 9$ will be three times more tightly clustered around the value of 10 than are the *individual package weights,* where we measure the degree of clustering by the standard deviation. Another example should help to clarify these points.

EXAMPLE 1

Suppose we took 50 samples of size $n = 36$ from a population consisting of savings accounts and examined the account balances. These samples might be made up of 36 randomly selected account balances taken at the close of business each working day for ten weeks. Suppose further that the mean balance in the population of accounts (which will be considered to be very large) is $3900 and that the standard deviation of the population is $1200. That is,

$$\mu = \$3900 \qquad \text{and} \qquad \sigma = \$1200$$

After the 50 samples have been taken, there will be 50 · 36, or 1800, observations.

$$\text{Variance} (\bar{X}) = \frac{N-n}{N-1} \cdot \frac{\sigma^2}{n}$$

where N is the size of the population. The factor $(N - n)/(N - 1)$ is the *finite population correction factor.* In the preceding example, the population variance is:

$$\sigma^2 = E(X_i - 6)^2 = \frac{1}{5}(16 + 4 + 0 + 4 + 16) = 8$$

And the variance of $\bar{X}$ is:

$$E(\bar{X} - 6)^2 = 9 \times .1 + 4 \times .1 + 1 \times .2 + 1 \times .2 + 4 \times .1 + 9 \times .1 = 3$$

which agrees with the preceding formula for $N = 5$ and $n = 2$:

$$3 = \frac{5-2}{5-1} \cdot \frac{8}{2}$$

If the population size N is large in comparison with the sample size n, then the correction factor is nearly 1, so that the formula for Variance $(\bar{X})$ for a finite population is practically the same thing as Formula (7.3) in this case.

But there will be *only 50 means*, one for each sample. According to Formula (7.2) the mean of these 50 means will be approximately $3900, the mean of the population. (We say *approximately* $3900 since Formula (7.2) assumes that infinitely many samples, not just 50, have been taken.) Also, according to Formula (7.4) the standard deviation of the 50 mean values will be approximately:

$$\text{Standard deviation } (\overline{X}) = \frac{\$1200}{\sqrt{36}} = \frac{\$1200}{6} = \$200$$

That is, the distribution of the means of samples of size $n = 36$ will have a mean value of $3900 and a standard deviation of $200.

EXAMPLE 2

Consider the distribution shown in Figure 7.1. The variable whose distribution is presented there is the amount of time required for people to pass through a border inspection station. The distribution shows that many people pass with a very short delay or none at all. Also, many are detained a rather long period for inspections of baggage and personal belongings which might contain drugs or other contraband. Very few people are delayed a moderate amount of time. The mean and standard deviation of this distribution of times are $\mu = 8$ minutes and $\sigma = 6$ minutes. If 150 random samples of size $n = 64$ delay times at this inspection station were taken, then a total of $(150)(64) = 9600$ individual times would be available, but only 150 sample means would be calculated. According to Formula (7.2), these 150 means would have a mean value very close to the population mean $\mu = 8$ minutes. (The many sample means would have a mean of exactly 8 minutes in the long run, i.e., after infinitely many samples of $n = 64$ were taken.) According to Formula (7.4), these 150 sample means will have a standard deviation of approximately $\sigma/\sqrt{n} = (6 \text{ minutes})/\sqrt{64} = .75$ minutes. If we use the discussion points about the standard deviation mentioned in Section 3.12, we can say that virtually all of the 150 sample means will fall within ± 3 standard deviations of their central value, 8 minutes. That is, it would be most unlikely for us to observe the mean for any sample of $n = 64$ times outside the interval defined by $8 \pm 3(.75) = 8 \pm 2.25$, or 5.75 to 10.25 minutes. If someone were to tell us that they observed the amount of time that it took a randomly chosen sample of $n = 64$ people to pass through the inspection station and found they averaged $\overline{X} = 11$ minutes per person, then since 11 exceeds the highest mean we would expect to see, 10.25, we could reach one of three conclusions:

1. The person has observed a *very* unusual sample,
2. The person is not telling us the truth or,
3. For some reason the population mean time has shifted upward from $\mu = 8$ minutes.

If we trust the person who took the sample and if we believe that rare events don't happen very often, we would draw the third conclusion.

FIGURE 7.1   Distribution of time to pass a border inspection station

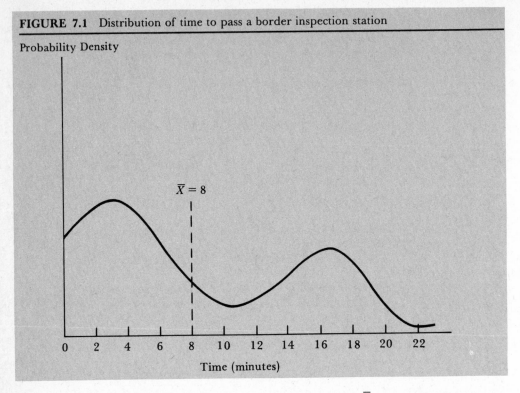

Time (minutes)

In addition to the mean and variance of $\bar{X}$, we can find the mean of the sample variance s^2. Recall from Chapter 3 that

$$s^2 = \frac{1}{n-1} \sum_{i=1}^{n} (X_i - \bar{X})^2 \tag{7.5}$$

For a sample from an *infinite* population, we have the formula*

$$E(s^2) = \sigma^2 \tag{7.6}$$

Formula (7.6) concerns the mean of the variance, whereas 7.3

---

*For the sake of simplicity, we derive this only for the case where μ is 0. By Formula (3.12),

$$\Sigma(X_i - \bar{X})^2 = \Sigma X_i^2 - \frac{1}{n}(\Sigma X_i)^2 = \Sigma X_i^2 - n\bar{X}^2$$

If $\mu = 0$, so that $E(X_i) = 0$ and $E(\bar{X}) = 0$, then $E(X_i^2) = \sigma^2$ and $E(\bar{X}^2) = \text{Variance}(\bar{X}) = \sigma^2/n$; therefore

$$E\left(\sum_{i=1}^{n} (X_i - \bar{X})^2 \right) = \sum_{i=1}^{n} \sigma^2 - n\frac{\sigma^2}{n} = (n-1)\sigma^2$$

which gives Formula (7.6).

concerns the variance of the mean. To keep these straight, the terms must be expanded: Formula (7.6) concerns the expected value of the sample variance, and 7.3 concerns the variance of the sample mean.

Formula (7.6) can be explained in terms of the 400 samples taken by the quality control engineer mentioned earlier. Each of the 400 samples would have a variance, which would be computed by dividing the sum of squared deviations from the $\bar{X}$ value by *eight*. These 400 sample variances would average out to be the population variance. That is, if all the 400 sample variances were added up and divided by 400, their value would be close to $\sigma^2 = (.6)^2 = .36$.

7.3   SAMPLING FROM NORMAL POPULATIONS

Let $\bar{X}$ be the sample mean for an independent sample of size n. As we saw in the last section, if the population mean and variance are μ and σ^2, then $\bar{X}$ has mean μ and variance σ^2/n, and this is true whatever the form of the parent population may be. Suppose now that the population is *normal*. In this case $\bar{X}$ is normally distributed with mean μ and variance σ^2/n. This is a fundamental fact about normal populations.

This characteristic is plainly seen in the results of a sampling exercise. Each of 158 statistics students took two random samples of five values each from a population approximately normal with mean 40 and variance 100. Each computed the two means and also the mean of the ten values combined into one sample. Table 7.1 shows the resulting distribu-

Table 7.1   Distributions of the Sample Means in Samples of 5 and 10

MEAN	FREQUENCY ($n = 5$)	FREQUENCY ($n = 10$)
25–27	1	1
27–29	1	1
29–31	4	2
31–33	18	0
33–35	18	5
35–37	37	14
37–39	47	36
39–41	60	43
41–43	41	26
43–45	45	19
45–47	27	9
47–49	12	1
49–51	1	1
51–53	4	0
Totals	316	158

FIGURE 7.2   Histogram for the means of samples of size 5

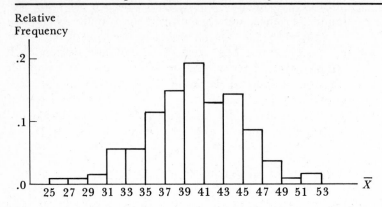

tions of sample means, and Figures 7.2 and 7.3 show the histograms.

The means for these empirical distributions are 40.06 for the 316 samples of size 5 and 39.98 for the 158 samples of size 10. The variances are 20.95 and 12.64. Even with relatively few means, the histograms show a tendency toward the symmetrical, bell-shaped normal curve; the means and variances agree well with the theoretical values $\mu = 40$, $\sigma^2/5 = 20$, and $\sigma^2/10 = 10$.

FIGURE 7.3   Histogram for the means of samples of size 10

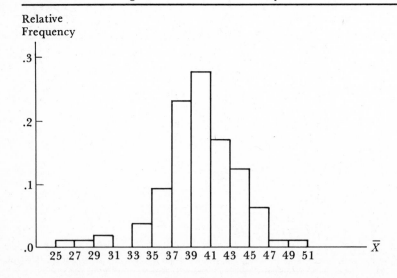

7.4   THE STANDARDIZED SAMPLE MEAN

Since $\bar{X}$ has mean μ and standard deviation $\sigma/\sqrt{n}$, the standardized variable

$$Z = \frac{\bar{X} - \mu}{\sigma/\sqrt{n}} \tag{7.7}$$

has mean 0 and variance 1; see Formulas (5.7) and (5.8) in Section 5.6.

If the population is normal, then Z, defined by Equation (7.7), is a standard normal variable and we may use the table of normal areas to calculate probabilities for the sample mean.

EXAMPLE 1

Problem: Given a normal distribution with mean 50 and variance 100, find the probability that the mean of a sample of 25 observations will differ from the population mean by less than four units.

Solution: We want

$$P(-4 < \bar{X} - \mu < 4)$$

If we divide each member of the inequality by $\sigma/\sqrt{n}$ we have*

$$P(-4 < \bar{X} - \mu < 4) = P\left(\frac{-4}{\sigma/\sqrt{n}} < \frac{\bar{X} - \mu}{\sigma/\sqrt{n}} < \frac{4}{\sigma/\sqrt{n}}\right)$$

$$= P\left(\frac{-4}{10/\sqrt{25}} < Z < \frac{4}{10/\sqrt{25}}\right)$$

$$= P(-2 < Z < 2)$$

$$= .9544$$

EXAMPLE 2

Problem: For the normal distribution of the preceding example find two values equidistant from the mean such that 90% of the means of all samples of size n equals 64 will be contained between them.

---

*This step can be taken since the same positive amount has been divided into every term in the inequality string. In general, if $a \leq b \leq c$, then $a/d \leq b/d \leq c/d$ as long as $d > 0$. Also, the same amount can be added to or subtracted from each term in an inequality string without changing the relationship. Thus, if $a \leq b \leq c$, then $a + d \leq b + d \leq c + d$. This fact is used in the solution to Example 2. Finally, the same positive amount can be multiplied by each term in an inequality string without changing the relationship. If $a \leq b \leq c$, then $ad \leq bd \leq cd$ as long as $d \geq 0$.

Solution: From the table of normal areas we have

$$.90 = P(-1.64 \leq Z \leq 1.64)$$

From Equation (7.7),

$$.90 = P\left(-1.64 \leq \frac{\bar{X} - \mu}{\sigma/\sqrt{n}} \leq 1.64\right)$$

Therefore, since $\sigma/\sqrt{n} = 10/8 = 1.25$

$$.90 = P\left(-1.64 \leq \frac{\bar{X} - 50}{1.25} \leq 1.64\right)$$

$$= P(-2.05 \leq \bar{X} - 50 \leq 2.05)$$

$$= P(-2.05 + 50 \leq \bar{X} \leq 50 + 2.05)$$

$$= P(47.95 \leq \bar{X} \leq 52.05)$$

The desired values are 47.95 and 52.05.

EXAMPLE 3

Problem: The lengths of individual machined parts coming off a production line are normally distributed around their mean of $\mu = 30$ centimeters. Their standard deviation around the mean is $\sigma = .1$ centimeter. An inspector just took a sample of $n = 4$ of these parts and found that $\bar{X}$ for this sample is 29.875 centimeters. What is the probability of getting a sample mean this low or lower if the process is still producing parts at a mean of $\mu = 30$?

Solution: We know that sample means from normal populations are normally distributed around μ with a standard deviation of $\sigma/\sqrt{n}$. Knowing this, we can standardize our sample mean of $\bar{X} = 29.875$ and use the normal table to find the answer we seek as follows:

$$P(\bar{X} \leq 29.875) =$$

$$P\left(Z = \frac{\bar{X} - \mu}{\sigma/\sqrt{n}} \leq \frac{29.875 - 30.0}{.1/\sqrt{4}}\right) =$$

$$P\left(Z \leq \frac{-.125}{.05}\right) =$$

$$P(Z \leq -2.50)$$

In the normal Table V in the Appendix, we find this value is:

$$.5000$$

$$\frac{-.4938}{.0062}$$

What could the inspector conclude from this result? Either a sample has been taken which occurs with very small probability or the assumption that $\mu = 30$ is not valid.

7.5 THE CENTRAL LIMIT THEOREM

The **central limit theorem** concerns the approximate normality of means of random samples or of sums of random variables; we accordingly state it in two forms.

Suppose $X_1, X_2, \ldots, X_n$ is a sample from an infinite population with mean μ and variance σ^2; the X_i are independent random variables.

First Form: **If n is large, then**

$$Z = \frac{\bar{X} - \mu}{\sigma/\sqrt{n}} \tag{7.8}$$

has approximately a standard normal distribution, or (what is the same) $\bar{X}$ has approximately a normal distribution with mean μ and standard deviation $\sigma/\sqrt{n}$.

We know that whatever the parent population may be, the Standardized Variable (7.8) has mean 0 and standard deviation 1; and we know that, if the parent population is normal, then Variable (7.8) has exactly a standard normal distribution. The remarkable fact is that, even if the parent population is not normal, the standardized mean is approximately normal if n is large. The importance of the theorem lies in the fact that it permits us to use normal theory for inferences about the population mean regardless of the form of the population, provided only that the sample size is large enough.

The importance of the central limit theorem lies in the fact that it opens up an entirely new class of problems that can be solved using the normal probability distribution and Table V from the Appendix. In Chapter 5, we found that the normal distribution could be used to find the probability that an *individual* measure would lie in a particular interval by using Table V, assuming that the measure was known to be normally distributed. However, the central limit theorem tells us we can use the normal distribution to find the probability that a *sample mean* will lie in a particular interval—regardless of the distribution followed by the measure in its population, as long as the sample size we use is large enough (usually in excess of $n = 30$). This second use of the normal distribution is demonstrated in Example 1.

EXAMPLE 1

Problem: The probability density function shown in Figure 7.4 is a triangular distribution showing the probability that a delicate new medical device will fail between 0 and 20 years after it is implanted in the human body. The mean time to failure is $\mu = 6.7$ years, and the standard deviation is $\sigma = 3$ years. Do the following:

a)   Verify that this density function has a total area under it of 1.0.
b)   Find the probability that an *individual* device will fail 15 years or more after implantation.
c)   Find the probability that in a sample of $n = 36$ of these devices the *sample mean* time to failure, $\bar{X}$, will be 8 years or less.

Solution:

a)   To verify that the total area under the density function is 1.0, we have to use the relationship for the area of a triangle: Area = (1/2)(base)(height). In this case, Area $(1/2)(20)(.1) = (10)(.1) = 1.0$.

FIGURE 7.4   Distribution of times to failure for a medical device

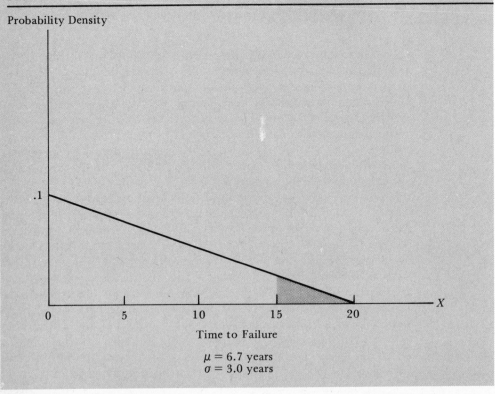

Probability Density

.1

0   5   10   15   20   X

Time to Failure

$\mu = 6.7$ years
$\sigma = 3.0$ years

b) Since the second question concerned the probability that an *individual* device will fail in 15 years or more, we must use the probability distribution for individual times to failure given in Figure 7.4. The probability we seek is given by the shaded area under the density function from 15 to 20 years. This is the area in a triangle of base $(20 - 15) = 5$ and unknown height. To find the height of the function at $X = 15$, we must find the equation of the function. A little examination of the problem convinces us that the intercept of this line is at .1, and the slope is $-.1/20 = -.005$. Thus, the equation of the line is $f(X) = .1 - .005X$. This means that the height of the line at $X = 15$ is

$$f(15) = .1 - .005(15) = .025,$$

and the area of the triangle from $X = 15$ out to $X = 20$ is

$$\text{Area} = \frac{1}{2}(5)(.025) = .0625.$$

Thus, the chance that one of these devices will last 15 years or longer is .0625, or one out of sixteen.

c) The third question asks about the probability that a sample mean will be 8 years or less. To answer this question, we must use the probability distribution shown in Figure 7.5. From Formula (7.2), we know that the mean of the $\bar{X}$'s for samples of size $n = 36$ is the population mean, $\mu = 6.7$ years. From Formula (7.4), we know that the distribution of the sample means has a standard deviation of

$$\sigma_{\bar{X}} = \sigma/\sqrt{n} = 3/\sqrt{36} = .5$$

The central limit theorem tells us that even though the population of times from which the sample was drawn was triangular in shape (Figure 7.4), the shape of the distribution of sample means is almost normal since the sample size, $n = 36$, is large. Thus, we seek the shaded area under the normal curve in Figure 7.5 to the left of $\bar{X} = 8$.

FIGURE 7.5 Distribution of means of samples of $n = 36$ devices

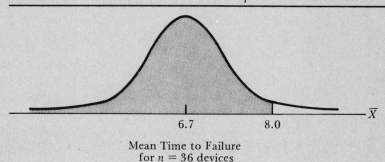

6.7   8.0 $\bar{X}$

Mean Time to Failure
for $n = 36$ devices

$$P(\overline{X} \le 8) = P\left(Z \le \frac{8 - 6.7}{3/\sqrt{36}}\right) = P\left(Z \le \frac{1.3}{.5}\right) = P(Z \le 2.6)$$

$$= .5000 + .4953$$

$$= .9953$$

Thus, the chances are nearly certain that the mean of a sample of size $n = 36$ will be 8 years or less.

To *prove* the central limit theorem would require a full use of mathematical probability. But we can illustrate it further by starting with a specific nonnormal distribution, a J-shaped exponential distribution with mean μ of 2 and variance σ^2 of 4. Figure 7.6(a) shows this distribution (solid line) and the normal distribution (dashed line) with the same mean and variance. The two distributions are very dissimilar.

For samples with size n equal to 4 from the exponential population of Figure 7.6(a), the sample mean $\overline{X}$ has mean 2 and variance 1. Figure 7.6(b) shows the exact distribution (solid line) of $\overline{X}$ together with the normal distribution (dashed line) with mean 2 and variance 1. The two curves are rather similar.

Figure 7.6(c) shows the same pair of curves for a sample size n of 12 (the mean and variance are now 2 and 1/3), and here the agreement is really quite close. These graphs show insufficient detail for the tails of the distribution; the normal approximation is usually better for values near the mean than for values far removed. As appears from this example, however, the sample need not be excessively large before we can feel reasonably safe in using the central limit theorem.

The material in this section deals with the probability distribution of sample means. The reader might legitimately ask, "Who cares about the distribution of sample means?" The answer is that many people do. And the reason that they do is that, in many probability applications, we want to determine the probability of getting a sample result like the one we have just observed, and the sample result we have observed is very often a sample mean, an $\overline{X}$. For instance, in Example 2 of Section 7.2, we considered a situation where the delay time for inspection of baggage at a border station had a mean of $\mu = 8$ minutes and $\sigma = 6$ minutes. The distribution of those times was very nonnormal and was shown in Figure 7.1. Assume that a representative of a particular minority group took a sample of $n = 64$ people from her minority and found that their mean time to get through the inspection station was $\overline{X} = 10$ minutes. Would the representative be justified in suggesting that the minority group people were being delayed longer than one would expect if the minority group people are being processed the

FIGURE 7.6   (a) Exponential and normal distributions with $\mu = 2$ and $\sigma^2 = 4$. (b) Distribution of $\overline{X}$ for samples of 4 and the normal distribution with the same mean and variance. (c) Distribution of $\overline{X}$ for samples of 12 and the normal distribution with the same mean and variance

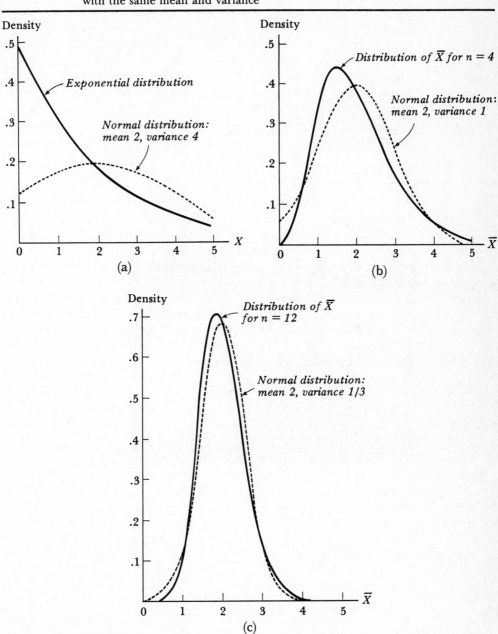

same as people in general? That is, we desire to know the probability that the sample mean, $\bar{X} \geq 10$, could occur by random chance alone. In Example 2 of Section 7.2, we suggested that a sample mean of 10 minutes or more would be a rather unusual result. But now we have the tools to specify exactly how unusual this event would be. Figure 7.7 shows the distribution of sample means for samples of size $n = 64$ from a population with $\mu = 8$ and $\sigma = 6$. Our knowledge of the central limit theorem tells us that, even though the population (shown in Figure 7.1) is not normal in shape, the distribution of sample means in Figure 7.7 is very close to normal in shape since the sample size of $n = 64$ is large. Thus

$$P(\bar{X} \geq 10) = P\left(Z \geq \frac{10 - 8}{6/\sqrt{64}}\right) = P\left(Z \geq \frac{2}{.75}\right) = P(Z \geq 2.67)$$

$$= .5000 - .4962 = .0038$$

Thus, our knowledge of the central limit theorem has allowed us to calculate exactly how unusual a sample mean the minority group representative has observed, *if the mean processing time for members of her group is the same as for people in general.* This value of .0037 is so unusual that we would likely conclude that people from the minority group are being processed in significantly longer time than the population mean of 8 minutes, and they might have some justification to complain about discrimination against them.

A second version of the central limit theorem concerns the sum ΣX_i of a set $X_1, X_2, \ldots, X_n$ of random variables all having the same distribution with mean μ and variance σ^2.

Second Form: If n is large, then

FIGURE 7.7   Distribution of sample means for samples of $n = 64$

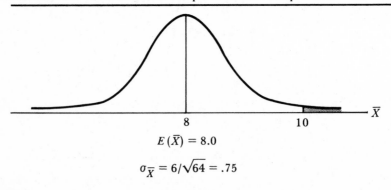

$$E(\bar{X}) = 8.0$$

$$\sigma_{\bar{X}} = 6/\sqrt{64} = .75$$

$$Z = \frac{\left(\sum_{i=1}^{n} X_i\right) - n\mu}{\sigma\sqrt{n}} \tag{7.9}$$

has approximately a standard normal distribution, or (what is the same) ΣX_i has approximately a normal distribution with mean $n\mu$ and standard deviation $\sigma\sqrt{n}$.*

This theorem (together with more general versions of it) is one of the reasons why the normal distribution often arises in nature. If one performs a complicated physical measurement, the measurement error is the sum ΣX_i of many small independent random errors X_i. The height of a plant is the sum ΣX_i of many small independent increments X_i. In such cases, the normal distribution at least roughly approximates the distribution of the sum.

EXAMPLE 2

Problem: The daily catch of a small tuna fishing fleet averages 130 tons. The fleet's log book shows that the weight of the catch varies from day to day, and this variation is measured by the standard deviation of the daily catch, $\sigma = 42$ tons. What is the probability that, during a sample of $n = 36$ fishing days, the total weight of the catch will be 4320 tons or more?

Solution: We don't know the shape of the distribution of weight for individual daily catches, but we don't need to know. The question concerns the probability that the *total* weight in 36 days is 4320 or more. Thus, we can use Expression (7.9) and the second form of the central limit theorem which says the sum of a large number of measurements is approximately normal. Figure 7.8 shows the distribution of the sum of $n = 36$ measurements from a population that has mean $\mu = 130$ and $\sigma = 42$. The probability we seek is in the shaded area.

$$P(\Sigma X \geq 4320) = P\left(Z \geq \frac{4320 - (36)(130)}{42\sqrt{36}}\right) = P\left(Z \geq \frac{4320 - 4680}{252}\right)$$

$$= P(Z \geq -1.43) = .5000 + .4236$$

$$= .9236$$

Thus, the chances are quite good that the total catch will be 4320 tons or more in a sample of $n = 36$ days. The owner of this fleet might thus be

---

*This theorem is really the same as the first one, because ΣX_i is $n\overline{X}$, so that the Variable (7.9) is just $(n\overline{X} - n\mu)/\sigma\sqrt{n}$, and algebra reduces this to the Variable (7.8).

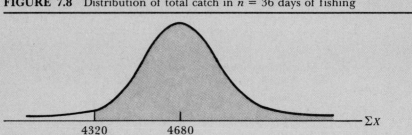

FIGURE 7.8 Distribution of total catch in $n = 36$ days of fishing

willing to sign a contract to deliver that much tuna to his buyers over the next 36 days of fishing if he is willing to assume that the next 36 days will be a random sample and will have no unusual deviations from the past fishing results.

As a final note in this section, we should point out that the normal approximation to the binomial, as treated in Section 6.3, really comes under the central limit theorem. In an infinite (or finite but large) population split into two categories, success and failure, label each element in the success category with a 1 and each element in the failure category with a 0. Now a random sample of size n from the population becomes converted into a set $X_1, X_2, \ldots, X_n$, of independent random variables, each having value either 0 or 1. The number of successes is the number of 1's, and this is ΣX_i, so the central limit theorem applies. In this case [see Formulas (5.16) and (5.17)], $n\mu = np$ and $\sigma \sqrt{n} = \sqrt{np(1-p)}$. Variable (7.9) is the same thing as the standardized number of successes; it has approximately a standard normal distribution.

SUMMARY. In Chapters 5 and 6, we reviewed a chart which showed the logic of how to solve a probability problem. That chart was presented as Figures 5.18 and 6.3. Now we can use the knowledge we have gained concerning the Central Limit Theorem to expand the chart. The revised Figure 6.3 is presented on page 215. The bottom leg of the chart summarizes the material covered by the Central Limit Theorem.

FIGURE 6.3 (Revised)   Probability Summary Chart

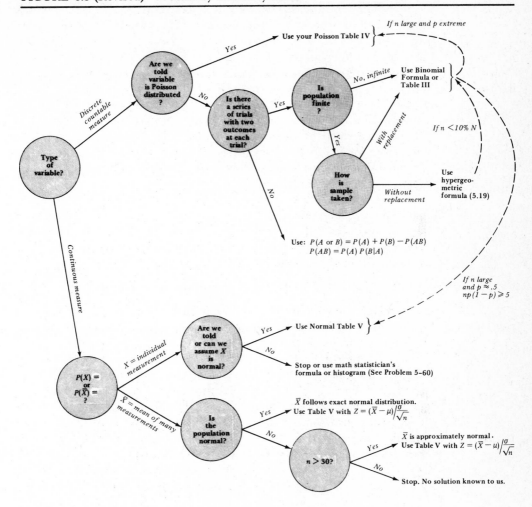

■ 7.6   SAMPLE RESULTS FROM TWO POPULATIONS

There are many times when we are interested in comparing the results of samples that come from two populations. For instance, if a company has two different sources which supply it with raw materials, it might be interested in which source, on the average, provides higher quality

■The material in this section is required only in other optional sections.

material. If a personnel manager has two different training programs, he or she might be interested in which program, on the average, produces more efficient workers. The manager of a factory may have two production lines that produce the same goods. The manager may be interested in knowing which line, on the average, gives higher output.

In order to make comparisons like those just suggested, we often take two samples—one from each of the two populations we are interested in comparing. It is often useful to compare the differences between the two samples' mean values. That is, we may have two populations which have means μ_1 and μ_2. If we take samples of n_1 items from the first population and n_2 items from the second population, we can calculate two sample means, $\bar{X}_1$ and $\bar{X}_2$. These two sample means can then be compared.

If we wish to make probability statements about the differences that can arise between two sample means, then we must study the distribution of the difference $\bar{X}_1 - \bar{X}_2$. The distribution of the difference between sample means has many of the same characteristic features as the distribution of single sample means. First,

$$E(\bar{X}_1 - \bar{X}_2) = \mu_1 - \mu_2 \tag{7.10}$$

For example, if we took a set of many, many pairs of samples from two populations that have means μ_1 and μ_2, then the difference between the sample means of the paired samples would average out to $\mu_1 - \mu_2$, in the long run. Table 7.2 gives an indication of what this statement means. This table shows ten pairs of samples where the sample sizes are $n_1 = 5$ and $n_2 = 5$. The third and fourth columns of the table give the ten pairs of sample means. The fifth column shows the difference between the two sample means. The populations from which these samples were drawn is the set of integers from 0 through 9. Thus, both populations have the same mean. That is, $\mu_1 = \mu_2 = (0 + 1 + 2 + \ldots + 8 + 9)/10 = 4.5$. To obtain the pairs of samples we merely took the first twenty sets of five random digits from Table II in the Appendix.

Since $\mu_1 = \mu_2$ for this example, Formula (7.10) tells us that in the long run (after *infinitely many pairs* of $\bar{X}_1$ and $\bar{X}_2$ have been calculated) the mean value of the fifth column in Table 7.2 should be $\mu_1 - \mu_2 = 4.5 - 4.5 = 0$. After taking only *ten pairs* of samples, this difference averages out to .76 as is shown.

Consider another example. Suppose that two chemical processes are being run at the same plant, side by side. The first process produces $\mu_1 = 2.00$ ounces of impurities per gallon of product. The second process produces $\mu_2 = 1.25$ ounces of impurities per gallon of output. Further, assume that on each shift there is one quality control engineer

TABLE 7.2   Distribution of differences in sample means for pairs of samples from uniform populations

FIRST SAMPLE	SECOND SAMPLE	$\bar{X}_1$	$\bar{X}_2$	$d = \bar{X}_1 - \bar{X}_2$
3,9,5,9,1	4,6,3,0,4	5.4	3.4	2.0
9,9,5,4,7	0,6,7,4,3	6.8	4.0	2.8
6,9,5,6,8	9,1,2,4,0	6.8	3.2	3.6
3,5,2,4,9	9,7,4,5,8	4.6	6.6	-2.0
3,8,9,8,0	1,0,7,5,0	5.6	2.6	3.0
3,6,2,4,7	7,0,9,9,4	4.4	5.8	-1.4
9,9,6,3,8	7,2,0,5,5	7.0	3.8	3.2
2,4,0,3,8	7,4,9,7,6	3.4	6.6	-3.2
3,5,5,5,3	3,5,6,7,6	4.2	5.4	-1.2
7,4,8,1,5	4,5,2,4,6	5.0	4.2	0.8

Mean $(\bar{X}_1 - \bar{X}_2) = .76$
Variance $(\bar{X}_1 - \bar{X}_2) = 6.28$
Standard Deviation $(\bar{X}_1 - \bar{X}_2) = 2.51$

assigned to each process. Each day the engineers take random samples of twenty-four gallons (three each hour of their eight-hour day) from the line to which they are assigned, and they measure the weight of impurities in their sample gallons. At the end of a day, each engineer will have 24 measurements, which are then averaged to get $\bar{X}_1$ for the mean impurities per gallon coming off the first line that day, and $\bar{X}_2$ for the mean impurities per gallon coming off the second line. After many, many working days, Equation (7.10) tells us that the difference between the mean daily impurities measures will average out to $\mu_1 - \mu_2 = 2.00 - 1.25 = .75$ ounces.

Theoretically, we could take infinitely many pairs of independent random samples from two populations like those shown in Figure 7.9(a). Again theoretically, we could plot the difference between the infinitely many pairs of sample means. (Imagine a set of values like Column 5 of Table 7.2, except that the column of numbers would be infinitely long.) An example of what the distribution of these values might look like is given in Figure 7.9(b). Formula (7.10) tells us that the mean

FIGURE 7.9(a)

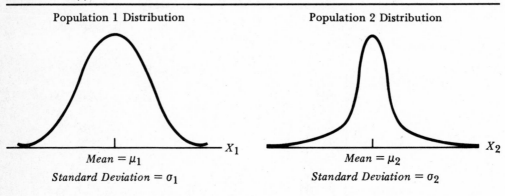

Population 1 Distribution

X_1

Mean $= \mu_1$

Standard Deviation $= \sigma_1$

Population 2 Distribution

X_2

Mean $= \mu_2$

Standard Deviation $= \sigma_2$

FIGURE 7.9(b)

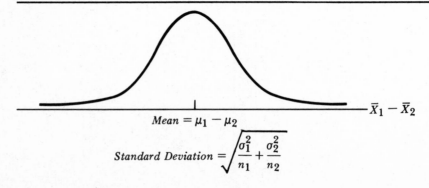

$\bar{X}_1 - \bar{X}_2$

Mean $= \mu_1 - \mu_2$

Standard Deviation $= \sqrt{\dfrac{\sigma_1^2}{n_1} + \dfrac{\sigma_2^2}{n_2}}$

of the distribution in Figure 7.9(b) is $\mu_1 - \mu_2$. But if we want to make probability statements when we compare two sample means, we need to know the standard deviation and shape of the distribution in Figure 7.9(b).

According to Equation (5.10), Section 5.7, if X and Y are independent random variables, then their variances add: variance $(X + Y) =$ variance $(X) +$ variance (Y). Now the distribution curve for $-Y$ is the mirror image of that for Y, so their measures of spread are the same: variance $(-Y) =$ variance (Y). Hence $X - Y$, or $X + (-Y)$, has for its variance the sum variance $(X) +$ variance $(-Y) =$ variance $(X) +$ variance (Y). Thus we have the formula:

$$\text{Variance}(X - Y) = \text{Variance}(X) + \text{Variance}(Y)$$

(One might at first expect a minus on the right, but that could even give a negative variance.)

Since $\bar{X}_1$ and $\bar{X}_2$ are the means of independent random samples, they are independent of one another. They have variances σ_1^2/n_1 and σ_2^2/n_2, respectively; therefore, the variance of their difference $d = \bar{X}_1 - \bar{X}_2$ is:

$$\sigma_d^2 = \frac{\sigma_1^2}{n_1} + \frac{\sigma_2^2}{n_2} \tag{7.11}$$

Thus, the standard deviation of the distribution shown in Figure 7.9(b) is the square root of the value in Formula (7.11):

$$\sigma_d = \sqrt{\frac{\sigma_1^2}{n_1} + \frac{\sigma_2^2}{n_2}} \tag{7.12}$$

Let us return to Table 7.2 to demonstrate the meaning of Formulas (7.11) and (7.12). These formulas tell us what the variance and standard deviation of the fifth column in Table 7.2 would be after infinitely many pairs of samples had been taken. Since the population from which the samples are being drawn is the uniform distribution of digits between 0 and 9, we could calculate that $\sigma_1^2 = \sigma_2^2 = 9.2$. The long run variance of Column 5 would then be given by Formula (7.11) as

$$\sigma_d^2 = \frac{9.2}{5} + \frac{9.2}{5} = 3.68$$

and $\sigma_d = \sqrt{3.68} = 1.92$. The variance and standard deviation of Column 5 after only ten pairs of samples are 6.28 and 2.51. These differ from the long run values, and the error is due to taking only ten pairs of samples instead of infinitely many. If more and more pairs of samples were taken, Column 5 of the table would get longer and longer. Its mean would get closer to zero, and its variance would converge to 3.68.

Finally, we ask: what is the shape of the distribution of differences in sample means shown in Figure 7.9(b)? The answer is given by extension of the concepts discussed in Section 7.5.

1. If the populations from which the samples are drawn are normal in shape, then the distribution of $\bar{X}_1 - \bar{X}_2$ will be normal in shape.
2. If the populations from which the samples are drawn are not normal in shape, then the distribution of $\bar{X}_1 - \bar{X}_2$ will be approximately normal if the sample sizes, n_1 and n_2, are both large.

Since the populations in Figure 7.9(a) appear to be normal in shape, the first point applies, and we can assume that the distribution of $\bar{X}_1 - \bar{X}_2$ values in Figure 7.9(b) is normal. That is, $[(\bar{X}_1 - \bar{X}_2) - (\mu_1 - \mu_2)]/\sigma_d = Z$ follows the normal distribution.

The implication of the two points listed is that when we desire to make probability statements about the difference between two sample means, there are many situations in which we can use Table V of the Appendix to do so. Example 1 demonstrates how this can be done.

EXAMPLE 1

Problem: The daily production of ore from Mine 1 averages $\mu_1 = 150$ tons and is normally distributed with a standard deviation $\sigma_1 = 20$ tons. Mine 2's daily production is also normally distributed, but the mean is $\mu_2 = 125$ tons and the second standard deviation is $\sigma_2 = 25$ tons. A sample of five randomly selected daily production figures is taken from each of the mines. What is the probability that the sample mean production for Mine 1 will be less than or equal to the sample mean production for Mine 2? That is, what is $P(\bar{X}_1 - \bar{X}_2 \leq 0)$?

Solution: From the discussion in this section, we know that the difference $\bar{X}_1 - \bar{X}_2$ for samples taken from these two mines should average out to $\mu_1 - \mu_2 = 150 - 125 = 25$ tons. Also, the standard deviation of the distribution for many pairs $\bar{X}_1 - \bar{X}_2$ is

$$\sigma_d = \sqrt{\frac{\sigma_1^2}{n_1} + \frac{\sigma_2^2}{n_2}} = \sqrt{\frac{(20)^2}{5} + \frac{(25)^2}{5}} = 14.32$$

The distribution of the differences of sample means taken from the two mines is shown in Figure 7.10. The shape of this distribution is normal since the two populations are normal in shape, and the shaded area shows the probability we seek. The Z value associated with this area is:

FIGURE 7.10 Distribution of differences of sample means from two normal populations

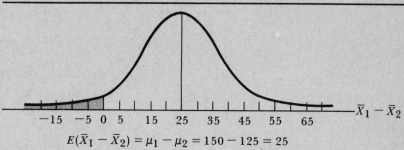

$$E(\bar{X}_1 - \bar{X}_2) = \mu_1 - \mu_2 = 150 - 125 = 25$$

$$Z = \frac{0 - (150 - 125)}{14.32} = \frac{-25}{14.32} = -1.75$$

That is, the shaded area begins at 1.75 standard deviations to the left of the center of the distribution. The area to the left of 1.75 standard deviations is found through Table V to be $.5000 - .4599 = .0401$. Thus, there is only a 4% chance that the first sample's mean will be smaller than the second sample's mean. Thus, if the owner of the two mines found $\bar{X}_1 = 130$ tons and $\bar{X}_2 = 135$ tons in samples of five randomly selected days from each mine, he would suspect that either the sampling was faulty or that the difference in the mines' mean daily outputs had changed.

EXAMPLE 2

Problem: Two production processes are, on the average, identical. Both use an average of $\mu_1 = \mu_2 = 500$ kilograms of raw material per day. Both have the same standard deviation of daily usage $\sigma_1 = \sigma_2 = 9$ kilograms per day. Thus, the daily usage of material may vary for the two processes, but on the average they are the same. What is the probability that two samples of $n_1 = 100$ and $n_2 = 100$ randomly selected daily usage figures would show the sample means $\bar{X}_1$ and $\bar{X}_2$ differing by less than one-half kilogram. That is, what is

$$P(-.5 < \bar{X}_1 - \bar{X}_2 < .5)$$

Solution: Since the two populations have the same mean, Formula (7.10) tells us that the distribution of differences between the sample means is centered at $E(\bar{X}_1 - \bar{X}_2) = 500 - 500 = 0$. The standard deviation of the distribution is given by Formula (7.12)

$$\sigma_d = \sqrt{\frac{9^2}{100} + \frac{9^2}{100}} = \sqrt{1.62} = 1.27$$

The distribution of the differences $\bar{X}_1 - \bar{X}_2$ for this problem is shown in

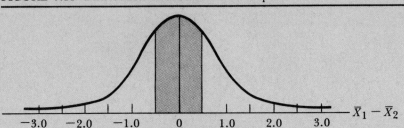

FIGURE 7.11 Distribution of differences of sample means for material usage

Figure 7.11. The shape of this distribution is approximately normal. This is because we don't know the shape of the distributions of material usage in the two populations, but we do know that the samples are large ($n_1 = 100$ and $n_2 = 100$). Thus, the differences in the sample means are *approximately* normal at a minimum, and we can use Table V to answer the probability question posed in this problem. The probability we seek is shown by the shaded area in Figure 7.11. Note that the difference may be positive or negative, as long as the two sample means differ by no more than .5 kilograms. Due to the symmetry of the normal curve, we need only find the area to the right of zero and double it to get the answer we seek.

$$P(0 < \bar{X}_1 - \bar{X}_2 < .5) = P[0 < Z = (\bar{X}_1 - \bar{X}_2)/\sigma_d < (.5 - 0)/1.27]$$

$$P(0 < Z < .39) = .1517$$

Doubling this figure gives .3034, the answer to this problem.

One might wonder what happens if we take pairs of *small* samples from two *nonnormal* populations. What is the shape of the $\bar{X}_1 - \bar{X}_2$ distribution? The answer is that we don't know and thus cannot make probability statements about the $\bar{X}_1 - \bar{X}_2$ differences. The only choice we would have in a situation like that would be to continue sampling until we obtained large samples (which might be expensive) or to use a completely different approach to the problem.

■ 7.7 TWO SAMPLES FROM BINOMIAL POPULATIONS

There are many instances when we want to know about the differences in the proportion of successes in two binomial populations. We may have two production processes. If the first one produces defectives

---

■The material in this section is required only in other optional sections.

at a rate p_1 and the second produces defectives at a rate p_2, we may wish to know about the distribution of $f_1 - f_2$, the difference between the rates of defectives found in two independent samples taken from the two processes. A company may have two products which are bought by both men and women. If p_1 is the proportion of male buyers for the first product and p_2 is the proportion of male buyers for the second product, we may wish to know about the chance of getting a value of $f_1 - f_2$, the difference in proportion of male buyers in two independent samples of these products' customers.

It should not be surprising to the reader that

$$E(f_1 - f_2) = p_1 - p_2 \tag{7.13}$$

That is, if one population has a success rate of $p_1 = .50$ and another has a success rate of $p_2 = .35$, then, on the average, two independent samples will yield a difference in sample success rates of $.50 - .35 = .15$.

Since we are assuming that two *independent* random samples are drawn from the two binomial populations, the variance of the differences $f_1 - f_2$ is the sum of their individual variances:

$$\sigma^2_{f_1 - f_2} = \frac{p_1(1 - p_1)}{n_1} + \frac{p_2(1 - p_2)}{n_2} \tag{7.14}$$

By extending the normal approximation to the binomial from Section 6.3, we know that for *large* n_1 and n_2 or (by the rule of thumb given in Section 6.3) when $n_1 p_1 (1 - p_1) \geq 5$ *and* $n_2 p_2 (1 - p_2) \geq 5$ then the shape of the $f_1 - f_2$ distribution is approximately normal. Thus, the term

$$Z = \frac{(f_1 - f_2) - (p_1 - p_2)}{\sqrt{\dfrac{p_1(1 - p_1)}{n_1} + \dfrac{p_2(1 - p_2)}{n_2}}} \tag{7.15}$$

has approximately a standard normal distribution. That means we can use Table V in the Appendix to answer probability questions about the differences $f_1 - f_2$. Example 1 demonstrates how this can be done.

EXAMPLE 1

Problem: A company has two sales outlets. At both outlets 40% of the customers charge their purchases. That is, $p_1 = p_2 = .40$. In doing an audit, the company accountant took random samples of $n_1 = 100$ and $n_2 = 100$ sales slips from each outlet. $r_1 = 41$ and $r_2 = 36$ charge customers were found in the two samples. What is the probability that a result would be achieved where the first outlet's proportion of charge customers exceeds the second outlet's proportion of charge customers by this much or more?

Solution: In this example, $f_1 = r_1/n_1 = 41/100 = .41$, and $f_2 = r_2/n_2 = 36/100 = .36$. Since $p_1 = p_2$, on the average we would expect $f_1 - f_2$ to be zero, according to Formula (7.13). The probability we seek can be written as $P(f_1 - f_2 \geq .41 - .36)$ or $P(f_1 - f_2 \geq .05)$. Using Expression (7.15), we can find this probability as follows:

$$P(f_1 - f_2 \geq .05) = P\left(Z \geq \frac{.05 - 0.0}{\sqrt{\dfrac{(.4)(.6)}{100} + \dfrac{(.4)(.6)}{100}}} \right) = P\left(Z \geq \frac{.05}{.069} \right)$$

$$= P(Z \geq .72)$$

Thus, we wish to find the area under the normal curve to the right of $Z = .72$. This is $.5000 - .2642 = .2358$. This means there is about a 24% chance of getting two independent random samples that differ by .05 or more in the rate of charge customers from the *first outlet* over the rate of charge customers from the *second outlet*. Thus, the accountant did not obtain an unusual difference. In fact, due to the symmetry of the problem, we can say that in approximately 48% of the samples taken from these two outlets' sales slips, *one* of the outlet's charge rate will exceed *the other* by .05 or more.

EXAMPLE 2

Problem: Assume your company receives parts sent to you by two suppliers. Historical records show the first supplier's goods are rejected at a rate of $p_1 = .08$ and the second supplier's goods have a rejection rate of $p_2 = .05$. Your production operation uses $n_1 = 150$ items from the first supplier and $n_2 = 300$ items from the second supplier each day. On what proportion of the days will the difference in the rejection rates of the two suppliers, $f_1 - f_2$, be 1% or less?

Solution: The probability we want is represented by the shaded area of Figure 7.12. Since the daily usages of the parts constitute *large* random samples of output from the two suppliers, the differences $f_1 - f_2$ each day will be approximately normally distributed around $p_1 - p_2 = .08 - .05 = .03$. By Formula (7.15) we want the area to the left of Z:

$$Z = \frac{.01 - .03}{\sqrt{\dfrac{(.08)(.92)}{150} + \dfrac{(.05)(.95)}{300}}} = \frac{-.02}{\sqrt{.0006}} = -.82$$

This value is $.5000 - .2939 = .2061$. Thus, on about 21% of the days the difference in the rejection rates of the two suppliers, $f_1 - f_2$, will be 1% or less.

FIGURE 7.12 Distribution of differences in sample rejection rates

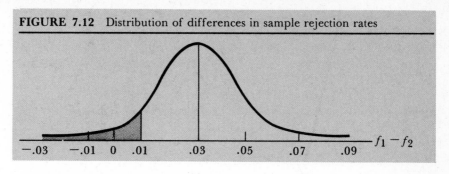

We should note that in order to solve problems like the preceding two examples, we must have large samples. When n_1 and n_2 are both large, then the normal approximation to the binomial holds, and the differences of sample proportions of success, $f_1 - f_2$, are approximately normally distributed. We noted previously that a good rule of thumb for determining when this approximation can be used is that both $n_1 p_1 (1 - p_1)$ and $n_2 p_2 (1 - p_2)$ should be greater than or equal to five.

7.8 SUMMARY

Chapters 4 through 7 of the text have dealt with the subject of probability. Subsequent chapters will cover the area of statistics known as statistical inference. It is in this latter area that most of the realistic problems of statistics lie. Many of the examples in this and previous chapters have been somewhat contrived. It is not often that a manager or administrator can work problems of the type we have shown here.

This is due to the fact that the probability problems involve a knowledge of what the population looks like. That is, in order to work a probability problem we have to know μ, σ, or p of the populations from which we sample. Then we can ask questions about the probability of obtaining certain sample results.

In most realistic situations, we do not know the population parameters μ, σ, or p. We usually have sample results like $\bar{X}$, s, or f. From these sample results, we usually want to estimate the population characteristics. However, to do so we must use the knowledge of probability we have gained in these chapters. Thus, before we proceed to the subject of statistical inference, it is important that the reader have a good grasp of the material covered in this chapter.

Table 7.3 summarizes four important probability situations which will be used often in the next two chapters. These concepts were covered in this chapter and in Section 6.3. These four concepts are basically outgrowths of the central limit theorem, the most important theorem in all of statistics.

TABLE 7.3   Summary of Four Important Probability Situations

SAMPLING SITUATION	SAMPLE RESULT	EXPECTED VALUE OF SAMPLE RESULT	STANDARD DEVIATION OF SAMPLE RESULT OVER MANY SAMPLES	SHAPE OF THE SAMPLE RESULT'S PROBABILITY DISTRIBUTION
1. One sample of n items from a continuous population with mean μ and standard deviation σ.	$\bar{X}$ Sample Mean	$E(\bar{X}) = \mu$	$\sigma_{\bar{x}} = \dfrac{\sigma}{\sqrt{n}}$	a. Exactly normal if the population is normal in shape. b. Approximately normal if the population is not normal but n is large.
2. One sample of n items from a binomial population with a proportion of successes p.	$f = r/n$ Sample Proportion of Successes	$E(f) = p$	$\sigma_f = \sqrt{\dfrac{p(1-p)}{n}}$	Approximately normal if n is large. RULE: $np(1-p) \geq 5$.
3. Two independent samples of n_1 items and n_2 items from continuous populations which have means μ_1 and μ_2 and standard deviations of σ_1 and σ_2, respectively.	$\bar{X}_1 - \bar{X}_2$ Difference in Sample Means	$E(\bar{X}_1 - \bar{X}_2) = \mu_1 - \mu_2$	$\sigma_d = \sqrt{\dfrac{\sigma_1^2}{n_1} + \dfrac{\sigma_2^2}{n_2}}$	a. Exactly normal if both populations are normal in shape. b. Approximately normal if the populations are not normal but n_1 and n_2 are large.
4. Two independent samples of n_1 items and n_2 items from binomial populations which have p_1 and p_2 proportions of success, respectively.	$f_1 - f_2$ Difference in Sample Success Rates	$E(f_1 - f_2) = p_1 - p_2$	$\sigma_{f_1-f_2} =$ $\sqrt{\dfrac{p_1(1-p_1)}{n_1} + \dfrac{p_2(1-p_2)}{n_2}}$	Approximately normal if n_1 and n_2 are large. RULE: $n_1 p_1(1-p_1) \geq 5$ and $n_2 p_2(1-p_2) \geq 5$.

PROBLEMS

1. If many samples of size 100 (that is, each sample consists of 100 items) were taken from an infinite normal population with mean 10 and variance 16, what would be the mean and variance of these many samples' means?

2. If many samples of size 64 were taken from a continuous nonnormal population with mean of 50 and standard deviation of 20, what would be the mean and standard deviation of these many samples' means?

3. What differences, if any, will there be between the distributions of the sample means in Problems **1** and **2**?

4. In a certain manufacturing process, the standard length of a machined part is 100 cm. It is known from past measurements that when the process is in adjustment, this measurement is really a random variable with a mean of 100 cm and a standard deviation of .5 cm. The individual measurements are normally distributed around their mean.

   a. What proportion of the parts will be longer than 100.2 cm?

   b. What is the probability that a randomly selected part will have a length longer than 100.2 cm?

   c. What proportion of the parts will have lengths between 99.6 and 101.2 cm?

   d. If many samples of size 25 were taken from this process, what would be the mean and standard deviation of these many samples' means?

   e. What probability distribution would the distribution of these sample means follow?

   f. Did you need your knowledge of the central limit theorem to answer Parts **d** and **e**? Explain.

   g. If a single sample of size 25 is taken, what is the probability that the mean length of the parts in this sample will exceed 100.2 cm?

   h. What is the difference between the problems in Parts **b** and **g**?

5. The Bills Processing Company makes an item which it sells to the U.S. Navy. The item is called the KM-2. During the manufacturing process, the individual KM-2's are placed in a baking oven. The time it takes to bake the items is normally distributed around a mean of 64 minutes with a standard deviation of 5 minutes. Thus the population of baking times is normal in shape.

   a. If 100 KM-2's are baked, what is the probability that the mean baking time will be 64 minutes and 45 seconds or greater?

   b. What proportion of the individual KM-2's bake in 57 minutes or less?

   c. Did you use the Central Limit Theorem in either of Parts **a** and **b**? If so, which one(s)?

6. A certain chemical company claims that its major product contains on the average 4.0 fluid ounces of caustic materials per gallon. It further states that the distribution of caustic materials per gallon is normal and has a standard deviation of 1.3 fluid ounces.

   a. What proportion of the individual gallon containers for this product will contain more than 5.0 fluid ounces of caustic materials?

   b. A government inspector randomly selects 100 gallon-size containers of the product and finds the mean weight of caustic material to be 4.5 fluid ounces per gallon. What is the probability of finding the mean of a sample of 100 that is 4.5 or greater? Do you think the production process was producing its usual level of caustic materials when this sample was taken?

7. A certain type of electron tube has a mean transconductance equal to 10,000 μmhos (micromhos). The variance of the transconductances is 3600. If we take samples of 25 tubes each and for each sample we find the mean transconductance, between what limits (symmetric with respect to the mean) would 50% of the sample means be expected to lie?

8. If the population of times measured by 3-minute egg timers is normally distributed with μ equal to 3 minutes and σ equal to .2 minutes, and if we test samples of 25 timers, find the time that would be exceeded by 95% of the sample means.

9. In Example 1 of Section 7.5, find the probability that in a sample of $n = 81$ devices the mean time to failure, $\overline{X}$, will be six years or less.

10. A food processing company packages a product which is periodically inspected by the Food and Drug Administration (FDA). The FDA has ruled that the company's product may have no more than 2.0 grams of a certain toxic substance in it. Past records of the company show that packages of this product have a mean weight of toxic substance equal to 1.25 grams per package and that the weights are normally distributed around 1.25 grams with a standard deviation of .50 grams.

   a. What proportion of the individual packages exceed the FDA limit?

   b. What is the probability of selecting an individual package which has between 1.75 and 2.00 grams of toxic substance it it?

   c. A team of FDA inspectors is on its way to inspect a random sample of the company's output. They plan to take a preliminary sample of 25 packages. If they find that the mean weight of toxic substance in this sample exceeds 1.6 grams, they will close down the plant and have an extensive inspection of the company's entire inventory. What is the probability they will close the plant?

11. Fifty-five percent of a company's customers are women. If many samples of $n = 200$ were taken from this company's customer list,

    a. What is the expected proportion of women customers in these samples over the long run?

    b. What is the standard deviation of f, the proportion of women customers found in these many samples?

12. A production line produces 10% defective items. If a sample of $n = 64$ items is taken, what is the probability that 5 or fewer items in this sample will be defective? (*Hint:* This is a Chapter 6 problem, which uses a special case of the central limit theorem. Use the continuity correction.)

13. A light bulb manufacturer claims that 90% of the bulbs it produces meet tough, new standards imposed by the Consumer Protection Agency. You just received a shipment containing 400 bulbs from this manufacturer. What is the probability that 375 or more of the bulbs in your shipment meet the new standards? (See the hint in Problem **12.**)

14. Seventy-five percent of a school's law class passes the state bar examination

on the first attempt. If a randomly selected group of 60 of this school's law graduates take the examination, what is the probability that 80% or more of them will pass on the first attempt?

15. Two populations of measurements are normally distributed with $\mu_1 = 57$ and $\mu_2 = 25$. The two populations' standard deviations are $\sigma_1 = 12$ and $\sigma_2 = 6$. Two independent samples of $n_1 = n_2 = 36$ are taken from the populations.

   a. What is the expected value of the difference in the sample means, $\bar{X}_1 - \bar{X}_2$?

   b. What is the standard deviation of the distribution of $\bar{X}_1 - \bar{X}_2$?

   c. What is the shape of the distribution of $\bar{X}_1 - \bar{X}_2$? How do you know?

16. What proportion of the time will the means of the samples described in Problem 15 differ by +35 or more? That is, find $P(\bar{X}_1 - \bar{X}_2 \geq 35)$.

17. For Example 2 in Section 7.6, find the probability that two samples of $n_1 = 81$ and $n_2 = 36$ will produce sample means, $\bar{X}_1$ and $\bar{X}_2$, which differ by no more than 1.0 kilogram. (*Hint:* "Differ by no more . . ." implies that you must consider the absolute value of the difference.)

18. There are two classes of land investment in a certain company's investment plans. Both have five-year mean yields of 220% return on investment. That is, $\mu_1 = \mu_2 = 220$. They also have the same variances of returns for the investment purchases within the two classes: $\sigma_1^2 = \sigma_2^2 = 30$. If the company purchases $n_1 = 50$ pieces of land in the first class and $n_2 = 75$ pieces of land in the second class, what is the probability that the mean yields of these two sample portfolios will differ by less than 2.3%? That is, find $P(|\bar{X}_1 - \bar{X}_2| \leq 2.3)$.

19. A weight reduction clinic has offices in New York and Chicago. It has found that the proportion of people signed up for its weight reducing classes who actually complete their entire program is $p_1 = .80$ in New York and $p_2 = .72$ in Chicago. A class of $n_1 = 80$ participants just started the program in New York and a class of $n_2 = 60$ just started the program in Chicago. If you consider these two groups to be independent random samples,

   a. What is the expected value of the difference in completion rates, $f_1 - f_2$, between these New York and Chicago groups?

   b. What is the standard deviation of this difference over many pairs of classes of this size?

20. What is the probability that the difference $f_1 - f_2$ will be 10% or larger in Problem 19?

21. Two different salespeople have the same sales success rate when they call on customers: $p_1 = p_2 = .30$. During the past month, each salesperson has called on the same number of randomly selected customers: $n_1 = n_2 = 210$.

   a. What is the expected difference in their sales success rates?

   b. What is the standard deviation of the difference in success rates over many months where each salesperson visits 210 customers?

   c. What is the probability that the sales success rates will differ by less than 2% this month? That is, what is $P(|f_1 - f_2| \leq .02)$?

22. Two large national companies in the same industry differ in the proportions of women in their production labor forces. The first company has a proportion of women that is $p_1 = .30$. The second company's proportion of women is $p_2 = .18$. Randomly selected groups of $n_1 = 80$ and $n_2 = 70$ production workers are being sent by their companies to an industry-sponsored training program in Miami. What is the probability that the proportion of women in the first company's group will exceed the proportion of women in the second company's group by 20% or more? That is, what is $P(f_1 - f_2 \geq .20)$?

SAMPLE DATA SET QUESTIONS: *Refer to the 113 applicants for credit listed in the Sample Data Set Appendix of this book.*

a. Find the probability that the difference in the mean ages of applicants who were extended credit and those who were denied credit would differ by as much or more (in either direction) than they do in the samples of $n_1 = 63$ applicants granted credit and $n_2 = 50$ applicants denied credit, assuming the mean ages are no different in the population. (Assume that the variances computed from the samples are equal to the population variances: $s_i^2 = \sigma_i^2$ where $i = 1$ or 2.)

b. Assume that the proportion of women both granted and denied credit is 10% in the population of all applicants. Find the probability that the sample difference $f_2 - f_1$ or a larger difference in the proportions of women denied and granted credit could have occurred by pure chance.

REFERENCES

7.1   Hoel, Paul G. *Elementary Statistics,* 3rd edition. John Wiley & Sons, New York, 1971. Chapter 6.

7.2   Hogg, Robert V., and Allen T. Craig. *Introduction to Mathematical Statistics,* 3rd edition. Macmillan, New York, 1970. Chapter 4.

ESTIMATION

Chapters 4 through 7 of this text dealt with probability topics. We mentioned in Chapter 1 that all probability problems have one feature in common. We take a sample from a population (whose characteristics we know) and make statements about the likelihood of the sample's having certain characteristics.

In this chapter, we begin discussing the third major area of statistics—the area called *statistical inference*. The problems in this area of statistics all have some features in common. First, a statistician collects data by experiment, sample, or sample survey in the hope of drawing conclusions about the phenomenon under investigation. From his experimental results or sample values he wants to pass to *inferences* about the underlying population. He may use his data for the *estimation* of the values of·unknown parameters or for *tests of hypotheses* concerning these values. In Section 7.1, we discussed a personnel manager who takes a sample from a population of employees divided into those who belong to a union and those who don't. The unknown parameter is the proportion p who belong to a union. The personnel manager may estimate p (try to guess its value), or he may test an hypothesis about p—test the hypothesis, say, that $p \geq 1/2$. In either case he infers something about p.

This chapter deals with methods of estimation and the principles underlying them. Hypothesis testing is taken up in Chapter 9.

8.1 ESTIMATORS

We have at hand a sample from a population involving an unknown parameter; the problem is to construct a sample quantity that will serve to estimate the unknown parameter. Such a sample quantity we call an **estimator;** the actual numerical value obtained by evaluating an estimator in a given instance is the **estimate.** For example, the sample mean $\bar{X}$ is an estimator for the population mean μ; if for a specific sample the sample mean is 10.31, we say 10.31 is our estimate for μ. Notice that an estimator must be a statistic; it must, of course, depend only on the sample and not on the parameter to be estimated.

Now the sample mean $\bar{X}$ is an estimator for the population mean μ, but then so is the quantity $\bar{X} + 1000$. We will all agree that, as an estimator for μ, $\bar{X}$ is more reasonable than $\bar{X} + 1000$, but why? Because $\bar{X}$, unlike $\bar{X} + 1000$, is equal to μ *on the average:*

$$E(\bar{X}) = \mu \tag{8.1}$$

for all μ (see Equation (7.2) in Section 7.2). This is a statement typical of statistical theory. Putting ourselves in the position of the experimenter, we do not know μ and doubtless never will; given the sample, we do know $\bar{X}$; and *whatever* the unknown μ may be, $\bar{X}$ balances out at μ in the sense that $E(\bar{X})$ is equal to μ.

This discussion leads us to the following definition:

> **Unbiased Estimator:** A sample estimator of a population parameter is said to be unbiased if it has expected value equal to the population parameter. That is,

$$E(\text{sample estimator}) = \text{population parameter} \tag{8.2}$$

For example, $\bar{X}$ is an unbiased estimator of μ, and s^2 is an unbiased estimator of σ^2. Expression (8.2) means that an estimator is *unbiased* if, on the average, it gives the parameter's value. We saw in Chapter

7 that the distribution of sample means has a mean equal to the population mean. Thus, $\bar{X}$ has this desirable property. So has the sample variance. If $X_1, X_2, \ldots, X_n$ is an independent sample, then the sample variance

$$s^2 = \frac{1}{n-1} \sum_{i=1}^{n} (X_i - \bar{X})^2 \tag{8.3}$$

is an unbiased estimator of σ^2:

$$E(s^2) = \sigma^2 \tag{8.4}$$

(See Equation (7.6), Section 7.2.) Notice that s^2 is a statistic; it does not depend on the unknown μ and σ^2. The point of Equation (8.4), the condition for unbiasedness, is not that we can somehow check it for the true values of μ and σ^2—we do not know what these true values are. The point is that Equation (8.4) holds *whatever* values μ and σ^2 may happen to have. The same remark applies to any unbiased estimator.

It was exactly in order to make the sample variance unbiased that, in the original Definition (3.14) in Section 3.11, we divided by the then-mysterious $n - 1$; division by n would have introduced bias. As an estimator of σ, the sample standard deviation

$$s = \sqrt{\frac{1}{n-1} \sum_{i=1}^{n} (X_i - \bar{X})^2} \tag{8.5}$$

is ordinarily used, even though it is somewhat biased. That is, on the average s does not equal σ.

Returning to the mean, consider a sample $X_1, X_2, \ldots, X_{100}$ of size 100. Here $\bar{X}$, or

$$\frac{1}{100} \sum_{i=1}^{100} X_i \tag{8.6}$$

is an unbiased estimator of μ. If we throw out the last 50 observations, the remaining ones, $X_1, X_2, \ldots, X_{50}$, form a sample of size 50 by themselves, and their mean

$$\frac{1}{50} \sum_{i=1}^{50} X_i \tag{8.7}$$

is another unbiased estimator of μ. Given the full sample of 100, we will all agree that Expression (8.6) is a better estimator of μ than Expression (8.7) is, but why? Each of the two estimators is unbiased, but one feels that (8.6) is stronger than (8.7) because there is somehow more information in it. This idea can be made precise by using variances.

Expression (8.6) has variance $\sigma^2/100$ (see Equation (7.3); here σ^2 is the population variance, also unknown to us), where Expression (8.7) has the larger variance $\sigma^2/50$. Since variance measures spread, of two unbiased estimators we naturally prefer the one with smaller variance. The ideal, the so-called *minimum-variance unbiased estimator,* is an estimator which not only is unbiased but also has smaller variance than *any other* unbiased estimator.

In this section, we have discussed point estimators, sample quantities which estimate population parameters with specific numerical values. More often, however, we estimate a parameter of a population by constructing a range, or interval, within which we think the parameter might lie. The following sections of this chapter show how these ranges, or confidence intervals, are constructed.

8.2   CONFIDENCE INTERVALS FOR NORMAL MEANS: KNOWN VARIANCE

We have seen that the sample mean $\bar{X}$ is an unbiased estimator of the population mean μ. Now if the parent population is *normal,* it is even possible to show that $\bar{X}$ is a minimum-variance unbiased estimator of μ. In this sense, we may say that, for a given sample size, $\bar{X}$ is the *best* estimator of μ.

To prove that $\bar{X}$ is best or optimal in the sense of having the smallest possible variance lies outside the purpose of this book (see Reference 8.2 for a proof). It is simple and natural to estimate μ by $\bar{X}$, and we do know that $\bar{X}$ is unbiased and has variance σ^2/n even if we have not proved this variance to be minimal.

This section deals with the problem of estimating the mean μ of a normal population when the variance σ^2 is known. This is a rather artificial circumstance: If one is ignorant of the mean he is usually ignorant of the variance as well. Assuming σ^2 known, however, serves to simplify the reasoning and make clear the principles underlying estimation; it also serves to introduce the more complicated and more realistic case, treated in the next two sections, where σ^2 is unknown.

EXAMPLE 1

Ten patients were given a soporific drug. In almost every case, the patient slept longer under the effect of the drug than usual, and Table 8.1 shows the amount of the increase in each case.

TABLE 8.1   Additional Hours Sleep Gained by Using Drug

PATIENT	1	2	3	4	5	6	7	8	9	10
INCREASE	1.2	2.4	1.3	1.3	0.0	1.0	1.8	0.8	4.6	1.4

$$\bar{X} = 1.58 \text{ hours}$$

Let us assume we know from experience that the increase is normally distributed with some mean μ and that the variance σ^2 is 1.66. The estimator $\bar{X}$ then has a variance $\sigma^2/10$, or .166, and a standard deviation $\sigma/\sqrt{10}$, or .408. Knowing this and the fact that $\bar{X}$ is normally distributed, we can use the methods of Section 7.4 to calculate the probability that $\bar{X}$ is within .8, say, of the population mean μ. We do not know μ, but whatever value it may have, the standardized variable $(\bar{X} - \mu)/.408$ has a standard normal distribution and hence (see Figure 8.1)*

$$P(\mu - .8 \le \bar{X} \le \mu + .8) = P\left(\frac{-.8}{.408} \le \frac{\bar{X} - \mu}{.408} \le \frac{.8}{.408}\right)$$

$$= P\left(-1.96 \le \frac{\bar{X} - \mu}{.408} \le 1.96\right) = .95$$

Thus, the chance is 95% that $\bar{X}$ is off by .8 or less. Since $\mu - .8 \le \bar{X} \le \mu + .8$ is the same as $\bar{X} - .8 \le \mu \le \bar{X} + .8$ (both being the same as $|\bar{X} - \mu| \le .8$), we can say that

$$P(\bar{X} - .8 \le \mu \le \bar{X} + .8) = .95 \tag{8.8}$$

FIGURE 8.1 Standard normal curve

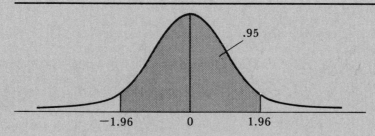

*The first step in the computation that follows was accomplished by subtracting μ from each term in the inequality string and then dividing each term by .408. These are both legitimate operations, as was explained in the footnote at the beginning of Section 7.4.

whatever μ may be. Now $\bar{X}$ for our data in Table 8.1 has the value 1.58, and it is tempting to replace the $\bar{X}$ in Equation (8.8) by 1.58:

$$P(1.58 - .8 \leq \mu \leq 1.58 + .8) = .95. \tag{8.9}$$

That is, it is tempting to conclude that there is probability .95 that the unknown μ lies between $.78 = (1.58 - .8)$ and $2.38 = (1.58 + .8)$. But this would be incorrect. Although μ is unknown to us, it is not a random variable, but a fixed number. Thus $.78 \leq \mu \leq 2.38$ is simply either true or false; the probability on the left in Equation (8.9) is accordingly either 1 or 0—and we do not know which, since we do not know μ.

Although Equation (8.9) is wrong as it stands, the idea lying behind it can be made sense of by means of **confidence intervals.** The first thing to understand is the source of the error in passing from Equation (8.8) to Equation (8.9). Consider a related but simpler case. Let Y be the total obtained in rolling a pair of fair dice. As we know from Chapter 4, the chance of rolling a 7 is $1/6$; $P(Y = 7) = 1/6$. Suppose we actually roll the dice and obtain a total of 3; if in the equation $P(Y = 7) = 1/6$ we replace Y by 3, we get $P(3 = 7) = 1/6$, which is nonsense. The probability of $3 = 7$ is 0, not $1/6$. Or suppose we happen to roll a 7; if in the equation $P(Y = 7) = 1/6$ we replace Y by 7, we get $P(7 = 7) = 1/6$, again nonsense. The probability of $7 = 7$ is 1, not $1/6$.* In passing from Equation (8.8), which is true, to Equation (8.9), which is false, we have made the same error, that of replacing inside a probability statement the random variable $\bar{X}$ by the specific value 1.58 that it happened to assume when the experiment was carried out.

For the preceding example, the terms L and R, where

$$L = \bar{X} - .8 \quad \text{and} \quad R = \bar{X} + .8$$

are 95 percent *confidence limits* for μ, and the interval bounded "on the left" by L and "on the right" by R is a 95 percent *confidence interval* for μ. Now L and R are random variables, and the interval they determine is a random interval. What is the chance that the confidence limits surround μ, or that the confidence interval includes μ?

Now $L \leq \mu \leq R$ if $\bar{X} - .8 \leq \mu \leq \bar{X} + .8$, and this is the same as $|\bar{X} - \mu| \leq .8$ or $-.8 \leq \bar{X} - \mu \leq .8$. Since $(\bar{X} - \mu)/.408$ has

---

*It is perhaps illuminating to observe also that in the expected value $E(Y)$ the random variable Y cannot be replaced by a numerical value. In fact this expected value is 7: $E(Y) = 7$. If we roll a 3, substitution yields the equation $E(3) = 7$, nonsense once more. The illegitimacy of these substitutions becomes clearer if we keep in mind that Y is a name for the total number of dots "before the dice are rolled."

a standard normal distribution, whatever μ may be,

$$P(L \leq \mu \leq R) = P(-.8 \leq \bar{X} - \mu \leq .8)$$

$$= P\left(-\frac{.8}{.408} \leq \frac{\bar{X} - \mu}{.408} \leq \frac{.8}{.408}\right)$$

$$= P\left(-1.96 \leq \frac{\bar{X} - \mu}{.408} \leq 1.96\right) = .95$$

Thus, the terms $L = \bar{X} - .8$ and $R = \bar{X} + .8$ have the property that the probability of $L \leq \mu \leq R$ is *always* .95, no matter *what* value the unknown μ may happen to have.

Since the limits L and R have a 95 percent chance of enclosing the true μ between them whatever μ may be, they are called 95 percent confidence limits. For the data in Table 8.1, L is .78 and R is 2.38, and we say we are 95 percent confident that .78 $\leq \mu \leq$ 2.38. This is not to be interpreted as saying μ is random and has a 95 percent chance of lying between .78 and 2.38; it is really only a rephrasing of Equation (8.8). The confidence we have in the limits .78 and 2.38 really derives from our confidence in the *statistical procedure* that gave rise to them.* The procedure gives random variables L and R that have a 95 percent chance of enclosing the true μ; whether their specific values .78 and 2.38 actually enclose μ we have no way of knowing.

To construct confidence intervals for the general case, we need some auxiliary concepts. If Z is a standard normal variable, the quantity Z_ξ (where the subscript letter is Greek xi), defined by the relationship

$$P(Z > Z_\xi) = \xi$$

(see Figure 8.2), is the *upper percentage cutoff point* of the standard

FIGURE 8.2   Standard normal curve

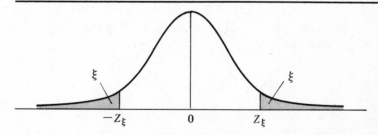

---

*This accords with at least one everyday use of the word "confidence": What confidence a layman has in a physician's advice very likely derives from what confidence he or she has in the physician.

normal distribution corresponding to a probability of ξ. It is the point on the Z scale such that the probability of a *larger* value is equal to ξ. The upper 2.5 percent point for the standard normal distribution is denoted by $Z_{.025}$ and is the value on the Z scale such that the area to the *right* of it is .025. By the table of normal areas, Table V, we find that $Z_{.025}$ is equal to 1.96. We may say that at $Z = 1.96$, 2.5 percent of the area under the normal curve is *cut off* in the curve's tail.

Because of the symmetry of the standard normal distribution, its lower percentage cutoff points are equal in absolute value to the corresponding upper ones but are negative. Thus $-Z_\xi$ is the point on the Z scale such that the probability of a *smaller* value is ξ; the area to the *left* of $-Z_\xi$ is ξ. There is probability .025 that a standard normal variable is less than $-Z_{.025}$, or -1.96. In terms of cutoff points, we have, by definition,

$$P(-Z_{\alpha/2} \le Z \le Z_{\alpha/2}) = 1 - \alpha \qquad (8.10)$$

Note that we use the $(1/2)\alpha$ cutoff points in order to have $100(1 - \alpha)\%$ of the area between them; each *tail* has area $(1/2)\alpha$ so the two together have area α (Greek alpha).

Now consider a normal population with unknown mean μ and known variance σ^2. Since the standardized variable $(\overline{X} - \mu)/(\sigma/\sqrt{n})$ has the standard normal distribution, it follows that

$$P\left(-Z_{\alpha/2} \le \frac{\overline{X} - \mu}{\sigma/\sqrt{n}} \le Z_{\alpha/2}\right) = 1 - \alpha \qquad (8.11)$$

Using the algebraic rules of inequalities,* it can be shown that Formula (8.11) is the same as

$$P\left(\overline{X} - Z_{\alpha/2}\frac{\sigma}{\sqrt{n}} \le \mu \le \overline{X} + Z_{\alpha/2}\frac{\sigma}{\sqrt{n}}\right) = 1 - \alpha \qquad (8.12)$$

The $100(1 - \alpha)\%$ confidence interval is the interval bounded by the confidence limits

$$L = \overline{X} - Z_{\alpha/2}\frac{\sigma}{\sqrt{n}} \quad \text{and} \quad R = \overline{X} + Z_{\alpha/2}\frac{\sigma}{\sqrt{n}} \qquad (8.13)$$

By Equation (8.12), there is probability $1 - \alpha$ that the confidence interval will contain the unknown μ, whatever μ may be.

In Example 1,

---

*These rules are given in a footnote at the beginning of Section 7.4.

$$1 - \alpha = .95 \qquad \alpha = .05 \qquad \frac{\alpha}{2} = .025 \qquad Z_{.025} = 1.96$$

Since $\sigma = \sqrt{1.66}$ and $n = 10$, and since the data give $\bar{X} = 1.58$,

$$L = \bar{X} - 1.96 \frac{\sqrt{1.66}}{\sqrt{10}} = 1.58 - .8 = .78$$

and

$$R = \bar{X} + 1.96 \frac{\sqrt{1.66}}{\sqrt{10}} = 1.58 + .8 = 2.38$$

Thus we can feel 95 percent confident that the population mean lies between .78 and 2.38.

The reason that "we can feel 95 percent confident" is this: if we were to take 100 different samples from the same population and calculate the confidence limits for each sample, then we would expect about 95 of these 100 intervals would contain the true value of μ, and 5 would not contain the true value of μ. Since we usually only have one sample and hence one confidence interval, we don't know whether our interval is one of the 95 or one of the 5. In this sense, then, we are "95 percent confident."

EXAMPLE 2

Problem: It is known that a package-filling process fills bags to an average weight of μ but that μ changes from time to time as the process adjustment changes through wear. However, even though the mean changes over time, the variance of the weights is a constant 9 pounds. A sample of 25 bags has just been taken, and their mean weight $\bar{X}$ was found to be 150 pounds. Assume that the weights of the individual bags are normally distributed around μ, and find the 90% confidence limits for μ.

Solution: Here,

$$1 - \alpha = .90 \qquad \alpha = .10 \qquad \frac{\alpha}{2} = .05 \qquad Z_{.05} = 1.645$$

By computation,

$$L = 150 - 1.645 \frac{3}{\sqrt{25}} = 150 - .987 = 149.013$$

$$R = 150 + 1.645 \frac{3}{\sqrt{25}} = 150 + .987 = 150.987$$

We are 90% confident that μ lies between 149.013 and 150.987 pounds, in that the procedure just followed will give limits enclosing μ 90% of the time.

EXAMPLE 3

Problem: For the data of Example 2 we are to calculate 95% confidence limits.

Solution: Here $Z_{\alpha/2}$ is 1.96, as in Example 1, and so

$$L = 150 - 1.96 \frac{3}{\sqrt{25}} = 150 - 1.176 = 148.824$$

$$R = 150 + 1.96 \frac{3}{\sqrt{25}} = 150 + 1.176 = 151.176$$

The width of the 95 percent confidence interval in this example is 2 × 1.176, or 2.352; the width of the 90 percent confidence interval computed in Example 2 is 2 × .987, or 1.974. The 90 percent confidence interval is better in that it is narrower; it apparently gives greater precision. But of course we have less confidence in it (90 as opposed to 95 percent). Confidence has been traded for precision, and neither interval can really be said to be better than the other.

8.3   THE t DISTRIBUTION

The interval estimates of the last section, where σ^2 was assumed *known*, were based on the fact that, in normal sampling, the standardized mean

$$Z = \frac{\bar{X} - \mu}{\sigma/\sqrt{n}} \tag{8.14}$$

has a distribution that does not depend on μ and σ^2, namely, the standard normal distribution. In trying to construct a confidence interval for the case of *unknown* variance, we need to know what happens if in Expression (8.14) we merely replace σ by its estimator s, the sample standard deviation. We then have the statistic

$$t = \frac{\bar{X} - \mu}{s/\sqrt{n}} \tag{8.15}$$

Expression (8.14) varies from one sample to the next due to the fact that each sample has a different $\bar{X}$. Thus, the Z of Expression (8.14) has one source of variation. Expression (8.15), however, differs in that it varies from one sample to the next due to *two* sources of variation. The sample mean, $\bar{X}$, *and* the sample standard deviation, s, change from sample to sample. Thus, the term on the right side of Expression (8.15) follows a sampling distribution different from the normal distribution, which is the distribution followed by the term on the right side of Expression (8.14). In order to use Expression (8.15), however, the *population* from which the *n* sample items was drawn must be normally distributed.

Expression (8.15) is the first of a number of *t statistics* we shall encounter. The *t distributions* form a family of distributions dependent on a parameter, the degrees of freedom. For the *t* variable (8.15) the degrees of freedom are $n - 1$, where n is the sample size. In general, the degrees of freedom for a *t* statistic are the degrees of freedom associated with the sum of squares used to obtain an estimate of a variance.

The *t* distribution is a symmetric distribution with mean zero. Its graph is similar to that of the standard normal distribution, as Figure 8.3 shows. There is more area in the tails of the *t* distribution, and the standard normal distribution is higher in the middle. The larger the number of degrees of freedom, the more closely the *t* distribution resembles the standard normal. As the number of degrees of freedom increases without limit, the *t* distribution approaches the standard normal distribution, and it is convenient to regard the standard normal distribution as a *t* distribution with an infinite number of degrees of freedom.

Table VI of the Appendix gives percentage cutoff points of the *t* distribution, those points on the *t* scale such that the probability of a larger *t* is equal to a specified value. The percentage cutoff point t_ξ is defined as that point at which

FIGURE 8.3   The *t* distribution with 5 degrees of freedom and the standard normal distribution

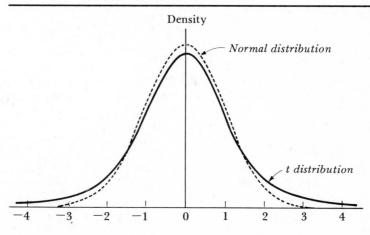

$$P(t > t_\xi) = \xi$$

Since the distribution is symmetric about zero, only positive t values are tabulated. The lower ξ cutoff point is $-t_\xi$, because

$$P(t < -t_\xi) = P(t > t_\xi) = \xi$$

In general, we denote a cutoff point for t by $t_{\xi,\nu}$, where ξ is the probability level and ν is the degrees of freedom.

To find a cutoff point in Table VI, we locate the row of the table that corresponds to the given degrees of freedom ν and then take the value in that row that is also in the column headed by the given probability level ξ. For example, $t_{.01,15}$ is the .01 percentage point of t with 15 degrees of freedom. Using Table VI, we find that it is equal to 2.602: If 15 is the degrees of freedom for t, then $P(t > 2.602) = .01$. The bottom line of the table corresponds to an infinite number of degrees of freedom, that is, to the standard normal distribution. Thus, the .05 cutoff point for the standard normal, $Z_{.05}$, is $t_{.05,\infty} = 1.645$, and the .025 cutoff point, $Z_{.025}$, is $t_{.025,\infty} = 1.96$.

8.4 CONFIDENCE INTERVALS FOR NORMAL MEANS: UNKNOWN VARIANCE

By the definition of percentage points, the following holds

$$P(-t_{\alpha/2,\nu} \leq t \leq t_{\alpha/2,\nu}) = 1 - \alpha$$

if t has a t distribution with ν degrees of freedom. The fact underlying the construction of confidence intervals in the case where σ^2 is *unknown* is that the Statistic (8.15) has a t distribution with $n - 1$ degrees of freedom, where n is the sample size and the population is assumed normal. Therefore,

$$P\left(-t_{\alpha/2,n-1} \leq \frac{\overline{X} - \mu}{s/\sqrt{n}} \leq t_{\alpha/2,n-1}\right) = 1 - \alpha \qquad (8.16)$$

If we multiply each member of the expression within the parentheses by $s/\sqrt{n}$, subtract $\overline{X}$ from each, and then multiply through by -1, we arrive at

$$P\left(\overline{X} - t_{\alpha/2,n-1}\frac{s}{\sqrt{n}} \leq \mu \leq \overline{X} + t_{\alpha/2,n-1}\frac{s}{\sqrt{n}}\right) = 1 - \alpha \qquad (8.17)$$

On the basis of Equation (8.17), we define the $100(1 - \alpha)\%$ confidence interval as the interval bounded by the limits

$$L = \bar{X} - t_{\alpha/2, n-1} \frac{s}{\sqrt{n}} \quad \text{and} \quad R = \bar{X} + t_{\alpha/2, n-1} \frac{s}{\sqrt{n}} \qquad \textbf{(8.18)}$$

The point is that, whereas the Confidence Limits (8.13) involve the population standard deviation σ, assumed known in Section 8.2 but not here, the Confidence Limits (8.18) involve instead the sample standard deviation s, an estimate of σ. Thus these limits can be computed from the sample alone when:

1. The population standard deviation is not known, and
2. The population from which the sample was taken is normally distributed.

By Equation (8.17), the probability that the confidence interval includes μ—the probability that the confidence limits L and R surround μ—is just $1 - \alpha$. The interpretation of these limits is just as for those in Section 8.2.

EXAMPLE 1

Problem: A random sample of twenty homeowners in a particular city showed that the mean mortgage payment being made by the people in the sample was $200 per month. The sample standard deviation was $60 per month. Find an interval within which you can be 98% confident that the mean monthly mortgage payment of all the people in the population will lie.

Solution: From the problem statement we have:

$$\bar{X} = \$200/\text{mo.} \qquad s = \$60/\text{mo.} \qquad n = 20$$

The 98% confidence limits can be found if we note two things. First, we must assume the population of payments from which we are sampling is normally distributed. Second, since we don't know the population standard deviation, we must use the confidence limits involving t and Expression (8.18), which incorporates the sample standard deviation, s. Thus, the limits are:

$$L = \bar{X} - t_{.01,19} \frac{s}{\sqrt{n}} = 200 - 2.539 \times \frac{60}{\sqrt{20}} = 200 - 34 = 166$$

and

$$R = \bar{X} + t_{.01,19} \frac{s}{\sqrt{n}} = 200 + 2.539 \times \frac{60}{\sqrt{20}} = 200 + 34 = 234$$

Thus we can be 98% confident that the mean monthly mortgage payment of the people in the population from which we sampled is between $166 and $234.

EXAMPLE 2

Let us return to Example 1 of Section 8.2 where ten patients took a soporific drug, but this time without the unrealistic assumption that we know the population variance. The sample variance s^2 for the data in Table 8.1 is 1.513, so the sample standard deviation s is $\sqrt{1.513}$, or 1.23, and $s/\sqrt{n}$ is $s/\sqrt{10}$, or .389. The degrees of freedom $n - 1$ is 9. If we are to compute 95% confidence limits ($\alpha = .05$, $\alpha/2 = .025$), the appropriate cutoff point is $t_{.025,9}$, or 2.262. Since $\bar{X}$ is equal to 1.58, the 95% confidence limits are

$$L = \bar{X} - 2.262\,\frac{s}{\sqrt{10}} = 1.58 - 2.262 \times .389 = 1.58 - .88 = .70$$

and

$$R = \bar{X} + 2.262\,\frac{s}{\sqrt{10}} = 1.58 + 2.262 \times .389 = 1.58 + .88 = 2.46$$

We are 95% confident that μ lies between .70 and 2.46, in the sense that the random variables L and R (which for our data take values .70 and 2.46) have probability .95 of enclosing μ between them, whatever μ may be.[*]

EXAMPLE 3

A small electronics firm claims that it has developed a new, cheap process for manufacturing color television picture tubes. Assume they have approached you about investing in their company so that they can begin mass production of the picture tubes. One of the things you might want to know is how durable the tubes manufactured by the cheaper process may be. To answer this question the company tested 16 picture tubes and found the following:

Mean length of life for 16 tubes: 3220 hours

Standard deviation of life length in sample: 120 hours

Construct a 90% confidence interval for the mean length of life for the new picture tubes. Since $n = 16$, there are 15 degrees of freedom, and the t value associated with 5% in each tail of the t distribution is 1.753. Thus

$$L = \bar{X} - 1.753\,\frac{120}{\sqrt{16}} = 3220 - 52.6 = 3167.4$$

$$R = \bar{X} + 1.753\,\frac{120}{\sqrt{16}} = 3220 + 52.6 = 3272.6$$

Thus we can be 90% confident that the mean length of life of all picture tubes manufactured under the cheaper process is between 3167.4 and 3272.6 hours. This interval is relatively small; it is only 2×52.6, or 105.2 hours wide.

---

[*]The t distribution was discovered in 1908 by W. S. Gosset, who wrote under the name "Student." The data in the example are his; see Reference 8.1.

8.5   ESTIMATING BINOMIAL p

In Section 5.9, we defined a binomial population as a population whose elements are classified as belonging to one of two classes conventionally labeled success and failure. We represent the proportion of the population that belongs to the first class by p and the proportion that belongs to the second class by $1 - p$. In general, p is unknown and the problem is to estimate it. That is, someone may ship us a large order of goods, and we wish to estimate p, the proportion of the order that meets our standards.

Suppose we have at hand an independent sample of size n; let r be the number of successes in the sample (those that meet standards), and let f (equal to r/n) be the sample's proportion of successes. As stated in Section 5.9, r has mean np and variance $np(1 - p)$, and f has mean, variance, and standard deviation

$$E(f) = p$$

$$\text{Variance}(f) = p(1 - p)/n$$

$$\text{Standard deviation}(f) = \sqrt{p(1 - p)/n}$$

The first equation shows that f is an unbiased estimator of p. The second gives the variance of this estimator. Its value, of course, cannot be found without knowledge of the value of the parameter p to be estimated. But we can get an idea of the variance by replacing p by its estimator f. When we do this, the variance is approximated by $f(1 - f)/n$.

We now turn our attention to the problem of finding confidence interval estimates for the proportion p. The calculation of "exact" confidence intervals based on the binomial probabilities is complex; therefore, tables and graphs of these intervals have been constructed for various confidence probabilities. Table VII of the Appendix gives the 95 percent confidence intervals for p for a number of sample sizes from 10 to 1000. If our sample size is one of those tabulated and is less than 100, we find the number observed in the leftmost column and then move horizontally into the column corresponding to our sample size. The two numbers we find there are the endpoints of the interval and may be expressed as either percentages or relative frequencies. For example, suppose we take a sample of 30 students and find that 8 of them do not smoke. In the column for $n = 30$ we find the figures 12 and 46 opposite the observed number 8. We may then state that, on the basis of this sample, we are 95 percent confident that the percentage of nonsmoking students is between 12 and 46 percent. Alternatively, the confidence interval for the relative frequency of nonsmoking students is .12 to .46.

For sample sizes greater than or equal to 100, we use the column for observed relative frequency rather than that for the observed number of successes. If n is 250 and r is 75, then the observed fraction is 75/250, or .3. We locate .3 in the *fraction observed* column, and in the column for values of n equal to 250 we find the corresponding pair of values to be 24 and 36. This means our 95 percent confidence interval for p is from .24 to .36.

If the observed fraction f is greater than .50 we use $1 - f$ and subtract each limit from 100. For example, out of 1000 heads of families questioned it was found that 930 owned one or more television sets. The observed fraction, .93, is greater than .50. We subtract .93 from 1.00 to get .07. The corresponding figures in the column for $n = 1000$ are 6 and 9. We subtract these from 100 to get 94 and 91, respectively. The 95 percent confidence interval for the percentage of families that own TV sets is 91 to 94 percent.

Interpolation may be used in this table and is straightforward for the three largest sample sizes. For example, if n is 500 and r is 150, then the observed fraction f is 150/500, or .30. For $n = 250$ the confidence limits would be .24 and .36. For $n = 1000$ the limits would be .27 and .33. The proportion we want is

$$\frac{500 - 250}{1000 - 250} = \frac{250}{750} = \frac{1}{3}$$

Therefore, the lower limit is

$$.24 + \frac{1}{3}(.27 - .24) = .25$$

and the upper limit is

$$.36 + \frac{1}{3}(.33 - .36) = .35$$

For values of n from 10 to 50, inclusive, the interpolation is not quite as simple, for we must calculate for each sample size the observed number that corresponds to a given observed fraction. Let n be 25 and r be 15. Then f is 15/25, or .60. For $n = 20$ the observed number corresponding to $f = .60$ is 20(.60), or 12, and the confidence limits are .36 and .81. For $n = 30$ the value of X used to enter the table is 30(.60), or 18, and the confidence limits are .40 and .77. Since our sample size, 25, is midway between $n = 20$ and $n = 30$, our confidence limits are

$$.36 + \frac{1}{2}(.40 - .36) = .38$$

and

$$.81 + \frac{1}{2}(.77 - .81) = .79$$

The confidence limits we obtain from Table VII are based on exact binomial probabilities. However, the table is of limited use since it can only be used to find 95 percent confidence intervals. Another approach must be used if we have no tables or if we want something other than a 95 percent confidence interval. The normal approximation to the binomial provides a method of finding approximate confidence limits for large sample sizes. A knowledge of the method is of practical importance and also deepens our understanding of binomial estimation.

As stated in Section 6.3, the standardized fraction of successes

$$Z = \frac{f - p}{\sqrt{p(1 - p)/n}} \tag{8.19}$$

has for large n approximately a standard normal distribution; see Equation (6.2), Section 6.3, and the example following it. If in Expression (8.19) we replace each p in the denominator by its estimator f, we get a ratio

$$Z \approx \frac{f - p}{\sqrt{f(1 - f)/n}} \tag{8.20}$$

which turns out also to have approximately a normal distribution for large n. The rule of thumb is that $nf(1 - f) \geq 5$ if we desire a good approximation.

By the definition of the cutoff points $Z_{\alpha/2}$ for the standard normal distribution, we therefore have the approximate equation

$$P\left(-Z_{\alpha/2} \leq \frac{f - p}{\sqrt{f(1 - f)/n}} \leq Z_{\alpha/2}\right) \cong 1 - \alpha$$

If we operate on the inequalities here just as we did in constructing the confidence intervals in Sections 8.2 and 8.4, we arrive at

$$P\left(f - Z_{\alpha/2}\sqrt{\frac{f(1 - f)}{n}} \leq p \leq f + Z_{\alpha/2}\sqrt{\frac{f(1 - f)}{n}}\right) \cong 1 - \alpha$$

Therefore, the random variables

$$L = f - Z_{\alpha/2}\sqrt{\frac{f(1 - f)}{n}} \quad \text{and} \quad R = f + Z_{\alpha/2}\sqrt{\frac{f(1 - f)}{n}} \tag{8.21}$$

have approximate probability $1 - \alpha$ of containing between them the true value of p, whatever it may be; they therefore can be used as approximate $100(1 - \alpha)\%$ confidence limits if n is large. As another rough rule of thumb most statisticians feel that n should be at least 30 before this approximation method is valid.

EXAMPLE 1

Problem: A sample of 100 voters in a town contained 64 persons who favored a bond issue. Between what limits can we be 95% confident that the proportion of voters in the community who favor the issue is contained?

Solution: Here,

$$n = 100 \qquad r = 64 \qquad f = .64 \qquad Z_{.025} = 1.96$$

Hence

$$\sqrt{\frac{f(1-f)}{n}} = \sqrt{\frac{.64 \times .36}{100}} = .048$$

is our best approximation for the standard deviation of f and

$$L = .64 - 1.96 \times .048 = .64 - .094 = .546$$

$$R = .64 + 1.96 \times .048 = .64 + .094 = .734$$

We are 95% confident the true proportion is covered by the interval from .55 to .73. The 95% confidence limits from Table VII are .54 and .73; the normal approximation is quite accurate.

EXAMPLE 2

Problem: A company has received a shipment of several thousand parts. A random sample of 81 parts is selected, and 8 are found to be defective. Find the 90% confidence limits for the proportion of defects in the entire shipment.

Solution: Here,

$$n = 81 \qquad r = 8 \qquad f = .099 \qquad Z_{.05} = 1.645$$

Hence

$$\sqrt{\frac{f(1-f)}{n}} = \sqrt{\frac{.099 \times .901}{81}} = .033$$

approximates the standard deviation of f and

$$L = .099 - 1.645 \times .033 = .099 - .054 = .045$$

$$R = .099 + 1.645 \times .033 = .099 + .054 = .153$$

We are 90% confident that the proportion of defects in the shipment is between .045 and .153.

■ 8.6 SAMPLE SIZE

In all of the examples in the previous two sections, we assumed that a sample *had been taken*. That is, we looked at the data *after* the sample results had been obtained. There are many situations, however, when we know *before* we take the sample that we will be constructing a confidence interval with the results we obtain. In these situations, we often ask the question: how big a sample should we take? Three formulas will be given in this section for determining the sample size needed to obtain confidence intervals for the population mean, μ, and the population proportion of successes, p.

In order to determine the sample size needed to construct a confidence interval *before* the sample is taken, we need three pieces of information:

1. The confidence level (usually 90, 95, or 98 percent).
2. The desired maximum width of the confidence interval that is to be constructed. That is, the value $w = R - L$. The amount of error we can tolerate in our estimate of the population mean is thus $w/2$.
3. The standard deviation associated with the population from which we are about to sample. That is, we need to know σ if our interval is estimating μ, or σ_f if our interval is estimating p.

If we desire a confidence interval to estimate the population mean μ, then the interval is formed by finding a sample mean, $\overline{X}$, and adding and subtracting the following term.

$$Z_{\alpha/2} \frac{\sigma}{\sqrt{n}}$$

Since this amount is both added and subtracted, the width of the confidence interval will be twice this amount:

$$w = 2Z_{\alpha/2} \frac{\sigma}{\sqrt{n}} \tag{8.22}$$

If we know the three items of information listed previously we can use Formula (8.22) to solve for n, the sample size we need to obtain the confidence interval specified by the three features. That is, we must specify w, the width of the confidence interval we want to construct; Z, which indicates the confidence level; and σ, the population standard

---

■ The material in this section is not required in the sequel.

deviation. Once these items are specified, we can solve Formula (8.22) for n and obtain the following formula:

$$n = \frac{4Z_{\alpha/2}^{2}\sigma^{2}}{w^{2}} \tag{8.23}$$

EXAMPLE 1

Problem: Suppose that we know the standard deviation of daily output on a production line is $\sigma = 10$ tons. Suppose further that we want to estimate the mean daily output of the production line, μ, to within ± 2.5 tons with 95% confidence. What sample size is required?

Solution: In this problem, $w = 5$ tons (since "± 2.5 tons" implies a total width of $2 \times 2.5 = 5$). We also know from the problem statement that $\sigma = 10$ and $Z_{.025} = 1.96$, since the confidence level is 95%. Using Formula (8.23) yields a sample size:

$$n = \frac{4(1.96)^{2}(10)^{2}}{(5)^{2}} = 61$$

Thus a sample size of $n = 61$ will give us an interval that allows us to estimate the population mean daily output to within 2.5 tons with 95% confidence.

This example is somewhat artificial since there will be few actual applications where one wishes to estimate the population mean μ and can use the population standard deviation σ in Formula (8.23). That is, if we don't know μ, we probably don't know σ either. Thus, we are often forced to replace σ^{2} in Equation (8.23) with some reasonable estimate, s^{2}, which has been obtained from a previous sample of the population or from a preliminary sample. But we learned in Section 8.4 that confidence intervals constructed with sample standard deviations (since σ is not known) require the use of $t_{\alpha/2,n-1}$ instead of $Z_{\alpha/2}$.

These intervals were formed in Expression (8.18) by adding and subtracting the amount

$$t_{\alpha/2,n-1} \frac{s}{\sqrt{n}}$$

around the sample mean. Thus, this expression is half the width of the confidence interval and

$$w = 2t_{\alpha/2,n-1} \frac{s}{\sqrt{n}} \tag{8.24}$$

When Formula (8.24) is solved for n, we obtain Formula (8.25), which is exactly parallel to (8.23), the sample size formula used when the population standard deviation σ is known.

$$n = \frac{4\left(t_{\alpha/2, n-1}\right)^2 s^2}{w^2} \tag{8.25}$$

The use of this formula is demonstrated in the following example.

EXAMPLE 2

Problem: A company manager wishes to estimate the mean length of time, μ, it takes company crews to do a certain type of job. She wishes to estimate μ to within ± 5 minutes with 90% confidence. Since she has no idea about the value of σ, the population standard deviation, she took a preliminary sample of $n = 15$ jobs and found the 15 job completion times had a standard deviation of $s = 20$ minutes. How much larger should she make her sample to obtain the confidence interval she desires?

Solution: From the problem statement we know:

$$w = 10 \qquad s = 20 \qquad \alpha = .10$$

Since the population standard deviation is unknown, we must use Formula (8.25) to determine the sample size. We will use a t value of $t_{\alpha/2, n-1} = t_{.05, 14} = 1.761$. Then

$$n = \frac{4\,(1.761)^2 (20)^2}{(10)^2} = 50$$

Thus, the manager must take $50 - 15 = 35$ additional observations to complete her total sample.

Two important points should be observed about this example. First, it demonstrates a process known as *two-stage sampling*. The first stage of the sampling was the preliminary step involving fifteen observations. This step allowed the manager to obtain a sample standard deviation s to use in Formula (8.25). The second stage of the sampling involved thirty-five observations to fill out the total sample to the full fifty items.

The second point about the example concerns the use of the t value. We chose to use $t_{.05, 14}$. Technically, however, the final confidence interval following the second stage of sampling will be constructed with the *total* sample's standard deviation, based on fifty observations and a t value with $n - 1 = 49$ degrees of freedom. Since $t_{.05, 49}$ is about 1.678, the actual confidence interval obtained will likely be narrower than the ± 5 minutes specified in the problem. Thus, Formula (8.25) and the method demonstrated in Example 2 usually give a conservatively large sample size—one that is as large *or larger* than what will be needed to obtain the desired confidence interval.*

EXAMPLE 3

Problem: A claims manager for an insurance company would like to know the mean size of the automobile insurance repair claims paid by his company. Thus, he took a sample of $n = 20$ claims and found $\bar{X} = \$900$ and $s = \$300$. How much larger should his sample be if he wants to estimate the mean payment to within $\pm\$100$ with 95% confidence?

Solution: From the problem statement we have $w = 200$, $s = 300$, and $t_{.025, 19} = 2.093$ to be used in Formula (8.25) (since we only know the sample standard deviation):

$$n = \frac{4 (2.093)^2 (300)^2}{(200)^2} = 39$$

Thus, if he examines $39 - 20 = 19$ more payments, the claims manager will be able to obtain the interval he desires (or one slightly narrower).

The last three examples have concerned methods of determining the sample size needed to estimate μ, the population mean. Comparable methods can be used to estimate p, the proportion of successes in a binomial population. According to Formula (8.21), a confidence interval used to estimate p is formed by taking a sample, obtaining the sample proportion of successes f, then adding to f and subtracting from f the following term:

$$Z_{\alpha/2} \sqrt{\frac{f(1 - f)}{n}}$$

Thus, any confidence interval is twice this wide and

$$w = 2 Z_{\alpha/2} \sqrt{\frac{f(1 - f)}{n}} \tag{8.26}$$

We can solve Formula (8.26) for the value n. This gives us

$$n = \frac{4 Z^2_{\alpha/2} f (1 - f)}{w^2} \tag{8.27}$$

---

*There is an iterative method for determining the sample size more exactly than Formula (8.25) does. However, such a method is beyond the scope of this material. Our method does not *always* give an interval that is too large, however, since when the sample is filled out, the sample standard deviation, s, may become large and offset the increased sample size.

This formula can be used to find the sample size necessary to obtain a confidence interval as long as we can specify the level of confidence desired (and thus the $Z_{\alpha/2}$ value), the width of the desired interval (and thus w), and some estimate of the proportion of successes that will be obtained from the sample (and thus f).

The value of f inserted in Formula (8.27) can be obtained in several ways. First, it might be estimated using a preliminary sample in a two-stage sampling procedure like that described earlier. Also, a person experienced with the situation where the sampling is being done might be able to supply a good estimate of what f will be. Finally, where there is no suggestion as to the possible value for f and when two-stage sampling is not practical, we should let $f = .5$. This is because when f is assigned the value of .5, the product $f(1 - f)$ has a higher value than when any other value is used for f. Table 8.2 shows how the value of $f(1 - f)$ changes with various assignments for f. When $f = .5$, the product term reaches the maximum value of .25. Thus, when this product is inserted in Formula (8.27), the largest numerator of the fraction is obtained, and the largest possible value of n results. In other words, assigning $f = .5$ is the most conservative choice we can make since it will yield a sample size as large as, or possibly a little larger than, we need for the confidence interval we seek.

TABLE 8.2 The Product $f(1 - f)$ for Selected Values of f

f	$(1 - f)$	$f(1 - f)$
.1	.9	.09
.2	.8	.16
.3	.7	.21
.4	.6	.24
.5	.5	.25
.6	.4	.24
.7	.3	.21
.8	.2	.16
.9	.1	.09
1.0	.0	.00

EXAMPLE 4

A quality control engineer would like to estimate the proportion of defects being produced on a production line to within ±.05 with 95% confidence. In a preliminary sample of 25 items, the engineer found four defectives. How much larger a sample needs to be taken? From the problem statement, we know:

$$w = .10 \qquad Z_{\alpha/2} = 1.96 \qquad \text{and} \qquad \text{preliminary} f = \frac{4}{25} = .16$$

From Formula (8.27) we find

$$n = \frac{4(1.96)^2(.16)(.84)}{(.10)^2} = 207$$

Thus, the quality control engineer needs to examine $207 - 25 = 182$ more items.

We can use this example to point out the value of a preliminary sample. If the preliminary sample had not been taken, the engineer would have had to assume, conservatively, that the defect rate, p, was close to .5. In that case, the engineer would propose a sample of

$$n = \frac{4(1.96)^2(.5)(.5)}{(.10)^2} = 384$$

Thus, the preliminary sample resulted in a total sample that is $384 - 207 = 177$ smaller than the sample that would have been taken without it.

EXAMPLE 5

Problem: The credit manager of a department store would like to know what proportion of the charge customers take advantage of the store's "deferred payment plan" each year. The manager would like to estimate this figure to within 10% at a 90% confidence level, but has no idea right now about what this proportion might be.

Solution: From the problem statement we can find:

$$w = .20 \qquad \text{and} \qquad Z_{\alpha/2} = 1.645$$

Since there is no preliminary sample and the credit manager has no idea about the possible f value, we should assign $f = .5$. Then from Formula (8.27) we obtain

$$n = \frac{4(1.645)^2(.5)(.5)}{(.20)^2} = 67$$

This seems like a reasonable sample to take, but the credit manager could probably save a little work by taking a preliminary sample first to get an estimate of f. If this estimate is very much different from .5, it should be used in Formula (8.27) again to reestimate the sample size, which would necessarily be less than 67.

Formulas (8.23), (8.25), and (8.27) can be examined to yield a few generalizations about the size of samples needed to obtain confidence intervals. First, the more confidence we desire, all other things being constant, the larger will be the sample size (since the Z and t values in these formulas are larger). Second, the larger the variation in the population (as measured by σ or s or indirectly by f in these formulas), the larger the sample. And finally, the most influential element of these formulas is the confidence interval width, w. The narrower the confidence interval, the larger the sample size will be. Since w^2 appears in the denominator of all these formulas, a general rule is that cutting the width of the confidence interval in *half* increases the sample size by *four* times.

In many sampling situations such as auditing, the population from which the sample is taken is finite. That is, the population of invoices being audited may have some finite size we will call N. In that case the standard deviation of the sample result is smaller by a factor which is called the *finite population correction factor*:

$$\sqrt{\frac{N - n}{N - 1}}$$

When sampling is done from a finite population, then the confidence interval widths in Formulas (8.22), (8.24), and (8.26) are all reduced in size by multiplying this factor by the terms on the right sides of the equations. This factor then has an impact on the sample size Formulas (8.23), (8.25), and (8.27). For instance, when the finite population correction factor is included, Formula (8.27) becomes

$$n = \frac{4Z^2_{\alpha/2}f(1 - f)N}{(N - 1)w^2 + 4Z^2_{\alpha/2}f(1 - f)} \qquad \textbf{(8.27 revised)}$$

This revised equation seems quite complicated, and we will not use it in the problems at the end of this chapter. This is because the impact of the finite population correction factor is quite minimal in terms of reducing the required sample size unless the sample becomes more than about 10 percent of the population. Since there are few realistic situations in which the sample size turns out to be more than 10 percent of the population, we will ignore the revised sample size equations and stay with the simpler versions of (8.23), (8.25), and (8.27). However, auditors sometimes use standard tables for determining sample sizes, and the standard tables were developed using the more complicated formulas.

■ 8.7   CONFIDENCE INTERVALS FOR $\mu_1 - \mu_2$

We have looked at some of the inferences that are possible when we have a single random sample from a normal population, and, also, when our random sample is assumed to have been taken from a binomial population. We now consider some of the inferences that may be made when we have two independent random samples, one from each of two populations. We will examine some of the methods used when the data are continuous, and then turn our attention to inferences based on samples from two binomial populations.

When we have two samples, an observation X_{ij} has two subscripts; the first denotes the sample, and the second designates the particular element within the sample. Thus X_{24} is the fourth value in the second sample. Let $X_{11}, X_{12}, \ldots, X_{1n_1}$ represent the observed values in a sample of size n_1 taken from a normal population with mean μ_1 and variance σ^2, and let $X_{21}, X_{22}, \ldots, X_{2n_2}$ be the observations in a sample of size n_2 taken from a second normal population with mean μ_2 and variance σ^2 (the same σ^2 as for the first population). Such a two-sample situation is characterized by three basic assumptions:

1.   Independent random samples.
2.   Normal populations.
3.   Common variance for both populations.

The methods to be presented in this section are valid only if these assumptions are satisfied.

Data that satisfy our three assumptions and that fit exactly into the framework of the two-sample situation may be obtained as the result of an experiment for comparing the effects of two treatments. In statistical usage, *treatments* are any procedures, methods, or stimuli whose effects we want to estimate and compare. Treatments in one situation might represent different machines, but in another they could be different operators; they could be different chemicals, or different rates of application of one chemical; they could be different advertisements; and so on.

After the data have been obtained, either by selecting the samples or by performing an experiment, the information they contain may be summarized in tabular form, as shown in Table 8.3. The pooled sum of squares, pooled SS, can be obtained by adding together the sums of squares for the individual samples in the last column of the table. In practical problems these sums of squares would ordinarily be obtained by means of the machine formulas:

■ The material in this section is not required in the sequel.

TABLE 8.3   Summary Table for Two Samples from Continuous Populations

SAMPLE	SIZE	DEGREES OF FREEDOM	MEAN	SUM OF SQUARES
1	n_1	$n_1 - 1$	$\overline{X}_1$	$\Sigma(X_{1j} - \overline{X}_1)^2$
2	n_2	$n_2 - 1$	$\overline{X}_2$	$\Sigma(X_{2j} - \overline{X}_2)^2$
Totals	$n_1 + n_2$	$n_1 + n_2 - 2$		Pooled SS

$$\Sigma(X_{1j} - \overline{X}_1)^2 = \Sigma X_{1j}^2 - \frac{(\Sigma X_{1j})^2}{n_1}$$

$$\Sigma(X_{2j} - \overline{X}_2)^2 = \Sigma X_{2j}^2 - \frac{(\Sigma X_{2j})^2}{n_2}$$

After the information contained in the data has been summarized in the table, we turn our attention to the estimation of the parameters of interest. It was explained in Section 7.6 that the best unbiased estimators for the population means μ_1 and μ_2 are the respective sample means $\overline{X}_1$ and $\overline{X}_2$. The best unbiased estimator for the difference between the means, $\delta = \mu_1 - \mu_2$, is the difference between the sample means, $d = \overline{X}_1 - \overline{X}_2$. The latter statistic is of particular interest because two populations are involved, and our primary concern is making inferences about the difference between the means. In experimental situations, the difference between the means is also the difference between the effects of the two treatments.

Since the variance is assumed to be the same for both populations, we could obtain an unbiased estimate for σ^2 from each sample. But the best unbiased estimate is obtained by pooling the information contained in both samples to get what we call the *pooled estimate of variance*. This pooled estimate of variance is an estimate of the common population variance σ^2 and can be found in two equivalent ways. The first (less obvious) method involves pooling (or adding) the sum of squares from each of the samples and then dividing by the pooled degrees of freedom $(n_1 + n_2 - 2)$:

$$s^2 = \frac{\text{pooled SS}}{\text{pooled df}} = \frac{\Sigma(X_{1j} - \overline{X}_1)^2 + \Sigma(X_{2j} - \overline{X}_2)^2}{n_1 + n_2 - 2} \tag{8.28}$$

It is easy to show that this pooled estimate of variance is equivalent to that found as the weighted mean of the individual sample estimates, where the weights are the degrees of freedom. That is, the pooled estimate of variance using the second method is:

$$s^2 = \frac{(n_1 - 1)s_1^2 + (n_2 - 1)s_2^2}{n_1 + n_2 - 2} \tag{8.29}$$

where
$$s_1^2 = \frac{\Sigma(X_{1j} - \bar{X}_1)^2}{n_1 - 1}$$
$$s_2^2 = \frac{\Sigma(X_{2j} - \bar{X}_2)^2}{n_2 - 1}$$

In order to form a confidence interval that estimates $\mu_1 - \mu_2$, we find $d = \bar{X}_1 - \bar{X}_2$ and then add and subtract either a t or Z value times σ_d to obtain the right and left confidence limits. The procedure is exactly like those described in Formulas (8.13), (8.18), and (8.21) which were used to form confidence intervals for μ and p. Thus, the confidence interval for d, the difference in the means of two normally distributed populations is

$$d \pm Z_{\alpha/2}\sigma_d \qquad \text{or} \qquad d \pm t_{\alpha/2, n_1 + n_2 - 2}s_d \tag{8.30}$$

The expression in 8.30 needs a little explanation. First, we found in Section 7.6 that σ_d has the value given in Equation (7.12), which is repeated here.

$$\sigma_d = \sqrt{\frac{\sigma_1^2}{n_1} + \frac{\sigma_2^2}{n_2}} \tag{7.12}$$

However, in this section we must assume that the two populations have the same variance: $\sigma^2 = \sigma_1^2 = \sigma_2^2$. When we know this value of the common variance; we can obtain σ_d using Formula (7.12) and find the confidence interval estimating $\mu_1 - \mu_2$ using $Z_{\alpha/2}$ and the first form of Expression (8.30).

In those cases where σ^2 is not known, the two sample variances can be pooled using (8.28) or (8.29). Then s^2 can be factored out of the numerator in Expression (7.12) and the value of σ_d can be estimated as

$$s_d = \sqrt{s^2\left(\frac{1}{n_1} + \frac{1}{n_2}\right)} \tag{8.31}$$

In these cases the second form of (8.30) must be used. Note that the t value involved has $n_1 + n_2 - 2$ degrees of freedom. As always, the degrees of freedom for t are those associated with the sum of squares used in finding our unbiased estimate of σ^2. The most common form of the confidence limits estimating $\mu_1 - \mu_2$ is expressed as

$$L = d - t_{\alpha/2, n_1 + n_2 - 2}s_d$$
$$R = d + t_{\alpha/2, n_1 + n_2 - 2}s_d \tag{8.32}$$

EXAMPLE 1

Problem: To compare the durabilities of two paints for highway use, 12 four-inch-wide lines of each paint were laid down across a heavily traveled road. The order was decided at random. After a period of time, reflectometer readings were obtained for each line. The higher the readings, the greater is the reflectivity and the better is the durability of the paint. The data are as follows:

PAINT A	12.5	11.7	9.9	9.6	10.3	9.6
	9.4	11.3	8.7	11.5	10.6	9.7
PAINT B	9.4	11.6	9.7	10.4	6.9	7.3
	8.4	7.2	7.0	8.2	12.7	9.2

Solution: For these sets of data we find

PAINT A	PAINT B
$\Sigma X_{1j} = 124.8$	$\Sigma X_{2j} = 108.0$
$\overline{X}_1 = 10.4$	$\overline{X}_2 = 9.0$
$\Sigma X_{1j}^2 = 1312.00$	$\Sigma X_{2j}^2 = 1010.64$
$(1/n_1)(\Sigma X_{1j})^2 = 1297.92$	$(1/n_2)(\Sigma X_{2j})^2 = 972.00$
$\Sigma(X_{1j} - \overline{X}_1)^2 = 14.08$	$\Sigma(X_{2j} - \overline{X}_2)^2 = 38.64$

The summary table is:

PAINT	n	df	$\overline{X}$	SUM OF SQUARES
A	12	11	10.4	14.08
B	12	11	9.0	38.64
	24	22		52.72

Hence

$$d = \overline{X}_1 - \overline{X}_2 = 1.40$$

The pooled estimate of variance is

$$s^2 = \frac{52.72}{22} = 2.40$$

the estimated variance of the difference between the means is

$$s_d^2 = s^2\left(\frac{1}{n} + \frac{1}{n}\right) = 2.40\left(\frac{2}{12}\right) = 0.40$$

and the estimated standard deviation of d is

$$s_d = \sqrt{0.40} = 0.63$$

If we desire a 95% confidence interval, we must obtain $t_{.025,22} = 2.074$ from

Table VI of the Appendix. The 95% confidence interval can now be formed using Formula (8.32).

$$L = 1.40 - (2.074)(.63) = .09$$

$$R = 1.40 + (2.074)(.63) = 2.71$$

We can be 95 percent confident that the difference between the actual means is between .09 and 2.71.

Note that this confidence interval is for $\mu_1 - \mu_2$. If we wanted the interval for $\mu_2 - \mu_1$, the estimator would be $\overline{X}_2 - \overline{X}_1$, or -1.40, and the confidence limits would be -2.71 and $-.09$ (the variance of $\overline{X}_2 - \overline{X}_1$ is the same as the variance of $\overline{X}_1 - \overline{X}_2$). Thus, there is no difference in these two confidence intervals except for the algebraic signs on the confidence limits.

EXAMPLE 2

Problem: Random samples of weights were taken from two package filling processes. The first sample consisted of $n_1 = 160$ bags, and the second consisted of $n_2 = 200$ bags. The two sample means and standard deviations are as follows.

$$\overline{X}_1 = 10.2 \text{ kilograms} \qquad \overline{X}_2 = 9.9 \text{ kilograms}$$

$$d = (10.2 - 9.9) = .3 \text{ kilograms}$$

$$s_1 = .3 \text{ kilograms} \qquad s_2 = .4 \text{ kilograms}$$

Find an interval within which we can be 90% sure that $\mu_1 - \mu_2$ lies.

Solution: Since the samples are large, even if the populations of weights in the two processes are not normally distributed, the difference between the means will be normally distributed according to Section 7.6. The first step in solving this problem is to find s^2, the estimate of the common variance of the two processes. Since we know the sample standard deviations, we can use Formula (8.29) to obtain:

$$s^2 = \frac{(160 - 1)(.3)^2 + (200 - 1)(.4)^2}{160 + 200 - 2} = .13$$

Now we can find s_d using Formula (8.31).

$$s_d = \sqrt{.13 \left(\frac{1}{160} + \frac{1}{200} \right)} = .038$$

Finally, we use Expression (8.30) to form the confidence interval. Technically, since we do not know the populations' standard deviation, we must use the second form of expression 8.30. However, $t_{.05, 358}$ is the t value with 358 degrees of freedom. We noted earlier that for many degrees of freedom the t distribution approaches the normal distribution. Thus we seek the t value from the .05 column of Table VI and the last row, where the normal distribution values are found. So $t_{.05,358} \approx 1.645$, and the confidence interval is given by:

$$R = .3 + (1.645)(.038) = .363$$

$$L = .3 - (1.645)(.038) = .237$$

Thus we can be 90% sure that the first process fills packages to a mean weight that is between .237 and .363 kilograms heavier than the mean of the second process. Equivalently, we could say we are 90% sure that the difference in the population mean weights is between 237 and 363 grams.

8.8   TWO SAMPLES FROM BINOMIAL POPULATIONS

We now suppose that we have two independent random samples, one of size n_1 from a binomial distribution with a proportion of successes equal to p_1, the other of size of n_2 from a binomial population with proportion of successes equal to p_2. We let r_1 and r_2 be the observed number of successes in the first and second samples, respectively. Our primary objectives will be to estimate the difference $p_1 - p_2$. To achieve this objective we make use of the normal approximation to the binomial.

The proportion p_i is estimated unbiasedly by the corresponding sample fraction f_i (that is, r_i/n_i), and the difference $p_1 - p_2$ is estimated unbiasedly by $f_1 - f_2$. The variance of f_i is estimated by

$$s_{f_i}^2 = \frac{f_i(1 - f_i)}{n_i}$$

which is the sample estimate of the formula in Example 4 of Section 6.3. Since f_1 and f_2 are independent, the variance of their difference, $f_1 - f_2$, is the sum of their variances and is estimated by

$$s_{f_1 - f_2}^2 = \frac{f_1(1 - f_1)}{n_1} + \frac{f_2(1 - f_2)}{n_2}$$

For large n_1 and n_2, the quantity $f_1 - f_2$ is approximately normally distributed, we learned in Section 7.7. So the figure

$$\frac{(f_1 - f_2) - (p_1 - p_2)}{\sqrt{\dfrac{p_1(1 - p_1)}{n_1} + \dfrac{p_2(1 - p_2)}{n_2}}} \tag{7.15}$$

has approximately a standard normal distribution. If we replace the p_i in the denominator by their estimates f_i, we obtain

$$\frac{(f_1 - f_2) - (p_1 - p_2)}{\sqrt{\dfrac{f_1(1 - f_1)}{n_1} + \dfrac{f_2(1 - f_2)}{n_2}}}$$

and this too has approximately a standard normal distribution. This leads to the approximate confidence limits

$$L = (f_1 - f_2) - Z_{\alpha/2} \sqrt{\frac{f_1(1 - f_1)}{n_1} + \frac{f_2(1 - f_2)}{n_2}}$$

and   (8.33)

$$R = (f_1 - f_2) + Z_{\alpha/2} \sqrt{\frac{f_1(1 - f_1)}{n_1} + \frac{f_2(1 - f_2)}{n_2}}$$

EXAMPLE 1

Problem: Two different methods of manufacture, casting and die forging, were used to make parts for an appliance. In service tests of 100 of each type, 10 castings failed but only 3 forged parts were found to be defective. Find 95% confidence limits for the difference between the proportions of the cast and forged parts that would fail under similar conditions.

Solution: Here

$$f_1 = \frac{10}{100} = .10 \qquad f_2 = \frac{3}{100} = .03$$

$$s_{f_1}^2 = \frac{.10(.90)}{100} \qquad s_{f_2}^2 = \frac{.03(.97)}{100}$$

$$s_{f_1 - f_2}^2 = \frac{1}{100}[.10(.90) + .03(.97)] = .001191$$

$$s_{f_1 - f_2} = \sqrt{.001191} = .0345$$

From the last row of Table VI $Z_{.025}$ is found to be 1.96, and approximate confidence limits for $p_1 - p_2$ are given by

$$L = (f_1 - f_2) - Z_{.025} s_{f_1 - f_2} = .07 - 1.96(.0345) = .002$$

and

$$R = (f_1 - f_2) + Z_{.025} s_{f_1 - f_2} = .07 + 1.96(.0345) = .138$$

We are approximately 95% confident that the difference between the proportions of failures is between .00 and .14. Thus, the two different methods might differ in failure rates by as little as zero ($p_1 - p_2 = .00$) or the first method might produce failures at a rate as high as 14% greater than the second method ($p_1 - p_2 = .14$).

8.9 SUMMARY

The topic of this chapter has been statistical estimation. The procedures described in the previous sections are used in the situation where we are interested in taking a sample in order to estimate some characteristic of the population. When we deal with a population of continuous measurements, that population is characterized by a mean μ and a standard deviation σ. These characteristics are estimated by the sample mean $\overline{X}$ and the sample standard deviation s. When we deal with a population composed only of successes and failures, a binomial population, then the population is characterized by the proportion of successes p in the population. The sample proportion of successes f is the best single estimator of p.

However, a single point estimate of a population characteristic is almost surely off the mark to some extent. Thus, in estimating population characteristics we often construct confidence intervals that show a range of values within which the population characteristic might lie. All our $100(1 - \alpha)\%$ confidence intervals have the same general form:

$$L = \left(\begin{matrix} \text{Sample} \\ \text{Estimate} \end{matrix}\right) - (Z_{\alpha/2} \text{ or } t_{\alpha/2, \nu}) \, \sigma_{\text{Sample Estimate}}$$

$$R = \left(\begin{matrix} \text{Sample} \\ \text{Estimate} \end{matrix}\right) + (Z_{\alpha/2} \text{ or } t_{\alpha/2, \nu}) \, \sigma_{\text{Sample Estimate}}$$

That is, the left confidence limit is found by calculating the sample estimator and then subtracting a Z or a t value times the standard error of the sample estimator. The right confidence limit is found by adding the same quantity from the sample estimator. The Z value is used when we have large samples or when we have small samples from populations of known variance (which is seldom the case in actual practice). The t value is used when we have small samples from normally distributed populations whose variance is unknown. When the population characteristic we are estimating is p or $p_1 - p_2$, then the populations we deal with are binomial (very much not normal), and the t value cannot be used. The number of degrees of freedom associated with the t value is ν, and its value is determined by the number of degrees of freedom in the standard error of the sample estimator.

Table 8.4 summarizes the four types of confidence intervals dis-

TABLE 8.4 Summary of Confidence Interval Formulas

POPULATION CHARACTERISTIC BEING ESTIMATED	FORMULA FOR THE $100(1-\alpha)\%$ CONFIDENCE INTERVAL	COMMENTS
1. μ, the Mean of a Population of Continuous Measurements	$\bar{X} \pm (Z_{\alpha/2} \text{ or } t_{\alpha/2,\nu}) \dfrac{\sigma}{\sqrt{n}}$	$\nu = n - 1$ when t is used. If σ is unknown, use s and the appropriate t value.
2. p, the proportion of Successes in a Binomial Population	$f \pm (Z_{\alpha/2}) \sqrt{\dfrac{f(1-f)}{n}}$	The formula applies when $nf(1-f) \geq 5$. For small n and selected large n use Table VII.
3. $\mu_1 - \mu_2$, the Difference in the Means of Two Populations of Continuous Measurements	$(\bar{X}_1 - \bar{X}_2) \pm (Z_{\alpha/2} \text{ or } t_{\alpha/2,\nu}) \sqrt{\dfrac{\sigma^2}{n_1} + \dfrac{\sigma^2}{n_2}}$	$\nu = n_1 + n_2 - 2$ when t is used. The two populations must have the same variance. If σ^2 is unknown, pool the sample variances and use the appropriate t value.
4. $p_1 - p_2$, the Difference in the Proportion of Successes in Two Binomial Populations	$(f_1 - f_2) \pm (Z_{\alpha/2}) \sqrt{\dfrac{f_1(1-f_1)}{n_1} + \dfrac{f_2(1-f_2)}{n_2}}$	The formula applies for large samples. Rule: $n_1 f_1 (1 - f_1) \geq 5$ and $n_2 f_2 (1 - f_2) \geq 5$. When sample sizes are small, no confidence interval of this type can be constructed.

cussed in this chapter. Note that the formulas given in the middle column of that table all have the general form for confidence intervals indicated previously. When the "plus" is used in the plus-or-minus operation, the right confidence limit R is obtained. L is obtained when the "minus" is used.

In addition, this chapter dealt with the question of how large a sample size is needed to construct confidence intervals for a population mean μ and for a population proportion of successes p. In order to determine the sample size one must specify three pieces of information: the confidence level desired, the maximum width of the confidence interval to be constructed, and an estimate of the population dispersion. The third item is often estimated using a preliminary sample, which is the first stage in two-stage sampling. Formulas (8.23), (8.25), and (8.27) can be used to determine sample sizes once the appropriate parameters have been identified.

PROBLEMS

1. For the following results from samples from normal populations, what are the best estimates for the mean, the variance, the standard deviation, and the standard deviation of the mean?

   a. $n = 9$, $\Sigma X_i = 36$, $\Sigma (X_i - \bar{X})^2 = 288$
   b. $n = 16$, $\Sigma X_i = 64$, $\Sigma (X_i - \bar{X})^2 = 180$
   c. $n = 9$, $\Sigma X_i = 450$, $\Sigma (X_i - \bar{X})^2 = 32$
   d. $n = 25$, $\Sigma X_i = 500$, $\Sigma X_i^2 = 12{,}400$
   e. $n = 16$, $\Sigma X_i = 320$, $\Sigma X_i^2 = 6640$

2. For each of the following samples from normal populations find the best estimates for μ, σ^2, σ, and standard deviation $(\bar{X})$.

   a. 6, 15, 3, 12, 6, 21, 15, 18, 12
   b. 4, 10, 2, 8, 4, 14, 10, 12, 8
   c. 2, 5, 9, 11, 13
   d. 1, 3, 3, 5, 6, 6
   e. −4, 2, −6, 0, −4, 6, 2, 4, 0
   f. 6676, 6678, 6681, 6680, 6681, 6678

3. Given the following results from random samples from binomial populations, find the best estimates for p, the variance and standard deviation of r, and the variance and standard deviation of the observed fraction f.

   a. $n = 25$, $r = 5$ b. $n = 64$, $r = 32$
   c. $n = 400$, $r = 144$ d. $n = 100$, $r = 64$

4. In a sample of 100 small castings, 98 were not defective. What are the best estimates for (a) the proportion of nondefectives in the lot from which the sample was taken, and (b) the standard deviation of the estimator?

5. Assuming random samples from populations with known variance, find

confidence intervals for the means that have the specified degree of confidence.

a. $n = 9,$ $\overline{X} = 20,$ $\sigma^2 = 9,$ confidence $= 90\%$.
b. $n = 16,$ $\overline{X} = 52,$ $\sigma^2 = 64,$ confidence $= 98\%$.
c. $n = 25,$ $\overline{X} = 120,$ $\sigma^2 = 400,$ confidence $= 95\%$.

6. Assuming samples from normal populations with known variance find:
    a. The degree of confidence used if $n = 16$, $\sigma = 8$, and the total width of a confidence interval for the mean is 3.29 units.
    b. The sample size when $\sigma^2 = 100$ and the 95% confidence interval for the mean is from 17.2 to 22.8.
    c. The known variance when $n = 100$ and the 98% confidence interval for the mean is 23.26 units in width.

7. Find the value of t such that:
    a. The probability of a larger value is .005 when ν (degrees of freedom) is equal to 28.
    b. The probability of a smaller value is .975 when $\nu = 24$.
    c. The probability of a larger value, sign ignored, is .10 when $\nu = 20$.

8. Assuming unknown variances, find 95% confidence intervals for the means based on the sample results given in Problem 1.

9. Assuming unknown variances, and using the data given in Problem 2, find:
    a. 95% confidence intervals for the means for Parts a and b.
    b. 99% confidence intervals for the means for Parts c, d, and e.
    c. 98% confidence intervals for Part f.

10. Sensitivity tests were made on 25 randomly selected tubes of a given type. The mean was 3.2 microvolts, and the estimated variance was .20. Find a 95% confidence interval for the mean sensitivity of this type of tube.

11. A soft drink bottler took a sample of nine families and found a mean consumption of his drink of 100 ounces per week with a standard deviation of 18 ounces. Find a 98% confidence interval for the mean weekly consumption of this drink in the entire population.

12. A company has just installed a new automatic milling machine. The time it takes the machine to mill a particular part is recorded for a sample of nine observations. The mean time is found to be $\overline{X} = 8.50$ seconds, and $s^2 = .0064$. Find a 90% confidence interval for the true unknown mean time for milling this part.

13. A state highway department inspector is interested in knowing the mean weight of commercial vehicles traveling on the roads in his state. He takes a sample of 100 randomly selected trucks passing through state weigh stations and finds the mean gross weight $\overline{X} = 15.8$ tons and a sample standard deviation $s = 4.2$ tons. Construct a 95% confidence interval for the mean gross weight of commercial vehicles traveling the highways of this state.

14. Twenty secretaries were given a spelling test. They were then given a special short course designed to improve spelling ability and were tested again at the end of the course. The differences between the first and

the second scores had a mean equal to 4 (that is, there was a 4-point improvement). The variance of the improvement was 16. Find a 90% confidence interval for the mean of the score improvement if this course were given to all the secretaries in the company.

15. A certain stimulus was tested for its effect on blood pressure. Twenty men had their blood pressure measured before and after the stimulus. The results are given as follows. Find a 95% confidence interval for the mean change in blood pressure.

$$8, 7, 1, 9, -8, -3, 1, 2, -8, 2, 3, 8, 1, 7, -5, -4, 0, 7, -1, 5$$

16. Cattle are sold by weight. A sample of six animals from one herd yielded the following weights (in pounds):

$$692, 800, 685, 790, 695, 793$$

Find a 95% confidence interval for the mean weight of all animals in the herd.

17. Use the data of Problem 9, Chapter 3, to find a 95% confidence interval for the mean weight of feedlot cattle.

18. Use the data of Problem 10, Chapter 3, to find a 99% confidence interval for the mean billing error value.

19. Use the data of Problem 15, Chapter 3, to find a 90% confidence interval for the mean difference in DC9 engine thrust.

20. Find a 95% confidence interval for the mean based on the following distribution. Here, $n = 60$.

CLASS	RELATIVE FREQUENCY
10 but less than 20	.10
20 but less than 30	.30
30 but less than 40	.40
40 but less than 50	.20

21. Use the data of Problem 25, Chapter 3, to find a 99% confidence interval for the mean time between malfunctions for all equipment of the type under discussion.

22. The mean height of 16 children whose parents bought a certain style of bicycle was 56 inches. The standard deviation in the sample of 16 was 8 inches. Find a 90% confidence interval for the true mean height of children whose parents buy this bicycle.

23. A sample of 25 tanks holding compressed air had a mean loss of pressure over a year of 15 pounds per square inch (psi). The standard deviation of these losses was 4 psi. Find a 95% confidence interval for the mean yearly pressure loss for tanks of this type.

24. A consumer protection agency took a random sample of a product whose package was marked as weighing one pound. The 16 packages in the sample had a mean weight of 1.10 pounds and a standard deviation of .36 pound.

    a. Find a 95% confidence interval for the mean weight of the one-pound packages.

b. On the basis of this interval would the agency be justified in challenging the company's contention that the mean weight of packages is one pound?

c. On what ground might the agency challenge the company's designation of a one-pound weight on its packages?

25. A random sample of 28 families in a city revealed that the children in these families average 130 minutes of television viewing each day. The standard deviation of the sample was 50 minutes. Find a 98% confidence interval for the mean daily viewing time of children in this city.

26. Use the table of binomial confidence intervals to find 95% confidence intervals for p when:

    a. $n = 10, r = 7$ **b.** $n = 50, r = 0$

    c. $n = 100, r = 23$ **d.** $n = 100, r = 64$

    e. $n = 250, r = 150$ **f.** $n = 1000, r = 50$

27. Use the table of binomial confidence intervals to find 95% confidence intervals for p when:

    a. $n = 15, r = 12$ **b.** $n = 30, r = 30$

    c. $n = 100, r = 87$ **d.** $n = 250, r = 125$

    e. $n = 1000, r = 120$ **f.** $n = 250, r = 200$

28. Use interpolation in the table of binomial confidence intervals to find 95% confidence intervals for p in each of the following situations:

    a. $n = 40, r = 10$ **b.** $n = 60, r = 18$

    c. $n = 750, r = 300$ **d.** $n = 175, r = 70$

29. Use interpolation in the table of binomial confidence intervals to find 95% confidence intervals for p when:

    a. $n = 500, r = 50$ **b.** $n = 150, r = 45$

    c. $n = 12, r = 6$ **d.** $n = 25, r = 0$

30. Read the case at the back of the book entitled "Mountain States Feed Exchange" and find 95% confidence intervals for the proportion of sick animals associated with each feed. Assume the sample design is valid.

31. Read the case at the back of the book entitled "Statistics 101" and find a 95% confidence interval for the proportion of students in the class who checked out the tapes.

32. In a certain city, a sample of 1000 people were interviewed. It was found that 290 drove Fords. Give the 95% confidence interval for the percentage of people in this city who drive Fords.

33. In a sample of 100 urban families, it was found that 64 had checking accounts. Find a 95% confidence interval for the proportion of families who have checking accounts. Use the normal approximation to the binomial.

34. A local governmental agency has just purchased a new computer. The agency chief is interested in determining how many of the local government's 17,000 employees have computer programming experience. He takes a sample of 50 people and finds 3 that have had this experience. Construct a 95% confidence interval for the number of people who have had this experience:

    a. Using the table of 95% confidence intervals.

    b. Using the normal approximation to the binomial.

35. If 10% of a sample of 400 items were found to be defective, what is the 99% confidence interval for the proportion defective in the lot from which the sample was taken? Use the normal approximation to the binomial.

36. In a sample of 100 high-pressure castings, 20 were found to have chill folds resulting from improper heating of the mold. Find the 90% confidence interval for the true percent of all castings which have chill folds.

37. One-fourth of 300 persons interviewed were found to be opposed to the policies of a certain county school superintendent. Calculate a 95% confidence interval for the fraction of the population who are opposed to these policies. Use the normal approximation.

38. If a normal population has a known standard deviation σ of .75, how large a sample must be taken in order that the total width w of the 95% confidence interval for the population mean will not be greater than .10?

39. If X is a normal variable with known variance equal to 625, how large a sample must we take to be 95% confident that the sample mean will not differ from the true mean by more than ±6 units?

40. If a normal population is known to have σ equal to 5, how large a sample should we take in order to be 95% confident that the sample mean will not differ from the population mean by more than ±.8 units?

41. A preliminary sample of 10 ball bearings made by a certain process yielded a standard deviation of the diameters that was $s = .3$ millimeters. How much larger a sample is needed to estimate the mean diameter of all ball bearings made by this process to within ±.01 millimeters and 95% confidence?

42. A large oil company wishes to estimate the mean amount of water that has seeped into the oil storage tanks·at its refineries in the southern United States. A preliminary sample of $n = 21$ tanks showed that $s = 45$ gallons. How much larger should the sample be in order to estimate the mean water content of the tanks to within ±10 gallons with 95% confidence?

43. How large a sample should we take from a binomial population in order to be about 95% sure that the sample proportion of successes f will be within ±.04 of p? Assume that we suspect that p is about .20.

44. A sales manager wants to know what proportion of her accounts are inactive. A preliminary sample of $n = 25$ accounts showed that $r = 4$ were inactive. How many more accounts should she examine if she wants her confidence interval to be no more than $w = .08$ wide with 90% confidence?

45. A national polling organization wishes to determine which of two presidential candidates will win an election. They decide to find out what proportion of the voters favor candidate A, and they desire to be accurate to within ±.01. Find the sample size they need, assuming that the candidates seem to be running about even in the race at this point, i.e., p is about .5, and assuming 95% confidence.

46. A typewriter manufacturing firm wishes to determine what proportion of typing teachers uses its brand of machines. Find the sample size needed to be about 95% confident that their sample results f will not differ from p by more than $\pm.05$. Assume they believe they currently have about 30% of the market.

47. Two samples, one from each of two normal populations having the same variance, gave the following results.
    a. What are the best estimates for $\mu_2 - \mu_1$ and σ_d^2?
    b. Find the 95% confidence interval for $\mu_2 - \mu_1$.

SAMPLE	n	$\overline{X}_i$	$\Sigma(X_{ij} - \overline{X}_i)^2$
1	11	150	5600
2	11	120	5400

48. Given the following summary table, and assuming common variance:
    a. Find the best estimates for $\mu_1 - \mu_2$ and σ_d^2.
    b. Find the 90% confidence interval for $\mu_1 - \mu_2$.

SAMPLE	n	$\overline{X}_i$	$\Sigma(X_{ij} - \overline{X}_i)^2$
1	6	30	300
2	4	20	180

49. In an experiment designed to compare the mean service lives of two types of tires it was found that the difference between the means was 2250 miles, and that the pooled estimate of variance was equal to 625,000. There were 10 tires of each type. Find a 95% confidence interval for the difference between the mean lives of these types of tires.

50. Fifteen of each of two types of fabricated wood beams were tested for breaking load with the following results. Find a 95% confidence interval for the difference between the mean breaking loads.

TYPE	n	$\overline{X}_i$	s_i^2
I	15	1560	3500
II	15	1600	2500

51. An employee of a certain company wishes to know if managerial salaries are higher for those who have worked their way up from nonmanagerial positions in the company or for those who were hired from outside the company. He randomly selects 10 of each type of manager. The mean annual income for those hired from outside he computes to be $31,750, while the mean figure for the insiders is $27,000. The corresponding standard deviations are $6350 and $3050. Find a 90% confidence interval for the difference in annual income of the two groups. Should this interval lead the employee to conclude that there is a real difference?

52. Two assembly lines produce the same item. The amount of time to produce one finished piece varies due to the speed of the operators and the number of times the lines must be stopped for malfunctions or repairs. The amount of time (in minutes) it took to finish each piece in a sample of 9 items on the first line and 16 items on the second line is presented in the following summary table.
    a. Find 95% confidence intervals for $\mu_2 - \mu_1$.

b. Would the confidence interval for $\mu_2 - \mu_1$ lead you to conclude that the mean times per piece for the two lines are different? Why?

SAMPLE	n	$\overline{X}_i$	$\Sigma(X_{ij} - \overline{X}_i)^2$
1	9	52	68
2	16	55	24

53. Samples from two normal populations with the same variance gave the following results. Find a 95% confidence interval for $\mu_1 - \mu_2$.

SAMPLE	n	$\overline{X}$	s^2
1	9	96	24
2	16	80	34

54. Given the following summary table, find a 98% confidence interval for $\mu_2 - \mu_1$.

SAMPLE	n	$\overline{X}_i$	$\Sigma(X_{ij} - \overline{X}_i)^2$
1	15	25	55
2	12	20	45

55. In a sample of 100 from one binomial population there were 28 successes. A sample of 200 from a second binomial population contained 92 successes.
  a. Find estimates for p_1, p_2, and $p_2 - p_1$.
  b. Estimate the standard deviation of $f_2 - f_1$.
  c. Find a 98% confidence interval for $p_2 - p_1$.

56. Samples of 200 were taken from each of two binomial populations. They contained 104 and 96 successes, respectively.
  a. Find the estimated standard deviation of $p_1 - p_2$.
  b. Find a 90% confidence interval for $p_1 - p_2$.

57. A sample of 500 people were classified as being "athletic" or "nonathletic." Among 300 classified as athletic it was found that 60 regularly eat a certain breakfast food. Among the 200 nonathletic persons there were 50 who regularly use the product. Find a 95% confidence interval for the difference in the proportions that use the product.

58. In a sample of 50 turnings from an automatic lathe, 8 were found to be outside specifications. The cutting bits were then changed and the machine restarted. A new sample of 50 contained 3 defective turnings. Find a 90% confidence interval for the decrease in the proportion defective after changing bits.

SAMPLE DATA SET QUESTIONS: *Refer to the 113 applicants for credit listed in the Sample Data Set Appendix of this book.*

a. Find an interval within which you can be 95% confident that the population mean value of debt owed by successful credit applicants lies.

b. Discuss whether you must assume the variable in Question **a** has a normal distribution in order for the confidence interval you constructed to be a valid one. Do the data look normally distributed?

c. Assume you wanted to know the population mean value of TOTBAL for successful applicants to within ±$100 with 90% confidence. How much larger would your sample have to be?

d. Find the 95% confidence interval for the proportion of all applicants in the population who are married:
1. Using interpolation on the values in Table VII.
2. Using Formula (8.21).

REFERENCES

8.1 Fisher, R. A. *Statistical Methods for Research Workers,* 13th edition. Hafner, New York, 1963.

8.2 Hogg, Robert V., and Allen T. Craig. *Introduction to Mathematical Statistics,* 3rd edition. Macmillan, New York, 1970. Chapters 6, 7, and 8.

8.3 Mosteller, Frederick, Robert E. K. Rourke, and George B. Thomas, Jr. *Probability with Statistical Applications,* 2nd edition. Addison-Wesley, Reading, 1970. Chapter 12.

8.4 Snedecor, George W., and William G. Cochran. *Statistical Methods,* 6th edition. Iowa State University Press, Ames, 1967. Chapters 1, 2, 3, and 4.

TESTS
OF
HYPOTHESES

Turning from estimation to the theory of testing hypotheses, let us consider an example that exhibits in concrete form the problems that arise in the general case.

A "triangle test" provides one way of checking whether a candidate for a food technician's consumer taste panel can detect subtle differences in taste. In a single trial of this test the subject is presented with three food samples, two of which are alike, and is asked to select the odd one. Except for the taste difference, the samples are as similar as possible, and the order of presentation is random. In the absence of any ability at all to distinguish tastes, the subject has a one-third chance of correctly distinguishing the odd food sample, and the question is whether the candidate can do better than that.

To check the subject's ability, we present him with a series of triangle tests, and the order of presentation is randomized within each trial to eliminate any consistent bias and to make the trials independent of one another. Let p be the probability that the subject correctly identifies the odd food sample in a single trial. Then $p > 1/3$ if the subject has an ability in taste discrimination, and $p = 1/3$ if he has none. (Since it is hard to see how he could do worse than chance, we rule out the possibility $p < 1/3$ on a priori grounds.)

9.1   ESTABLISHING HYPOTHESES

To place this problem in a standard framework, we establish two hypotheses:

$$\textit{Null hypothesis } H_0 : p = 1/3$$

$$\textit{Alternative hypothesis } H_a : p > 1/3$$

We call the first hypothesis the **null hypothesis.** In this case, the null hypothesis is that the subject has *no* ability in taste discrimination. In general, the null hypothesis usually is established in such a way that it states *nothing is different* from what it is supposed to be, or is claimed to be, or has been in the past.

We normally assume that the null hypothesis is correct unless we see strong evidence to the contrary. When this happens, we *reject* the null hypothesis. That is, we no longer consider it to be true. But in rejecting the null hypothesis, we automatically *accept* something else. This "something else" is the **alternative hypothesis.** In the case of the subject whose taste discrimination ability is being tested, if we reject the hypothesis that he has *no* ability, then we must accept the hypothesis that he has *some* ability, and this is what the alternative hypothesis $p > 1/3$ states.

Some hypothesis-testing situations are presented below. The null hypothesis and alternative hypothesis are given for each one.

SITUATION 1.   A canning company is required by law to make sure that its products have 1% or less of a certain chemical in them. Several of the company's cans may be examined by the Food and Drug Administration to test the null hypothesis against the alternative hypothesis:

$$H_0 : p \leq .01$$

$$H_a : p > .01$$

If many cans are found with more than 1% of the chemical

in them, then the FDA will reject the null hypothesis, accept the alternative hypothesis, and perhaps recall some of the company's product from the market.

SITUATION 2. The Skills Unlimited Company specializes in training people for jobs in the building trades. The company's advertising claims that 90% or more of their graduates are placed in jobs immediately. Someone wishing to test this claim might take a sample of recent graduates and find out what proportion of them were placed in jobs. The hypotheses involved would be:

$$H_0: p \geq .90$$

$$H_a: p < .90$$

If the sample showed that a large number had not been placed in jobs, then we would reject the company's claim.

SITUATION 3. Hammond Manufacturing, Inc., has two manufacturing plants which produce special photographic lenses. The proportion of lenses that fail to pass final inspection has historically been the same at the two plants. If a quality control engineer desires to check if this is still the case, he could sample the inspection records at each plant and compare the proportion of rejects at each. If he calls the proportions at the two plants p_1 and p_2, then he could look at the hypotheses:

$$H_0: p_1 = p_2 \quad \text{or} \quad p_1 - p_2 = 0$$

$$H_a: p_1 \neq p_2 \quad \text{or} \quad p_1 - p_2 \neq 0$$

If the engineer found the proportions of rejects in his samples to be quite different, then he would reject the null hypothesis.

Hypotheses do not always involve proportions like p in the preceding situations. Sometimes we test hypotheses about what a population mean μ might be. For instance, consider the following situations.

SITUATION 4. The Stretchy Rubber Band Company claims that its number 7 box of rubber bands contains an average of "600 or more little office helpers" (rubber bands, that is). If a competitor wished to test this claim, the hypotheses involved would be:

$$H_0: \mu \geq 600$$

$$H_a: \mu < 600$$

If the competitor took a sample of number 7 boxes and found that the mean number of little office helpers per box was much less than 600, they might accuse Stretchy of stretching a point.

SITUATION 5.   Production workers at the American Standard Products Company have been trained in their jobs using two different training programs. The company's training director would like to know if there is a difference in the output of workers trained in the two programs. She might take a sample of workers and compare the output levels of those trained under program 1 with the output of those trained under program 2. She would be testing the hypotheses

$$H_0 : \mu_1 = \mu_2 \quad \text{or} \quad \mu_1 - \mu_2 = 0$$

$$H_a : \mu_1 \neq \mu_2 \quad \text{or} \quad \mu_1 - \mu_2 \neq 0$$

If she found that the mean output of one of her sample groups was quite different from the mean output of the other, she would be justified in rejecting the null hypothesis.

In each of the preceding situations we have suggested that the null hypothesis is rejected if we find strong evidence that it is not true. But what constitutes "strong" evidence in hypothesis testing? A formal procedure for hypothesis testing and methods for determining the strength of evidence supporting or contradicting the null hypothesis are presented in the remainder of this chapter.

9.2   TESTING HYPOTHESES

Let us return to the example in which a food technician wishes to test a subject's ability to distinguish tastes. We confront the problem of testing the null hypotheses $H_0 : p = 1/3$ against the alternative $H_a : p > 1/3$. We assume a priori that either H_0 is true or else H_a is true.

If we make n trials with the subject, and if r is the number of correct identifications made, then r has the binomial distribution with parameters n and p:

$$P(r) = \binom{n}{r} p^r (1 - p)^{n - r} \qquad r = 0, 1, 2, \ldots, n \qquad (9.1)$$

This probability can be found for selected values of n and p in Table

III of the Appendix. We want to use the statistic r to decide whether or not the alternative hypothesis H_a is more reasonable or plausible than the null hypothesis H_0—that is, whether or not the subject has ability. And we want to set up in advance of experimentation a rule for making the decision; we want to set up a statistical procedure in the form of a *test*. This parallels what we do in constructing confidence intervals, for a given problem settling on confidence limits $\overline{X} \pm 2.262(s/\sqrt{n})$, say, in advance of sampling, and letting the sample give to these limits actual numerical values. In the present problem we want to set up a *rejection region* or *critical region*, a set R of values for r that will lead us to reject H_0 and prefer H_a. If the experiment gives to r a numerical value lying in R, we reject H_0; otherwise, we do not.

To illustrate, suppose that n is 10. Consider two rejection regions:

$$R_5 = \{5, 6, 7, 8, 9, 10\}$$

and

$$R_7 = \{7, 8, 9, 10\}$$

If we use R_5, the rule or test is: If $r \geq 5$, reject H_0 and accept H_a (decide that the subject has ability); if $r < 5$, we do not reject H_0 (decide that the subject has no ability). If we use R_7, the rule is: If $r \geq 7$, reject H_0 and accept H_a; if $r < 7$, do not reject H_0. (Of course, we could use some cutoff point other than 5 or 7.)

Each of these two rules makes sense. We decide in favor of the hypothesis H_a that $p > 1/3$ when X is large. It is contrary to common sense to use a rule that says to decide in favor of H_a in case $r \leq 3$, for example, or a rule that says to decide in favor of H_a in case X is even. But of the sensible regions R_5 and R_7, which is better? The answer to this question depends on the kinds of errors these rules lead us to, as is shown in the next section.

9.3   TYPE I AND TYPE II ERRORS

Applying an hypothesis test may lead to the wrong conclusion. There are in fact two kinds of error possible, called **Type I** and **Type II errors.**

> *Type I error:* We reject H_0 but H_0 is true
>
> *Type II error:* We fail to reject H_0 but H_0 is false

Note: We can make a Type I error only when we reject H_0 and can make a Type II error only when we fail to reject H_0. We can assess the strength of a test or rejection region by calculating the probabilities of these two kinds of errors.

$$\alpha = P(\text{reject } H_0 \mid H_0 \text{ is true}) = P(\text{Type I error})$$

$$\beta = P(\text{accept } H_0 \mid H_0 \text{ is false}) = P(\text{Type II error})$$

Naturally we want α and β to be small. The probability α of a Type I error is also called the *size* of the test or the *level of significance*. And $1 - \beta$ is called the *power*.

TRUE SITUATION	CONCLUSION:	
	H_0 *is true*	H_0 *is false*
H_0 *is true*	No error; probability $= 1 - \alpha$	Type I error; probability $=$ size $= \alpha$
H_0 *is false*	Type II error; probability $= \beta$	No error; probability $=$ power $= 1 - \beta$

For the rejection region R_5 in our taste-discrimination example, a Type I error occurs when the subject has no skill but scores five or more successful identifications just by luck. The chance of this is:

$$\alpha \quad \text{for} \quad R_5 = P(r \geq 5) = \sum_{r=5}^{10} \binom{10}{r} \left(\frac{1}{3}\right)^r \left(\frac{2}{3}\right)^{10-r} = .213$$

This value cannot be obtained directly from Table III of the Appendix but can be approximated using interpolation of the table. For the rejection region R_7, the chance of a Type I error is:

$$\alpha \quad \text{for} \quad R_7 = P(r \geq 7) = \sum_{r=7}^{10} \binom{10}{r} \left(\frac{1}{3}\right)^r \left(\frac{2}{3}\right)^{10-r} = .020$$

Now $P(r \geq 7)$ is smaller than $P(r \geq 5)$. As far as Type I error is concerned, then, R_7 is a *better* rejection region than R_5.

But what about β, the chance of a Type II error? If H_0 is false, so that $p > 1/3$, then β depends on which alternative value of p we check. We should write β_p to indicate this dependence on p. Let us check for a p of .65 (any other value of p that exceeds .33 would also illustrate the calculation). For the region R_5,

$$\beta_{.65} \quad \text{for} \quad R_5 = P(r < 5) = \sum_{r=0}^{4} \binom{10}{r} (.65)^r (.35)^{10-r} = .095$$

This time we use .65 as the value for p and .35 as a value for $1 - p$ in the binomial formula because we are computing the chance that we will erroneously accept the null hypothesis that the subject is devoid of skill when in fact he or she can guess right 65% of the time. For R_7,

$$\beta_{.65} \quad \text{for} \quad R_7 = P(r < 7) = \sum_{r=0}^{6} \binom{10}{r} (.65)^r (.35)^{10-r} = .486$$

Now $P(r < 7)$ is greater than $P(r < 5)$. As far as Type II error is concerned, R_7 is a *worse* rejection than R_5.

We can summarize our computations in a table:

	α	$\beta_{.65}$
R_5	.213	.095
R_7	.020	.486

The α column makes us prefer R_7, but the β column makes us prefer R_5. This is inevitable: As α goes down, β goes up, and vice versa. If we refuse to allow a Type I error at all, so that α is 0, we must never reject H_0, whatever r may be (so the rejection region R is empty); but then if H_0 should be false, we always make an error of Type II, so that β is 1. Or if we demand that β equal 0, we must always reject H_0, whatever r may be (so the rejection region R is $\{0, 1, 2, \ldots, 10\}$); in this case, if H_0 is true we always make an error of Type I, so that α is 1. It is impossible to arrange that both α and β are equal to 0. For whatever consolation it may be, it is equally impossible to arrange that both α and β are equal to 1. It should be noted that α and β do not ordinarily sum to unity (see the preceding table for R_5 and R_7). The only thing we can say is that α and β are inversely related to one another in that changing the rejection region to lower one of them will automatically raise the other. It is only under a very special circumstance that $\alpha + \beta = 1$.

The procedure in a court of law affords a parallel. Let the null hypothesis be that the defendant is innocent and let the alternative hypothesis be that he or she is guilty. To condemn an innocent person is to commit a Type I error; to acquit a guilty person is to commit a Type II error. The rules that govern a trial are loosely analogous to a statistical test or a rejection region. Any rule that decreases the chance α of a Type I error (a rule, say, that the defendant need not testify against him- or herself) necessarily increases the chance of β of a Type II error. And any rule that decreases the chance β of a Type II error (a rule allowing a split jury to convict, say) necessarily increases the chance α of a Type I error. It is impossible to arrange that both α and β are equal to 0. But it is equally impossible to arrange that both α and β are 1.

In most hypothesis testing, as in a court of law, the null hypothesis is chosen so that the Type I error is a more serious error than the Type II error (convicting someone who is innocent is generally viewed as more serious than letting a guilty person go). Thus, the most important probability in hypothesis tests is usually α, the probability that the hypothesis test will lead to a Type I error—rejection of the null hypothesis when it is true. It is somewhat more difficult to deal with β since, to calculate this figure, one must know or assume the true value of μ or p for the population. We will leave discussions of β, the Type II error probability, to more advanced texts. One can always be safe, however, with using large samples whenever possible. Under these circumstances, both α and β are small.

9.4   THE STEPS OF HYPOTHESIS TESTING

In the sequel, we shall encounter many testing problems. They share certain features with our taste-discrimination example, and, like it, they can be conveniently analyzed according to a standard sequence of steps.

STEP 1.   Formulate the null hypothesis H_0 in statistical terms.
STEP 2.   Formulate the alternative hypothesis H_a in statistical terms.
STEP 3.   Set the level of significance α and the sample size n.
STEP 4.   Select the appropriate statistic and the rejection region R.
STEP 5.   Collect the data and calculate the statistic.
STEP 6.   If the calculated statistic falls in the rejection region R, reject H_0 in favor of H_a; if the calculated statistic falls outside R, do not reject H_0.

Each of these steps calls for comment:

STEP 1.   The null hypothesis must be stated in statistical terms. It would not suffice in our example merely to hypothesize that the subject cannot distinguish subtle differences in taste; we need the specific binomial model, together with the identification of $p = 1/3$ as representing absence of skill.

STEP 2.   The alternative hypothesis is essential to the problem. Suppose that a soap manufacturer claims that $1/3$ of all soap purchasers in a certain city buy his brand. Thus, as in the triangle test, the hypothesis to be tested is $H_0: p = 1/3$, that one-third of the buyers prefer his brand. Should the alternative hypothesis again be $H_a: p > 1/3$? Obviously not, since either more *or less* than one-third of the buyers might buy the brand in question. In the taste experiment, an excessively large r leads us to reject H_0 in favor of H_a. In the soap example, we might interview 3000 soap purchasers and find that 1900 of them prefer the brand in question. Since this is 900 over the expected number of 1000, we would probably reject the hypothesis that $p = 1/3$ and conclude that $p > 1/3$. On the other hand, if only 100 say they prefer this brand, then we would probably reject H_0 and conclude that $p < 1/3$. Thus, for this case the alternative hypothesis should be $H_a: p \neq 1/3$. Since the form of the alternative hypothesis determines the rejection region we will choose in Step 4, the alternative hypothesis must be carefully stated to represent the problem at hand.

STEP 3.   Since the null hypothesis is usually set up in such a way that we want strong evidence against it before we reject it, the level of α is usually rather small. Values often used for α are .01, .02, .05, and .10. Researchers seldom allow more than a 10 percent chance of a Type I error in hypothesis tests. Larger values of α are sometimes tolerated when a Type I error is not as serious as a Type II error.

STEP 4.   In the taste experiment, the statistic is the number of successes and the rejection region has the commonsense form $\{r, r + 1, \ldots, n\}$. In all our testing problems, common sense will dictate the proper form of the rejection region; choosing the exact region (choosing the actual value of r in our example) requires knowing the distribution of the statistic (the binomial distribution in our example). In general, when the alternative hypothesis is a strict inequality, as in the triangle test where $H_a: p > 1/3$, then the rejection region consists of one set of values such as $R_7 = \{7, 8, 9, 10\}$. But when the alternative hypothesis is of the "not equal to" type, as in the soap example where $H_a: p \neq 1/3$, then the rejection region consists of two sets of values, either very high *or* very low numbers of people saying they prefer the manufacturer's brand.

STEP 5.   The sampling procedure must accord with the model, the binomial model in our example.

STEP 6.   The hypotheses H_0 and H_a are not treated in a symmetric fashion. That is, if we reject H_0, we accept H_a. However, if we *do not reject* H_0, that does not mean we *accept* H_0. Just because we do not have enough evidence to reject a hypothesis does not mean that we must "accept" it. There is a parallel situation in the law. A juror in a court case may feel that there is not enough evidence to convict the defendant (rejecting the null hypothesis of innocence), but that does not mean the juror must accept the notion that the defendant is "innocent." The juror will vote "not guilty," which is a middle ground between "guilty" and "innocent." In the same way, when we do not have enough evidence to reject the null hypothesis, we simply say we "cannot reject H_0" rather than saying we "accept" it. This too is a middle-ground position.

Sometimes, tests of hypothesis are called *tests of significance*. If the α we use in such a test is .05 and if we reject the null hypothesis,

we sometimes say "the hypothesis was rejected at the 5 percent significance level." This means that the sample results we obtained were
significantly different from those we would expect to see if the hypothesis
were true. In fact, the sample result would have only a 5 percent
chance or less of occurring *if* the hypothesis is true, and thus the
result is significantly different from the hypothesized or expected result.

In what follows, we shall construct a number of tests of hypotheses.
In each case we must actually construct the rejection region R, and
these constructions will clarify the general principles set out previously.
Tests of hypotheses are used extensively in research, though not
necessarily in rigid accord with the theory. The theoretical framework
holds in that it makes it possible to understand what test is appropriate
to what problem and what the strength of the test will be.

9.5   HYPOTHESES ON A NORMAL MEAN

Consider the null hypothesis that a normal population has a specified
mean μ_0, the variance σ^2 being unknown. Under the null hypothesis
$H_0: \mu = \mu_0$, the statistic

$$t = \frac{\bar{X} - \mu_0}{s/\sqrt{n}} \tag{9.2}$$

has a t distribution with $n - 1$ degrees of freedom (see Section 8.3),
which can be made the basis of a test.

Suppose first that the alternative hypothesis is $H_a: \mu > \mu_0$. It
is intuitively clear that if we take a sample to test the hypothesis H_0,
then we ought to reject H_0 in favor of H_a when $\bar{X}$ is too large—in
fact, when $\bar{X}$ exceeds μ_0 by too much. Since the t statistic of Equation
(9.2) increases when $\bar{X}$ increases, we can just as well reject H_0 when
t is excessively large. How large? Recall that the upper α percentage
cutoff point $t_{\alpha,n-1}$ is defined so that $P(t > t_{\alpha,n-1})$ is equal to α. Therefore,
if we adopt the rule of rejecting H_0 when t exceeds $t_{\alpha,n-1}$, then the
chance of rejecting H_0 when it is true is just α. The testing problem
and the rule are

$$\begin{cases} H_0: \mu = \mu_0 \\ H_a: \mu > \mu_0 \\ R: t > t_{\alpha,n-1} \end{cases} \tag{9.3}$$

The rejection region is specified by the inequality $t > t_{\alpha,n-1}$. The shaded
area in Figure 9.1 represents α, and all t values to the right of $t_{\alpha,n-1}$
form the rejection region. If we substitute the ratio in Equation (9.2)

FIGURE 9.1

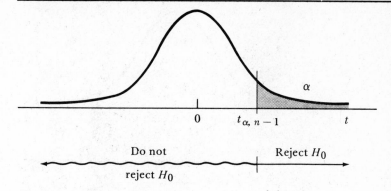

for t here, multiply by $s/\sqrt{n}$ and then add μ_0, the inequality becomes $\bar{X} > \mu_0 + t_{\alpha,n-1}(s/\sqrt{n})$. Thus we reject the hypothesis that $\mu = \mu_0$ when $\bar{X}$ exceeds μ_0 by too much, $t_{\alpha,n-1}(s/\sqrt{n})$ providing the proper measure of "too much."

If μ is equal to μ_0, the distribution of the Ratio (9.2) corresponds to the density curve (a) in Figure 9.2. If instead μ is equal to μ_1, say, where $\mu_1 > \mu_0$, then the ratio has the density curve (b); it has the same shape as (a) but is displaced to the right, its mean being $\mu_1 - \mu_0$ instead of 0. The cutoff point $t_{\alpha,n-1}$ is where the area of the right tail of curve (a) equals the level α of the test; the area of the left tail of curve (b) is the probability β of a Type II error if the mean μ is really μ_1.

Similarly, if the alternative is $\mu < \mu_0$, the sensible thing is to reject the hypothesis that $\mu = \mu_0$ when t is small:

$$\begin{cases} H_0 : \mu = \mu_0 \\ H_a : \mu < \mu_0 \\ R : t < -t_{\alpha,n-1} \end{cases} \tag{9.4}$$

FIGURE 9.2 Distributions corresponding to $H_0: \mu = \mu_0$ and $H_a: \mu = \mu_1$

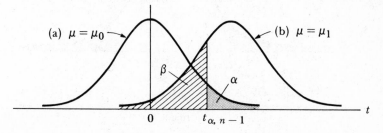

FIGURE 9.3

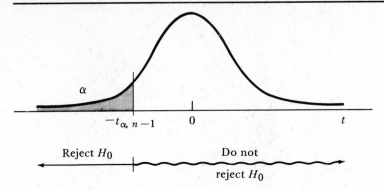

Figure 9.3 shows the rejection values of t for this hypothesis test, and the shaded area represents the probability of a Type I error, α. Notice again that $t < -t_{\alpha,n-1}$ has probability α if μ does equal μ_0; and notice that $t < -t_{\alpha,n-1}$ is the same thing as $\bar{X} < \mu_0 - t_{\alpha,n-1}(s/\sqrt{n})$.

The alternative hypotheses in Setups (9.3) and (9.4) are onesided, and so are the corresponding rejection regions. A two-sided alternative, including both possibilities $\mu > \mu_0$ and $\mu < \mu_0$, calls for a two-sided rejection region:

$$\begin{cases} H_0 : \mu = \mu_0 \\ H_a : \mu \neq \mu_0 \\ R : t > t_{\alpha/2,n-1} \quad \text{or} \quad t < -t_{\alpha/2,n-1} \end{cases} \tag{9.5}$$

The cutoff points are $(1/2)\alpha$ percentage points, since the test is two-tailed; as Figure 9.4 indicates, the total area of the two tails taken together is α. The rejection rule can be restated as $|\bar{X} - \mu_0|/(s/\sqrt{n}) > t_{\alpha/2,n-1}$.

FIGURE 9.4

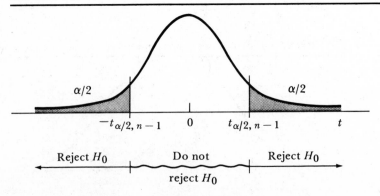

We reject H_0 if $\overline{X}$ is excessively far from μ_0 in either direction, i.e., many standard deviations from the hypothesized value μ_0.

EXAMPLE 1

Problem: A manufacturer of small electric motors asserts that on the average they will not draw more than .8 amperes under normal load conditions. A sample of 16 of the motors was tested, and it was found that the mean current was .96 amperes with a standard deviation of .32 amperes. Are we justified in rejecting the manufacturer's assertion?

Solution: Here the null hypothesis that the assertion is correct is $H_0: \mu \leq$ .8. If we are willing to take a risk of 1 in 20 of rejecting the assertion if it is true, we set α at .05. Since we want to reject the assertion only if the evidence indicates that the mean current consumption (the population value) is greater than .8, we have a one-sided alternative $H_a: \mu > .8$. Under the assumption of normality, we use a t statistic with 15 degrees of freedom and a one-tailed test; the cutoff point is $t_{.05,15}$, or 1.753:

$$\begin{cases} H_0: \mu \leq .8 \\ H_a: \mu > .8 \\ R: t > 1.753 \end{cases}$$

For our data the calculated t is:

$$t = \frac{\overline{X} - \mu_0}{s/\sqrt{n}} = \frac{.96 - .80}{.32/4} = 2.00$$

Since this exceeds 1.753, we reject the null hypothesis and conclude that the mean current consumption is greater than .8 amperes. The chance that our conclusion is in error is less than 5%.*

EXAMPLE 2

Problem: In an attempt to determine whether or not special training increases the speed with which assembly-line workers can do a certain job, 25 workers are timed doing this job. Then they are given a special course designed to increase their assembly efficiency. At the end of the course they are timed doing the same job. The differences between the first and the second times are recorded; the mean reduction for the 25 workers is found to be 3 minutes, and the sample standard deviation is 9 minutes. Has the training reduced the time required on this job?

---

*In reality, the chance that we are in error is either 0 or 1 since our conclusion must be either right or wrong. However, it is useful to think of α as the level of risk we expose ourselves to in rejecting the null hypothesis, and little is lost if we think of α as the probability of being in error.

Solution: The null hypothesis that the training has no effect is that the population mean reduction is $\mu = 0$. We will reject H_0 only if we think the mean reduction is positive, so the alternative is $H_a : \mu > 0$. We need a one-tailed test, and if we set α at .05, the cutoff point is $t_{.05,24}$, or 1.711:

$$\begin{cases} H_0 : \mu = 0 \\ H_a : \mu > 0 \\ R : t > 1.711 \end{cases}$$

The calculated t is:

$$t = \frac{\overline{X} - 0}{s/\sqrt{25}} = \frac{3 - 0}{9/5} = 1.667$$

Thus, the sample evidence and the null hypothesis are consistent with one another since the reduction of 3 minutes is not large enough to cause rejection of H_0. However, note that not rejecting does not mean we have accepted H_0 and concluded that $\mu = 0$. We simply do not have enough evidence to reject the suggestion that $\mu = 0$.

EXAMPLE 3

Problem: Last year's retail sales records show that the average monthly expenditure per person for a certain food product was $5.50. We should like to know whether there has been any significant change in this average during the first quarter of this year. Thus, we are looking for both positive and negative changes.

Solution: To make a comparison, we sample 30 families and find that the mean expenditure is $5.10 with a standard deviation of $.90. Here the null hypothesis of no change is $H_0 : \mu = 5.50$. The two-sided alternative, $H_a : \mu \neq 5.50$, represents a change in either direction, up or down. Therefore, we want a two-tailed test; if α is set at .01, the cutoff is the $(1/2)\alpha$, or .005, percentage point for 29 degrees of freedom: $t_{.005,29}$ is 2.756. Hence

$$\begin{cases} H_0 : \mu = 5.50 \\ H_a : \mu \neq 5.50 \\ R : t > 2.756 \quad \text{or} \quad t < -2.756 \end{cases}$$

Since

$$t = \frac{\overline{X} - 5.50}{s/\sqrt{30}} = \frac{5.1 - 5.50}{.9/\sqrt{30}} = -2.434$$

the data do not allow us to reject the null hypothesis.

In each of our three examples we observe these features:

1. The null hypothesis always contains the equality statement, and the alternative hypothesis is determined by the question implicit in the statement of the problem.
2. The calculated statistic is based on the difference between the observed mean $\overline{X}$ and the mean μ_0 under the null hypothesis. Since the difference between them is meaningful only in relation to the amount of variation in the data, the statistic is the ratio of the difference to its estimated standard deviation.
3. When the difference between $\overline{X}$ and μ_0 is large enough that the calculated t value falls into one of the tails of the t distribution (the appropriate tail or tails are determined by the alternative hypothesis), we reject the null hypothesis because the probability of such a t value, given that the null hypothesis is true, is too small to be attributed to normal sampling variation.

Whenever we use the t distribution for calculating confidence intervals or for testing hypotheses, we are assuming that the data are a random sample from a normal population. The first assumption, randomness, is necessary if we want to make probability statements about the results we obtain; therefore it cannot be relaxed.

The second assumption, normality, is not too strong and may be relaxed. Because of the central limit theorem, we may say that, even though the population has a distribution that is not normal, the t statistic may be used and the probabilities will not be greatly affected, provided we have a sufficiently large sample. What size sample may be considered sufficiently large depends on how much the population concerned departs from normality.

Finally, if for some reason we are fortunate enough to know the standard deviation σ of the population for which the hypothesis test is being conducted, then σ can be substituted for s in Formula (9.2). With this change the ratio in (9.2) is exactly normally distributed as the standardized Z variable.

$$Z = \frac{\overline{X} - \mu_0}{\sigma / \sqrt{n}} \qquad (9.6)$$

Hypothesis tests using Expression (9.6) are conducted just as those using (9.2), although it is rather rare that such tests are conducted since we seldom know the population standard deviation. In this rare case the normal table would be used to determine the rejection region.

EXAMPLE 4

Problem: A cutting machine is designed so that the lengths of cut parts turned out by the machine may be adjusted to different settings. However, the variance of the measurements, once a setting is chosen, is always $\sigma^2 = .5$. The machine was set to cut at a mean length of 30. Use the following sample data to test the null hypothesis that the mean length of the cuts is 30 against the alternative hypothesis that μ is not 30.

Solution:

$$n = 16$$

$$\overline{X} = 30.25$$

$$s = .6$$

Since we know the population standard deviation is $\sigma = \sqrt{.5} = .707$, we will use the ratio in Formula (9.6) and ignore the $s = .6$ value. If we assume that $\alpha = .05$, then the Z value that defines our rejection region is $Z_{.025}$. This figure is found in the normal table or in the last row of the t table, and its value is 1.96.

$$\begin{cases} H_0 : \mu = 30 \\ H_a : \mu \neq 30 \\ \quad R : Z > 1.96 \quad \text{or} \quad Z < -1.96 \end{cases}$$

The calculated Z is:

$$Z = \frac{30.25 - 30}{.707/\sqrt{16}} = 1.41$$

This value of Z does not fall in the rejection region so we cannot reject the hypothesis that the mean length of cuts is $\mu = 30$.

9.6 HYPOTHESES ON A BINOMIAL p

If we take a sample from a binomial population with the intention of using the information it contains to test the hypothesis that p has a specified value p_0, we can use an exact test or an approximate test. To use the exact test we must either have tables of the binomial probabilities like those which form Table III in the Appendix or be willing to compute them. Often, then, we use an approximate test based on the central limit theorem.

Under the null hypothesis $H_0 : p = p_0$, the ratio

$$Z = \frac{r - np_0}{\sqrt{np_0(1 - p_0)}} \tag{9.7}$$

has approximately a standard normal distribution, where r is the number of successes in the sample (see Section 6.3). The denominator in the ratio is the standard deviation of r, we know its value under the null hypothesis, and there is no need to estimate it from the sample.

Testing H_0 is like testing for a normal mean when there are infinitely many degrees of freedom—that is, with the normal distribution in place of the t distribution. The rejection region will be one-tailed or two-tailed depending on whether the alternative hypothesis is one-sided or two-sided.

Dividing numerator and denominator in Ratio (9.7) by the sample size n gives

$$Z = \frac{r - np_0}{\sqrt{np_0(1 - p_0)}} = \frac{f - p_0}{\sqrt{p_0(1 - p_0)/n}} \tag{9.8}$$

where f is r/n, the fraction of successes in the sample. Clearly, the test may be performed either with r or with f, as convenience dictates. We must, however, be sure to use the right standard deviation in the denominator.

EXAMPLE 1

Problem: In a sample of 400 bushings there were 12 whose internal diameters were not within the tolerances. Is this sufficient evidence for concluding that the manufacturing process is turning out more than 2% defective bushings? Let α be .05.

Solution: The null hypothesis that the process is in control is $H_0: p \le .02$. The alternative hypothesis that it is out of control is $H_a: p > .02$. The upper .05 percentage cutoff point $Z_{.05}$ for the standard normal distribution is 1.645. We should reject if we calculate larger Z values, since a one-sided test calls for a one-sided rejection region:

$$\begin{cases} H_0: p \le .02 \\[2mm] H_a: p > .02 \\[2mm] R: \dfrac{r - 400 \times .02}{\sqrt{400 \times .02 \times .98}} > 1.645 \end{cases}$$

Our data give

$$\frac{r - 400 \times .02}{\sqrt{400 \times .02 \times .98}} = \frac{12 - 8}{2.8} = 1.429$$

The data favor the hypothesis that the process is in control; 12 defective bushings out of 400 is not excessively large, and we cannot reject the null hypothesis (nor can we "accept" it).

EXAMPLE 2

Problem: A package designer wishes to determine if there is any difference in the way customers react to two new package designs. She places equal numbers of packages on the shelf in a supermarket and observes how many of each package are purchased each day. She rotates the position of the two package types from one day to the next to eliminate the effect of shelf position and makes sure that there are always equal numbers of the two package types on the shelf. After 500 purchases have been made, 150 of the first package design and 350 of the second design have been purchased. Use $\alpha = .05$.

Solution: Let p equal the proportion of purchasers of the first design in the population (we could as easily have chosen p to be those preferring the second design). We have observed a sample value $f = 150/500$, or .30. The null hypothesis here would be that there is no difference in the two package designs and $H_0: p = .5$. If we test this hypothesis with α of .05, then the cut-off points are $\pm Z_{.025}$, or ± 1.96, on a two-tailed test. This is a two-tailed test since very large values of f indicate preference for the first design, while very small values of f indicate preference for the second. Thus:

$$\begin{cases} H_0: p = .5 \\ H_a: p \neq .5 \\ R: \dfrac{f - .5}{\sqrt{(.5 \times .5)/500}} > 1.96 \quad \text{or} \quad < -1.96 \end{cases}$$

Since $f = .3$, we have

$$Z = \frac{.3 - .5}{\sqrt{(.5 \times .5)/500}} = \frac{-.2}{.02236} = -8.94$$

This figure falls well into the lower tail of the rejection region, so we reject the null hypothesis and conclude that the first package design is inferior to the second. Our chance of error in this conclusion is less than 5%. In fact, since Z is *so* negative the chance of error is far less than .05.

9.7   A CONNECTION BETWEEN TESTING HYPOTHESES AND ESTIMATION

There is a connection between hypothesis testing and interval estimation which illuminates both:

Given a level-α test, we can construct a level-α confidence interval by collecting together all the null hypotheses not rejected by the test.

For example, in a two-sided, level-α t test of $\mu = \mu_0$, the condition for *not* rejecting the null hypothesis [see Formula (9.5)] is:

$$-t_{\alpha/2,n-1} < \frac{\overline{X} - \mu_0}{s/\sqrt{n}} < t_{\alpha/2,n-1} \tag{9.9}$$

By the usual algebraic manipulations, this pair of inequalities is the same as:

$$\overline{X} - t_{\alpha/2,n-1}\frac{s}{\sqrt{n}} < \mu_0 < \overline{X} + t_{\alpha/2,n-1}\frac{s}{\sqrt{n}} \tag{9.10}$$

But the extreme terms here are the level-α confidence limits for a normal mean with variance unknown; see Formula (8.17) in Section 8.4. The confidence interval consists exactly of those μ_0 having the property that the test does not reject the null hypothesis $H_0 : \mu = \mu_0$. The point is that the Conditions (9.9) and (9.10) are equivalent and each has probability $1 - \alpha$ if the mean is μ_0; Condition (9.9) ensures that the test has probability α of a Type I error, and Condition (9.10) ensures that the confidence interval has probability $1 - \alpha$ of covering the true mean.

The principle works equally well the other way around:

Given level-α confidence limits, we can construct a level-α test by the rule of rejecting the null hypothesis if the test statistic falls outside the confidence limits.

As an illustration, consider estimating a normal mean when the variance σ^2 is known. The level-α confidence limits (see Formula (8.13) in Section 8.2) are:

$$L = \overline{X} - Z_{\alpha/2}\frac{\sigma}{\sqrt{n}} \quad \text{and} \quad R = \overline{X} + Z_{\alpha/2}\frac{\sigma}{\sqrt{n}} \tag{9.11}$$

Now μ_0 lies outside these limits if

$$\mu_0 < \overline{X} - Z_{\alpha/2}\frac{\sigma}{\sqrt{n}} \quad \text{or} \quad \mu_0 > \overline{X} + Z_{\alpha/2}\frac{\sigma}{\sqrt{n}}$$

and this is equivalent to

$$Z_{\alpha/2} < \frac{\overline{X} - \mu_0}{\sigma/\sqrt{n}} \quad \text{or} \quad -Z_{\alpha/2} > \frac{\overline{X} - \mu_0}{\sigma/\sqrt{n}}$$

Thus, if we know the variance σ^2 and want a two-sided, level-α test of the null hypothesis $H_0 : \mu = \mu_0$, the rule is to reject if

$$\left| \frac{\overline{X} - \mu_0}{\sigma / \sqrt{n}} \right| > Z_{\alpha/2} \tag{9.12}$$

This is a new test.

Keeping in mind this parallel between interval estimation and testing will help to unify the concepts and techniques of the subsequent chapters, particularly when we note that our tests and estimates have been based on just a few statistics. First is the standardized mean which was given in Chapter 7 to be

$$Z = \frac{\overline{X} - \mu}{\sigma / \sqrt{n}} \tag{9.13}$$

From this we construct the confidence limits of Formula (9.11) and the test in Condition (9.12) appropriate when σ is known. If σ is unknown, we replace it in the denominator in the Ratio (9.13) by its natural estimate s. This gives the t statistic

$$t = \frac{\overline{X} - \mu}{s / \sqrt{n}} \tag{9.14}$$

and the tests and confidence intervals appropriate when σ is unknown.

We have also used the standardized number of successes and the fraction of successes in binomial trials. The standardized ratio was presented in Chapter 6.

$$Z = \frac{r - np}{\sqrt{np(1 - p)}} = \frac{f - p}{\sqrt{p(1 - p)/n}} \tag{9.15}$$

By using the approximate normality of this statistic, we can test for a particular value of p. If we replace p in the denominator by its natural estimator f, we get the statistic

$$Z = \frac{f - p}{\sqrt{f(1 - f)/n}} \tag{9.16}$$

also approximately normally distributed. It is from this statistic that we construct confidence intervals for p.

In the following chapters we shall encounter further statistics that arise from standardizing some random variable by its mean and standard deviation and then replacing the standard deviation by its natural estimate. A standard notation will help in this connection. The standard deviation of a statistic Y we denote σ_Y; a sample estimate of this standard deviation we denote s_Y. Thus, the standard deviation of $\overline{X}$ is $\sigma_{\overline{X}}$, equal to $\sigma / \sqrt{n}$, of which the estimate $s_{\overline{X}}$ is $s / \sqrt{n}$. The number r of successes in binomial trials has standard deviation σ_r, equal to $\sqrt{np(1 - p)}$, and the estimate s_r of this is $\sqrt{nf(1-f)}$. Finally, the fraction f of successes has standard deviation σ_f, equal to $\sqrt{p(1-p)/n}$, and the estimate s_f of this is $\sqrt{f(1-f)/n}$.

■ 9.8 TESTING THE HYPOTHESIS $\mu_1 = \mu_2$

To test the hypothesis that the difference between the means is equal to a given value δ_0, we rely on the quantity defined as follows, for under the hypothesis $H_0 : \mu_1 - \mu_2 = \delta_0$, the statistic

$$t = \frac{\overline{X}_1 - \overline{X}_2 - \delta_0}{s_d} \tag{9.17}$$

has a t distribution with $n_1 + n_2 - 2$ degrees of freedom. The test is a simple t test except that we must assume the populations have the same variance. Thus, to obtain s_d for use in the denominator of Expression (9.17), we must pool the sample variances just as we did in constructing confidence intervals for the value of $\mu_1 - \mu_2$, using either Equation (8.28) or Equation (8.29).

An hypothesis of particular interest in the two-sample case or in the case of comparing two treatments in a completely randomized experiment is the one that states that there is no difference between the means: $H_0 : \mu_1 = \mu_2$. Under this hypothesis, δ_0 is 0 and the test statistic as given in Equation (9.17) reduces to

$$t = \frac{\overline{X}_1 - \overline{X}_2}{s_d} \tag{9.18}$$

Here, as with other t tests, the critical region depends upon the alternative hypothesis. If it states simply that the means are not equal ($H_a : \mu_1 \neq \mu_2$), we use the two-tailed critical region $|t| \geq t_{\alpha/2, n_1+n_2-2}$. If the alternative is one-sided, i.e., $H_a : \mu_1 > \mu_2$ or $H_a : \mu_1 < \mu_2$, we use the one-tailed critical regions $t > t_{\alpha, n_1+n_2-2}$ or $t < -t_{\alpha, n_1+n_2-2}$.

For a one-tailed test, there is often some doubt as to which tail should be used. This difficulty is resolved if we always write our t, Equation (9.18), and the alternative hypothesis in such a way that the subscripts are in the same order. If the alternative is $H_a : \mu_2 > \mu_1$, and if we write the numerator of our t statistic as $\overline{X}_2 - \overline{X}_1$, then the inequality in H_a is an arrowhead pointing to the right and we use the upper tail. If the alternative hypothesis is written as $H_a : \mu_1 < \mu_2$, we would use $\overline{X}_1 - \overline{X}_2$. The inequality sign then points to

---

■ The material in this section is not required in the sequel.

the left, showing that we use the lower tail for the critical region.

The determination of the proper tail to use can also be seen this way: For the testing problem

$$\begin{cases} H_0 : \mu_1 = \mu_2 \\ H_a : \mu_1 > \mu_2 \end{cases}$$

we ought to reject H_0 in favor of H_a if $\overline{X}_1$ (which estimates μ_1) is greater than $\overline{X}_2$ (which estimates μ_2) by an excessive amount. Since this happens exactly when $\overline{X}_1 - \overline{X}_2$ is greater than 0 by an excessive amount, we reject H_0 if the value of the Statistic (9.18) is greater than $t_{\alpha, n_1 + n_2 - 2}$. Similarly, for the problem

$$\begin{cases} H_0 : \mu_1 = \mu_2 \\ H_a : \mu_1 < \mu_2 \end{cases}$$

we should prefer H_a to H_0 if $\overline{X}_1$ is greatly less than $\overline{X}_2$, or if $\overline{X}_1 - \overline{X}_2$ is greatly less than 0, so we reject H_0 if the value of the Statistic (9.18) is less than $-t_{\alpha, n_1 + n_2 - 2}$.

EXAMPLE 1

Problem: For the following summary table for a two-sample situation, is there any reason to believe that the mean of the first population is less than that of the second? Let $\alpha = .05$ in testing the hypothesis that the population means are equal against the alternative that the second population mean is larger.

SAMPLE	n	df	$\overline{X}$	$\Sigma(X_{ij} - \overline{X}_i)^2$
1	15	14	21	1224
2	9	8	29	756
		22		1980

$$s^2 = \frac{1980}{22} = 90 \qquad s_d^2 = 90\left(\frac{1}{15} + \frac{1}{9}\right) = 16$$

Solution: The hypothesis is $H_0 : \mu_1 = \mu_2$. The alternative hypothesis is $H_a : \mu_2 > \mu_1$; therefore, we calculate

$$t = \frac{\overline{X}_2 - \overline{X}_1}{s_d} = \frac{29 - 21}{4} = 2$$

The critical region is $t > t_{.05,22} = 1.717$. The calculated t is greater than the tabular value; therefore, we reject the hypothesis and conclude that $\mu_2 > \mu_1$.

EXAMPLE 2

To test the hypothesis that in the example of Section 8.7 the two paints are equal in reflectivity we test the hypothesis $H_0: \mu_1 = \mu_2$ against the alternative $H_a: \mu_1 \neq \mu_2$. The t statistic is

$$t = \frac{\overline{X}_1 - \overline{X}_2}{s_d} = \frac{1.40}{.63} = 2.222$$

The critical region for α of .05 is $|t| \geq t_{.025,22} = 2.074$. Our calculated t falls into the upper tail of the critical region; therefore, we reject the null hypothesis and conclude that the means are not equal.

When we first introduced statistical analysis involving two populations, we stated three assumptions that must be met for the analyses to be valid. Let us now reconsider those assumptions and see which, if any, can be relaxed. The first, randomness, must be satisfied if we are to make probability statements in connection with our inferences.

As is frequently the case, the second assumption, normality, is not a strong assumption. This is because of the dictate of the central limit theorem that if the samples are sufficiently large, even large departures from normality will not affect the probabilities to any great extent.

The third assumption, common or homogeneous variance, cannot be relaxed. If the population variances are not the same, a significant result for the t test may be due to the different variances and not to different means. If this assumption is not satisfied there is no exact test of the hypothesis $H_0: \mu_1 - \mu_2 = \delta_0$. However, there are several approximate tests available; see Reference 9.6.

■ 9.9 PAIRED COMPARISONS

In the completely randomized experiment for comparing two treatments, the treatments are randomized over, or randomly assigned to, the whole set of experimental units. That is, a random device is used to determine which of the two treatments is applied to any given experimental unit. It is usual to restrict the randomization to the extent that each treatment is applied to the same number of units. As we have seen, the data from this type of experiment can be analyzed by using the two-sample techniques discussed in the previous section and in Section 8.7.

■ The material in this section is not required in the sequel.

For this type of experiment the efficiency and the precision of estimation are inversely proportional to the *pooled* estimate of variance, s^2; the smaller the variance, the greater the precision. To increase the precision of an experiment with a fixed amount of experimental material, we have to alter the design of the experiment in order to reduce the unexplained variation in the data as measured by s^2. Since s^2 is based on the variation among experimental units treated alike, either we must reduce this variation by using more homogeneous experimental material, or we must eliminate some of this variation by using a priori information about the responses of the experimental units. It is the latter approach we now consider.

If the experimental units occur in pairs or can be grouped into pairs in such a way that the variation in the responses between the members of any pair is less than the variation between members of different pairs, we can improve the efficiency of our experiment by randomizing the two treatments over the two members of each pair. We restrict our randomization so that each treatment is applied to one member of each pair; hence, we obtain a separate estimate of the difference between the treatment effects from each pair, and the variation between, or among, the pairs is not included in our estimate of the variance. If the variation among the pairs of units is large relative to the variation within the pairs, the variance will be smaller than if a completely randomized design were used.

For example, suppose we want to compare the efficiencies of two different barnacle-resistant paints and there are ten ships available for participation in the test. We could randomly assign the two paints to five ships each and then, after a suitable length of time, obtain a measure of the weight of barnacles clinging to each ship. Since the variance would be based on the variation among ships painted alike we could expect it to be large because the ships would very likely be sailing in different waters for different periods of time.

On the other hand, where the port side of a hull goes, the starboard goes also; therefore, we would probably have a much more precise experiment if we were to paint one side of each hull with one paint, and the other side with the second paint, tossing a coin to decide which side gets which paint. Each ship would then provide a measure of the difference in effectiveness of the two paints, and the variation due to other factors, such as time at sea and the parts of the world where the ships sailed, would be eliminated from our estimate of variance.

These same considerations lead to the use of identical twins or of littermates in animal experiments, and to grouping experimental units into pairs according to some factor such as age, weight, chemical composition, environment, etc. If the pairing is successful, much of the variation due to the factor on which it is based is eliminated from the estimate of variance and the efficiency of the experiment is increased accordingly. The paired experiment is the simplest example of a class of experimental designs known as the *randomized complete block* designs.

The analysis of paired comparisons reduces a problem to the single-sample techniques discussed in the early sections of Chapter 8 and this chapter. The data for a paired comparison may be represented as shown in the following table. In this table, a single observation is denoted by X_{ij}, where the first subscript refers to the treatment name and the second to the pair name. Thus X_{23} refers to the observation produced by applying the second treatment in the third pair.

			PAIR:			
TREATMENT	1	2	3	4	•••	n
1	X_{11}	X_{12}	X_{13}	X_{14}	. . .	X_{1n}
2	X_{21}	X_{22}	X_{23}	X_{24}	. . .	X_{2n}
DIFFERENCE	D_1	D_2	D_3	D_4	. . .	D_n

To estimate the mean difference and to test hypotheses about the difference between the treatment effects, we use the differences between the members of each pair,

$$D_j = X_{1j} - X_{2j} \qquad (9.19)$$

and treat these as a random sample from the population of differences. We then have

$$\bar{D} = \frac{1}{n} \Sigma D_j \qquad (9.20)$$

and

$$s_D^2 = \frac{1}{n-1} \Sigma (D_j - \bar{D})^2 \qquad (9.21)$$

as unbiased estimates for μ_D, the mean difference, and for σ_D^2, the variance of the differences, respectively. The variance of the mean difference $\bar{D}$ is estimated unbiasedly by

$$s_{\bar{D}}^2 = \frac{s_D^2}{n} \qquad (9.22)$$

Since the differences D_j constitute a single random sample from a normal population, the appropriate methods for finding confidence intervals and for testing hypotheses are the methods for a single normal population as considered in the early sections of Chapter 8 and this chapter. Thus $100(1 - \alpha)\%$ confidence limits for μ_D are given by

$$L = \bar{D} - t_{\alpha/2, n-1} s_{\bar{D}}$$

(9.23)

$$R = \bar{D} + t_{\alpha/2, n-1} s_{\bar{D}}$$

and the test of the hypothesis $H_0: \mu = \mu_0$ is the single-sample t test with $n - 1$ degrees of freedom based on the statistic

$$t = \frac{\bar{D} - \mu_0}{s_{\bar{D}}}$$

(9.24)

The test of the hypothesis of no difference in the effects of the two treatments is the special case $H_0: \mu_D = 0$.

The assumptions are the same as for the group comparisons: independent random observations, normal populations of responses, and common within-treatment variance.

EXAMPLE 1

In order to determine whether or not a particular heat treatment is effective in reducing the number of bacteria in skim milk, counts were made before and after treatment on 12 samples of skim milk, with the following results. The data are in the form of log DMC, the logarithms of direct microscopic counts.

	LOG DMC:		
SAMPLE	Before Treatment	After Treatment	DIFFERENCE $D_j = X_{1j} - X_{2j}$
1	6.98	6.95	.03
2	7.08	6.94	.14
3	8.34	7.17	1.17
4	5.30	5.15	.15
5	6.26	6.28	−.02
6	6.77	6.81	−.04
7	7.03	6.59	.44
8	5.56	5.34	.22
9	5.97	5.98	−.01
10	6.64	6.51	.13
11	7.03	6.84	.19
12	7.69	6.99	.70
			3.10

For these data:

$$\Sigma D_j = 3.10 \qquad\qquad \bar{D} = .258$$

$$\Sigma D_j^2 = 2.1990 \quad (\Sigma D_j)^2/n = .8008$$

$$\Sigma (D_j - \bar{D})^2 = \Sigma D_j^2 - (\Sigma D_j)^2/n = 2.1990 - .8008 = 1.3982$$

$$s_D^2 = \frac{1.3982}{11} = .12711$$

$$s_{\bar{D}}^2 = \frac{.12711}{12} = .0106$$

$$s_{\bar{D}} = .103$$

For the hypothesis of no effect, $H_0: \mu_D = 0$, against $H_a: \mu_D > 0$,

$$t = \frac{.258}{.103} = 2.50$$

From the tables, $t_{.05,11}$ equals 1.796; therefore, we reject the null hypothesis at the 5% level and conclude that the heat treatment has reduced the number of bacteria.

EXAMPLE 2

Problem: The training director of a large chain of department stores wants to know if a special one-day training program will help the chain's sales personnel to get along with customers better and reduce the number of complaints customers turn in on the salespeople.

Solution: In order to make a test of the program's effectiveness, the training director could present the program to 100 salespeople. She could then compare the number of complaints these people get in the following month to the number of complaints received on another 100 salespeople who have not had the program presented to them. However, she can receive more information about the effectiveness of the program by making a paired comparison of the first 100 salespeople's complaint records *before and after* the training program. Suppose that this was done and the D_i values were recorded for the 100 salespeople as the difference between the number of complaints received in the month before and the month after the program. This pairing eliminates the variations in complaint records between salespeople. The data showed $\bar{D} = 1.2$; that is, the group averaged 1.2 more complaints in the month prior to the training program. Also, $s_D^2 = 144$. Thus

$$s_{\bar{D}}^2 = \frac{144}{100} = 1.44 \quad \text{and} \quad s_{\bar{D}} = 1.2$$

For the hypothesis of no effect by the training program, $H_0: \mu_D = 0$, against the alternative $H_a: \mu_D > 0$, we will use an α of .05.

$$t = \frac{1.2}{1.2} = 1.0$$

Table VI in the Appendix does not give $t_{.05,99}$, but interpolation of the .05 column between 60 and 120 degrees of freedom gives a value for t of 1.664. Since our t value is only 1.0, we cannot conclude that the training program reduces the number of complaints on salespeople.

It should be noted that in this problem there is an unstated assumption that the salespeople served approximately the same number of customers in the two months for which the data were gathered. If this were not true and if, say, the second month were much busier than the first, then the smaller number of complaints was spread over a larger number of customers and the reduction in complaints might be significant after all. Under these circumstances it would be more appropriate to do one of two things. First, the *number of complaints per customer* might be compared before and after the training program. But if records on the numbers of customers served are not available, then the complaint records of the trained 100 should be compared with those for another 100 untrained personnel over the same sales period.

■ 9.10   GROUPS VERSUS PAIRS

Paired comparisons can give greater precision of estimation than group comparisons, but only if the pairing is effective. To be effective the pairs must be such that the variation among the pairs is greater than the variation between the units within the pairs. The degrees of freedom for the t test based on paired comparisons are equal to $n - 1$, compared with $2(n - 1)$ for the group comparison based on the same number of experimental units, and since degrees of freedom are like money in the bank they should not be invested unless a suitable return can be expected.

With pairing, then, as compared with grouping, we lose $n - 1$ degrees of freedom. But if the variance is reduced enough to more than compensate for their loss, then a gain in efficiency is achieved. If, however, the experimental material is nearly homogeneous, we have no justification for pairing, because the variation among pairs will be but little greater (if at all) than that within the pairs. We would squander degrees of freedom without getting a sufficient reduction in the variance.

For data from experiments involving two treatments, it is important that the proper method of analysis be employed. If the data are from a paired experiment and if we ignore the pairing and use the group analysis of Sections 8.7 and 9.8, the estimate of variance is inflated by the variation among pairs and will be too large. A difference between the treatment effects that is actually significant might not be detected because of the inflated variance estimate.

To illustrate what can happen when the two-sample techniques

■ The material in this section is not required in the sequel.

are used to analyze paired data, we use the example of Section 9.9 concerning the reduction due to a heat treatment of the number of bacteria in skim milk. The summary table is as follows:

SAMPLE	n	df	$\overline{X}$	SUM OF SQUARES
Before	12	11	6.721	8.0997
After	12	11	6.463	4.7750
	24	22		12.8747

$$s^2 = \frac{12.8747}{22} = .5852$$

$$s_d^2 = .5852 \left(\frac{1}{12} + \frac{1}{12} \right) = .0975 \text{ [by Formula (8.31)]}$$

$$s_d = .312$$

For the hypothesis that $\mu_d = 0$,

$$t = \frac{\overline{X}_1 - \overline{X}_2}{s_d} = \frac{.258}{.312} = .827$$

which is not significant.

Using this analysis we would conclude that the treatment had no effect. Notice that the standard deviation of the difference between the means is over three times as large as when these paired data were analyzed by the proper techniques.

Just as we have run into trouble using group techniques to analyze paired data, we are also in trouble if the data from a group experiment are paired. Random pairing would result in a loss of degrees of freedom and, at the same time, the variance estimate could be either larger or smaller than it should be and could lead to the wrong conclusion.

Suppose we were to systematically pair the results of a completely randomized experiment by ranking both sets of responses from high to low and then pairing the largest of each group, the next largest, and so on. Such a procedure would generally result in a variance estimate that is much smaller than it should be and, therefore, in t values that are too large. We might well detect a difference that does not exist.

The design of the study dictates the method of analysis that should be used. The reader will have an opportunity to examine situations where either group or paired techniques are appropriate at the end of this chapter.

■ 9.11   TESTING THE HYPOTHESIS THAT $p_1 = p_2$

When we have two binomial populations, it is often necessary to test the hypothesis that the proportion of successes is the same in both. For instance, we may wish to test the hypothesis that the rate of defectives produced by two production lines is the same. To test the hypothesis that $p_1 = p_2$, we obtain independent random samples of n_1 items from the first production line and n_2 items from the second. The first sample yields $f_1 = r_1/n_1$, which is the proportion of defects in that sample. The second sample yields $f_2 = r_2/n_2$. Then we use a test based on the standard normal distribution. The hypothesis states that the proportions are equal to each other, but does not specify the common value; therefore, we must estimate the variance of the difference under the assumption that the proportions are the same.

If $p_1 = p_2 = p$, both f_1 and f_2 are unbiased estimates for p, but the best estimate will be obtained by pooling the two samples into one sample of size $n_1 + n_2$ with the pooled number of successes $r_1 + r_2$. The observed fraction for the combined samples,

$$f = \frac{r_1 + r_2}{n_1 + n_2} \tag{9.25}$$

is our best estimate for the common value of p.

If $p_1 = p_2$, the estimated variance of $f_1 - f_2$ is

$$\frac{f(1-f)}{n_1} + \frac{f(1-f)}{n_2}$$

as was described in Section 7.7. This expression can be reduced to

$$f(1-f)\left(\frac{1}{n_1} + \frac{1}{n_2}\right) \tag{9.26}$$

Under the null hypothesis, the quantity

$$Z = \frac{f_1 - f_2}{\sqrt{f(1-f)\left(\dfrac{1}{n_1} + \dfrac{1}{n_2}\right)}} \tag{9.27}$$

has approximately a standard normal distribution if n_1 and n_2 are large. This, then, provides a basis for an approximate test procedure. The hypothesis test consists of computing a Z value from the sample results and comparing the computed Z to the standard normal Z which is exceeded with probability α (or $\alpha/2$ if the test is two-tailed).

■ The material in this section is not required in the sequel.

EXAMPLE 1

Problem: In order to test the effectiveness of the approach and layout of two direct-mail brochures, a marketing manager mailed out 150 copies of each brochure and recorded the number of responses generated by each. There were 30 responses generated by the first brochure and 10 generated by the second. Can the marketing manager conclude that the first brochure is more effective? Let $\alpha = .05$.

Solution: Here we want to test the hypothesis $H_0: p_1 = p_2$ against $H_a:$ $p_1 > p_2$, where p_1 is the proportion of responses that will be generated from a nationwide mailing of the first brochure and p_2 is the proportion of responses generated by the second brochure.

$$f_1 = \frac{30}{150} = .200 \qquad f_2 = \frac{10}{150} = .067$$

$$f = \frac{r_1 + r_2}{n_1 + n_2} = \frac{30 + 10}{300} = \frac{40}{300}$$

$$s_{f_1-f_2}^2 = \frac{40}{300}\left(\frac{260}{300}\right)\left(\frac{1}{150} + \frac{1}{150}\right) = .001541$$

$$s_{f_1-f_2} = .0392$$

Our calculated Z is

$$Z = \frac{f_1 - f_2}{s_{f_1-f_2}} = \frac{.133}{.0392} = 3.40$$

Since the critical region in this test is $Z > Z_{.05} = 1.645$, we reject the null hypothesis and conclude that the first brochure is more effective in generating responses.

EXAMPLE 2

Problem: A brick-making firm makes its bricks by two different processes. In samples of 200 bricks from the first process and 300 bricks from the second process, it was found that 20 of the first type broke during baking in the kiln and 45 of the second type broke in the kiln. Test the hypothesis that the kiln breakage rates are the same against the alternative hypothesis that they are different. Let α be .02.

Solution: For this problem we have the following:

$$H_0: p_1 = p_2$$

$$H_a: p_1 \neq p_2$$

$$R: Z > Z_{.01} = 2.326 \quad \text{or} \quad Z < -Z_{.01} = -2.326$$

$$f_1 = \frac{20}{200} = .10 \quad f_2 = \frac{45}{300} = .15$$

The pooled value of f, the best estimate of the breakage rate in the two populations, assuming these rates are equal, is

$$f = \frac{20 + 45}{200 + 300} = \frac{65}{500} = .13$$

Then

$$s^2_{f_1 - f_2} = (.13)(.87)\left(\frac{1}{200} + \frac{1}{300}\right) = .000943$$

and finally,

$$s_{f_1 - f_2} = \sqrt{.000943} = .031$$

Our calculated Z is

$$Z = \frac{f_1 - f_2}{s_{f_1 - f_2}} = \frac{.10 - .15}{.031} = -1.61$$

Since this value is not within the rejection region, we cannot reject the null hypothesis that the breakage rates of the two processes are the same. However, due to considerations mentioned previously, we do not "accept" the hypothesis either.

9.12   TESTING THE HYPOTHESIS THAT $\sigma_1^2 = \sigma_2^2$ AND THE F DISTRIBUTION

In Sections 9.8 and 9.9 (which readers skipping the optional material will not have read), we noted that in order to test the hypothesis H_0: $\mu_1 = \mu_2$, we had to assume three conditions existed:

1. The samples taken for the hypothesis test were random and independent.
2. The populations from which the samples were drawn were normally distributed.
3. The two populations had the same variance, σ^2.

We can insure the independence and randomness of samples by using careful sampling design and random number tables to obtain the samples. We will discuss in Chapter 11 how a new probability distribution called the chi-square distribution can be used to test a set of sample data for normality. But how, one might ask, can we test this third assumption: that two populations have the same variance. Also, there are numerous situations in which we might want to test

$H_0: \sigma_1^2 = \sigma_2^2$ completely outside the context of the problems in Sections 9.8 and 9.9. This section introduces the F probability distribution and demonstrates how it can be used to test the hypothesis that two populations have equal variances.

If two populations have the same variance, then $\sigma_1^2 = \sigma_2^2$. If we take a sample from each of these populations, the best estimates we will obtain for the population variances are s_1^2 and s_2^2. If these two values differ from one another a great deal, it is evident that the two population variances are not equal. We can compare the sample variances by computing their ratio s_1^2/s_2^2. When repeated independent random samples are drawn from two normal populations with the same variance, this ratio follows the F probability distribution. If this ratio is very large or very small, then the hypothesis that the two populations have the same variance is probably not true.

In order to perform this hypothesis test, we proceed in the following way:

STEP 1.   Set up the null hypothesis $H_0: \sigma_1^2 = \sigma_2^2$ and the alternative hypothesis (which is usually that the populations do not have equal variances).

STEP 2.   Calculate the F ratio as $F = s_1^2/s_2^2$. However, since the designation of which population is the first and which is the second is somewhat arbitrary, it is convenient if we always place the *largest* sample variance in the numerator of this ratio. If the null hypothesis is true, we expect this ratio to be 1.0. Calculated F values a great deal larger than 1.0 cause us to reject the null hypothesis of equal variances.

The big question is, of course, how big does the F ratio of s_1^2/s_2^2 have to get before we can safely reject the hypothesis that the two populations have the same variance? In order to answer this question, we must compare the F ratio calculated from our data to a theoretical F ratio located in Table VIII in the Appendix.

Each F value has two different degrees-of-freedom figures associated with it. We call v_1 the number of degrees of freedom in the larger sample variance, s_1^2, and v_2 the number of degrees of freedom in the smaller sample variance, s_2^2. Thus,

$$v_1 = n_1 - 1$$

$$v_2 = n_2 - 1$$

In order to determine if a calculated $F = s_1^2/s_2^2$ is so large that we should reject the hypothesis that the two population variances are equal, we must consult the F table, Table VIII of the Appendix. In that table, we find the F values that are exceeded by pure chance with only 5 percent probability for various combinations of degrees of freedom in the numerator and the denominator of the ratio. For example, if $\nu_1 = n_1 - 1 = 4$ and $\nu_2 = n_2 - 1 = 7$, from Table VIII we find that an F ratio of 4.1203 is exceeded with only a 5 percent probability when 4 and 7 degrees of freedom are involved. Thus if $n_1 = 5$ and $n_2 = 8$ and if the level of α (probability of a Type I error that we are willing to tolerate) for testing the hypothesis H_0: $\sigma_1^2 = \sigma_2^2$ were .05, then we would reject the hypothesis if s_1^2/s_2^2 exceeded 4.1203. Thus, the third step in the hypothesis test is:

STEP 3. To determine if the calculated ratio is so large as to cause us to reject the hypothesis that $\sigma_1^2 = \sigma_2^2$, compare the calculated F value with the F value in Table VIII using $n_1 - 1$ for ν_1 and $n_2 - 1$ for ν_2, remembering that the sample with the larger variance is designated by the subscript 1. Reject the hypothesis H_0: $\sigma_1^2 = \sigma_2^2$ if the calculated F ratio exceeds the table value of F for $\alpha/2$ (since the test is two-tailed). Table VIII contains F values for α levels of .05, .025, .01, and .005. The values in this table are designated F_{α, ν_1, ν_2}.

EXAMPLE 1

Problem: A small company offers a service in which they collect packages to be sent by air freight from businesses in downtown Washington, D.C., to New York City. The company likes to send out its last delivery to the air freight office at National Airport in Washington in time to make the 6:00 P.M. flight to New York. Since they must leave for the airport during the rush hour, the drivers are interested in which of two routes is the fastest. Thus, they took Route A for 5 days one week and Route B for 5 days the next week. The means and standard deviations of the samples' travel times follow.

$$\overline{X}_A = 34.2 \text{ minutes} \qquad \overline{X}_B = 33.8 \text{ minutes}$$

$$s_A = 6.1 \text{ minutes} \qquad s_B = 16.4 \text{ minutes}$$

Solution: In order to proceed with this problem, we have to assume that the route travel times are normally distributed, the samples are random weeks, and the weeks' times are independent of one another. Route B looks slightly better than Route A since the sample mean time is smaller (but not by much). In order to test to see if the difference in population mean times for the two routes are the same, H_0: $\mu_A = \mu_B$, we would have to be able to assume that the two population variances are the same, $\sigma_A^2 = \sigma_B^2$, and the sample data suggest that this is not the case. Thus, we should begin with testing the hypothesis that the two routes have the same travel time variance.

$$H_0 : \sigma_A^2 = \sigma_B^2$$

$$H_a : \sigma_A^2 \neq \sigma_B^2$$

The calculated F value is found by taking the ratio of the larger to the smaller sample variance:

$$F = (16.4)^2 / (6.1)^2 = 7.23$$

We now compare this value with the Table VIII F value for an $\alpha/2$ value of .05 and $\nu_1 = 4$ and $\nu_2 = 4$ and find that

$$7.23 > F_{.05,4,4} = 6.3883$$

Thus we reject the hypothesis that the two routes have the same travel time variances.

This conclusion restricts us from using a t test to test the hypothesis that the population mean travel times are the same. (It should be noted that if the hypothesis $H_0 : \sigma_1^2 = \sigma_2^2$ had not been rejected, then we would need to take a separate sample to test $H_0 : \mu_1 = \mu_2$. This is because the sample to check an assumption underlying a test and the sample to actually perform the test should be independent—although in practice some statisticians ignore this restriction.) In this example, we would use common sense and probably advise the company to use Route A. The route *appears* to have almost the same travel time, on the average, and it is more reliable. That is, the smaller sample standard deviation for this route indicates that each trip on this route takes about the same time. The large sample variance for Route B indicates that on this route some of the times are very fast and some are very slow (perhaps due to having to wait at a railroad crossing). Since Route A is more predictable, it would likely be the preferred route.

EXAMPLE 2

Problem: A company has two production lines that mill parts to a predetermined length. Quality control engineers are interested in knowing if the two lines have similar variances around the predetermined length. Thus, they took a sample of 16 items from the older line and 20 items from the newer line. They found:

$$s_{older} = 12 \quad \text{and} \quad s_{newer} = 9$$

Test the hypothesis that the two lines have the same variance from the predetermined length. Let $\alpha = .05$.

Solution: The null and alternative hypotheses are

$$H_0: \sigma^2_{\text{older}} = \sigma^2_{\text{newer}}$$

$$H_a: \sigma^2_{\text{older}} \neq \sigma^2_{\text{newer}}$$

We can now calculate the sample F ratio, remembering to put the larger sample variance in the numerator of the ratio. Thus,

$$F = \frac{s^2_{\text{older}}}{s^2_{\text{newer}}} = \frac{(12)^2}{(9)^2} = \frac{144}{81} = 1.778$$

We must compare this calculated value of F to the theoretical value of F which is exceeded by chance with only .025 probability (since this is a two-tailed test). To find this value we enter the second section of Table VIII where $\nu_1 = 16 - 1 = 15$ and $\nu_2 = 20 - 1 = 19$ and find $F_{.025,15,19} = 2.6171$. Since our calculated F value is smaller than this, we cannot reject the hypothesis that both lines have the same variance from the predetermined length.

9.13   SUMMARY

In this chapter, we have discussed several types of hypothesis testing. The first two tests involved single populations and required our comparing the sample evidence from one sample to the hypothesized value of the population characteristic. In the case of a single population with a continuous measure, we computed the sample $\overline{X}$ and compared it to the hypothesized value of μ. In the case of a single binomial population, we computed the sample frequency of success, f, and compared it to the hypothesized value of p. Summary Table 9.1 presents the statistics and assumptions used in these cases as the first two rows of the table.

When one wishes to test the hypothesis that two populations are the same in some regard, there are four hypothesis tests to keep in mind. The first involves the comparison of two populations with the same continuous measure and the test of $H_0: \mu_1 = \mu_2$. Two samples are taken—one from each population—and the sample means, $\overline{X}_1$ and $\overline{X}_2$, are compared using the test shown in the third row of the Summary Table.

In some cases where two populations are being compared, the sample items (people, accounts, firms, or automobiles) differ more from each other within each population than they do between the populations. This is often the case in experiments involving a comparison of performance or attitudes "before and after" some type of action is taken. Then paired comparisons are often better in identifying any real differences in the two populations. The fourth row of the Summary

Table illustrates the test of $H_0: \mu_1 = \mu_2$ using the paired comparison approach.

The fifth row of the Table 9.1 presents the approach to testing $H_0: p_1 = p_2$ for two binomial populations. Again, two samples are taken from each population, and the sample frequencies of success, f_1 and f_2, are compared.

Finally, we discussed the hypothesis test which checks the assumption used in the third test mentioned—that is, $H_0: \sigma_1^2 = \sigma_2^2$. There are also situations where we wish to test this hypothesis on its own and not as a part of checking the assumptions of another hypothesis test. In testing this hypothesis, we take samples from each of two populations and obtain s_1^2 and s_2^2 from the sample data. The significance in the different sizes of these two sample variances can be tested as is shown in the last line of Table 9.1.

PROBLEMS

1. A box contains four marbles, some of which are white. The others are black. To test the hypothesis that there are two of each color we select two *without* replacement and conclude that there are not two of each if both marbles selected are of the same color.

   a. What is the probability we will come to the wrong conclusion if, in fact, there are two of each color in the box? That is, what is the probability of selecting either two whites or two blacks?

   b. Assume that there are two of each type marble in the box. If you select two from the box and find they are both black, what type of error will you incur, Type I or Type II?

2. An employer has always assumed that the company's employees are honest. However, there have been many shortages from the cash register lately. There is only one employee who could have taken any money from the register during these periods. Realizing that the shortages might have resulted from the employee's inadvertently giving incorrect change to customers, the employer does not know whether to forget the situation or accuse the employee of theft.

   a. What is the null hypothesis in this problem?

   b. What are the Type I and Type II errors?

3. A tuna fish canning company suspects that several thousand cases of recently canned tuna may be contaminated with a substance causing nausea and vomiting in those who eat the tuna. The company decides to examine several randomly selected cases. Their null hypothesis is that the cases are contaminated.

   a. What are the Type I and Type II errors in this problem?

   b. Would the company want a small α or a small β if it had to choose one or the other to be small?

4. Given each of the following sets of values, test the indicated hypothesis.

TABLE 9.1 Summary of Hypothesis Tests

POPULATION TYPE	HYPOTHESIS TO BE TESTED	TEST STATISTIC	COMMENTS AND ASSUMPTIONS
ALL HYPOTHESIS TESTS →			All the tests described assume random samples have been taken. When two samples are taken for the same test, the two samples must be independent, except in Situation 4 below.
1. One population, continuous measure	$H_0: \mu = \mu_0$	$t = \dfrac{\bar{X} - \mu_0}{s/\sqrt{n}}$	a. If $n < 30$, then this test is valid only if the population is normally distributed. b. If σ is known and used rather than s in this calculation, then the ratio is a Z value. c. When $n > 30$, then $t \approx Z$ even if the population is not normally distributed. d. The t value has $n - 1$ degrees of freedom.
2. One binomial population	$H_0: p = p_0$	$Z = \dfrac{r - np_0}{\sqrt{np_0(1 - p_0)}}$ or $Z = \dfrac{f - p_0}{\sqrt{p_0(1 - p_0)/n}}$	a. Use the first form of the test statistic when the sample results are presented in terms of "number of successes." Use the second form when sample results are presented as "proportion" or "fraction of successes." b. This test is valid only for large samples ($n > 30$, but $n = 100$ is better).
3. Two populations with continuous measures X_1 and X_2	$H_0: \mu_1 = \mu_2$	$t = \dfrac{\bar{X}_1 - \bar{X}_2}{s_d}$ where $s_d = \sqrt{s^2\left(\dfrac{1}{n_1} + \dfrac{1}{n_2}\right)}$	a. If either sample is small ($n_i < 30$), then we must assume the populations are normally distributed. b. Both populations must have the same variance, which is estimated by s^2, the weighted average of s_1^2 and s_2^2: $s^2 = \dfrac{(n_1 - 1)s_1^2 + (n_2 - 1)s_2^2}{n_1 + n_2 - 2}$

			c. The t value has $n_1 + n_2 - 2$ degrees of freedom.
4. Two populations with continuous measures X_1 and X_2	$H_0: \mu_1 = \mu_2$	$t = \dfrac{\bar{D}}{s_D/\sqrt{n}}$ where s_D is the standard deviation of the *paired* differences.	a. $D_j = X_{1j} - X_{2j}$, the difference of paired observations. b. n is the number of pairs. c. If the sample size is small ($n < 30$), then we must assume the populations are normally distributed. d. Differences in sample items in the same population must exceed differences in paired items across the two populations before this test is preferred to Test 3. e. The t value has $n - 1$ degrees of freedom.
5. Two binomial populations	$H_0: p_1 = p_2$	$Z = \dfrac{f_1 - f_2}{\sqrt{f(1-f)\left(\dfrac{1}{n_1} + \dfrac{1}{n_2}\right)}}$	a. Both sample sizes must be large ($n_i > 30$). b. f is the pooled estimate of the common value for p_1 and p_2. It is found as follows: $f = $ (total number of successes) $/(n_1 + n_2)$
6. Two populations with continuous measures X_1 and X_2.	$H_0: \sigma_1^2 = \sigma_2^2$	$F = \dfrac{s_1^2}{s_2^2}$	a. The populations must be normally distributed. b. The larger sample variance is always placed in the numerator. The H_a is $\sigma_1^2 \neq \sigma_2^2$. c. The F value has two degrees-of-freedom values: $\nu_1 = n_1 - 1$ $\nu_2 = n_2 - 1$ d. If this test is used to check the assumption in test 3, a new sample is needed before Test 3 can be conducted.

    a. $n = 25$, $\overline{X} = 28$, $s = 3$, $H_0: \mu = 20$, $H_a: \mu > 20$, $\alpha = .05$

    b. $n = 25$, $\overline{X} = 50$, $s^2 = 100$, $H_0: \mu = 55$, $H_a: \mu \neq 55$, $\alpha = .01$

    c. $n = 9$, $\overline{X} = 329.3$, $s^2 = 9$, $H_0: \mu \leq 327$, $H_a: \mu > 327$, $\alpha = .10$

5. For each of the following, test the indicated hypothesis.

    a. $n = 16$, $\overline{X} = 1550$, $s^2 = 12$, $H_0: \mu = 1500$, $H_a: \mu > 1500$, $\alpha = .01$

    b. $n = 9$, $\overline{X} = 10.1$, $s^2 = .81$, $H_0: \mu = 12$, $H_a: \mu \neq 12$, $\alpha = .05$

    c. $n = 49$, $\overline{X} = 17$, $s = 1$, $H_0: \mu \geq 18$, $H_a: \mu < 18$, $\alpha = .05$

6. Sensitivity tests were made on 18 randomly selected transistors of a given type. The mean was 2.5 microvolts and the sample variance, s^2, was .48. Would we conclude that the mean sensitivity of transistors of this type in the given circuit is greater than 2.0 microvolts? That is, use $H_a: \mu > 2.0$. Let $\alpha = .05$.

7. The estimated variance based on four measurements of a spring tension was .25 gram. The mean was 37 grams. Test the hypothesis that the true value is 35 grams. Use $\alpha = .10$ and $H_a: \mu > 35$.

8. With a random sample from a normal population with n equal to 8 and $\overline{X}$ equal to 62, and given that the variance of the population, σ^2, is known to be equal to 2, can we reject the null hypothesis $H_0: \mu \leq 60$? Let $\alpha = .10$.

9. A population has a variance σ^2 of 100. A sample of 25 from this population had a mean equal to 17. Can we reject $H_0: \mu = 21$ in favor of $H_a: \mu \neq 21$? Let $\alpha = .05$.

10. A sample for which n was 25 had a mean equal to 33 and a sample variance equal to 100. Would we have reason to reject the null hypothesis $H_0: \mu \geq 37$? Let $\alpha = .025$.

11. Given the following information, test the stated hypothesis at the given level.

    a. $n = 9$, $\overline{X} = 76$, $\Sigma(X_i - \overline{X})^2 = 32$, $H_0: \mu = 75$, $H_a: \mu \neq 75$, $\alpha = .10$

    b. $n = 25$, $\Sigma X_i = 500$, $\Sigma X_i^2 = 12,400$, $H_0: \mu \leq 17$, $H_a: \mu > 17$, $\alpha = .10$

12. Nine rafters were tested for breaking load, giving a mean breaking strength of 1500 pounds and a sample standard deviation, s, of 110 pounds. At the 5% level of significance, would we reject the null hypothesis that the mean breaking load is 1600 pounds or more?

13. Suppose that a sample of 15 rulers from a given supplier have an average length of 12.04 inches and that the sample standard deviation is .015 inches. If α is .02, can we conclude that the average length of rulers produced by this supplier is 12 inches or should we accept $H_a: \mu \neq 12.00$?

14. The manufacturer of a certain type of four-wheel drive vehicle claims that a driver can cover 150 miles or more over rough terrain in an 8-hour day. An independent testing agency was hired by the manufacturer's competitor to test this claim. The agency drove 12 of the vehicles an average of 145 miles over rough terrain during an 8-hour period. The standard deviation of the sample was 8.3 miles.

    a. What is the null hypothesis in this problem? (*Hint:* Give the manufacturer the benefit of the doubt.)

b. Test the hypothesis using 5% as the risk you are willing to accept for a Type I error.

15. It is known that mean time between malfunctions for a certain type of electronic equipment that uses electron tubes is 600 hours. It is claimed that a completely transistorized model will, on the average, operate *at least* three times as long without malfunctioning. A number of these supposedly superior equipments are put into service and give rise to 61 malfunctions with a mean time between malfunctions equal to 1350 hours. The sample standard deviation is 980 hours. At the 5% level, is the claim justified?

16. Five years ago, the State Department of Transportation did a study of automobile pollution at a particular location. During the month of November, they found that the mean pollution count was 132. During November of this year, the Department took a random sample of $n = 8$ days and found that the sample mean pollution count was 120 and the sample standard deviation was $s = 10$. Test the hypothesis that the population mean count is still 132 against the alternative that the mean is now less than 132. Use $\alpha = .025$.

17. The keypunch operation in a large computer department claims that it gives its customers a turnaround time of 6.0 hours or less. In order to test this claim, one of the customers took a sample of thirty-six jobs and found that the sample mean turnaround time was $\overline{X} = 6.5$ hours with a sample standard deviation of $s = 1.5$ hours. Use $H_0: \mu = 6.0$ and $H_a: \mu > 6.0$ and $\alpha = .10$ to test the keypunch operation's claim.

18. The drained weights in ounces for a sample of 15 cans of fruit are as follows. At the 5% level of significance, test the hypothesis that on the average a 12-ounce drained weight standard is being maintained. Use $H_a: \mu \neq 12.0$ as the alternative hypothesis.

12.1	12.1	12.3	12.0	12.1
12.4	12.2	12.4	12.1	11.9
11.9	11.8	11.9	12.3	11.8

19. The following are the times in seconds that it took the sand in a sample of timers to run through. At the 10% level can we conclude that the mean for timers of this type is not equal to the nominal three minutes? Use $H_a: \mu \neq 180$ as the alternative hypothesis.

190	198	180	181	208	198
199	176	174	183	188	165

20. In a sample of 400 seeds, 326 germinated. At the 2.5% level would we reject the null hypothesis $H_0: p \geq .90$?

21. In a sample of 50 die castings, we found 8 with defects. On the basis of this sample have we any reason to believe that more than 8% of all such castings would show defects? Use a 5% level of significance to test $H_0: p \leq .08$.

22. In a sample of 144 voters, it was found that 84 were in favor of a bond issue. Test the hypothesis that opinion is equally divided on this issue. Let $\alpha = .10$ and $H_a: p \neq .50$.

23. On 384 out of 600 randomly selected farms, it was discovered that the

farm operator was also the owner. Is there reason to believe we should reject $H_0: p \le .60$ if we use $\alpha = .10$?

24. A U.S. automobile manufacturer asked each of 50 randomly selected drivers to compare the ride in his car with the smoothness of ride in a \$20,000 European touring car. The number that preferred the ride in the American car was 38. Would the manufacturer be justified in making the claim in advertising that his car has the smoother ride? That is, can he reject $H_0: p = .50$ in favor of $H_a: p > .50$ where p is the proportion of people preferring the American car's ride. Let $\alpha = .05$.

25. The study mentioned in Problem 16 showed that the proportion of cars tested failing to meet state pollution control standards was .37 in the study of 5 years ago. In a sample of $n = 100$ cars this year, the proportion not meeting the standards was .28. Test the hypothesis $H_0: p = .37$ against the alternative $H_a: p < .37$ using $\alpha = .025$.

26. The rate of defectives produced by a certain production line is usually 15%. The quality control engineer took a sample of $n = 80$ items from this line and found $r = 18$ defectives. Test $H_0: p = .15$ against $H_a: p > .15$ using $\alpha = .10$.

27. An aircraft manufacturer receives large lots of wing struts from a subcontractor who guarantees that no more than 5% of each lot will fail to meet design specifications. One hundred struts are inspected in each lot.
   a. What are the null and alternative hypotheses?
   b. How many struts in the sample should fail to meet specifications before the hypothesis is rejected, if we desire an α of about .05? (*Note:* This is a special problem in which you are asked to find the rejection region.)

28. Independent random samples from two normal populations with common variance gave the following results. Can we reject the null hypothesis $H_0: \mu_1 = \mu_2$ with $\alpha = .02$? (*Hint:* Use interpolation in the t table.)

SAMPLE	n	$\bar{X}_i$	$\Sigma(X_{ij} - \bar{X}_i)^2$
1	10	62	340
2	20	70	302

29. A mining company owns two mines. The number of tons of ore taken from each of the mines each day is normally distributed with a common variance. The amount of ore extracted was recorded over 10 days at mine 1 and 15 days at mine 2. Given the following summary of data, will the company be justified in concluding that the mean daily yield of the first mine is equal to that of the second? That is, test $H_0: \mu_1 = \mu_2$ against $H_0: \mu_1 \ne \mu_2$ using $\alpha = .10$.

SAMPLE	n	$\bar{X}_i$	$\Sigma(X_{ij} - \bar{X}_i)^2$
1	10	25	250
2	15	30	302

30. Independent random samples from normal populations with the same variance gave the following results. Can we conclude that $H_0: \mu_1 = \mu_2$

should be rejected in favor of $H_a: \mu_1 < \mu_2$ when α is .025?

SAMPLE	n	MEAN	SUM OF SQUARES
1	10	17	106
2	6	21	124

31. Independent random samples from normal populations with the same variance gave the following results. Can we conclude that the difference between the means, $\mu_1 - \mu_2$, is less than 5? That is, test $H_0: \mu_1 - \mu_2 \geq 5$ with $\alpha = .05$.

SAMPLE	n .	MEAN	STANDARD DEVIATION
1	15	22	9
2	9	25	7

32. Independent random samples from normal populations with the same variance gave the following results. Test the hypothesis that $\mu_1 - \mu_2 = 10$ at the 2% level of significance. Use $H_a: \mu_1 - \mu_2 \neq 10$. [See Equation (9.17).]

SAMPLE	n	MEAN	STANDARD DEVIATION
1	12	62.3	10
2	20	77.1	21

33. The following are yields in bushels per acre for two oat varieties. Each was tried on eight different plots. Can we conclude that the yields are the same for the two varieties, A and B? Or is the alternative hypothesis more reasonable: $H_a: \mu_A \neq \mu_B$ if we use $\alpha = .10$?

A	81.2	72.6	56.8	76.9	42.5	49.6	62.8	48.2
B	56.6	58.6	45.4	39.1	42.8	65.2	40.7	49.9

34. The following are percentages of fat found in samples of two types of meat. Do the meats have different fat contents? That is, test $H_0: \mu_A = \mu_B$ against $H_a: \mu_A \neq \mu_B$ using $\alpha = .10$.

MEAT A	30	26	30	19	25	37	27	38	26	31
MEAT B	40	34	28	29	26	36	28	37	35	42

35. The daily catch of two fishing boats was recorded on a random basis. The results of two independent random samples are as follows. Test $H_0: \mu_1 = \mu_2$ against the alternative $H_a: \mu_1 \neq \mu_2$ with $\alpha = .02$.

BOAT 1	108	110	103	100	107	107	101
BOAT 2	113	110	108	98	111	112	110

36. To test the effect of two different sales approaches, each of 16 commercial cleaning compound salespeople used alternately both of the two approaches for the same period of time and the same number of sales contacts. The sales of the compound in pounds for each salesperson (**1–16**) are as follows.

    a. Why would paired comparisons be expected to provide a better test in this problem?

    b. Test the hypothesis that the mean improvement in sales is zero against the alternative that $\mu_1 - \mu_2 > 0$ using $\alpha = .05$.

APPROACH 1				APPROACH 2			
1	130	**9**	73	**1**	44	**9**	110
2	120	**10**	56	**2**	62	**10**	38
3	61	**11**	65	**3**	77	**11**	66
4	111	**12**	71	**4**	58	**12**	120
5	93	**13**	109	**5**	88	**13**	81
6	56	**14**	122	**6**	101	**14**	54
7	25	**15**	85	**7**	42	**15**	31
8	123	**16**	131	**8**	57	**16**	11

37. To generate some data for illustrating the differences between grouped and paired comparisons, 10 egg timers were timed in two positions: vertical, and at 20 degrees from vertical. The values in seconds are as follows. Use $H_a: \mu_1 \neq \mu_2$ and $\alpha = .10$.

    a. Treat the data as two groups and calculate the t for testing the hypothesis that the mean time is the same in both positions.

    b. Use paired comparison methods to find the value of t for testing the same hypothesis. Note that when differences among timers are eliminated from the variance, as is the case with paired comparisons, the t value is larger.

    c. What conclusions would be reached in these two analyses?

POSITION	*1*	*2*	*3*	*4*	*5*	*6*	*7*	*8*	*9*	*10*
Vertical	170	191	205	181	210	192	183	205	185	216
Tipped	160	197	175	181	163	172	177	185	183	177

TIMER:

38. To compare the average weight gains of pigs fed two different rations, nine pairs of pigs were used. The pigs within each pair were littermates, the rations were assigned at random to the two animals within each pair, and they were individually housed and fed. The gains, in pounds, after 30 days are as follows. Test the hypothesis that the mean gains are the same against the alternative that feed A produces a larger gain. Let $\alpha = .01$.

RATION	*1*	*2*	*3*	*4*	*5*	*6*	*7*	*8*	*9*	SUM
A	60	38	39	49	49	62	53	42	58	450
B	53	39	29	41	47	50	56	47	52	414

LITTER:

39. The management in a large assembly plant wishes to make changes in assembly techniques, but union members are afraid that the changes will result in lower wages for their workers, who are paid on the basis of output. Thus, management set up a test using a randomly selected group of 14 employees who had varying levels of experience in the assembly process. The assembly process was run using the old assembly techniques, and then, after a suitable training period, it was run using the new techniques. The hourly wages earned by the 14 employees using the two techniques are as follows. Is there sufficient evidence to indicate that the labor force will be able to make more money using the new techniques? Test the hypothesis that the mean wage change is zero or less against the alternative hypothesis that the change is positive. Use paired comparisons and let the probability of a Type I error be .10.

| | TECHNIQUE: | | | | TECHNIQUE: | |
EMPLOYEE	Old	New	EMPLOYEE		Old	New
1	3.96	4.61	8		7.35	8.12
2	6.13	6.33	9		4.13	4.75
3	8.21	7.38	10		6.37	6.43
4	6.05	6.87	11		7.08	7.38
5	5.21	6.62	12		7.56	6.45
6	5.25	6.82	13		6.24	6.16
7	6.32	6.87	14		8.05	8.14

40. A random sample of 900 persons included 467 smokers and 433 nonsmokers. Of the smokers, 18 were found to have required hospitalization in the past year, while only 2 nonsmokers had required hospitalization. From these data, could a health insurance company reject the hypothesis $H_0: p_s = p_{ns}$ in favor of $H_a: p_s > p_{ns}$, where p_s is the proportion of smokers requiring hospitalization and p_{ns} is the proportion of nonsmokers requiring hospitalization. Use $\alpha = .01$.

41. In a test of alloy applicability, parts for an appliance were manufactured first from one alloy and then from another. One hundred samples of parts made from each type of alloy were subjected to shock testing. Defects developed in 18 of those made from alloy I and in 26 of those made from alloy II. Can we reject $H_0: p_I = p_{II}$ in favor of $H_a: p_I < p_{II}$ using $\alpha = .025$? Note that p represents the defective rate.

42. An official of the U.S. Savings and Loan League feels that savings and loan associations should be more active in soliciting the savings of minority-group members. To support this position, the official took a random sample of 1000 people in various midwestern cities. Of the 100 minority-group members in this sample, only 20 had accounts with a savings and loan association. But 480 of the 900 other people in the sample had S&L accounts. Test the hypothesis that the proportion of account holders is the same in the two groups against the alternative hypothesis that the proportion of account holders is less among minority group members. Let $\alpha = .10$.

43. In a sample of 200 males with MBA degrees it was found that 120 had

received promotions within two years on the job. However, a sample of 90 women with MBA degrees showed only 45 had received promotions within two years on the job.

  a. What are the appropriate null and alternative hypotheses assuming a one-tailed test is desired?

  b. Perform the test using $\alpha = .05$.

44. In Section 7.5, there is a discussion of a minority group whose members were being delayed in crossing a border. A sample of $n = 64$ showed that the sample mean delay time for members of the minority group was $\overline{X} = 10$ minutes. If the minority group representative mentioned in that discussion were to test the null hypothesis $H_0: \mu = 8.0$ against the alternative hypothesis $H_a: \mu > 8.0$ and were to reject the null, what would be the chance of a Type I error, α?

45. See Example 2 in Section 7.6. Assume that samples of $n_1 = 100$ and $n_2 = 100$ randomly selected days are taken from the two processes with the result that $\overline{X}_1 - \overline{X}_2 = .5$ kilograms. If you decide to reject the hypothesis that the two processes have the same mean usage, what is the probability of a Type I error if:

  a. You are using a one-tailed test?

  b. You are using a two-tailed test?

46. Test the hypothesis that the two populations in Problem 31 have the same variance. Let $\alpha = .05$.

47. Test the hypothesis that the two populations in Problem 32 have the same variance. Let $\alpha = .10$. What does your conclusion suggest about the validity of the test in Problem 32?

48. Test the hypothesis that the two populations in Problem 33 have the same variance. Let $\alpha = .01$.

49. Two economic forecasting services have the same mean error in predicting the gross national product (GNP) over the past ten years. However, the standard deviation of the first service's error is $10 billion, and the other service's standard deviation is $15 billion.

  a. Test the hypothesis that the two services have the same variance of error using $\alpha = .05$.

  b. True or false: Since the mean error figures are the same, you would be satisfied with the prediction of either service as to what next year's GNP will be. Why?

50. Two different package designs were tested for a product. The shelf lives of the contents were measured for a sample of $n_1 = 25$ packages of the first design and $n_2 = 25$ packages of the second design. Use the following sample variances to test the hypothesis that the two package designs have the same variance of shelf life. Use $\alpha = .05$. What do you have to assume about the way the population values are distributed around their means?

$$s_1^2 = 720 \text{ hrs.}^2$$

$$s_2^2 = 1455 \text{ hrs.}^2$$

SAMPLE DATA SET QUESTIONS: *Refer to the 113 applicants for credit listed in the Sample Data Set Appendix of this book.*

a. Test the hypothesis that the population means for JOBINC are the same for those who were granted credit and those who were denied credit. Leave out those applicants who did not list a JOBINC value when you make this comparison. As your alternative hypothesis, use "applicants in the two groups do not have equal incomes." Use an α value of .02.

b. Test the hypothesis that the proportion of women is the same for the populations of people who were granted and denied credit. As an alternative hypothesis, use "the proportions of women in the two groups are not the same." Use an α value of .05.

c. Test the hypothesis that the variance of AGE is the same in the groups where credit was granted and where it was denied. Let $\alpha = .05$.

REFERENCES

9.1 Croxton, Frederick E., Dudley J. Cowden, and Ben W. Bolch. *Practical Business Statistics,* 4th edition. Prentice-Hall, New York, 1969. Chapters 11 and 12.

9.2 Hoel, Paul G. *Elementary Statistics,* 3rd edition. John Wiley & Sons, New York, 1971. Chapter 8.

9.3 Hogg, Robert V., and Allen T. Craig. *Introduction to Mathematical Statistics,* 3rd edition. Macmillan, New York, 1970. Chapter 9.

9.4 Li, Jerome C. R. *Statistical Inference,* I, Edwards Brothers, Ann Arbor, 1964. Chapters 10, 11, and 21.

9.5 Mosteller, Frederick, Robert E. K. Rourke, and George B. Thomas, Jr. *Probability with Statistical Applications,* 2nd edition. Addison-Wesley, Reading, Massachusetts, 1970. Chapter 9.

9.6 Snedecor, George W., and William G. Cochran. *Statistical Methods,* 6th edition. Iowa State University Press, Ames, 1967. Chapters 1, 2, 3, and 4.

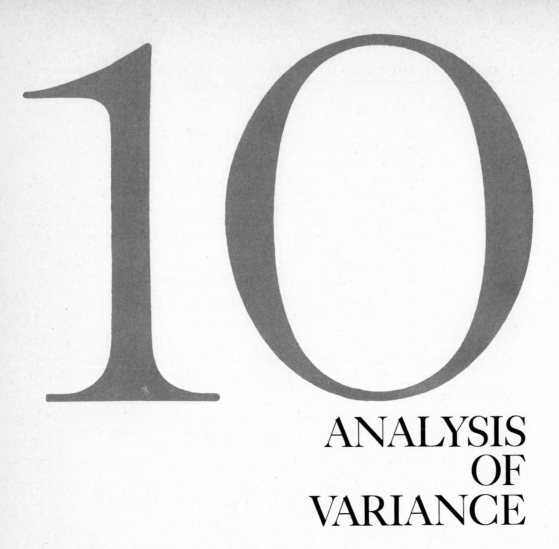

ANALYSIS OF VARIANCE

10.1 REVIEW OF HYPOTHESIS TESTING

In Section 9.5, we discussed testing the hypothesis that a population mean has a particular value. We stated the hypothesis as $H_0: \mu = \mu_0$ and performed the test using the statistic in Expression (9.2)

$$t = \frac{\bar{X} - \mu_0}{s/\sqrt{n}}$$

where s is the sample standard deviation; the population from which the sample was taken is normally distributed, and t has $n - 1$ degrees of freedom.

In Section 9.8, we extended this t test to include the test that two populations have the same mean value. The hypothesis was stated as $H_0: \mu_1 = \mu_2$ and was tested by calculating a t value from two samples as follows:

$$t = (\bar{X}_1 - \bar{X}_2) \bigg/ \sqrt{\frac{s^2}{n_1} + \frac{s^2}{n_2}}$$

where s^2 was the pooled variance from the two samples. In order for this t ratio to follow the t distribution, we had to assume that the two samples were randomly and independently selected from two normally distributed populations. In addition, we assumed that the two populations had the same variance, σ^2.

Analysis of variance is an extension of this two-sample test of hypothesis for continuous populations in Chapter 9. The analysis of variance hypothesis is: $H_0: \mu_1 = \mu_2 = \mu_3 = \ldots = \mu_k$, and the alternative hypothesis is H_a: one or more of the population means is not equal to the others.

10.2   THE ANALYSIS OF VARIANCE COMPUTATIONS

To test the hypothesis that several populations have the same mean, we undertake a set of computations using several variances. Thus, this procedure is called *analysis of variance*. To conduct an analysis of variance, we must take samples from each of the populations and obtain sample means and sample variances (or standard deviations). Let us assume we have taken four samples of size $n = 3$ from four populations that we think *might* have the same mean. The sample values are shown in Table 10.1 and represent the water pollution counts for a production cooling system when it was fitted with four different types of filters. Three random water samples were taken for each filter. We want to test the hypothesis that the mean pollution count is the same for all four filters. That is, we wish to test $H_0: \mu_1 = \mu_2 = \mu_3 = \mu_4$ against the alternative hypothesis that one or more of the filters produce a different pollution count.

TABLE 10.1   Numerical Example: Four Groups of Size 3, pollution counts.

	GROUP:		
1	*2*	*3*	*4*
10	11	13	18
9	16	8	23
5	9	9	25

We can summarize those sample results as follows:

$n_1 = 3$	$n_2 = 3$	$n_3 = 3$	$n_4 = 3$	
$\overline{X}_1 = 8$	$\overline{X}_2 = 12$	$\overline{X}_3 = 10$	$\overline{X}_4 = 22$	$\overline{\overline{X}} = 13$
$s_1^2 = 7$	$s_2^2 = 13$	$s_3^2 = 7$	$s_4^2 = 13$	

The $\overline{\overline{X}}$ figure is the "grand mean" of all twelve sample values in the table. The last row of figures is the variances computed from each of the four samples. For instance, s_3^2 is found by determining the variance of the three values in sample 3 around their mean of $\overline{X}_3 = 10$. That is,

$$s_3^2 = [(13 - 10)^2 + (8 - 10)^2 + (9 - 10)^2]/(3 - 1) = 14/2 = 7$$

The other sample variances were determined in the same way.

The hypothesis that all four of these samples came from populations with the same mean can be tested under the following conditions:

1. All the samples were randomly selected and independent of one another.
2. The populations from which the samples were drawn are normally distributed.
3. All the populations have the same variance σ^2.

Testing the hypothesis that all the filters produce the same mean is achieved by examining the third condition more closely. The following paragraphs outline a general procedure for testing H_0: $\mu_1 = \mu_2 = \mu_3 = \mu_4$.

If the third condition can be assumed true, then we might ask: What is the value of σ^2? What is the value of the variance that is common to all k populations? There are two ways of estimating the population variance σ^2.

The first estimate of the common population variance is found by using the now familiar relationship:

$$\text{Standard deviation } (\overline{X}) = \sigma_{\overline{X}} = \sigma/\sqrt{n}$$

That is, if many samples of size n were taken from a population of variance σ^2 and if the many sample means were examined, we would find that the standard deviation of the many sample means would be equal to the standard deviation of the population divided by the square root of the sample size.

In our numerical example, we did not take *many* samples, we took only *four* samples. But from those four samples, we got four means. The four means can be used to estimate the value $\sigma_{\overline{X}}$. If we have this value, then we can estimate the population variance σ^2 because:

$$\sigma_{\bar{X}} = \sigma / \sqrt{n}$$

or, squaring both sides of the equation, we get

$$\sigma_{\bar{X}}^2 = \sigma^2 / n$$

and

$$n\sigma_{\bar{X}}^2 = \sigma^2$$

Thus, n times the variance of the sample means around their grand mean is an estimate of the population variance of pollution counts. In formula form we can say:

$$s_B^2 = n \left[(\bar{X}_1 - \bar{\bar{X}})^2 + (\bar{X}_2 - \bar{\bar{X}})^2 + \ldots + (\bar{X}_k - \bar{\bar{X}})^2 \right] / (k - 1) \qquad (10.1)$$

$$= n \sum_{j=1}^{k} (\bar{X}_j - \bar{\bar{X}})^2 / (k - 1) \approx n\sigma_{\bar{X}}^2 = \sigma^2$$

where k is the number of populations being tested. This s_B^2 variance estimate is sometimes called the "between groups" variance estimate. This is because it is based on the variation from one sample mean to the next, or the variation *between* the samples. The s_B^2 figure has $k - 1$ degrees of freedom.

In our numerical example, we would calculate this variance estimate as follows:

$$s_B^2 = 3 \left[(8 - 13)^2 + (12 - 13)^2 + (10 - 13)^2 + (22 - 13)^2 \right] / (4 - 1)$$

$$= 348/3$$

$$= 116$$

We will return to this value when we complete the test of H_0: $\mu_1 = \mu_2 = \mu_3 = \mu_4$.

The second method of estimating the population variance is rather simple and straightforward. Since all the samples have sample variances, s_j^2, these variances should be good estimates of the common population variance. Thus, we can take the weighted average of all the sample variances. Note that when we pooled the sample variances in Chapters 8 and 9, we were just taking the weighted average of two samples' variances. Here we will proceed in the same fashion and estimate the common population variance as the weighted average of the sample variances, using the number of degrees of freedom in each sample, $n_j - 1$, as the weights.

$$s_W^2 = \sum_{j=1}^{k} (n_j - 1) s_j^2 / (n_1 + n_2 + \ldots + n_k - k) \qquad (10.2)$$

The value s_W^2 is the second estimate of the population variance, and it is sometimes called the "within groups" variance estimate (that is the reason for the subscript W). This is because the sample variances measure the variation *within* each sample, and this estimate is the weighted average of those within-groups estimates. The s_W^2 figure has $(n_1 + n_2 + \ldots + n_k - k)$ degrees of freedom.

In our numerical example, this within-groups variance estimate would be:

$$s_W^2 = [(3 - 1)(7) + (3 - 1)(13) + (3 - 1)(7)$$
$$+ (3 - 1)(13)] / (3 + 3 + 3 + 3 - 4)$$
$$= (14 + 26 + 14 + 26)/8 = 80/8 = 10$$

10.3   TESTING THE HYPOTHESIS OF SEVERAL EQUAL MEANS

The key to testing the hypothesis that all the k populations have the same mean lies in the second estimate's relationship to the first. We can think of the second estimate, s_W^2, as being a more reasonable estimate of the variance of all the populations. This is because we assumed all the populations had the same variance, and the weighted average of the sample variances is a good estimate of that common variance's value.

However, the first estimate of the population's variance, s_B^2, is somewhat suspect since it is based on the notion that all the populations have the same mean. That is, the estimate s_B^2 is a good estimate of the population variance *only if* the hypothesis is true and all the populations' means are equal: $\mu_1 = \mu_2 = \ldots = \mu_k$.

Let's assume for a moment that, unknown to us, all the population means are not equal. In fact, let's assume that the populations' means, the μ_j values, are *radically* different from one another. Then the sample means, the $\overline{X}_j$ values, will most likely be radically different from each other too! This will have a marked effect on the first variance estimate, s_B^2, of Formula (10.1). That is, the $\overline{X}_j$ values will vary a great deal and the $(\overline{X}_j - \overline{\overline{X}})^2$ terms will be large. Thus, if the population means are *not* all equal, then the s_B^2 variance estimate will be large relative to the s_W^2 estimate. That is to say, if the between-groups variance estimate is large relative to the within-groups variance estimate, it indicates that the hypothesis that all the populations have the same mean is not likely to be true.

The big question is, of course, how large is "large"? Also, how do we measure the relative sizes of the two variance estimates? The answer to these questions is given to us by the F distribution introduced in Section 9.12. If k samples of $n_j (j = 1, \ldots, k)$ items each are taken from k normal populations that have equal variances and for which the hypothesis $H_0: \mu_1 = \mu_2 = \ldots = \mu_k$ is true, then the ratio of the between-groups estimate of variance to the within-groups estimate of variance follows the F probability distribution. Here we use

$$F = s_B^2 / s_W^2 \tag{10.3}$$

Each F value has two different degrees-of-freedom figures associated with it: ν_1 is the number of degrees of freedom in the numerator's variance estimate, and ν_2 is the number of degrees of freedom in the denominator's variance estimate. Thus, for the F ratio in Formula (10.3):

$$\nu_1 = k - 1, \quad \text{and}$$
$$\nu_2 = (n_1 + n_2 + \ldots + n_k - k) \tag{10.4}$$

A set of samples gives a large F value only if s_B^2 is large relative to s_W^2. But this condition exists only if the sample means, the $\overline{X}_j$ values, differ greatly from one another. And this condition usually exists only if the population means, the μ_j values, differ greatly from one another. In sum then, a large F value usually indicates that the null hypothesis $H_0: \mu_1 = \mu_2 = \ldots = \mu_k$ is false.

In our example figures presented in Table 10.1, the F ratio is $116/10 = 11.6$. In order to determine if this figure is so large that we should reject the hypothesis that the four population means are equal, we must consult the F probability distribution table, Table VIII of the Appendix. In that table, the reader will recall, we find the F values that are exceeded by pure chance with only 5 percent probability for various combinations of degrees of freedom in the numerator and the denominator. In our example, $\nu_1 = k - 1 = 3$ and $\nu_2 = n_1 + n_2 + n_3 + n_4 - k = 8$. From Table VIII, we find that an F ratio of 4.0662 is exceeded with only a 5 percent probability when 3 and 8 degrees of freedom are involved. Thus if the α value (probability of a Type I error that we are willing to accept) for this problem were .05, we would reject the hypothesis that the four populations have the same mean. In fact, Table VIII has additional subsections which show us the F values that are exceeded by chance with only 2.5, 1.0, and .5 percent probabilities, respectively. We can see on page A-24 under $\nu_1 = 3$ and $\nu_2 = 8$ that an F value of 9.5965 is exceeded with only a .5 percent probability. Thus, our calculated F value of 11.6 would cause us to reject the hypothesis that the four filters produce the same mean pollution counts and the α value is less than .005. That is:

$$F_{.005,3,8} = 9.5965 < 11.6$$

so we reject

$$H_0: \mu_1 = \mu_2 = \ldots = \mu_k$$

We must conclude that one or more of the filters produces a pollution count different from the others.

We can see why this procedure for testing the hypothesis that several populations have the same mean is called "analysis of variance." This is because we reach our conclusion about the hypothesis by *analyzing* the *variances* within the samples and between the samples.

The results of an analysis of variance are usually summarized in a table like Table 10.2. This table shows the two "sources" of variation as that between the groups and that within the groups. The number of degrees of freedom in each of the two variance estimates is shown in the second column of the table. Variance estimates are obtained when we divide a sum of squares by the number of degrees of freedom in that sum of squares. Thus the SS column in Table 10.2 shows the sum of squares associated with each of the variance estimates. That is, the SS values are the numerators of Equations (10.1) and (10.2), respectively. The fourth column shows two variance figures. These figures are simply the s_B^2 and s_W^2, respectively, and can be found by dividing the df column into the SS column. The sample F value can be taken from this table very conveniently by computing the ratio of the two variance estimates. This table also conveniently displays the number of degrees of freedom in each of the estimates, which facilitates looking up the theoretical F value in Table VIII.

TABLE 10.2   Analysis of Variance Table

SOURCE	df	SS	VARIANCES
Between groups	3	348	116
Within groups	8	80	10
Total	11	428	

10.4   PROCEDURES FOR HANDLING UNEQUAL SAMPLE SIZES

Another example might be useful to help clarify and solidify the concepts we have just covered. The following example deals with a situation where there are three populations but the samples taken from each are different in size.

EXAMPLE 1

A regional fabric store chain consisted of 6 stores. Three years ago the chain acquired two other groups of stores. One of these, constituting group 2 of the enlarged chain, contained 4 stores and the other (group 3) contained 5 stores. Turnover of salespeople is a problem at all 15 stores. The number of salespeople who quit at each of the stores last year is shown in the following table. Assume, for simplicity, that all the stores have the same size sales force to start with. The employee compensation and benefit policies of the three groups of stores have not been standardized, and each group of stores has retained its old policies. The personnel manager of the entire 15-store chain is interested in knowing if there is a real difference in the turnover of personnel in the three groups of stores. If there is, he will consider standardizing compensation and benefits soon.

	GROUP:	
1	*2*	*3*
10	6	14
8	9	13
5	8	10
12	13	17
14		16
11		

This problem can be attacked using the analysis-of-variance method for unequal-sized groups. The hypothesis we will test is: H_0: $\mu_1 = \mu_2 = \mu_3$. That is, the mean annual turnover is the same for all three groups of stores. Before we proceed, we must assume that:

1. The turnover figures are normally distributed (methods of testing for the normality of data are presented in Chapter 11).
2. The three groups have equal population variances (methods for testing this assumption were discussed in Section 9.12).
3. The samples are random and independent (that is, last year was a typical, random year in terms of turnover, and turnover experiences in the three groups of stores were in no way related).

The first step in performing our analysis of variance is to find the grand mean turnover for the fifteen stores and the three samples' means and variances. These figures are:

$$\overline{\overline{X}} = 11.07$$

$$\overline{X}_1 = 10.0 \qquad \overline{X}_2 = 9.0 \qquad \overline{X}_3 = 14.0$$

$$s_1^2 = 10.0 \qquad s_2^2 = 8.67 \qquad s_3^2 = 7.50$$

The between-groups estimate of the common population variance cannot be found using Formula (10.1). That formula was developed for a situation where many samples, all of the same n, were taken. In the preceding example, however, each sample has a different size. Thus, we must modify Formula (10.1) to allow for sample sizes that are different. Since we have used differing sample sizes, we can call the jth sample size n_j. Then in Formula (10.1) we could bring the n_j figure inside the summation sign to yield the following formula:

$$s_B^2 = \sum_{j=1}^{k} n_j (\bar{X}_j - \bar{X})^2 / (k - 1) \qquad (10.5)$$

In this formula, each sample mean's squared deviation from the grand mean is weighted by that sample's size. When we apply this formula to the data in our example of the fabric stores, we obtain the following between-groups estimate of the common population variance:

$$s_B^2 = [6(10.0 - 11.07)^2 + 4(9.0 - 11.07)^2 + 5(14.0 - 11.07)^2] / (3 - 1)$$

$$= 66.93/2 = 33.47$$

This estimate has 2 degrees of freedom. The within-groups (store groups in this case) variance estimate is found using Equation (10.2) as follows:

$$s_W^2 = [(6 - 1)(10.0) + (4 - 1)(8.67) + (5 - 1)(7.5)] / (6 + 4 + 5 - 3)$$

$$= 106/12 = 8.83$$

This estimate has 12 degrees of freedom. The ratio of the two variance estimates given by Formulas (10.5) and (10.2) is our calculated F figure. A large ratio will cause us to reject the hypothesis that the three groups of stores have equal mean annual turnover. From Table VIII we find that $F_{.05,2,12} = 3.8853$. Thus, if we are willing to accept a 5 percent chance of rejecting the hypothesis of equal means when, in fact, it is true, then we will reject if our calculated F value exceeds 3.8853. Our F value is:

$$F = s_B^2 / s_W^2 = 33.47/8.83 = 3.79$$

which is less than 3.8853. Thus, we do not reject the hypothesis. The personnel manager may not want to consider standardizing compensation and benefits solely on the basis of these data. (Note that if he were willing to accept a slightly higher chance of a Type I error, he would probably reject the hypothesis of equal mean annual turnover since the calculated and Table F values are quite close.) The analysis-of-variance table which summarizes the calculations in this problem follows.

SOURCE	df	SS	VARIANCES
Between groups	2	66.93	33.47
Within groups	12	106.00	8.83
Total	14	172.93	

Analysis of variance is often applied to situations in which experiments have been conducted and the results are being evaluated. For instance, a chemical company may wish to conduct an experiment in which four different types of fertilizer are tested to determine if they produce differences in crop yields. If the company has 100 acres on which it might run tests, it could plant four twenty-five acre plots and treat each plot with a different fertilizer. Thus, each plot could be viewed as a sample of $n_j = 25$ one-acre sample elements. The results would be expressed in terms of four different means: $\bar{X}_j$ = mean crop yield per acre in the jth plot, $j = 1, \ldots, 4$.

In another example of experimentation, a company might wish to test the effectiveness of three different sales presentations. The sales manager could train three teams of salespeople in these presentation methods and then let them call on customers for a period of time. At the end of that time, the sales manager could examine the results of the test by comparing the means produced by the three sales teams: $\bar{X}_j$ = mean number of units purchased per customer exposed to the jth sales presentation.

In both of these examples, we have situations where test results are produced by applying different "treatments" (such as fertilizers or sales presentations) to different groups. The analysis of variance approach outlined in Sections 10.2 and 10.3 can be used just as it was described there to test the hypothesis that the treatment groups have equal means. The s_B^2 figure in experimental situations is often called the "treatments" variance estimate, and the s_W^2 figure is called "error" variance estimate. The reason for calling s_W^2 the error variance estimate is that it measures the variations of measurements for all units treated alike (all the units in the same group). Thus, even if the different treatments produce no impact on the measurements, we would expect to see a variation in measurements as large as s_W^2 simply due to natural random variation within the experimental units and to errors in our measurements. That is, s_W^2 measures random error or variation in our measurements.

10.5 ANOTHER VIEW OF ANALYSIS OF VARIANCE

Some statisticians like to approach the topic of analysis of variance from a different point of view. In this approach, they examine the

TABLE 10.1   **Numerical Example: Four Groups of Size 3**

	GROUP:			
1	2	3	4	
10	11	13	18	
9	16	8	23	$\bar{\bar{X}} = 13$
5	9	9	25	
$\bar{X}_1 = 8$	$\bar{X}_2 = 12$	$\bar{X}_3 = 10$	$\bar{X}_4 = 22$	

sum of squares of the individual items in a set of groups from their grand mean. The data from the example in Section 10.2 will be used to demonstrate this. Thus, Table 10.1 is reproduced here for convenient reference.

Figure 10.1 shows graphically how the total sum of squares around the grand mean would be obtained. Each of the twelve items in Table 10.1 is scattered around the grand mean of $\bar{\bar{X}} = 13$. The deviations of the individual items from this mean value are represented by the vertical deviations drawn in the graph. Deviations above the mean are

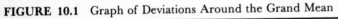

FIGURE 10.1   Graph of Deviations Around the Grand Mean

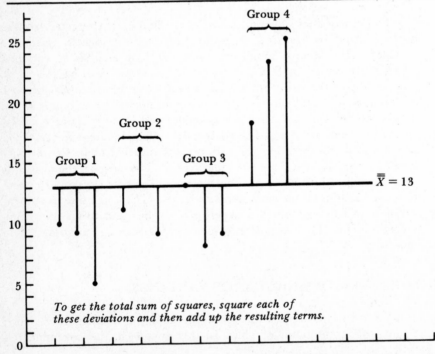

To get the total sum of squares, square each of
these deviations and then add up the resulting terms.

positive, and those below the mean are negative. To get the total sum of squares, we square each of these deviations from the mean of $\bar{\bar{X}} = 13$ and add these squared values.

For the data presented in Table 10.1 and Figure 10.1, this total sum of squares would be:

$$\text{Total SS} = (10 - 13)^2 + (9 - 13)^2 + (5 - 13)^2 + (11 - 13)^2$$
$$+ \ldots + (25 - 13)^2 = 428.$$

Note that *all* of the items in the four samples had their squared deviation from the *grand mean* added into this sum.

In general, the formula for computing the total sum of squares is:

$$\text{Total SS} = \sum_{j=1}^{k} \sum_{i=1}^{n_j} (X_{ij} - \bar{\bar{X}})^2 \tag{10.6}$$

where X_{ij} is the ith item in the jth sample. (Thus, $X_{23} = 8$ since it is the second item in the third sample.)

This total sum of squares can be written as the sum of two separate terms.* That is,

$$\sum_{j=1}^{k} \sum_{i=1}^{n_j} (X_{ij} - \bar{\bar{X}})^2 = \sum_{j=1}^{k} \sum_{i=1}^{n_j} (X_{ij} - \bar{X}_j)^2 + \sum_{j=1}^{k} n_j (\bar{X}_j - \bar{\bar{X}})^2 \tag{10.7}$$

The total sum of squares on the left has been divided on the right into two sums of squares. The first term on the right is a double

---

*The decomposition of the sum of squares into two terms can be shown as follows. If we add and subtract the mean of the jth group inside the parentheses for the sum of squares formula, we have

$$\sum_{j}^{k} \sum_{i}^{n_j} (X_{ij} - \bar{X}_j + \bar{X}_j - \bar{\bar{X}})^2$$

If we now square the quantity contained within the parentheses we get

$$\sum_{j}^{k} \sum_{i}^{n_j} (X_{ij} - \bar{X}_j)^2 + 2 \sum_{j}^{k} \sum_{i}^{n_j} (X_{ij} - \bar{X}_j)(\bar{X}_j - \bar{\bar{X}}) + \sum_{j}^{k} \sum_{i}^{n_j} (\bar{X}_j - \bar{\bar{X}})^2$$

The middle term is zero, since it can be written as

$$2 \sum_{j}^{k} \left[(\bar{X}_j - \bar{\bar{X}}) \sum_{i}^{n_j} (X_{ij} - \bar{X}_j) \right]$$

and $\Sigma_i^{n}(X_{ij} - \bar{X}_j)$ is zero for all j (the sum of the deviations of a set of values about their arithmetic mean is equal to zero). Since the expression in the right term has no i value in it, the Σ_i^{n} results in just n_j times the expression. It follows that the total sum of squares can be written as shown in Formula (10.7).

summation. We must remember that a double summation is evaluated from the inside out. Thus, it begins with the sum of squared deviations of the individual items in each sample from that sample's mean. This is the summation that goes from $i = 1$ to n_j. Then these sums for each sample are added for all the samples, and this is the summation that goes from $j = 1$ to k.

The result of this double summation operation is called the "within-groups sum of squares." The name comes from the fact that the sum of squared deviations *within* each group (the first summation operation of the double sum) is taken before the terms are added up for each group.

Figure 10.2 can be used to show graphically what the within-groups sum of squares represents. The figure shows that each sample in Table 10.1 has a different mean. The individual items in each sample are scattered around their mean. The deviations of the individual items

FIGURE 10.2   Graph of Deviations Used to Obtain Within Groups Sum of Squares

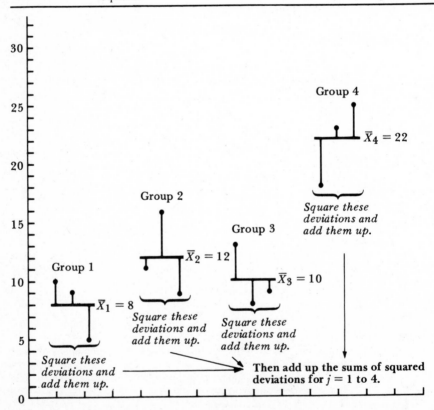

from the mean are represented by the vertical deviations drawn in the graph. Deviations above the mean are positive, and those below the mean are negative. If we were to add up these deviations for any one sample, the rules of summation found in Chapter 3 indicate that the positive and negative deviations would cancel each other out. That is, the sum of the deviations would be zero for all four samples.

If we square each of the deviations before we add them, we will get four different sums of squares. These are the four sums of squares we get by doing the first summation in Formula (10.7), the summation from $i = 1$ to n_j. Then we can add the four sums of squares to obtain the within-groups sum of squares. This second summation is the one that goes from $j = 1$ to k.

For the data in Table 10.1 this first term on the right side of Equation (10.7) is found as follows:

$$\text{Within Groups SS} = [(10 - 8)^2 + (9 - 8)^2 + (5 - 8)^2] +$$
$$[(11 - 12)^2 + (16 - 12)^2 + (9 - 12)^2] +$$
$$[(13 - 10)^2 + (8 - 10)^2 + (9 - 10)^2] +$$
$$[(18 - 22)^2 + (23 - 22)^2 + (25 - 22)^2]$$
$$= 80.$$

The second term on the right side of Equation (10.7) is a weighted sum. Each term is weighted by the sample size, $n = 3$. However, in the general case each sample has a different sample size, n_j. This second term is often called the "between groups sum of squares." The name derives from the fact that there is a different $(\bar{X}_j - \bar{\bar{X}})^2$ term for each group and thus the squared deviations differ "between the groups."

Figure 10.3 gives a graphical presentation of what the between-groups sum of squares represents. The deviations of the four groups' means from the grand mean of $\bar{\bar{X}} = 13$ are shown. However, the three boxes out at the ends of the deviations represent the fact that each deviation is to be squared and then counted $n = 3$ times. If one of the groups had contained four items, then four boxes would have been placed at the end of the deviation for that group.

For the data in Table 10.1, the mathematical calculation of this between-groups sum of squares is:

$$\text{Between Groups SS} = 3(8 - 13)^2 + 3(12 - 13)^2 + 3(10 - 13)^2$$
$$+ 3(22 - 13)^2 = 348.$$

Note that the (Total SS) = (Within Groups SS) + (Between Groups SS) in that $428 = 80 + 348$.

FIGURE 10.3   Graph of Deviations Used to Obtain Between Groups Sum of Squares

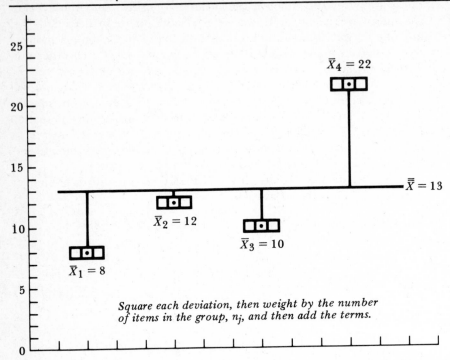

Square each deviation, then weight by the number of items in the group, n_j, and then add the terms.

It was mentioned in Chapter 3 that each sum of squares has a degree of freedom figure associated with it. The within-groups sum of squares has $[(n_1 - 1) + (n_2 - 1) + \ldots + (n_k - 1)]$ degrees of freedom. This figure is simply $(n_1 + n_2 + \ldots + n_k - k)$, or the sum of the samples' sizes minus the number of populations from which the samples were taken. The between-groups sum of squares has $k - 1$ degrees of freedom. The total sum of squares has just one less than the combined sample size.

It was mentioned in the previous section that the results of an analysis of variance are usually summarized in a table such as Table 10.2. Table 10.2 is presented again for convenient reference. Now the column labels on the table may be more clear. The SS column shows the sum of squares as they were developed in the previous paragraphs of this section. The MS column stands for "Mean Squares" and shows the "mean of the squares," if we consider division by the number of degrees of freedom as a sort of "averaging" process. That is,

TABLE 10.2   Analysis-of-Variance Table

SOURCE	df	SS	MS
Between groups	3	348	116
Within groups	8	80	10
Total	11	428	

Between Groups MS = (Between Groups SS)/(Between Groups df)
$$= 348/3 = 116$$
Within Groups MS = (Within Groups SS)/(Within Groups df)
$$= 80/8 = 10$$

It should be noted that these are exactly the same values as those that were obtained in Section 10.2 where we tried to estimate the common population variance for the four groups. That is, $s_B^2 = 116$ and $s_W^2 = 10$. A careful examination of formulas for these values will reveal that the between-groups variance estimate of Section 10.2 is the same as the between-groups mean square. Also, the within-groups variance estimate of Section 10.2 is the same as the within-groups mean square. That is,

$$s_B^2 = \text{Between Groups MS}$$

$$s_W^2 = \text{Within Groups MS}$$

(10.8)

Since this relationship holds, to test the hypothesis that several groups have the same mean, $H_0: \mu_1 = \mu_2 = \mu_3 = \ldots = \mu_k$, we can calculate the F ratio

$$F = (\text{Between Groups MS})/(\text{Within Groups MS}) \qquad \textbf{(10.9)}$$

and compare it with the F value in Table VIII for the appropriate level of α and number of degrees of freedom. The results of such an hypothesis test will be the same regardless of whether one approaches it by trying to estimate the common population variance as shown in Section 10.2 or by dividing the total sum of squares into two parts as shown in this section. The calculated F value and the conclusions reached will be the same regardless of the method used.

Even though an analysis-of-variance hypothesis testing problem can be solved by calculating the F ratio by either method, the second method, which involves dividing the total sum of squares into two parts, is useful in additional applications. One of these applications will be discussed in Chapter 12.

■ 10.6   ANALYSIS OF VARIANCE AND TWO-SAMPLE TESTS

In Section 10.1 the following statement was made:

> Analysis of variance is an extension of the two-sample test of hypothesis for continuous populations in Chapter 9. . . . The analysis of variance hypothesis is: $H_0: \mu_1 = \mu_2 = \mu_3 = \ldots = \mu_k$, and the alternative hypothesis is H_a: one or more of the population means is not equal to the others.

This statement implied that we use the t test of Chapter 9 to test the hypothesis $H_0: \mu_1 = \mu_2$ and the analysis-of-variance method with the F value to test the hypothesis that three or more population means are equal to one another.

From this relationship, one might assume that the two-sample t test is related to analysis of variance for the special case of testing $H_0: \mu_1 = \ldots = \mu_k$ where $k = 2$. This assumption is correct and is demonstrated in the following example.

EXAMPLE 1

A department store chain operates two outlets of the store in the same city. In an effort to determine if the outlets are serving different types of customers, the marketing manager of the chain took samples of the charge account customers at each outlet. She found the mean monthly billing for each customer in the two samples. The sample results follow.

$$n_1 = 31 \qquad \overline{X}_1 = \$42.16 \qquad s_1 = \$8.93$$

$$n_2 = 31 \qquad \overline{X}_2 = \$35.55 \qquad s_2 = \$9.78$$

Test the hypothesis that there is no difference between the mean monthly billings of the two outlets' charge account customers. Let the alternative hypothesis be that the two outlets' population means are not the same. Use an α value of .05.

Solution Introduction
This problem is going to be solved using the two-sample test of Chapter 9 *and* using analysis of variance.

$$H_0: \mu_1 = \mu_2$$

$$H_a: \mu_1 \neq \mu_2$$

Solution Using Two-Sample Methods of Chapter 9
The t test of Chapter 9 requires that we pool the sample variances. The best estimate of that common variance is given by Formula (8.29):

■This material is not required in the sequel.

$$s^2 = [(n_1 - 1)s_1^2 + (n_2 - 1)s_2^2] / (n_1 + n_2 - 2)$$
$$= [(30)(8.93)^2 + (30)(9.78)^2] / (31 + 31 - 2)$$
$$= 5261.80/60 = 87.70$$

We can now use this value in Formula (8.31) to obtain an estimated variance of $\overline{X}_1 - \overline{X}_2$:

$$s_d^2 = s^2(1/n_1 + 1/n_2) = 87.70(1/31 + 1/31) = 5.66$$

Then the hypothesis is tested using the t ratio of Formula (9.18):

$$t = (\overline{X}_1 - \overline{X}_2)/s_d = (42.16 - 35.55)/\sqrt{5.66}$$
$$= 6.61/2.38$$
$$= 2.78$$

Since the alternative hypothesis is two-tailed and since α is .05, we compare the calculated $t = 2.78$ to the Table VI value of $t_{.025,60} = 2.000$. Since the computed t exceeds the table t value, we will reject the null hypothesis and accept the alternative.

Solution Using Analysis of Variance

In order to use analysis of variance, we need two estimates of the population variance. The first is found using Formula (10.1). But that formula involves the grand mean, $\overline{\overline{X}}$. Since the two sample sizes were the same, the grand mean is merely the mean of the means:

$$\overline{\overline{X}} = (42.16 + 35.55)/2 = 38.86$$

Now Formula (10.1) gives us:

$$s_B^2 = 31[(42.16 - 38.86)^2 + (35.55 - 38.86)^2]/(2 - 1)$$
$$= 31(21.85)/1$$
$$= 677.23$$

The s_W^2 estimate of the population variance is given by Formula (10.2). However, this value was computed in the first solution approach as the pooled estimate of variance $s^2 = 87.70$. Thus

$$s_W^2 = 87.70$$

The F ratio for this problem is then

$$F = s_B^2/s_W^2 = 677.23/87.70 = 7.72$$

The numerator of this ratio has one degree of freedom. The denominator has $(n_1 + n_2 - 2) = 60$ degrees of freedom. In the 5 percent section of Table VIII we obtain $F_{.05,1,60} = 4.0012$, which is less than our calculated F value of 7.72. Therefore, we reject the hypothesis of equal mean monthly billing using the analysis of variance approach.

The two approaches arrived at the same conclusion, that we should reject the null hypothesis, but they have more in common than the conclusion. The parallel results of these two approaches are presented in the following summary:

TEST NAME	CALCULATED STATISTIC	TABLE VALUE	α	CONCLUSION
Two-sample t	$t = 2.78$	$t_{.025,60} = 2.000$	.05	Reject H_0
Analysis of variance	$F = 7.72$	$F_{.05,1,60} = 4.0012$	.05	Reject H_0

It should be noted in the second column that $(2.78)^2 \cong 7.72$ and that in the third column $(2.000)^2 \cong 4.0012$ with the exception of rounding errors. Thus, the two tests not only give the same conclusion but give virtually identical mathematical results.

In order to demonstrate the connection between analysis of variance and the t tests, we must examine the relationship between the t and the F distributions. It can be shown mathematically that

$$t^2 = F$$

for ν degrees of freedom in the t figure and for one and ν degrees of freedom in the numerator and denominator of the F ratio. However, since two t values ($-t$ and $+t$) result in the same $t^2 = F$ relationship, we can more precisely state the following relationship between t and F:

$$t^2_{\alpha/2,\nu} = F_{\alpha,1,\nu} \tag{10.10}$$

This means that we can square the t values in the $\alpha/2$ column of the t table (Table VI) and we will obtain the F values found in the first column (one degree of freedom in the numerator), α section of the F table (Table VIII). Conversely, the positive square root of the F values in Column 1 of the α section of Table VIII gives the t values in the $\alpha/2$ column of Table VI. (The reader may select a few values from each table to test this relationship.)

It follows, then, that a t test where the chance of a Type I error is $\alpha/2$ corresponds to an F test where the chance of this error is α. For the data in our example, both the calculated and table values of t^2 equal the corresponding F values and verify the relationship of 10.10. The reader will have additional opportunities to verify this relationship in the problems that follow.

10.7 SUMMARY

In this chapter, we have presented two different approaches to viewing analysis of variance as it is used to test $H_0: \mu_1 = \mu_2 = \ldots = \mu_k$. The two approaches produce identical results, and the reader may use that approach which seems most comfortable. We have also shown that the two-sample t test of Section 9.8 is simply a special case of analysis of variance where $k = 2$. The important points of this chapter are summarized in Table 10.3.

TABLE 10.3 Summary of Important Points in Chapter 10

HYPOTHESIS TO BE TESTED	TEST STATISTIC	COMMENTS AND ASSUMPTIONS
$H_0: \mu_1 = \mu_2 = \ldots = \mu_k$	$F = s_B^2 / s_W^2$ where $$s_B^2 = \frac{\sum_{j=1}^{k} n_j (\bar{X}_j - \bar{\bar{X}})^2}{k-1}$$ $$s_W^2 = \frac{\sum_{j=1}^{k} (n_j - 1) s_j^2}{\sum_{j=1}^{k} n_j - k}$$	a. The F value has two degrees of freedom values associated with it: $\nu_1 = k - 1$ and $\nu_2 = \sum_{j=1}^{k} n_j - k$. b. The populations are all assumed to be normally distributed. c. All the populations must have the same variance.
Alternative View of Test for $H_0: \mu_1 = \mu_2 = \ldots = \mu_k$	$F = \dfrac{\text{Between Groups MS}}{\text{Within Groups MS}}$ where Between Groups MS has same formula as s_B^2 above and Within Groups MS = $$\frac{\sum_{j=1}^{k} \sum_{i=1}^{n_j} (X_{ij} - \bar{X}_j)^2}{\sum_{j=1}^{k} n_j - k}$$	a. This method gives results identical to those obtained using the preceding method. b. All comments for the preceding method apply here.
Special Case of Analysis of Variance: $k = 2$ $H_0: \mu_1 = \mu_2$	F or the t value of Formula (9.18)	a. All comments for the first test hold. b. $F = t^2$ if the alternative hypothesis is $H_a: \mu_1 \neq \mu_2$.

PROBLEMS

1. The following table shows the sales figures (in hundreds of dollars) for three randomly selected days for four salespeople. Use analysis of variance to test the hypothesis that the mean daily sales figures are the same for all four salespeople. That is, test $H_0: \mu_1 = \mu_2 = \mu_3 = \mu_4$, and use $\alpha = .01$.

	SALESPERSON:			
	1	2	3	4
Day 1	15	9	17	13
Day 2	17	12	20	12
Day 3	22	15	23	17
Mean	18	12	20	14
s^2	13	9	9	7

2. Given the data in the following table, test the hypothesis that the three groups have the same population means. Let $\alpha = .05$.

GROUP:		
1	2	3
50	67	50
48	72	44
53	72	43
48	74	45
51	66	43

3. Three packaging materials were tested for moisture retention by storing the same food product in each of them for a fixed period of time and then determining the moisture loss. Each material was used to wrap 10 food samples. Given the following results,
    a. Construct the analysis-of-variance table.
    b. Can we reject the hypothesis that the materials are equally effective? Let $\alpha = .05$.

	MATERIAL:		
	1	2	3
Number of packages	10	10	10
Mean loss	224	232	228
Sample variance	30	40	38

4. To study the effects of different pressures on the yield of a dye, five lots were produced under each of three pressures with the following results.

PRESSURE	LOT:				
	1	2	3	4	5
200 mm	32.4	32.6	32.1	32.4	32.3
500 mm	37.8	38.2	37.9	38.0	37.8
800 mm	30.3	30.5	30.0	30.1	29.7

a.   What are the n_j values in this problem?
b.   What is the k value in this problem?
c.   Test the hypothesis that the mean yields are the same for all three pressure settings. Let the chance of a Type I error be .025.

5.   Given the following data,
a.   Construct the analysis-of-variance table.
b.   Estimate the population means and estimate the standard deviation of each group.
c.   Find a 99% confidence interval for the difference between the means of the first and third groups. (*Hint:* Pool all the variances for use in s_d^2.)
d.   Test the hypothesis of equal group means. Let $\alpha = .05$.

	GROUP:		
1	*2*	*3*	*4*
9	17	22	13
6	10	17	14
8	12	21	22
13			16
			10

6.   Three different automatic milling machines were set up to mill the same type of part. Observations were taken at random to find out how many parts were being produced per hour by each machine. Only four observations were taken on machine 3 since the inspector taking the observations became ill and had to go home before he could complete his work. Test the hypothesis of equal mean hourly output for each machine. Use $\alpha = .01$.

SAMPLE	MACHINE:		
	1	*2*	*3*
1	105	100	104
2	105	98	106
3	110	112	99
4	107	114	109
5	102	100	

7.   The following are the hourly rates of pay for samples of workers in three different types of firms. Test the hypothesis $H_0: \mu_1 = \mu_2 = \mu_3$. Let $\alpha = .01$.

HOURLY PAY	FIRM TYPE:		
	1	*2*	*3*
	2.30	2.00	1.65
	2.35	1.80	1.90
	2.25	2.15	1.85
	2.00	2.10	1.80
	1.95	1.90	1.75
	2.10	1.75	1.95
	2.40		2.00
			1.90

8. The manager of quality control at a sugar refinery was worried that two packaging production lines might be filling the packages to different weights. Thus, samples of size sixteen were taken from each of the production lines and the contents of 10-pound packages carefully weighed. The following sample results seem to indicate that the mean weights of the 10-pound packages from the two lines are the same, but there appears to be much more variation in the weights of the packages coming off the second line. Test the hypothesis that the two lines have the same variation in weights by testing H_0: $\sigma_1^2 = \sigma_2^2$ using $\alpha = .05$. Use H_a: $\sigma_1^2 \neq \sigma_2^2$. If the following data do not cause rejection of the null hypothesis, what must we do in order to test H_0: $\mu_1 = \mu_2$?

$$n_1 = 16 \qquad n_2 = 16$$

$$\overline{X}_1 = 10.15 \qquad \overline{X}_2 = 10.16$$

$$s_1 = .07 \qquad s_2 = .16$$

9. If we wish to test the hypothesis H_0: $\mu_1 = \mu_2 = \ldots = \mu_8$ and if we were to use the t test of Section 9.8 to conduct the hypothesis tests comparing two sample means at a time, how many of these two-sample t tests would be required?

10. If the hypothesis described in Problem **9** were tested and rejected using an analysis-of-variance F test, then which of the following tests might we use to identify which population(s) has (have) the different mean that caused rejection of the hypothesis:

   a. Another F test, dropping out the population with the most different sample mean.
   b. Pairwise t tests as described in Problem **9**.
   c. Either of these methods could be used to identify the population(s) causing the rejection of H_0.

11. The vice president for maintenance in a large national auto rental firm is concerned about the uniformity of wear given by tires put on the company's rental cars. If the tires on a car wear uniformly, then they can all be replaced at one time—when they all get to roughly the same level of tread wear. However, if they wear in a nonuniform manner, the car must be taken into the shop two, three, or four different times to change the one or two tires with the greatest wear. This latter situation is much more expensive in terms of maintenance costs. Thus, the vice president had the maintenance shops conduct a study in which they equipped 50 cars with Goodpoor tires and 50 cars with Goodmonth tires. Logs were kept on each of the cars, and when each car had traveled 20,000 miles from the time of the tire installation, the tread wear on each car's left rear tire was measured. Due to lost records, blowouts, and the sale of some cars, there were only 31 Goodpoor and 41 Goodmonth tire sets left at the end of the test. Problems 11–17 refer to the data given here.

TIRE TYPE	SAMPLE SIZE	SAMPLE MEAN	SAMPLE VARIANCE
Goodpoor	31	2.46	.65
Goodmonth	41	2.66	.98

What are the values of k and all the n_j figures in this problem?

12. Consider the following statement: Since the tire wear will likely vary more from one car to the next than it will vary between the tire types under similar driving conditions, a better experimental design for this problem would have involved putting Goodpoor tires on one side of the cars and Goodmonth tires on the other side and then comparing the differences in wear using the paired comparison test of Section 9.9. True or false?

13. Consider the data in Problem 11. If the hypothesis that both tire brands have the same mean tire wear were tested using the t test of Section 9.8, what theoretical t value would have to be exceeded by the calculated t in order to cause rejection of the hypothesis at the 5% level of significance? (Assume the alternative hypothesis is $H_a: \mu_1 \neq \mu_2$.)

14. If the hypothesis in Problem 11 were tested using analysis of variance, what theoretical F value would have to be exceeded by the calculated F in order to cause rejection of the hypothesis at the 5% level of significance?

15. Test the hypothesis that the mean tread wear in Problem 11 is the same for both brands of tire using the two-sample t test of Section 9.8 using $\alpha = .05$. (Assume $H_a: \mu_1 \neq \mu_2$.)

16. Test the hypothesis that the mean tread wear in Problem 11 is the same for both brands of tire using analysis of variance and an α value of .05.

17. Show that both the calculated and theoretical t and F values you used in Problems 15 and 16 are related. That is, verify that $t^2 = F$ for the values you used.

18. Use the data in Problem 33 of Chapter 9 to test the hypothesis H_0: $\mu_1 = \mu_2$ using the two-sample t test of Section 9.8 and $\alpha = .05$. Also, let the alternative be $H_a: \mu_1 \neq \mu_2$.

19. Use the data in Problem 33 of Chapter 9 to test the hypothesis H_0: $\mu_1 = \mu_2$ using analysis of variance with $k = 2$ and $\alpha = .05$.

20. Verify that the t values of Problem 18 are related to the F values of Problem 19 by the relationship $t^2 = F$.

21. The pollution measure for two midwestern cities was taken on 10 randomly selected days. The first city's mean measure was 14.6 with a sample standard deviation of 3.5. The second city's sample mean and standard deviation were 18.9 and 4.2, respectively.

   a. Test the hypothesis that the population means are the same for the two cities' pollution measures using the two-sample test approach. Use $\alpha = .05$.

   b. Perform the same test using analysis of variance and $k = 2$ and $\alpha = .05$.

SAMPLE DATA SET QUESTIONS: *Refer to the 113 applicants for credit listed in the Sample Data Set Appendix of this book.*

a. Use analysis of variance to test the hypothesis that the population means for JOBINC are the same for those who were granted credit and those who were denied credit. Leave out those applicants who did not list a JOBINC value. Let $\alpha = .05$.

b. Compare the results you obtained in Question **a** for the Sample Data Set questions of Chapter 9 with the results you obtained in Question **a** here.

REFERENCES

10.1   Cochran, W. G., and Gertrude M. Cox. *Experimental Designs,* 2nd edition. Wiley, New York, 1957. Chapters 2, 3, and 4.

10.2   Duncan, D. B. "Multiple range and multiple F tests," *Biometrics,* Vol. 11, 1955.

10.3   Kempthorne, Oscar. *The Design and Analysis of Experiments.* Wiley, New York, 1952. Chapters 6 and 7.

10.4   Ostle, Bernard. *Statistics in Research,* 2nd edition. Iowa State University Press, Ames, 1963. Chapters 10, 11, and 12.

10.5   Snedecor, George W., and William G. Cochran. *Statistical Methods,* 6th edition. Iowa State University Press, Ames, 1967. Chapters 10 and 11.

TESTS USING
THE CHI-SQUARE
DISTRIBUTION

In Chapter 9, we discussed testing hypotheses about two populations of continuous measurements. We showed that the hypothesis that two normal populations have the same mean can be stated as H_0: $\mu_1 = \mu_2$ and can be tested by calculating a t value from two samples as follows:

$$t = (\overline{X}_1 - \overline{X}_2) \Big/ \sqrt{\frac{s^2}{n_1} + \frac{s^2}{n_2}}$$

where s^2 is the pooled variance from the data in the two samples. In Chapter 10, we learned an important extension of this simple hypothesis test. We learned how to extend the test to several populations and test $H_0: \mu_1 = \mu_2 = \mu_3 = \ldots = \mu_k$ using analysis of variance. This test was accomplished using a new probability distribution, the F distribution.

In this chapter, we will also introduce a new probability distribution, the **chi-square distribution.** This distribution will allow us to extend some of the topics covered in Chapter 9 even further. First, it will allow us to extend the hypothesis test $H_0: p_1 = p_2$ to a general test $H_0: p_1 = p_2 = p_3 = \ldots = p_k$. This general test is called the multinomial test of hypothesis. This test and some of its extensions are discussed in the first three sections of this chapter.

The chi-square distribution will also be used to test one of the critical assumptions that has been made throughout much of the work of the previous chapters—that the populations from which we are sampling are normally distributed. Section 11.4 shows how this assumption can be tested as well as how the assumption that a population follows any particular distribution can be tested with the chi-square test for "goodness of fit."

A final use of this new distribution is demonstrated in Section 11.5, which discusses contingency table analysis. The following sections will make these three uses of the chi-square distribution more clear.

11.1   THE MULTINOMIAL DISTRIBUTION

In earlier chapters, we discussed the binomial distribution in some detail. Recall that it is the appropriate model when we are sampling from a population in which the elements belong to either one of two classes, and the sample is taken in such a manner that the probability of obtaining an element from a given class remains constant—that is, it is not affected by the sampling process. Not all actual statistical problems, of course, conform to this model. Frequently, statisticians must consider the more general case where the elements of the population are classified as belonging to one of k classes, where $k \geq 2$. We refer to such a population as a *multinomial* population, and, if the proportions of the elements belonging to each class are not changed by the selection of the sample, the appropriate model is the *multinomial distribution*. This condition typically exists when we are sampling from an infinite population which has multiple characteristics or from a similar finite population *with replacement*.

The multinomial distribution is the *joint* distribution for the random variables $Y_1, Y_2, \ldots, Y_k$, the numbers of elements in a sample of size n that belong to each of the k classes of the population, Y_i being the number that belong to the ith class. The corresponding sample fractions Y_i/n we denote by f_i. These quantities satisfy the conditions

$$\sum_{i=1}^{k} Y_i = n \qquad\qquad (11.1)$$

$$\sum_{i=1}^{k} f_i = 1 \qquad\qquad (11.2)$$

The parameters of the multinomial distribution are the sample size n and the proportions $p_1, p_2, \ldots, p_k$ of the elements in the population that belong to each of the k classes, p_i being the proportion that belong to the ith class. Since the k classes contain all of the elements,

$$\sum_{i=1}^{k} p_i = 1 \qquad\qquad (11.3)$$

As one might expect, each p_i is estimated unbiasedly by the corresponding sample fraction

$$f_i = \frac{Y_i}{n}$$

11.2   AN HYPOTHESIS ABOUT THE MULTINOMIAL PARAMETERS

The basic hypothesis to be considered here is that the proportions belonging to the k classes are equal to a set of specified values; thus, $H_0: p_i = p_{i0}$, where $i = 1, 2, \ldots, k$. An exact test for this hypothesis would be difficult to apply, particularly for large sample sizes; therefore, it is usually tested by a procedure involving approximations.

If the hypothesis is true, the mean, or *expected*, number of elements of the ith class in a sample of size n is:

$$E_i = np_{i0}$$

The *observed* number in the ith class is Y_i. If we calculate for each class the quantity

$$\frac{(Y_i - E_i)^2}{E_i}$$

we have the squared difference between the observed and expected numbers in that class relative to its expected number. The sum of these quantities over all classes, denoted by χ^2 (Greek chi, squared),

$$\chi^2 = \frac{(Y_1 - E_1)^2}{E_1} + \frac{(Y_2 - E_2)^2}{E_2} + \ldots + \frac{(Y_k - E_k)^2}{E_k}$$

or simply

$$\chi^2 = \sum_{i=1}^{k} \frac{(Y_i - E_i)^2}{E_i} \qquad (11.4)$$

is a measure of the lack of agreement between the data and the hypothesis. The idea is that if the null hypothesis is true, then the observed frequencies

$$Y_1, Y_2, \ldots, Y_k$$

ought not deviate too much from their respective expected values

$$E_1, E_2, \ldots, E_k$$

The χ^2 statistic gathers together the discrepancy between Y_i and E_i for all the values of i. To test the null hypothesis, we ask whether Statistic (11.4) has a larger value than can reasonably be accounted for by the workings of chance.

It can be shown that for sufficiently large sample sizes the distribution of the Statistic (11.4) can be approximated by a χ^2 distribution with $k - 1$ degrees of freedom. Note that the degrees of freedom are equal to one less than the number of *classes* and are not related to the size of the sample.

The χ^2 distribution, or **chi-square distribution,** is a continuous distribution ordinarily derived as the sampling distribution of a sum of squares of independent standard normal variables. It is a skewed distribution such that only nonnegative values of the variable χ^2 are possible, and it depends upon a single parameter, the degrees of freedom. The χ^2 distributions for degrees of freedom equal to 1, 4, and 10 are shown in Figure 11.1. It can be seen that the skewness decreases as the degrees of freedom increase. In fact, it can be shown that as the degrees of freedom increase without limit, the χ^2 distribution approaches a normal distribution.

Percentage points of the χ^2 distributions are given in Table IX of the Appendix and are defined by

$$P(\chi_\nu^2 \geq \chi_{\alpha,\nu}^2) = \alpha$$

—that is, $\chi_{\alpha,\nu}^2$ is that value for the χ^2 distribution with ν degrees of freedom such that the area to the *right*, the probability of a *larger* value, is equal to α. For example, the upper 5 percent point $\chi_{.05,20}^2$ for χ^2 with 20 degrees of freedom is 31.41.

To test the hypothesis H_0 that $p_i = p_{i0}$ for all i, we calculate the multinomial χ^2 Sum (11.4) and compare the calculated value with the percentage points of the χ^2 distribution with $k - 1$ degrees of freedom. Since good agreement between the observed and expected

FIGURE 11.1 χ^2 distributions with 1, 4, and 10 degrees of freedom

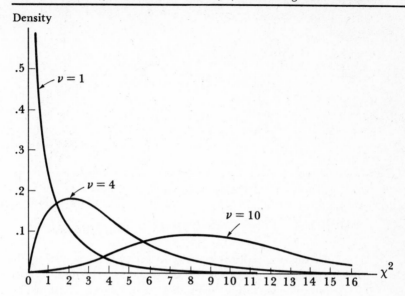

Density

numbers would result in a small χ^2 value and perfect agreement would give for χ^2 a value of 0, we are justified in rejecting the hypothesis only when χ^2 is large; hence, this test is always a one-tailed test on the upper tail of the χ^2 distribution. The critical region is $\chi^2 \geq \chi^2_{\alpha,k-1}$.

One note of caution should be observed. In order for the statistic defined in Formula (11.4) to approximate the chi-square distribution, the expected frequencies, the E_i values, should all be 5 or greater. If very small expected frequencies were allowed, then a small disagreement between Y_i and E_i could *appear* to be abnormally large when the squared difference $(Y_i - E_i)^2$ is divided by a small E_i value. If one category does not have an expected frequency of 5 or more, it is sometimes possible to combine two or more categories.

EXAMPLE 1

Problem: In a certain industry, the accounts receivable for companies are classified as being "current," "moderately late," "very late," and "uncollectable." Industry figures show that the ratio of these four classes is 9:3:3:1. Pratt Associates has 800 accounts receivable with 439, 168, 133, and 60 falling in each class. Are these numbers in agreement with the industry ratio? Let $\alpha = .05$.

Solution: The hypothesis is H_0: $p_1 = 9/16$, $p_2 = 3/16$, $p_3 = 3/16$, $p_4 = 1/16$.

The calculations may be conveniently arranged in tabular form. Note that all the E_i figures (second column) exceed 5, so we may proceed with the calculations.

CLASS	OBSERVED Y_i	EXPECTED $E_i = np_{i0}$	$Y_i - E_i$	$(Y_i - E_i)^2$	$(Y_i - E_i)^2/E_i$
Current	439	450	−11	121	.27
Moderately Late	168	150	18	324	2.16
Very Late	133	150	−17	289	1.93
Uncollectable	60	50	10	100	2.00
				Total	6.36

Thus, the calculated value of χ^2 is 6.36. From the table, $\chi^2_{.05,3}$ is found to be 7.81; therefore, since the calculated value is the smaller, we have no reason to reject the hypothesis. We conclude that the results are in agreement with the expected ratio.

In this example, we may be interested in testing the hypothesis that some, but not all, of the proportions are equal to specified values. For instance, we might be interested in testing whether the proportions in the first two classes are 9/16 and 3/16, respectively, regardless of the proportions in the last two. In such cases, the number of classes is reduced to conform with the hypothesis by combining the unspecified classes into one. For the accounts receivable example, they would be: "current," "moderately late," and "other." If we are interested in only one class, the problem falls into the framework of the binomial test, for we would have two classes: the class of interest and the class consisting of "all other classes."

EXAMPLE 2

Problem: A company sells its product in three primary colors: red, blue, and yellow. The marketing manager feels that customers have no color preference for the product. Thus, her null hypothesis is $H_0: p_1 = p_2 = p_3 = 1/3$. That is, each of the three colors (1 = red, 2 = blue, 3 = yellow) is preferred by one-third of the purchasers. To test this hypothesis, the manager set up a test in which 120 purchasers were given equal opportunity to buy the product in each of the three colors. The results were that 60 bought red, 20 bought blue, and 40 bought yellow. Test the marketing manager's null hypothesis using $\alpha = .05$.

Solution: To calculate the χ^2 statistic of Equation (11.4), we need to find the expected frequencies, the E_i values. Since the hypothesis states that one-third of the people prefer each of the colors, then $E_1 = E_2 = E_3 = (1/3)(120) = 40$. The calculation of χ^2 is facilitated by the following table. Note, again, that all the values in the E_i column exceed 5.

	OBSERVED Y_i	EXPECTED $E_i = np_{i0}$	$Y_i - E_i$	$(Y_i - E_i)^2$	$(Y_i - E_i)^2 / E_i$
Red	60	40	20	400	10
Blue	20	40	−20	400	10
Yellow	40	40	0	0	0
				Total	20

Note that the sum of the figures in the last column of the table is the χ^2 value for this problem. This figure is compared to $\chi^2_{.05,2}$ since $\alpha = .05$ and there are $\nu = 3 - 1 = 2$ degrees of freedom in the χ^2 calculation. Table IX yields a $\chi^2_{.05,2}$ value of 5.99. Since the calculated χ^2 value is 20 and exceeds 5.99, we should reject the hypothesis that people have no color preference with this product. It appears that red is the most popular color and blue is the least popular.

11.3 BINOMIAL DATA

In Section 9.6, we studied tests concerning the proportion p in a binomial population. The two-sided test considered in that section is a special instance of the χ^2 test for the case where k is 2. It is instructive to trace the connection.

We relabel the success category in the population as class 1 and the failure category as class 2. If p_0 is the proportion of successes in the population, then class 1 has proportion $p_{10} = p_0$ and class 2 has proportion $p_{20} = 1 - p_0$. And now the null hypothesis

$$H_0 : p_1 = p_0 \quad \text{and} \quad p_2 = 1 - p_0$$

is exactly the same thing as the null hypothesis

$$H_0 : p = p_0$$

considered in Section 9.6.

If r is the number of successes in a sample of size n, then the number Y_1 of observations falling in class 1 is r and the number Y_2 of observations falling in class 2 is $n - r$. Now the χ^2 Sum (11.4) for testing H_0 is

$$\chi^2 = \frac{(Y_1 - np_{10})^2}{np_{10}} + \frac{(Y_2 - np_{20})^2}{np_{20}}$$

$$= \frac{(r - np_0)^2}{np_0} + \frac{((n - r) - n(1 - p_0))^2}{n(1 - p_0)} \tag{11.5}$$

Algebra reduces this expression to

$$\chi^2 = \left[\frac{r - np_0}{\sqrt{np_0(1 - p_0)}} \right]^2 \qquad (11.6)$$

which is the square of Statistic (9.7) used to test the hypothesis $p = p_0$.

EXAMPLE 1

Problem: At one time, the sex ratio in a certain occupation was 8 women to every man. Suppose that in a recent survey we found that a random sample of 450 contained 68 men. Would we be justified in concluding that the ratio has changed?

Solution: If the ratio is 8 to 1, 1 out of every 9 persons is a male; therefore, if the ratio has not changed, the proportion p of males is $1/9$. We test the hypothesis $H_0: p = 1/9$ against the two-sided alternative, $H_a: p \neq 1/9$ at the 1% level of significance. Let r be the number of men in the sample; then

$$Y_1 = r = 68$$

$$Y_2 = n - r = 450 - 68 = 382$$

$$E_1 = \frac{1}{9}(450) = 50$$

$$E_2 = \frac{8}{9}(450) = 400$$

$$\chi^2 = \frac{(68 - 50)^2}{50} + \frac{(382 - 400)^2}{400} = 7.29$$

From Table IX of the Appendix, $\chi^2_{.01,1}$ is found to be 6.63. The calculated χ^2 is greater than the tabular value; therefore, we reject the null hypothesis and conclude that the ratio has changed.

In this calculation we have used Equation (11.5). Using Equation (11.6) instead gives

$$\chi^2 = \left[\frac{68 - 450 \times \dfrac{1}{9}}{\sqrt{450 \times \dfrac{1}{9} \times \dfrac{8}{9}}} \right]^2 = 2.7^2 = 7.29$$

which checks. The value 2.7 to be squared here is the value of Statistic

(9.7) used in Section 9.6 for a two-sided test of $H_0: p = p_0$. At the 1 percent level, we are to reject if the statistic has absolute value exceeding $Z_{.005}$, or 2.576; this is true of our value 2.7, so again we reject. The two procedures necessarily lead to the same conclusion because 6.63, the cutoff point for the χ^2 test, is 2.576^2.

Since it is true in general that $\chi^2_{\alpha,1}$ is equal to $Z^2_{\alpha/2}$, a χ^2 test based on Equation (11.5) is always the same as the two-sided test of Section 9.6. Note that the χ^2 test is always one-sided while the corresponding Z test is two-sided.

■ 11.4   A TEST FOR GOODNESS OF FIT

In the tests of hypotheses of the preceding chapters, we assumed that we knew the form of the distribution and we tested for the values of parameters. For example, we *assumed* the population was normal and tested the hypothesis $\mu = \mu_0$. But what if we want to check on the assumption of normality itself? The **multinomial χ^2 goodness-of-fit test** can be applied.

For an example, suppose we want to test at the 1 percent level the hypothesis that the 106 daily absence measurements of Table 2.3 come from some normal population. If we combine the first four categories and the last three, the data reduce to Table 11.1. The reason for combining the first few and last few classes is to ensure that the expected frequency for each class is at least 5. If this requirement is not met, then a small difference in Y_i and E_i could give an abnormally large figure when $(Y_i - E_i)^2$ is divided by a very small E_i figure. Now we must check whether the observed frequencies Y_i in the table agree well with a normal distribution having *some* mean and standard

TABLE 11.1   Distribution of Absences in 106 Days

CLASS	ABSENCES	NO. OBSERVATIONS Y_i
1	Less than 135.5	10
2	135.5 to 139.5	21
3	139.5 to 143.5	41
4	143.5 to 147.5	19
5	More than 147.5	15
		$n = 106$

■The material in this section is not required in the sequel.

deviation—a mean and standard deviation unspecified in advance. Our first step is to estimate the mean and standard deviation from the sample by $\overline{X}$ and s, which can be calculated from the grouped data of Table 2.2 as

$$\overline{X} = 141.77 \quad \text{and} \quad s = 5.41$$

The next step is to compute what probability a normally distributed random variable X with mean 141.77 and standard deviation 5.41 has of falling in each of the five classes represented in Table 11.1. To do this we standardize the class boundaries:

$$(135.5 - 141.77)/5.41 = -1.16$$

$$(139.5 - 141.77)/5.41 = -.42$$

$$(143.5 - 141.77)/5.41 = .32$$

$$(147.5 - 141.77)/5.41 = 1.06$$

The probability that such an X falls in class 2, for example, is the probability that the standardized variable $(X - 141.77)/5.41$ falls between the first two of these figures, namely -1.16 and $-.42$; by Table V in the Appendix this probability is $.3770 - .1628$, or $.2142$. Thus, we obtain the probabilities (rounded off) shown in Figure 11.2.

And now we compare the observed frequencies in Table 11.1 against these probabilities. Since the probability of class 2 is .214 and the total number of observations is 106, we expect on the average $106 \times .214$, or 22.68, observations to fall in class 2. Table 11.2 shows the remaining calculations.

The number of degrees of freedom in this example is 2. If we had not had to estimate the mean and standard deviation from the sample (by $\overline{X}$ and s), the degrees of freedom would have been (as

FIGURE 11.2   Areas for the standard normal curve

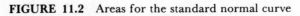

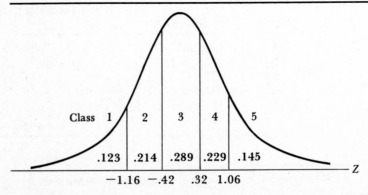

TABLE 11.2   Calculation of the χ^2 Goodness-of-Fit Test for the Data of Table 11.1

CLASS	PROBABILITY	E_i	Y_i	$Y_i - E_i$	$(Y_i - E_i)^2/E_i$
1	.123	13.04	10	−3.04	.71
2	.214	22.68	21	−1.68	.12
3	.289	30.63	41	10.37	3.51
4	.229	24.27	19	−5.27	1.14
5	.145	15.38	15	−.38	.01
Totals	1.000	106.00	106	0.00	$\chi^2 = 5.49$

in Section 11.2) one less than the number of categories: $k - 1$, or 4. But the rule is that we must subtract one *additional* degree of freedom for each parameter estimated. We have estimated two parameters, the mean and the standard deviation, so our degrees of freedom are $4 - 2$, or 2. We are to test at the 1 percent level the hypothesis that the data in Table 11.1 fit a normal curve. The 1 percent point on a χ^2 distribution with 2 degrees of freedom is $\chi^2_{.01,2}$, or 9.21. Our value, 5.49, is less than this; the data do fit well with a normal distribution.

The multinomial χ^2 may be used to test the goodness of fit of data with other distributions as well. We need only remember that each expected number should be at least 5, so that the calculated χ^2 statistic closely follows the χ^2 distribution, and that we lose one additional degree of freedom for every parameter estimated from the sample.

■ 11.5 CONTINGENCY TABLES

Many sets of countable data are such that they may be grouped according to two or more criteria of classification, and it is often the case that we should like to know whether or not these various criteria are independent of one another. For instance, we might like to know whether color preference for a product is independent of sex. In this case we could take a sample of people, record their sex and their favorite color, and classify their responses by sex and by color. In a similar manner we might want to test to see if brand preference and income level are independent.

If the data are to be classified according to two criteria, the only case we treat here, the classes based on one of them may be represented

---

■ The material in this section is not required in the sequel.

by the rows in a two-way table and the classes based on the other by the columns. A cell of the table is formed by the intersection of a row and column. We may then count the numbers of elements in the sample that belong to each cell of the table. The general two-way table with r rows and c columns is presented in Table 11.3. The notation used in Table 11.3 is as follows:

Y_{ij} = number observed to belong to the ith row and jth column. Note that the first subscript denotes the row, the second the column.

R_i = total observed number in the ith row; found by adding across the row.

C_j = total observed number in the jth column; found by adding down the column.

n = the sample size = $\Sigma R_i = \Sigma C_j$.

TABLE 11.3   The $r \times c$ Contingency Table

ROWS	COLUMNS: 1	2	3	$\cdots$	c	ROW TOTALS
1	Y_{11}	Y_{12}	Y_{13}		Y_{1c}	R_1
2	Y_{21}	Y_{22}	Y_{23}		Y_{2c}	R_2
3	Y_{31}	Y_{32}	Y_{33}		Y_{3c}	R_3
.						
.						
.						
r	Y_{r1}	Y_{r2}	Y_{r3}		Y_{rc}	R_r
Column totals	C_1	C_2	C_3		C_c	n

To test the hypothesis that rows and columns represent independent classifications, we compute an expected number E_{ij} for each cell and use the multinomial χ^2. By *independence* we mean that the proportion of each row total belonging in the jth column is the same for all rows (or that the proportion of each column total belonging in the ith row is the same for all columns). This is expressed mathematically by saying that the probability that a random element will belong in the (i,j)th cell is equal to the product of the probability that it belongs to the ith row times the probability that it belongs to the jth column. Symbolically, our hypothesis is

$$H_0: p_{ij} = p_{i \cdot} p_{\cdot j} \qquad i = 1, 2, \ldots, r; \quad j = 1, 2, \ldots, c$$

where p_{ij} = probability of ith row *and* jth column,
        $p_{i.}$ = probability of ith row ignoring columns,
        $p_{.j}$ = probability of jth column ignoring rows.

The null and alternative hypotheses are sometimes more understandable when they are stated as

H_0: The row classification and column classification are independent
H_a: The row and column classifications are dependent

The marginal probability $p_{i.}$ is estimated by the observed fraction in the ith row, R_i/n, and $p_{.j}$ by the observed fraction in the jth column, C_j/n. Under the hypothesis of independence, the value of p_{ij}, or of $p_{i.}p_{.j}$, is estimated by the product of these estimates $R_i C_j/n^2$; therefore, the expected number is the sample size n multiplied by the estimated probability:

$$E_{ij} = \frac{R_i C_j}{n}$$ (11.7)

We can now find the expected number for each cell in the table.

To perform the test we find the contribution of each cell to the multinomial χ^2. The contribution of the (i,j)th cell is

$$\frac{(Y_{ij} - E_{ij})^2}{E_{ij}}$$

There are a total of $r \times c$ such contributions, and the calculated χ^2 is their sum:

$$\chi^2 = \sum_{i,j} \frac{(Y_{ij} - E_{ij})^2}{E_{ij}}$$ (11.8)

This χ^2 has $(r-1)(c-1)$ degrees of freedom, the number of rows less one, multiplied by the number of columns less one.

If the calculated χ^2 exceeds the tabular value for probability α, we reject the hypothesis and conclude that rows and columns do not represent independent classifications; if the calculated χ^2 does not exceed the tabular value, we cannot reject the hypothesis.

We must observe the same precaution noted in earlier sections, however. Each expected frequency, E_{ij}, should be 5 or greater so that χ^2 is not artificially inflated. If the expected value in a cell is less than 5, one or more rows or columns should be combined with others.

The following simple example will serve to illustrate the computational procedure.

EXAMPLE 1

| | | COLUMNS: | | | ROW |
ROWS	1	2	3	4	TOTALS
1	36	16	14	34	100
2	64	34	20	82	200
3	50	50	16	84	200
Column totals	150	100	50	200	500

The expected numbers are:

$$E_{11} = \frac{(100)(150)}{500} = 30 \qquad E_{12} = \frac{(100)(100)}{500} = 20$$

$$E_{13} = \frac{(100)(50)}{500} = 10 \qquad E_{14} = \frac{(100)(200)}{500} = 40$$

$$E_{21} = \frac{(200)(150)}{500} = 60 \qquad E_{22} = \frac{(200)(100)}{500} = 40$$

$$E_{23} = \frac{(200)(50)}{500} = 20 \qquad E_{24} = \frac{(200)(200)}{500} = 80$$

$$E_{31} = \frac{(200)(150)}{500} = 60 \qquad E_{32} = \frac{(200)(100)}{500} = 40$$

$$E_{33} = \frac{(200)(50)}{500} = 20 \qquad E_{34} = \frac{(200)(200)}{500} = 80$$

Since all of the expected frequencies exceed 5 in value, we may proceed. The calculated χ^2 value has $(3 - 1)(4 - 1)$ degrees of freedom and is equal to 10.88:

$$\chi^2 = \frac{(36 - 30)^2}{30} + \frac{(16 - 20)^2}{20} + \frac{(14 - 10)^2}{10} + \frac{(34 - 40)^2}{40}$$

$$+ \frac{(64 - 60)^2}{60} + \frac{(34 - 40)^2}{40} + \frac{(20 - 20)^2}{20} + \frac{(82 - 80)^2}{80}$$

$$+ \frac{(50 - 60)^2}{60} + \frac{(50 - 40)^2}{40} + \frac{(16 - 20)^2}{20} + \frac{(84 - 80)^2}{80}$$

$$= 10.88$$

The tabular value for the case where α is .05 and there are 6 degrees of freedom is $\chi^2_{.05,6}$, or 12.59. Our calculated value is less than 12.59; therefore, at the .05 level of significance, we cannot reject the hypothesis that rows and columns are independent.

The simplest contingency table, the 2×2, can be used to test the hypothesis that two binomial populations have the same relative frequency of successes. Let p_1 be the proportion in the first population and p_2 the proportion in the second. Using the information contained in two random samples, one from each population, we can test the hypothesis $H_0: p_1 = p_2$.

The results of the two samples may be summarized in a two-way table as follows:

SAMPLE	SUCCESSES	FAILURES	SIZES
1	Y_1	$n_1 - Y_1$	n_1
2	Y_2	$n_2 - Y_2$	n_2
Totals	$Y_1 + Y_2$	$n_1 + n_2 - Y_1 - Y_2$	$n_1 + n_2$

In the preceding table, n_1 and n_2 are the two sample sizes, and Y_1 and Y_2 are the observed number of successes in the first and second samples, respectively.

If we test the hypothesis that the rows and columns of this table are independent we are, in fact, testing the hypothesis that the proportions of successes and failures are independent of the population. This could be true only if $p_1 = p_2$. The calculated χ^2 for the 2×2 table will, of course, have one degree of freedom.

EXAMPLE 2

Problem: In order to test the puncture resistance of a new type of off-the-road-vehicle tire, 120 of the new-type tires and 180 old-type tires were driven over the same special hazardous course for 24 hours. Just 6 of the new tires were punctured whereas 18 of the old tires had to be replaced due to punctures. Can we conclude that the new tires resist punctures better?

Solution: Here

$$n_1 = 120 \qquad n_2 = 180$$

$$Y_1 = 6 \qquad Y_2 = 18$$

$$n_1 - Y_1 = 114 \qquad n_2 - Y_2 = 162$$

and the two-way table is:

	PUNCTURED	NOT PUNCTURED	TOTALS
Old Tires	18	162	180
New Tires	6	114	120
Totals	24	276	300

The expected numbers are:

$$E_{11} = \frac{(180)(24)}{300} = 14.4 \qquad E_{12} = \frac{(180)(276)}{300} = 165.6$$

$$E_{21} = \frac{(120)(24)}{300} = 9.6 \qquad E_{22} = \frac{(120)(276)}{300} = 110.4$$

The calculated χ^2 is

$$\chi^2 = \frac{(18 - 14.4)^2}{14.4} + \frac{(162 - 165.6)^2}{165.6} + \frac{(6 - 9.6)^2}{9.6} + \frac{(114 - 110.4)^2}{110.4}$$

$$= 2.45$$

Since the calculated value is less than the tabular value for $\chi^2_{.05,1}$, which is 3.84, we have no reason, at the .05 level of significance, to reject the hypothesis of independence—no reason to believe that the puncture rates for the two groups are different.

As an interesting sidelight, the preceding example illustrates the dangers of the uncritical acceptance of numerical facts. If, on the basis of the data of Example 2, it were reported that, compared to old-type tires, only one-third as many new-type tires were punctured, this would be a statement of fact. However, before accepting the conclusion implicit in this statement one should look for additional information. Not only are the actual numbers of punctures in each group important; it is also necessary that the sample sizes be taken into account, since the observed numbers are meaningful only in relation to the size of the experiment.

Suppose that the number of tires in each group had been 20 but that the number of punctures, 6 and 18, remained the same. In this case the calculated χ^2 would be equal to 15, and the hypothesis H_0: $p_1 = p_2$ would be rejected. There is, however, no significant difference in the puncture rates when 6 out of 120 is compared with 18 out of 180.

11.6 SUMMARY

This chapter has dealt with the uses of the chi-square distribution. Three basic hypothesis tests were presented, and all of them involve chi-square computations. The first test involves the multinomial H_0: $p_1 = p_2 = \ldots = p_k$, which is an extension of the two-sample test of H_0: $p_1 = p_2$ discussed in Section 9.11. Of course, the special case of only one population which is assumed to contain the same proportion

TABLE 11.4 Summary of the Uses of the Chi-Square Distribution

TYPE OF PROBLEM	NULL HYPOTHESIS	TEST STATISTIC	COMMENTS
Multinomial Hypothesis	$H_0: p_1 = p_2 = \ldots = p_k$ or $H_0: p_i = p_{i0}$ $i = 1, 2, \ldots, k$	$\chi^2 = \sum_i \dfrac{(Y_i - E_i)^2}{E_i}$ where $E_i = np_{i0}$	a. The E_i values should all be 5 or more. If they are not, two or more classes can be combined. b. Note that this test can be used to test the hypothesis that several p values are equal to a predetermined set of values, not all necessarily equal. c. Degrees of freedom are $k - 1$.
Binomial Hypothesis (Special Case of Preceding Test where $k = 1$)	$H_0: p = p_0$	$\chi^2 = \left[\dfrac{r - np_0}{\sqrt{np_0(1 - p_0)}} \right]^2$	a. np_0 should be 5 or more. b. The chi-square figures have one degree of freedom.
Goodness of Fit	$H_0:$ The population is _____ distributed. (The blank can be filled in with the name of any theoretical distribution.)	$\chi^2 = \sum_i \dfrac{(Y_i - E_i)^2}{E_i}$	a. The E_i values should all be 5 or more. If they are not, two or more classes can be combined. b. Each E_i value is determined from the theoretical distribution being tested. c. The blank in the hypothesis statement can be filled in with any theoretical distribution. d. One additional degree of freedom is lost for each population parameter estimated.
Contingency Table Analysis	$H_0:$ Row and column classifications are independent	$\chi^2 = \sum_{i,j} \dfrac{(Y_{ij} - E_{ij})^2}{E_{ij}}$	a. The E_{ij} values should all be 5 or more. If they are not, some rows and/or columns can be combined. b. Each E_{ij} value is computed as $R_i C_j / n$. R_i = total in row i, and C_j = total in column j. c. Degrees of freedom are $(r - 1)(c - 1)$ where r is the number of rows in the table and c is the number of columns. d. The hypothesis $H_0: p_1 = p_2$ can be tested using a 2×2 contingency table.

of successes ($k = 1$) can be handled using the multinomial test or the methods of Section 9.6.

Many of the statistical procedures presented in earlier chapters of this text involve the assumption that the population from which samples are drawn is normally distributed. The chi-square, goodness-of-fit test can be used to test the assumption. It can also be used to test the assumption that a sample was drawn from a population with virtually any theoretical distribution: binomial, Poisson, hypergeometric, and others.

Finally, the chi-square distribution can be used to test the hypothesis that items classified according to two different characteristics exhibit independence in those characteristics. The table showing the classifications of the items by two characteristics is a contingency table, and the chi-square test here checks for dependence between the row and column classifications. Table 11.4 summarizes the uses of the chi-square distribution.

PROBLEMS

1. The following grades were given to a class of 100 students. The expected numbers are those corresponding to the "curve." Can we conclude that the instructor used the curve? Use $\alpha = .05$.

	A	B	GRADE: C	D	F
Given	10	30	40	16	4
Expected	7	24	38	24	7

2. An automobile manufacturer ships one model of its economy car in just four colors. The manufacturer's contract with its dealers states that the colors red, yellow, green, and blue are in the proportions 8:7:3:2. A sample of 400 contains 180 red, 120 yellow, 40 green, and 60 blue cars.
   a. Estimate the proportion of each color in the manufacturer's shipments.
   b. Test the hypothesis that one-tenth of the cars are blue. Let $\alpha = .01$.
   c. Test the hypothesis that the proportions are 8:7:3:2. Let $\alpha = .005$.

3. If a sample of 300 peak electrical loads for rural consumers should be distributed among three periods as follows:

2:00 AM–10:00 AM	80
10:00 AM–6:00 PM	90
6:00 PM–2:00 AM	130

   a. Estimate the proportions of peak loads that occur in each period.
   b. Test the hypothesis that the loads are uniformly distributed among the periods. That is, test $H_0: p_1 = p_2 = p_3 = (1/3)$. Let $\alpha = .05$.

4. The probabilities for the various outcomes when a pair of dice are tossed are as follows, with the frequencies observed when a pair of dice were tossed 1800 times.

   a. What is the estimated probability of throwing a 7 with these dice? of throwing a 12?

   b. At the 1% level of significance, are the observed frequencies in agreement with the theoretical values?

					TOTAL ON DICE r:						
	2	**3**	**4**	**5**	**6**	**7**	**8**	**9**	**10**	**11**	**12**
$P(r)$	$\dfrac{1}{36}$	$\dfrac{2}{36}$	$\dfrac{3}{36}$	$\dfrac{4}{36}$	$\dfrac{5}{36}$	$\dfrac{6}{36}$	$\dfrac{5}{36}$	$\dfrac{4}{36}$	$\dfrac{3}{36}$	$\dfrac{2}{36}$	$\dfrac{1}{36}$
Frequency	40	108	175	184	225	330	223	228	128	87	72

5. Fire Department records in an East Coast city show that the last 350 fires were distributed according to day of the week as follows:

Monday	35	Friday	65
Tuesday	35	Saturday	75
Wednesday	45	Sunday	55
Thursday	40	*Total*	350

   Test the hypothesis that fires are uniformly distributed throughout the week. That is, test $H_0: p_1 = p_2 = p_3 = p_4 = p_5 = p_6 = p_7 = (1/7)$. Use $\alpha = .01$.

6. A clothing manufacturer has always produced three sizes of sports shirts: small, medium, and large. The output of its shop has always been 25% small size, 50% medium size, and 25% large size. This breakdown was determined several years ago on the basis of a guess as to the proportion of small, medium, and large men in the population. The guess was made by the company president. Recently, the new company president questioned these proportions, so a random sample of 200 men was selected and measured for sport shirt size. it was found that 35 were small, 90 were medium, and 75 were large. Test the hypothesis that the proportions are $p_1 = .25$, $p_2 = .50$, and $p_3 = .25$ in the population using an α of .01.

7. The bank credit card operation of a large midwestern bank knows from credit applications that the education level of their card holders has the following distribution:

Some high school	$.02 = p_1$
High school complete	$.15 = p_2$
Some college	$.25 = p_3$
College complete	$.58 = p_4$
	1.00

   Of the 500 card holders whose cards have been called in for failure to pay their charges or for other abuses, the education distribution is as follows:

Some high school	50
High school complete	100
Some college	190
College complete	160
	500

Test the hypothesis that credit reliability is independent of education. That is, compare the distribution of the education of the 500 abusers to what would be expected for an education distribution in 500 random card holders. Let $\alpha = .01$.

8. An independent testing agency set up 6 color TV sets in the lobby of a Chicago hotel. The brand names of the sets were covered, the positions of the sets were changed each day, and the sets were adjusted daily by an independent TV repairman. Over a period of several weeks, 2700 people were asked: "Please look carefully at each TV screen and then indicate which color TV has the best picture." The number of people selecting each of the brands is as follows. Can we reject the hypothesis that all the brands have pretty much the same picture quality? Let $\alpha = .005$.

BRAND	NUMBER SELECTING BRAND AS "BEST"
U	1350
V	567
W	243
X	240
Y	162
Z	138
	2700

9. A polling agency reported that in a sample of 156 persons 81 were Democrats, 52 were Republicans, and 23 were independent or belonged to other parties. From these results can it be concluded that these political groups are equally strong in this area? That is, test the hypothesis H_0: $p_1 = p_2 = p_3$ using an α value of .05.

10. A television program rating service surveyed 600 families where the television was turned on during prime time on weeknights. They found the following numbers of people tuned to the various networks:

NBC		210
CBS	commercial	170
ABC		165
PBS		55
		600

a. Test the hypothesis that all 4 networks have the same proportion of viewers during this prime time period. Use $\alpha = .05$.

b. Eliminate the results for PBS and repeat the test of hypothesis for the 3 commercial networks using $\alpha = .05$ again.

11. Use the data in Problem 10 to test the hypothesis that each of the three major networks has 30% of the weeknight prime time market and PBS

has 10%. That is, test H_0: $p_1 = .30$, $p_2 = .30$, $p_3 = .30$, $p_4 = .10$ and let $\alpha = .005$. Would you reject the hypothesis if α were .05?

12. A random sample of 200 drivers contained 62 who had been involved in one or more accidents. Would we reject the hypothesis that the proportion of accident-free drivers is equal to .75? That is, test H_0: $p = .75$ against the alternative H_a: $p \neq .75$ using $\alpha = .05$.

   a. First perform the test using the method discussed in Chapter 9.

   b. Repeat the test using the χ^2 test shown in Section 11.3.

13. A market survey taken just prior to the introduction of a new product indicated that 5% of those who were offered the product would buy it. After the product had been on the market for three months, it had been offered to 6000 potential purchasers and 300 had bought it.

   a. Calculate the χ^2 value for this problem.

   b. Are the survey and sales results consistent with this χ^2 value?

14. The following is the distribution for the number of defective items found in 150 lots of a manufactured item. We want to test the hypothesis that r, the number of defectives, has a Poisson distribution, and we proceed as follows. Let $\alpha = .01$. (*Note:* Bracketed numbers indicate classes that were combined so the expected frequency of the combined class would exceed five.)

NUMBER DEFECTIVE r	OBSERVED FREQUENCY	EXPECTED FREQUENCY (ROUNDED)
0	23	20
1	39	41
2	43	41
3	23	27
4	10	14
5	7 ⎫	⎫
6	4 ⎬12	⎬7
7	1 ⎭	⎭

   a. The mean θ is 2 and is found by taking the weighted average of the r values with the observed frequencies as the weights. This estimates the parameter θ for the Poisson distribution:

$$P(r) = \frac{e^{-\theta}\theta^r}{r!} \qquad r = 0, 1, 2, \ldots$$

   b. Obtain the probabilities for each value of r from tables of Poisson probabilities (Table IV in the Appendix).

$$P(r = 0) = .135,$$

$$P(r = 1) = .270,$$

$$P(r = 2) = .270,$$

$$P(r = 3) = .180,$$

$$P(r = 4) = .090,$$

etc.

    c.   The expected number corresponding to a given value of r is $np(r)$.

    d.   We combine the last four observed and expected numbers in order that no expected number will be less than 5. Calculate χ^2. The degrees of freedom are equal to 4 since we had to estimate θ.

15.   Using the procedure outlined in the preceding problem, test the hypothesis that the following displayed data have a Poisson distribution. Let $\alpha = .025$. (*Hint:* The value of θ can be shown to be 1.3. Also, note that two or three classes might need to be combined.)

r	0	1	2	3	4	5
FREQUENCY	28	39	15	12	5	1

16.   The following is the distribution for the number of westbound cars arriving at a particular intersection in 60-second intervals. Can it be concluded that a Poisson distribution provides a suitable model? See Problem **14**. Let $\alpha = .025$. (Show that $\theta = 3.5$.)

NUMBER OF CARS	0	1	2	3	4	5	6	7	8	9	10
OBSERVED FREQUENCY	8	23	39	53	36	30	15	7	5	3	1

17.   Given the following distribution, test the hypothesis that the data came from a normal distribution. Let $\alpha = .05$ and use the procedure of Section 11.4. (*Hint:* Let the first class be "Under 50" and the last class "125 or more.")

CLASS	OBSERVED FREQUENCY
$25 \leq X < 50$	15
$50 \leq X < 75$	25
$75 \leq X < 100$	30
$100 \leq X < 125$	20
$125 \leq X < 150$	10

18.   Given the following distribution for the yields, in grams, obtained from 100 hills of corn, test that a normal distribution is an appropriate model. Let $\alpha = .05$ and use the procedure of Section 11.4. (*Hint:* Let the first class be "Under 300" and the last class "1500 or more.")

YIELD (GRAMS)	NUMBER OF HILLS
$100 \leq X < 300$	3
$300 \leq X < 500$	7
$500 \leq X < 700$	15
$700 \leq X < 900$	26
$900 \leq X < 1100$	22
$1100 \leq X < 1300$	13
$1300 \leq X < 1500$	9
$1500 \leq X < 1700$	5

19.   Questionnaires were mailed to graduates with degrees in statistics from

Iowa State University. The numbers who returned the questionnaires are as follows. Test the hypothesis that the proportion who did not return the questionnaire is independent of the level of degree earned. Let $\alpha =$ .05.

	B.S.	M.S.	Ph.D.
Returned	47	42	46
Did not return	13	13	8

20. Do the following data give us reason to believe that the fouling of boat bottoms can be reduced by the use of an antifouling paint? Let $\alpha =$ .01.

TREATMENT RESULT	PAINT: Antifouling	Standard
No fouling	40	25
Some fouling	50	50
Much fouling	10	45

21. Two chemical treatments were applied to random samples of seeds. After treatment, germination tests were conducted, with the following results. Do these data indicate that the chemicals differ in their effects on germination? Let the chance of a Type I error be .05.

CHEMICAL TREATMENT	NUMBER OF SEEDS	PERCENT GERMINATION
1	150	80
2	125	88

a. First perform the test using the method described in Chapter 9 to test $H_0: p_1 = p_2$ and $H_a: p_1 \neq p_2$, where p_1 is the proportion of seeds that germinated under treatment 1 and p_2 is the proportion that germinated under treatment 2.

b. Now perform the same test using the χ^2 test illustrated in Example 2 of Section 11.5.

c. Show that the calculated Z and χ^2 values are related to one another by $Z_{\alpha/2}^2 = \chi_{\alpha,1}^2$.

22. Two independent polls were taken to investigate public opinion with regard to a proposed recreational facility. One was conducted by a polling agency, the other by a local newspaper. Are the two polls homogeneous with respect to division of opinion? Let $\alpha =$ .025, and construct a three-row, two-column contingency table. Discuss anything unusual you see in the data.

	POLLING AGENCY	NEWSPAPER
Favor	52.5%	47.5%
Oppose	35.0%	37.5%
Undecided	12.5%	15.0%
Sample size	300	250

23. A sample of the employees of an industrial organization were asked to indicate a preference for one of three pension plans. Is there reason to believe that their preferences are dependent upon job classification? Let $\alpha = .01$.

	PLAN 1	PLAN 2	PLAN 3
Supervisory	29	13	10
Clerical	19	80	19
Labor	22	57	81

24. A random sample of 450 heads of households was taken in the trading area for a large financial institution. These people were asked to classify their own attitudes and their parents' attitudes toward borrowing money as being:
A: Borrow only for large purchases or emergencies.
B: Borrow whenever you want something and can't pay for it now.
C: Never borrow.
The heads of households were then classified as being BA, CB, etc. For example, a BA classification meant that the respondent had a B attitude toward debt while his or her parents had an A attitude. The 400 respondents' classifications are presented in the following table. Test the hypothesis that there is no relationship between debt attitudes of the respondents and how they perceive the debt attitudes of their parents. Let $\alpha = .05$.

	PARENTS:		
RESPONDENTS	A	B	C
A	120	20	10
B	90	60	20
C	90	20	20

25. A large metal fabrication shop operates three shifts. The following data give the distribution of accidents among the three shifts by type of accident. Test the hypothesis that shift and accident type are unrelated. Let $\alpha = .01$.

ACCIDENT TYPE	SHIFT:		
	Day	Swing	Night
Minor	130	95	48
Serious	40	35	12

26. An appliance manufacturer offers a product in four models: standard, deluxe, super deluxe, and majestic. The Randall Department Store has compiled data on the purchasers of this appliance relating the model of appliance purchased with the charge-account balance of the purchaser at the time of purchase. Is there evidence of a relationship between the charge-account balance and the model of appliance purchased? Let $\alpha = .025$.

ACCOUNT BALANCE	MODEL PURCHASED:			
	Standard	Deluxe	Super Deluxe	Majestic
Under $50	40	16	5	10
$50–100	24	12	15	8
Over $100	16	12	30	16

SAMPLE DATA SET QUESTIONS:

Refer to the 113 applicants for credit listed in the Sample Data set Appendix of this book.

a. Use the chi-square distribution and a 2×2 contingency table to test the hypothesis that the proportion of women is the same for the populations of people who were granted and denied credit. Use an α value of .10.

b. Compare the results you obtained in Question **b** of the Sample Data Set questions in Chapter 9 to the results you obtained in Question **a** here.

REFERENCES

11.1 Guttman, Irwin, and Samuel S. Wilks. *Introductory Engineering Statistics.* John Wiley & Sons, New York, 1965. Chapter 12.

11.2 Hoel, Paul G. *Elementary Statistics,* 3rd edition. John Wiley & Sons, New York, 1971. Chapter 11.

11.3 Snedecor, George W., and William G. Cochran. *Statistical Methods,* 6th edition. Iowa State University Press, Ames, 1967, Chapter 9.

12

REGRESSION AND CORRELATION

In most of our statistical procedures so far, we have been concerned with a single observation made on each element of the sample, that is, with a sample of values for a single variable X. We now consider the case where two measurements are made on each element of the sample: where the sample consists of pairs of values, one for each of the two variables X and Y. For example, consider the heights and weights of individuals. If we take a sample of individuals, obtain from each his height and weight, and then let the height be represented by X and the weight by Y, we obtain from the ith person the pair of numbers (X_i, Y_i). If there are n persons in the sample we have a sample of size n which consists of the n number pairs

$$(X_1, Y_1), (X_2, Y_2), \ldots, (X_n, Y_n)$$

Our object is to study the relationship between the variables X and Y.

One way to study this relationship is by means of regression. **Regression analysis** is the process of *obtaining a functional relationship, or equation,* between the variables being studied. For instance, a banker might be interested in the relationship between the mean current yield on corporate bonds and the rate of savings withdrawals at his bank. If he had an equation showing the relationship between these two variables and if he had forecasts for future current yields on corporate bonds, he could use the equation to predict savings withdrawals at his bank. These predictions would allow him to more easily cover the withdrawals. Regression analysis is the tool that would be used to develop the prediction equation.

Another means of studying the relationship between two variables is called correlation analysis. **Correlation analysis** involves *measuring the strength* of the relationship between two variables. This measurement takes a numerical form called the *correlation coefficient,* which will be discussed in Section 12.5. Some problems dealing with the relationships between two variables are concerned simply with determining whether there is a relationship and not with obtaining a prediction equation. For instance, researchers might study the relationship between an individual's educational level and his annual income. The researchers would not necessarily be interested in an equation relating these variables but might want to know if there is a strong relationship, or *correlation,* between the two.

In many problems, both regression analysis and correlation analysis are used. That is, an equation relating two variables is obtained, and the strength of the relationship between the variables is measured.

We shall study regression analysis first.

12.1 SIMPLE LINEAR REGRESSION

To use a regression analysis we must know or assume the functional form of the relationship between the variables. This is expressed in the form of a mathematical function in which Y, the dependent variable, is set equal to some expression which depends only on X, the independent variable, and on certain constants or parameters. For instance, the simplest form of the relationship between two variables would suggest that one variable is a constant multiple of the other. That is,

$$Y = BX \tag{12.1}$$

In this form, the value of Y is always B times X. A one-unit change in X is associated with B units change in Y. This form is a special case of the general linear equation

$$Y = A + BX \tag{12.2}$$

where A is called the intercept and B is called the slope of the equation. This equation implies that Y has a base value of A when $X = 0$ and that the value of Y changes B units for every unit change in X.

As an example we might consider the pricing formula for a company's product. The company's base price is \$2.00 per case, but the price drops two cents per case for every added case purchased up to forty cases. The pricing formula would be represented by

$$Y = 2.02 - .02X \quad \text{for} \quad 0 \le X \le 40$$

where Y is the price per case and X is the number of cases ordered. Note that the intercept of this equation must be 2.02 (the price of $X = 0$ cases) so that the price per case for $X = 1$ case will be $Y = 2.02 - (.02)(1) = 2.02 - .02 = 2.00$. The slope of the equation is $-.02$, which indicates that for every added case ordered (one unit increase in X) the price drops (since the sign is negative) two cents. With an order of $X = 40$ cases, the price per case would be $Y = 2.02 - (.02)(40) = 1.22$. The total cost would be $(1.22)(40) = \$48.80$.

Some variables are not related by the simple linear form just discussed. Two variables might be related according to a formula such as

$$Y = A + BX + CX^2 + 2^X \tag{12.3}$$

This is a rather complex relationship and not one we would expect to see often. Yet it raises the following question: If we are studying the relationship between two variables, how do we know if the form of the relationship is similar to that in Formulas (12.1) or (12.2), or some more complicated expression like Formula (12.3)? The answer is that in a given situation we may arrive at the functional form by either of two methods: (1) from analytical or theoretical considerations, or (2) by studying scatter diagrams.

Theoretical considerations, for instance, suggest that over a relatively narrow range of prices, the amount of a commodity supplied will have a linear relationship like that shown in Equation (12.2). That is,

$$P = A + BQ$$

is suggested by economists as the relationship between price, P, and quantity supplied, Q, in a perfectly competitive market. On the other hand, an aerospace engineer might suggest that the maximum theoretical speed, S, obtained by a new rocket is related to its weight, W, according to the following form:

$$S = A - e^{-(1/W^2)}$$

where $e = 2.718$ is the same mathematical constant found in the density function for the normal curve and the Poisson distribution. The preceding formula is not important to us. The only important point is that an aerospace engineer might be able to develop such a theoretical relationship from his knowledge of physics, aerodynamics, and the characteristics of the new rocket.

The second means of obtaining the functional form of the relationship between two variables involves the use of scatter diagrams (sometimes called scattergrams). A scatter diagram is obtained by gathering observations on pairs of (X, Y) values and then plotting the pairs as points in a plane, where Y is measured along the vertical axis and X along the horizontal axis. Examples of three scatter diagrams are presented in Figure 12.1.

After the points have been plotted, observation of the diagram may reveal a pattern to the points that indicates what functional form may be used for the purposes of the analysis. We shall be concerned in this chapter only with simple linear relationships in which the points in the scatter diagram appear to lie along a straight line, or in which theoretical reasoning leads us to conclude that two variables are linearly related. Thus, this chapter deals only with relationships like those shown in Part (a) of Figure 12.1. These types of relationships have the general linear form given by Equation (12.2).

As an example, a real estate investment firm may be considering the construction of a condominium complex in a certain city. It wishes to price its units in line with what is being charged by other builders in the area. Thus, the firm may wish to study the relationship between condominium selling prices and living area (measured in square feet) for the condominiums in this city. One would expect that the larger condominiums would sell at higher prices and the relationship between

FIGURE 12.1   Scatter diagrams that suggest (a) a linear relationship, (b) a curvilinear relationship, and (c) no relationship

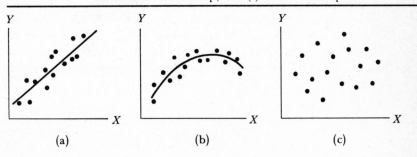

area, X, and price, Y, might be like that depicted in Figure 12.1(a).

No one would expect that there is a perfect relationship between price and area and that knowing area would allow us to predict a condominium's price exactly. Because of sampling variation and variations in price due to other factors such as land costs, view, and access to shopping, the observed values or points will not all lie on the line but will be scattered to some degree about the line. Therefore, we assume that for each X there is a distribution for Y rather than a single value and that our observed Y values corresponding to a given X are a random sample from that distribution. The regression curve or line is the curve or line that joins the *means* of the distributions corresponding to all possible values of X. Under these assumptions the relationship we want to estimate is

$$\mu_{Y|X} = A + BX \tag{12.4}$$

—that is, the mean of Y for a fixed X is equal to $A + BX$, where the constants A and B are the intercept and slope, respectively.* The intercept is the value of the mean of Y when X is equal to zero. The slope is the rate of change of $\mu_{Y|X}$ with X, the change in the mean of Y for a unit change in X. The real estate investment firm would be very interested in the value of B, the slope. This would tell the firm the mean increase in price of a condominium associated with an additional square foot of area. If $B = \$40$, this implies that every additional square foot of area is associated with an average of an additional \$40 in price for the condominiums sold in this city.

For any randomly selected individual condominium, the price, Y_i, would be

$$Y_i = A + BX_i + e_i$$

where e_i is the random variation of this particular condominium's price from the *mean* price, $\mu_{Y|X_i} = A + BX_i$, for all condominiums which have an area of X_i square feet. For any particular point in Figure 12.1, the value e_i is the amount of vertical distance between the point and the straight line. Thus, the e_i values are disturbances which would cause errors in our predictions of an individual price, Y_i, even if we knew the relationship $\mu_{Y|X} = A + BX$. In regression problems, we assume that the e_i values average out to zero. That is, $E(e_i) = 0$. Thus, the error value, e_i, is present in any prediction of condominium price since other variables that could help explain variations in price may

---

*In simple linear regression, α and β are usually used to denote the true intercept and slope, respectively. Here these symbols have been used for the probabilities of Type I and Type II errors; to avoid confusion, we use A and B for the regression parameters.

not be included in the prediction equation. This equation $Y_i = A + BX_i + e_i$ only states that a condominium's price is some base value, A, plus an average of B additional dollars for every additional square foot of area, plus some error value (which can be positive or negative) that cannot be predicted. This equation only shows how area is associated with price.

Note that the last sentence said "is associated with . . ." rather than "causes." That is, just because there is a relationship between two variables, we cannot assume that two variables are related by cause and effect. In the case of some physical relationships, however, the variables are causally related. For example, increases in temperature cause increases in the volume of a gas which is held under constant pressure. However, the price of a condominium, Y, may be only partially related to the area, X, of the unit, through cause and effect. It may be that the larger units also have many more frills which make them more expensive to purchase. Thus, if we find that $B = \$40$, it would be incorrect to say that an additional square foot of area "causes" the price to rise by an average of $40 because the price rise may be caused by numerous factors such as the frills put in the larger units *as well as* the added area. That is why, in regression analysis, we are careful to say that an additional unit of X "is associated with" an average of B additional units of Y.

12.2   FINDING THE SLOPE AND INTERCEPT OF A REGRESSION LINE

Up to this point, we have only discussed the form which regression equations might take. In this chapter, we will only deal with simple linear regression equations of the form $\mu_{Y|X} = A + BX$. In this section, we deal with the problem of how we use sample data to make estimates of the regression line's slope, B, and its intercept, A.

Let us assume that the real estate investment firm mentioned in the previous section took a random sample of $n = 5$ condominium units which recently sold in the city of interest. The sales prices, Y (in thousands of dollars), and the areas, X (in hundreds of square feet), for each unit are as follows.

$Y = SALES$ $PRICE$ (000)	$X = AREA$ $SQ.\ FEET$ (00)
36	9
80	15
44	10
55	11
35	10

This table shows that the first condominium in the sample had an area of 900 square feet and sold for $36,000. Figure 12.2 shows the scatter diagram for this sample. The figure shows that there is a definite positive slope to the points. The real estate investment firm would like to know $\mu_{Y|X} = A + BX$, the regression equation for the line that would pass through a scatter diagram of *all* the condominium sales in the city. That is, the firm would like to know the population regression line. However, there are sales prices and areas for only a small sample of five condominium sales. Using these data, the firm will be able to *estimate* the population regression equation $\mu_{Y|X} = A + BX$. In order to distinguish between a regression equation calculated from population data and one calculated from sample data, we indicate that the sample regression equation is

$$\hat{Y} = a + bX \qquad (12.5)$$

The sample data can be used to obtain the value a, which is a sample estimate of the population regression intercept A, and b, which is the sample estimate of the population regression line's slope B. The symbol $\hat{Y}$ indicates the sample regression line's estimate of $\mu_{Y|X}$, the mean

FIGURE 12.2   Scatter diagram for five condominium sales

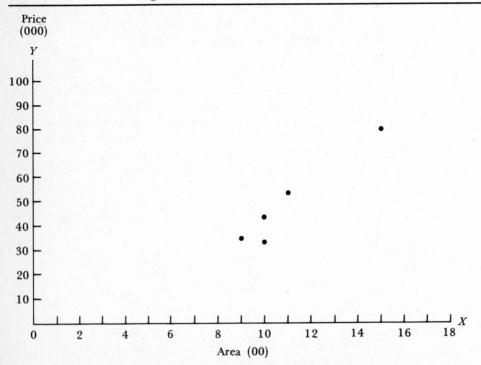

price of the population condominiums that have X hundred square feet of area.

In order to obtain values a and b from the sample data, we could draw a straight line through the points in Figure 12.2 in such a way that they pass close to all the points in the scatter diagram. Then with a ruler we could measure the intercept a and the slope b. However, this method would be very inexact, and two people drawing lines through the same set of points might obtain quite different estimates for the intercept and slope of the line. The method used by statisticians to find the intercept and slope of the regression line passing through a set of points is called the **least squares method.** The principle of least squares is illustrated in Figure 12.3. For every observed Y_i in a sample of points, there is a corresponding predicted value $\hat{Y}_i$, equal to $a + bX_i$ as given by Equation (12.5). The deviation of the observed Y from the predicted $\hat{Y}$ is $Y_i - \hat{Y}_i$, equal to $Y_i - a - bX_i$. The sum of squares of these deviations from the fitted line is

$$\Sigma e_i^2 = \Sigma(Y_i - \hat{Y}_i)^2 = \Sigma(Y_i - a - bX_i)^2 \tag{12.6}$$

where the estimators a and b are the functions of the sample values that make this sum of squares a minimum or *least* value.

It can be shown mathematically* that Expression (12.6) is minimum when the values for a and b are the ones given in Equations (12.7) and (12.8).

FIGURE 12.3   Sample points and estimated regression line

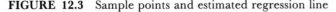
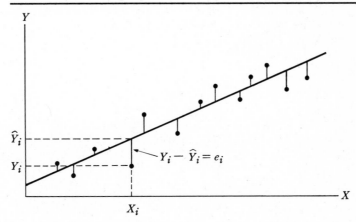

---

*A good understanding of calculus is necessary if one desires to know how Equations 12.7 and 12.8 were derived from Equation 12.6.

$$b = \frac{\Sigma(X_i - \bar{X})(Y_i - \bar{Y})}{\Sigma(X_i - \bar{X})^2} = \frac{\Sigma xy}{\Sigma x^2} \qquad (12.7)$$

and

$$a = \bar{Y} - b\bar{X} \qquad (12.8)$$

We shall not derive these estimates here; see Reference 12.4 for a derivation not requiring calculus. Notice, however, that Equation (12.8) implies that the regression line goes through the center of gravity of the points (X_i, Y_i). That is, it passes through the mean of the Y's and the mean of the X's.

In this chapter we make extensive use of the notation introduced in the rightmost member of Equation (12.7). That is, we represent the deviation of an observed value from its mean by the corresponding lowercase letter, thus:

$$x_i = X_i - \bar{X} \quad \text{and} \quad y_i = Y_i - \bar{Y} \qquad (12.9)$$

In this notation the sums of squares and cross products become:

$$\Sigma(X_i - \bar{X})^2 = \Sigma x^2$$
$$\Sigma(Y_i - \bar{Y})^2 = \Sigma y^2$$
$$\Sigma(X_i - \bar{X})(Y_i - \bar{Y}) = \Sigma xy$$

As an example of how Equations (12.7) and (12.8) are used to find the slope and intercept of a linear regression line, we will use the data from the five condominiums to obtain the sample regression line $\hat{Y} = a + bX$.

EXAMPLE 1

The following table has been prepared from the sample data in the condominium example. The mean price of the five units, $\bar{Y}$, was found to be 50, or \$50,000. The mean area, $\bar{X}$, was found to be 11, or 1100 square feet. The column headed y was formed by subtracting $\bar{Y}$ from each Y_i in the first column. The column headed x was formed by subtracting $\bar{X}$ from each value in the second column. Thus, the y and x columns are the columns of deviations from the means. The y^2, x^2, and xy columns are the columns of squared Y deviations, squared X deviations, and cross product deviations, respectively.

PRICE Y	AREA X	y	x	y^2	x^2	xy
36	9	−14	−2	196	4	28
80	15	30	4	900	16	120
44	10	− 6	−1	36	1	6
55	11	5	0	25	0	0
35	10	−15	−1	225	1	15
Sums 250	55	0	0	1382	22	169

$$\bar{Y} = 250/5 = 50$$

$$\bar{X} = 55/5 = 11$$

$$\Sigma xy = 169$$
$$\Sigma x^2 = 22$$
$$\Sigma y^2 = 1382$$

From the information given in the table we are prepared to use Formulas (12.7) and (12.8) to obtain the slope and intercept of the sample regression line.

$$b = \frac{\Sigma xy}{\Sigma x^2} = \frac{169}{22} = 7.68$$

$$a = \bar{Y} - b\bar{X} = 50 - (7.68)(11) = -34.48$$

Thus, the sample regression line is

$$\hat{Y} = -34.48 + 7.68X$$

This equation tells us that the least squares line passing through the points in the scatter diagram of Figure 12.2 has a Y-intercept of −34.48. This value can be thought of as a "starting point" for the regression line. (Some people might wish to attach a rather silly interpretation that a condominium with $X = 0$, that is, zero square feet of area, would sell for −$34,480!) The slope of 7.68 means that for condominiums in which the floor space ranges from 900 to 1500 square feet, each additional unit of X (100 square feet of area) is associated with 7.68 additional units of Y (in thousands of dollars). That is, 100 added square feet is associated with an added $7680 in price, on the average. The same result suggests that the average price increase is $76.80 per square foot.

It should be noted that this interpretation of the slope of the regression line is only valid for condominiums with areas in the range of 900 to 1500 square feet. This is the range of sizes for condominiums in the data set used to construct the line. We cannot extrapolate outside this range of X values and suggest that the average price increase is $76.80 per square foot for condominiums larger or smaller than those in our data set. The price change per square foot might be either greater or smaller than this figure—especially if the relationship between area and price is nonlinear outside the range of 900 to 1500 square feet.

After we have found numerical values for the estimators a and b using Equations (12.7) and (12.8), we have the regression equation which can be used to obtain predictions. For instance, if the real estate investment firm in the example is planning to build condominiums that have $X = 13$ (or 1300 square feet), they would predict a selling price of $\hat{Y} = -34.48 + 7.68(13) = 65.36$. Thus, a selling price of $65,360 would be consistent with the prices obtained for the five condominiums in the sample. This assumes, of course, that the firm's condominiums would have approximately the same type of amenities possessed by the condominiums used to construct the regression line.

It is interesting to note that the line $\hat{Y} = -34.48 + 7.68X$ passes through the points in Figure 12.2 in such a way that it is "close" to the points. Closeness is defined in a rather explicit way, however. That is, there is no other line that could be fitted to those points for which the sum of squares of vertical deviations from the line would be smaller. This is why the line is called the "least squares" line.

When the mean values $\bar{Y}$ and $\bar{X}$ are *not* whole numbers, as they were in our first example, it is often easier to obtain the sums of squares and cross products needed in a regression problem using the following machine formulas. The first two formulas were presented in Chapter 3, where we discussed the need to obtain the sums of squares of deviations around the mean in calculating the variance of a set of numbers. The third formula shows how to obtain the sum of cross product deviations without getting involved with unmanageable deviations produced by fractional means.

$$\Sigma x^2 = \Sigma X_i^2 - \frac{(\Sigma X_i)^2}{n}$$

$$\Sigma y^2 = \Sigma Y_i^2 - \frac{(\Sigma Y_i)^2}{n}$$

$$\Sigma xy = \Sigma X_i Y_i - \frac{(\Sigma X_i)(\Sigma Y_i)}{n}$$

The use of these equations is demonstrated in the following example.

EXAMPLE 2

The X values in the accompanying table represent the number of gasoline pumps operated by 7 different gas stations. The Y values represent the number of gallons, to the nearest 1000 gallons, that each of these stations sold during a specified period of time. We seek a regression equation relating the number of pumps to gasoline sales.

Y	X	Y^2	X^2	XY
1	2	1	4	2
3	3	9	9	9
5	4	25	16	20
8	5	64	25	40
9	6	81	36	54
11	7	121	49	77
12	8	144	64	96
Sums 49	35	445	203	298

$$\bar{X} = 35/7 = 5 \qquad \bar{Y} = 49/7 = 7$$

$$\Sigma y^2 = 445 - \frac{(49)^2}{7} = 102$$

$$\Sigma x^2 = 203 - \frac{(35)^2}{7} = 28$$

$$\Sigma xy = 298 - \frac{(49)(35)}{7} = 53$$

Thus, $b = 53/28 = 1.89$ and $a = 7 - 1.89(5) = -2.45$. The regression equation is thus:

$$\hat{Y} = -2.45 + 1.89X$$

The regression equation can be interpreted in the following way. The $b = 1.89$ value indicates that a one-unit change in X is associated with a 1.89-unit change in Y. That is, each additional pump was associated with a 1890-gallon increase in sales. This interpretation of the slope is valid, of course, only for stations with 2 to 8 pumps, inclusive. The $a = -2.45$ value is simply the intercept of the line with the Y axis. Literally, it means that a station with no pumps would sell -2450 gallons of gas during this period! However, this is meaningless, and the value of $X = 0$ is outside the range of 2 to 8 pumps for which the regression equation is a valid predictor. It is best to think of the a value as a starting point for the regression line in this problem. If another gas station is located in the same area as these seven, then this regression equation can be used to estimate its gasoline sales during the same period. For instance, assume this new station has 5 pumps. Our regression equation would give us an estimated sales volume of:

$$-2.45 + 1.89(5) = -2.45 + 9.45 = 7.0 \quad \text{or } 7000 \text{ gallons}$$

12.3   MEASURING THE ACCURACY OF PREDICTION

In most regression problems, the major reason for constructing the regression equation is to obtain a tool that is useful in predicting the value of Y, the dependent variable, from some known value of X, the associated independent variable. Since this is the case, we often wish to assess the accuracy of the regression line in predicting the Y values. In Section 12.1, we noted that the error in predicting a particular Y_i value is a term designated e_i. Thus, we can measure the accuracy of prediction for a regression equation by examining these e_i values. We mentioned earlier that the error terms average out to zero: $E(e_i) = 0$. But the variance of these error terms, called the mean squared deviation around the regression line, is used to measure how much the points cluster around the line. This value, denoted $\sigma^2_{Y|X}$, is simply the variance of the Y_i values around the regression line.

In the typical regression problem, we do not know the population of all Y_i values. We usually have a sample, and the mean squared deviation around the regression line is estimated from sample data as follows:

$$s^2_{Y|X} = \frac{1}{n-2} \Sigma \hat{e}_i^2 \qquad (12.10a)$$

or, since $\hat{e}_i = Y_i - \hat{Y}_i$,

$$s^2_{Y|X} = \frac{1}{n-2} \Sigma (Y_i - \hat{Y}_i)^2 \qquad (12.10b)$$

A third form of this equation, which is easier to compute, is

$$s^2_{Y|X} = \frac{1}{n-2} [\Sigma y^2 - b\Sigma xy] \qquad (12.10c)$$

The first form of Formula (12.10) shows why $s^2_{Y|X}$ is a good estimate of the accuracy of predictions using the sample regression equation. If the errors, the $\hat{e}_i$ values, are large, then the $s^2_{Y|X}$ figure will be large. If all the Y_i values lie right on the regression line, however, then each $\hat{e}_i = 0$ and $s^2_{Y|X} = 0$. The notation $s^2_{Y|X}$ refers to the estimated variance of the population of all Y_i values around their mean $\mu_{Y|X}$ for a given value of X. This notation is used to distinguish the variance around the regression line, $s^2_{Y|X}$, from the sample variance defined in Formula (3.14) in Chapter 3. That formula specifies

$$s^2_Y = \frac{1}{n-1} \Sigma (Y_i - \bar{Y})^2$$

Note the similarities in the second form of Formula (12.10) and Formula (3.14). This comparison shows the $s^2_{Y|X}$ is the variance of the individual Y_i values with respect to the regression line's predicted values, $\hat{Y}_i$. That is, $s^2_{Y|X}$ is the variance of the deviations shown in Figure 12.3. The sum of squared deviation around the regression line is divided by $n-2$, the number of degrees of freedom in the sum of squares. Two degrees of freedom were lost from the data set through the calculation of a and b for the regression line. The s^2_Y figure is similar to $s^2_{Y|X}$ except that its reference point is $\overline{Y}$ rather than the regression line.

Thus, s^2_Y can be thought of as the average squared deviation of the Y_i values from the mean while $s^2_{Y|X}$ can be thought of as the average squared deviation of the Y_i values from the regression line's predictions.

If we use Example 1 in the previous section, we find that

$$s^2_Y = \frac{1}{n-1} \Sigma(Y_i - \overline{Y})^2 = \frac{1}{n-1} \Sigma y^2$$

$$= \frac{1}{4}(1382)$$

$$= 345.50 \text{ (thousand dollars)}^2$$

and

$$s^2_{Y|X} = \frac{1}{n-2} \Sigma(Y_i - \hat{Y})^2 = \frac{1}{n-2}[\Sigma y^2 - b\Sigma xy]$$

$$= \frac{1}{3}[1382 - (7.68)(169)]$$

$$= 28.03 \text{ (thousand dollars)}^2$$

Thus, the variance of the condominium prices around the mean price, $\overline{Y}$, is 345.5 (thousand dollars)^2 while the variance of the condominium prices around the regression line's predicted values, $\hat{Y}_i$, is only 28.03 (thousand dollars)^2. Note that the variance around the regression line is quite a bit smaller than the variance around the mean. That is, the 28.03 figure is very much smaller than the 345.50 figure. This is usually the case in the regression equation since *the variation of Y around the regression line, $\Sigma(Y_i - \hat{Y}_i)^2$, is less than or equal to the variation around the mean, $\Sigma(Y_i - \overline{Y})^2$.*

This condition suggests a way of measuring the accuracy of a regression line's predictions. The accuracy has been defined by statisticians in terms of the squared error of prediction. The most commonly used measure of this accuracy is called the sample *coefficient of determination*, r^2. This value is defined as

$$r^2 = 1 - \frac{\Sigma(Y_i - \hat{Y}_i)^2}{\Sigma(Y_i - \overline{Y})^2} \qquad (12.11)$$

This formula needs a little explanation. The numerator of the fraction on the right is just Σe_i^2, which measures the squared error that remains in the predictions of Y when we use the sample regression equation. The denominator is the sum of squared deviations around the mean, $\overline{Y}$, and can be thought of as the sum of the squared errors in predicting the Y_i values if we were to use $\overline{Y}$ rather than the regression equation as our predictor. Thus, the fraction is the proportion of the squared errors of prediction around the mean that remains *unexplained* when we use the regression equation as a predictor. The r^2 figure is 1.0 minus the proportion of *unexplained* squared error; thus, r^2 is the proportion of squared error that the regression equation can *explain*, or eliminate, when we use it as the predictor rather than using $\overline{Y}$ as a predictor.

The sample coefficient of determination can be computed using another formula. This formula is more useful in calculations since the figures needed come directly from the table of values used to compute the regression equation's slope and intercept.

$$r^2 = \frac{b\Sigma xy}{\Sigma y^2} \qquad (12.12)$$

For our example on condominium prices, we find that

$$r^2 = \frac{(7.68)(169)}{1382} = .94$$

This figure can be interpreted in the following way. One person looking at the condominium price data in Example 1 of the previous section might forecast prices in this way: Every time someone asks the potential selling price of a condominium, he or she could say, "I suppose that $50,000 will be the selling price since that is the average selling price of the condominiums in my sample." Another person looking at the same data might forecast prices in this way: Every time someone asks the potential selling price of a condominium, he or she could say, "Tell me the number of square feet in the condominium, and I will plug that figure into the regression equation $\hat{Y} = -34.48 + 7.68X$ to get your predicted price." This second set of predictions would eliminate $r^2 = .94$ of the squared errors, or variation, in the first set of predictions. Thus, 94 percent of the squared errors, or variation, of prediction can be eliminated by using the regression equation to predict prices rather than using $\overline{Y}$ as the predictor.

The r^2 figure is a value that ranges between zero and unity. That is, if the regression equation is no better at predicting Y than the mean, $\overline{Y}$, then no variation or error has been eliminated, and $r^2 = 0$. However, if the regression equation is a perfect predictor of the Y_i values, then all the Y_i values in the scatter diagram will lie right on top of the regression line, and all of the squared error of prediction has been eliminated. Thus, $r^2 = 1$. Values of r^2 between zero and unity indicate the relative strength of the relationship between the X variable and the Y variable in the regression equation. The r^2 values that are close to unity indicate that there is a strong relationship and that the regression equation will give relatively accurate predictions of the Y_i figures once the associated X value is known and entered into the regression equation.

Since r^2 is the fraction of the total variation in Y that is accounted for by the regression, $1 - r^2$ must be the fraction of the variation in Y that is unaccounted for: the fraction associated with the errors of prediction. The latter quantity is sometimes called the *coefficient of alienation*.

The r^2 value associated with a regression equation can be used as a measure of the equation's accuracy regardless of the context of the problem. For instance, the r^2 figure associated with the regression equation developed in Example 2 of the previous section can be calculated using Equation (12.12) as

$$r^2 = \frac{(1.89)(53)}{102} = .98$$

Thus, the regression equation in Example 2 could be used to eliminate or explain ninety-eight percent of the variation in the amount of gasoline pumped at the stations. In this sense, the second example's regression equation is slightly more accurate in predicting gasoline sales (with r^2 equal to .98) than the first example's equation is in predicting condominium prices (with r^2 equal to .94).*

---

*Numerous computer programs which perform simple regression show an output called "Adjusted r^2." This figure is computed by the following formula:

$$\text{Adjusted} \quad r^2 = \frac{s_Y^2 - s_{Y|X}^2}{s_Y^2}$$

Adjusted r^2 is found by dividing the difference in variances, $s_Y^2 - s_{Y|X}^2$, by the sample variance around the mean. In our condominium example, the adjusted r^2 value is

$$\text{Adjusted} \quad r^2 = \frac{345.50 - 28.03}{345.50} = .92$$

The difference between the r^2 of Equation (12.11) and the adjusted r^2 is in the division of one sum of squared variations by $n - 1$ and the other by $n - 2$ in calculating the adjusted r^2 value. When n is large, the two r^2 figures are very close. In this text, we will use the r^2 figure of (12.11) exclusively.

In all the examples presented to this point, the slope of the regression line has been positive. That is, large values of X have been associated with large values of Y, and the regression line's graph has sloped from lower left to upper right in scatter diagrams like Figure 12.3. However, many regression lines have a negative slope like the one shown in Figure 12.4. This line slopes from upper left to lower right and indicates that large values of X are associated with small values of Y and vice versa. For example, in a particular industry, high inventory levels might be associated with low profitability. In certain job classifications, high pay might be associated with low employee turnover.

By examining Formula (12.7) we can see why the slope of the regression line will be negative when high values of one variable are associated with low values of the other. Since the denominator of Formula (12.7) is Σx^2, this figure is always positive. Thus, b is negative when Σxy is negative, or when $\Sigma(X_i - \bar{X})(Y_i - \bar{Y})$ is negative. This happens when many of the individual terms are negative, and negative terms occur when one member of the product is positive and one is negative. That is, when an above-average value of one variable (which gives

FIGURE 12.4   Scatter diagram and regression line with negative slope

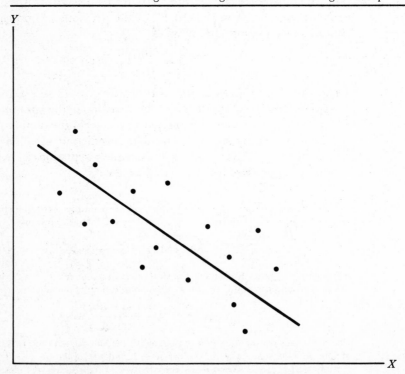

a positive deviation) is associated with a below-average value of the other variable (which gives a negative deviation), then the product is negative. Thus, high or above-average values of one variable associated with low or below-average values of the other variable gives a negative Σxy, which produces a negative slope in the regression line. Some of the problems at the end of this chapter involve regression lines with negative slope.

12.4 STATISTICAL INFERENCE IN REGRESSION ANALYSIS

At the beginning of Section 12.2, we noted that the real estate investment firm in our example would like to know the regression line showing the relationship between price and area for *all* the condominiums sold in the city of interest. That is, the firm wanted to know $\mu_{Y|X} = A + BX$, the population regression line. However, since the firm only had a sample taken from the population, we found the sample regression line $\hat{Y} = a + bX$. We also found the predicted price at which a condominium with $X = 13$ (1300 square feet of area) would sell: \$65,360. However, this estimate was made using the *sample* regression line, not the population regression line.

In Chapter 8, we discussed the topic of statistical inference and showed how sample results could be used to construct confidence intervals within which we think the population parameters will lie. Under certain conditions, we can use our sample regression information to make inferences about what the population regression parameters will be. That is, we can answer the following questions:

1. What is the slope, B, of the population regression line?
2. What is the intercept, A, of the population regression line?
3. What will the mean selling price $\mu_{Y|X_o}$ be for a condominium that has area of X_o?

Of course, any time a sample is taken from a population, we can perform hypothesis tests—if we have an idea of what the population parameters might be. Thus, we can also use the hypothesis testing methods of Chapter 9 to test hypotheses about:

1. The slope, B, of the population regression line,
2. The intercept, A, of the population regression line, and
3. The mean value of Y associated with a specific value X_o

if we have some preconceived notion about these values.

In order to construct confidence intervals or to test hypotheses like those just outlined, three conditions must be met by the population data. These are:

1. The X values are known, that is, nonrandom.
2. For each value of X, Y is normally and independently distributed with mean $\mu_{Y|X}$ equal to $A + BX$ and variance $\sigma^2_{Y|X}$, where A, B, and $\sigma^2_{Y|X}$ are unknown parameters.
3. For each X the variance of Y given X is the same; that is, $\sigma^2_{Y|X} = \sigma^2$ for all X.

The first condition implies that if we are to make predictions using a regression equation, we must know the value of X to plug into the regression equation—it will not materialize simultaneously with the Y value. The examples we have used satisfy this condition. We knew the number of square feet in the condominiums and the number of pumps at each gasoline station.

The second condition implies that for each value of X (or each area value for a condominium) the Y values (the prices of all the condominiums with that area) will be normally distributed around what the population regression line $\mu_{Y|X} = A + BX$ would predict for Y. Figure 12.5a illustrates the meaning of this condition. It shows that for a fixed value of X, the distribution of Y_i values around the regression line (or the distribution or errors in prediction) follow the normal distribution which has a variance $\sigma^2_{Y|X}$. The $\sigma^2_{Y|X}$ value is the population's equivalent of the sample's $s^2_{Y|X}$, which is the variance of the Y_i values around the sample regression line.

The second condition also states that the Y's should be *independently* distributed. Figure 12.5b shows an example where the Y values are not independently distributed. If we know the value of Y at a particular value of X, then we have a good idea of what the value of Y will be at the next value of X. In order for independence to exist, the Y values must be unrelated to one another at the different levels of X.

The third condition implies that the disturbances around the regression line must be the same for small values of X as they are for large values of X. Figure 12.6 illustrates two cases in which this condition is not met, diagrams (a) and (b), and one case where the condition is met, diagram (c). In the first two cases we sometimes say that the data are *heteroscedastic*. In the third case, the data are *homoscedastic*.

12.4.1   Inferences and Hypotheses Concerning B

If the three conditions previously discussed are met, then standard statistical inference procedures can be used to determine the population regression line's characteristics from the sample data. In Section 8.9

FIGURE 12.5   (a) Normal distribution of Y_i values around regression line at $X = X_0$.
(b) Example of dependent Y values in a regression problem.

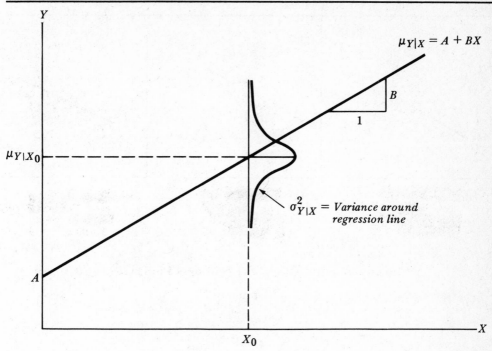

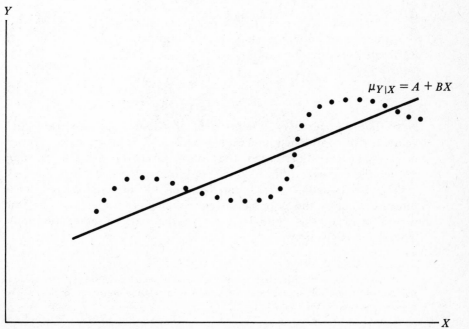

FIGURE 12.6   (a) Heteroscedastic data: variation large for large X values. (b) Heteroscedastic data: variation large for small X values. (c) Homoscedastic data: variation uniform for all X values.

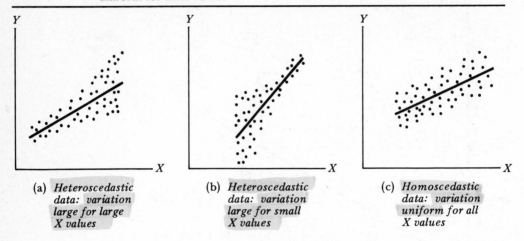

(a) *Heteroscedastic data: variation large for large X values*

(b) *Heteroscedastic data: variation large for small X values*

(c) *Homoscedastic data: variation uniform for all X values*

we noted that any $100(1 - \alpha)\%$ confidence interval has the form:*

$$L = \left(\begin{array}{c} \text{Sample} \\ \text{Estimate} \end{array} \right) - (t_{\alpha/2,\nu}) \; s_{\text{Sample Estimate}}$$

$$R = \left(\begin{array}{c} \text{Sample} \\ \text{Estimate} \end{array} \right) + (t_{\alpha/2,\nu}) \; s_{\text{Sample Estimate}}$$

If we desire to know the slope of the population regression line, B, we would form the interval

$$L = b - t_{\alpha/2,n-2} s_b$$

$$R = b + t_{\alpha/2,n-2} s_b$$

(12.13)

Note that the number of degrees of freedom associated with the t value in this confidence interval is $n - 2$. This is because two degrees of freedom were "lost" when the sample data were used to calculate the sample regression line's slope b and intercept a.

Before Equation (12.13) can be used to construct a confidence interval to locate the population regression line's slope, we must find the value of s_b, the standard error of the sample regression slope. That value is given by

---

*The $Z_{\alpha/2}$ value has been left out of the form presented here since the population standard deviation will almost never be known in a realistic regression problem. This requires that we use the t value.

$$s_b = \frac{s_{Y|X}}{\sqrt{\Sigma x^2}} \qquad (12.14)$$

The figure in the numerator of this expression is the standard error of estimate, Equation (12.10), which is the standard deviation of the Y_i values around the regression line.

EXAMPLE 1

Problem: Find an interval within which we can be 95% sure that the slope of the regression line will lie for all condominium sales in Example 1 of Section 12.2.

Solution: In that example we found the sample regression line to be $\hat{Y} = -34.48 + 7.68X$. Thus, the value of b in Formula (12.13) is 7.68. In that example we also found

$$s_{Y|X}^2 = 28.03 \qquad \text{so} \qquad s_{Y|X} = \sqrt{28.03} = 5.29$$

Thus, by applying Formula (12.14) we find that the standard error of b is

$$s_b = \frac{5.29}{\sqrt{22}} = 1.13$$

The number of degrees of freedom in that example was $n - 2 = 5 - 2 = 3$. From Table VI we find that $t_{.025,3} = 3.182$. Thus, the 95% confidence interval for B is:

$$L = 7.68 - (3.182)(1.13) = 4.08$$

$$R = 7.68 + (3.182)(1.13) = 11.28$$

Thus, we can be 95% sure that in the population of all condominiums in the city of interest to the real estate investment firm, an additional 100 square feet of area is associated with between 4.08 and 11.28 additional thousand dollars in price. This is the same as saying with 95% confidence that the population regression slope B is between $40.80 and $112.80 per square foot.

EXAMPLE 2

Problem: Find an interval within which we can be 90% sure that the population regression line's slope will lie for Example 2 of Section 12.2.

Solution: That example involved the following regression equation relating gasoline pumped to the number of pumps available: $\hat{Y} = -2.45 + 1.89X$. Thus, $B = 1.89$ and since seven gas stations were involved in the study, $n = 7$. Before we can apply Formula (12.13) we must find $s_{Y|X}$ and s_b. By Formula (12.10) and the data in the example we find

$$s^2_{Y|X} = \frac{1}{n-2} [\Sigma y^2 - b\Sigma xy] = \frac{1}{5} [102 - (1.89)(53)] = \frac{1.83}{5} = .366$$

or

$$s_{Y|X} = \sqrt{.366} = .605$$

From the table in the problem we know that $\Sigma x^2 = 28$ so

$$s_b = \frac{.605}{\sqrt{28}} = .11$$

Using $t_{.05,5} = 2.015$ from Table VI allows us to construct

$$L = 1.89 - (2.015)(.11) = 1.67$$

$$R = 1.89 + (2.015)(.11) = 2.11$$

Thus, we can be 90% sure that for the group of gasoline stations in the same area and affected by the same influences, each additional pump is associated with between 1670 and 2110 additional gallons in sales, on the average.

If we wish to test an hypothesis concerning the slope of the population's regression line, we proceed as follows. Under the null hypothesis that the slope is equal to a given value B_0, that is, H_0: $B = B_0$, the quantity

$$t = \frac{b - B_0}{s_b} \tag{12.15}$$

has a t distribution with $n - 2$ degrees of freedom. A common hypothesis is that B equals zero. We want to know whether or not there is a linear association between the variables, for if there is not there is nothing to be gained by using the X's, as they would contribute nothing to the analysis or prediction of the Y's. Under this hypothesis, H_0: $B = 0$, we have

$$t = \frac{b}{s_b} \tag{12.16}$$

We must remember, however, that this quantity is distributed as t only when B is zero. In this situation the alternative hypothesis is usually that B is not equal to zero. There are, of course, situations in which one is interested only in knowing whether or not there is a positive slope (or a negative slope). In these circumstances we would use single-sided alternatives.

To test the hypothesis H_0: $B = 0$ for the gas station example, we calculate the value of t, as follows:

$$t = \frac{b}{s_b} = \frac{1.89}{.11} = 17.18$$

In the t table we find that $t_{.005,5}$ equals 4.032; therefore, since the calculated t is greater than the tabular value, we reject the hypothesis that B is zero at the 1 percent level of significance. We conclude there is an underlying linear relationship. That is, if we conclude that B is not zero, then we run less than a 1 percent chance of being wrong.

12.4.2 Inferences and Hypotheses Concerning A

Let us now shift our attention to the second question introduced at the first of this section: What is the intercept, A, of the population regression line? The confidence interval for A has the following form:

$$L = a - t_{\alpha/2,n-2}s_a$$
$$R = a + t_{\alpha/2,n-2}s_a \tag{12.17}$$

In order to apply this formula we need to calculate the standard error of a according to the following formula:

$$s_a = s_{Y|X}\left[\frac{1}{n} + \frac{\bar{X}^2}{\Sigma x^2}\right]^{1/2} \tag{12.18}$$

where the exponent of $1/2$ in the expression indicates that the square root of the value in parentheses should be taken.

EXAMPLE 3

Problem: For the real estate example in Section 12.2, find an interval within which we can be 90% sure that the intercept, A, of the population regression line will lie.

Solution: The sample regression equation was $\hat{Y} = -34.48 + 7.68X$, n was 5, $\bar{X}$ was 11, Σx^2 was 22, and $s_{Y|X}$ was 5.29. Thus, the value of s_a is:

$$s_a = s_{Y|X}\left[\frac{1}{n} + \frac{\bar{X}^2}{\Sigma x^2}\right]^{1/2} = 5.29\left[\frac{1}{5} + \frac{(11)^2}{22}\right]^{1/2} = 12.63$$

and

$$L = -34.48 - (2.353)(12.63) = -64.20$$
$$R = -34.48 + (2.353)(12.63) = -4.76$$

Thus, we can be 90% confident that the intercept of the population regression line associating the area with the selling price of all condominiums in this city is between -4.76 and -64.20 thousand dollars.

The result found in this last example is not very meaningful since the intercept in this problem lacks concrete interpretation. This is the case in many regression problems in business and economics. Unless there are some negative values of X possible in the data set, the interpretation of the intercept of the regression line is usually meaningless. In these cases, the construction of confidence intervals for A is also meaningless.

The test of the hypothesis that the intercept A is equal to a specified value A_0 is an ordinary t test. Under $H_0: A = A_0$, the function

$$t = \frac{a - A_0}{s_a} \tag{12.19}$$

also has a t distribution with $n - 2$ degrees of freedom. Here, again, the hypothesis that A is equal to zero is of special interest, for if this is so the population regression line passes through the origin; that is, the mean of Y given that X equals zero is equal to zero. For the gas station example, if we test $H_0: A = 0$, we get

$$s_a = s_{Y|X} \left[\frac{1}{n} + \frac{\bar{X}^2}{\Sigma x^2} \right]^{1/2} = .605 \left[\frac{1}{7} + \frac{25}{28} \right]^{1/2} = .616$$

and

$$t = \frac{a}{s_a} = \frac{-2.45}{.616} = -3.98$$

At the 5 percent level of significance we would reject the hypothesis and conclude that A is not zero, since $t_{.025,5}$ is 2.571.

12.4.3   Inferences Concerning $\mu_{Y|X}$

Let us now turn our attention to the final question to be examined in this section: What will be the mean value of Y for some specific prediction using X_0? Rephrased in terms of our initial example, this question becomes: What will be the mean selling price $\mu_{Y|X}$ for a condominium that has area X_0? The confidence interval for this value is found by following the usual form:

$$\hat{Y}_0 \pm t_{\alpha/2, n-2} s_{\hat{Y}_0} \tag{12.20}$$

The standard error of the sample regression line has a rather complicated formula:

$$s_{\hat{Y}_0} = s_{Y|X} \sqrt{\frac{1}{n} + \frac{(X_0 - \bar{X})^2}{\Sigma x^2}} \tag{12.21}$$

The formula for $s_{\hat{Y}_0}$ is an estimate of how the regression line's estimate, $\hat{Y}_0$, would be scattered around the population estimate, $\mu_{Y|X}$, if many many sample regression lines were constructed. This formula has two terms in it. If we put the $s_{Y|X}$ under the square root sign, the formula becomes

$$s_{\hat{Y}_0} = \sqrt{\frac{s^2_{Y|X}}{n} + \frac{s^2_{Y|X}(X_0 - \bar{X})^2}{\Sigma x^2}} \qquad (12.22)$$

The term on the far right under the square root sign indicates the error in prediction that could be introduced by the possibility of the sample regression line's slope being in error. Note that the $(X_0 - \bar{X})^2$ factor in this term implies that the error in prediction gets larger for X values that are further from the mean, $\bar{X}$. That is, if the sample regression line's slope, b, is in error, the predicted values out toward the ends* of the sample regression line get "whipped" far from the actual values that would be given by the population regression line $\mu_{Y|X} = A + BX$, if we knew that line's equation. Figure 12.7 illustrates this point by showing that the width of the confidence interval defined

FIGURE 12.7   Confidence intervals for Y given X using Equation (12.20)

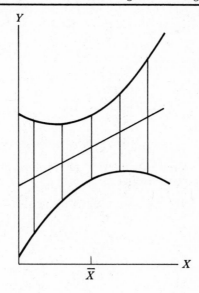

---

*Technically, the regression line has no "ends," but in practical regression problems the highest and lowest values of X in the sample data indicate approximately where we would locate artificial "ends" of the regression line.

in Expression (12.20) is narrow for X values close to $\overline{X}$ and wide at values of X far from $\overline{X}$.

The first term under the square root sign in Equation (12.22) indicates the prediction error that could be introduced by the possibility of the sample regression line's being located at the wrong mean height along the Y-axis, even if the slope were correct. That is, even if the sample slope, b, were exactly equal to the population slope, B, the intercept value might be in error. The $s^2_{Y|X}/n$ term is a measure of that error's contribution to the error in predicting $\mu_{Y|X}$ using the sample regression line with some specific X_0 value substituted in to give $\hat{Y}_0 = a + bX_0$.

EXAMPLE 4

Problem: If the real estate investment firm described in Example 1 of Section 12.2 plans to build condominiums with $X_0 = 13$, or area of 1300 square feet, find a 95% confidence interval for the mean selling price of 1300-square-feet condominiums in this city.

Solution: In Section 12.2 we found that the sample regression equation predicts a selling price of $\hat{Y} = -34.48 + (7.68)(13) = 65.36$, or $65,360. The standard error of this estimate can be figured using Formula (12.21) and values that can be obtained from the original example.

$$n = 5, \quad \overline{X} = 11, \quad \Sigma x^2 = 22, \quad \text{and} \quad s_{Y|X} = \sqrt{28.03} = 5.29$$

so

$$s_{\hat{Y}_0} = s_{Y|X} \sqrt{\frac{1}{n} + \frac{(X_0 - \overline{X})^2}{\Sigma x^2}} = 5.29 \sqrt{\frac{1}{5} + \frac{(13 - 11)^2}{22}}$$

$$= 3.27$$

Now the 95% confidence interval for the mean selling price can be calculated using Formula (12.20) and $t_{.025,3} = 3.182$.

$$\hat{Y}_0 \pm t_{\alpha/2, n-2} s_{\hat{Y}_0} = 65.36 \pm (3.182)(3.27)$$

or

$$54.96 \quad \text{to} \quad 75.76$$

Thus, the real estate investment firm can be 95% sure that the mean selling price for condominiums with 1300 square feet in area will lie between $54,960 and $75,760. This interval is quite wide, and does not provide a very precise interval, but part of the problem lies in the fact that the sample size of $n = 5$ is rather small and that only one variable, area of the condominium, was used to predict the price.

It should be remembered that the confidence intervals and
hypothesis tests demonstrated in this section are valid only if the variables
involved exhibit the three characteristics assumed at the first of the
section. That is, the X values must be nonrandom, the Y values must
be normally and independently distributed around the regression line,
and the scatter diagram should show homoscedasticity (common variance
around the regression line regardless of the X values). If these conditions
are not met, the least squares regression line may still be a somewhat
accurate predictor of the Y_i values, but the confidence intervals and
hypothesis tests described in this section are not valid. In order to
test the three assumptions to see if they hold, one could first make
sure that it is possible to secure a fixed value of X as a predictor
of Y in using the regression equation. Second, the normality assumption
can be tested using the goodness-of-fit test and the χ^2 distribution
discussed in Section 11.4, provided the sample size is large. Finally,
the assumption of homoscedasticity can usually be checked using the
scatter diagram of the $(X_i,\ Y_i)$ pairs to determine if the scatter of
the points around the regression line matches that shown in Figure
12.6(c).*

12.5 CORRELATION

In the preceding sections of this chapter we have been concerned with
the use of regression techniques to estimate the parameters of an assumed
linear relation between X and the mean of Y given X. We assumed
that the values of X were known and allowed them to be selected
and controlled by the experimenter; that is, we did not assume that
X was a random variable.

We now consider methods that are appropriate when it is assumed
that X and Y are both random variables and have a joint distribution.
We want to make inferences about the degree of linear relationship
between them without estimating the regression line.

One of the parameters of the joint distribution of X and Y is
the *product moment correlation coefficient* (or simply the *correlation coefficient*) ρ (Greek rho). The correlation coefficient ρ is a measure of the
linear covariation of the variables; that is, it measures the degree of
linear association between them. It is a dimensionless quantity that may
take any value between -1 and 1, inclusive. If ρ is either -1 or 1,

---

*Here we should repeat the note of caution in Section 9.12. The sample used
to validate the assumptions behind the model and the sample to which the model is
applied should be independent of one another.

the variables have a perfect linear relationship in that all of the points in a sample lie exactly on a line. If ρ is near -1 or 1, there is a high degree of linear association.

A *positive correlation* means that as one variable increases, the other likewise increases. A *negative correlation* means that as one variable increases, the other decreases. Heights and weights of humans are positively correlated, but the age of a car and its trade-in value are negatively correlated. If ρ is equal to zero, we say the variables are *uncorrelated* and that there is no linear association between them. Bear in mind that ρ measures only linear relationship. The variables may be perfectly correlated in a curvilinear relationship and ρ could still equal zero.

The population correlation coefficient, which measures the correlation between two sets of random variables, X and Y, can be obtained through a figure called the *covariance* of X and Y. The covariance of (X_i, Y_i) pairs in a population of N pairs is

$$\sigma_{XY} = \text{Cov}(X, Y) = \frac{\sum_{i=1}^{N} (X_i - \mu_x)(Y_i - \mu_y)}{N} \tag{12.23}$$

If the two variables move together, then the covariance is positive. However, if the two variables move in directly opposite directions, then the covariance is a negative value. Note that the population covariance is computed somewhat like the variance, which is denoted by σ^2. Unlike the variance, however, a covariance can be negative. Thus, the notation of σ_{XY} is not a squared figure. If the two variables are totally uncorrelated, then the covariance is zero.

Formula (12.23) defines the population covariance and assumes that the population is composed of N pairs that we have under examination. However, when the pairs we are examining form a sample, then we estimate the population covariance as

$$s_{XY} = \widehat{\text{Cov}}(X, Y) = \frac{\sum_{i=1}^{n} (X_i - \bar{X})(Y_i - \bar{Y})}{n - 1} = \frac{\Sigma xy}{n - 1} \tag{12.24}$$

where $\widehat{\text{Cov}}(X, Y)$ denotes the estimated covariance of X and Y. The divisor of $n - 1$ in Formula (12.24) is the number of degrees of freedom in the covariance estimate.

The covariance of pairs of values is a measure of how much they vary together in that a "big" positive covariance indicates that the variables move together, a "big" negative covariance indicates that they move inversely, and a "small" covariance indicates that they are uncorrelated with one another. However, the covariance's size depends on the units in which the variables X and Y are measured. In the real estate example of Section 12.2, the units of the covariance would be square feet-dollars, since X was measured in terms of square feet and Y was price. The size of the covariance could be changed by expressing X in terms of square yards and Y in terms of cents. Thus, if the covariance is to be used to measure the degree of correlation between two variables, it has to be adjusted. Thus, we define the population correlation coefficient, ρ, as

$$\rho = \frac{\sigma_{XY}}{\sqrt{\sigma_X^2 \sigma_Y^2}} \qquad (12.25)$$

The sample correlation coefficient, r, is an estimator for ρ and is defined as

$$r = \frac{s_{XY}}{\sqrt{s_X^2 s_Y^2}} \qquad (12.26)$$

Since $s_X^2 = \Sigma x^2 / (n - 1)$ and $s_Y^2 = \Sigma y^2 / (n - 1)$, Formula (12.26) can be written as

$$r = \frac{\Sigma xy / (n - 1)}{\sqrt{[\Sigma x^2 / (n - 1)][\Sigma y^2 / (n - 1)]}} \qquad (12.27)$$

A little algebraic manipulation can be used to show that the $(n - 1)$ figure in the numerator will cancel out the two $(n - 1)$ values under the square root sign in the denominator. Thus, the formula for the correlation coefficient becomes

$$r = \frac{\Sigma xy}{\sqrt{\Sigma x^2 \Sigma y^2}} \qquad (12.28)$$

While the correlation coefficient can be calculated using a number of different formulas,* Formula (12.28) is the most commonly used formula since it is so easily found from the sample data.

---

*The r^2 value defined in Formula (12.12) is the square of the r value defined in Formula (12.28). When 12.28 is squared, we obtain $r^2 = (\Sigma xy)^2 / \Sigma x^2 \Sigma y^2$. Since $b = \Sigma xy / \Sigma x^2$, we can substitute this expression into (12.12) for the value b and find that $r^2 = (\Sigma xy)^2 / \Sigma x^2 \Sigma y^2$.

EXAMPLE 1

Problem: An economist was interested in studying the relationship between the way in which families spend tax refunds and the size of their incomes. Thus, he took a sample of 6 families and looked at X = annual family income and Y = the percentage of the tax refund spent within 3 months of receipt. The following table of values shows the actual data for the families and the deviations used in a correlation problem.

X FAMILY INCOME (000)	Y PERCENT REFUND SPENT	x	y	x²	y²	xy
13	70	−7	20	49	400	−140
18	55	−2	5	4	25	−10
9	100	−11	50	121	2500	−550
25	40	5	−10	25	100	−50
36	15	16	−35	256	1225	−560
19	20	−1	−30	1	900	30
120	300	0	0	456	5150	−1280

$\overline{X} = 120/6 = 20$

$\overline{Y} = 300/6 = 50$

$\Sigma xy = -1280$

$\Sigma y^2 = 5150$

$\Sigma x^2 = 456$

Find the correlation coefficient using both Formulas (12.26) and (12.28).

Solution: Formula (12.26) requires that we find the covariance of X and Y and the variances of X and of Y. Using the sample data we may obtain

$$s_{XY} = -1280/(6-1) = -256$$

$$s_X^2 = 456/(6-1) = 91.2$$

$$s_Y^2 = 5150/(6-1) = 1030$$

Applying Formula (12.26) we find the value

$$r = \frac{-256}{\sqrt{(91.2)(1030)}} = -.835$$

We can apply Formula (12.28) using the data directly from the table without going through the covariance and variances first.

$$r = \frac{-1280}{\sqrt{(456)(5150)}} = -.835$$

Thus, regardless of which formula is used, $r = -.835$. This means that the two variables, family income and percent of tax refund spent in the first 3 months, have a rather strong negative correlation. That is, higher income families tend to spend a lower percentage of their refund in the first 3 months than low income families. This negative correlation is one of the reasons that economists recommend that tax cuts be aimed at lower income groups—they spend a higher percentage of the tax savings and give the economy a substantial boost.

Regression techniques and correlation methods are closely related, for r is the square root of the coefficient of determination. It is primarily in interpretation that the differences lie. In correlation, r is an estimator for the population correlation coefficient ρ. In regression, if X is not a random variable there is no correlation, and r^2 is simply a measure of closeness of fit.

When we want to make inferences about the population correlation coefficient ρ, we usually assume that the variables X and Y have a joint normal distribution; but even with the assumption of normality, if ρ is not equal to zero the sampling distribution of r is complicated and not at all easy to use. For this reason, tables and graphs have been made for finding confidence intervals for ρ. See Reference 12.1 or 12.6 since we will not treat these methods in this introductory text.

If we want to test the hypothesis that the variables are not linearly related, that is, that ρ is 0, we may use an ordinary t test, for when ρ is 0, the term

$$t = r \sqrt{\frac{n-2}{1-r^2}} \tag{12.29}$$

has the t distribution with $n - 2$ degrees of freedom. To test H_0: $\rho = 0$ we merely evaluate t by Equation (12.29) and compare it with the tabular t value for the given probability of Type I error.

EXAMPLE 2

Use the data in Example 1 to test the hypothesis that there is no correlation between the size of a family's income and the percentage of a tax refund spent in the first 3 months after receiving it. Use as the alternative hypothesis that there is a negative correlation between these two variables. Let $\alpha = .05$.

$$H_0 : \rho = 0$$

$$H_a : \rho < 0$$

Since the alternative hypothesis is of the "less than" type, this is a one-tailed test and

$$R : t < t_{.05,4} = -2.132$$

Using Formula (12.29) to calculate t, we obtain

$$t = -.835 \sqrt{\frac{6 - 2}{1 - (-.835)^2}} = -.835 \sqrt{12.21}$$

$$= -3.03$$

Since this value is beyond the tabular value of -2.132, we reject the null hypothesis and conclude that there is a negative correlation between the size of a family's income and the percentage of a tax refund spent in the first 3 months after receiving it.

The method just illustrated and Formula (12.29) cannot be used if the hypothesis is that the population correlation coefficient is something other than zero. That is, more complicated methods are needed to test a null hypothesis such as $H_0: \rho = .6$. These methods will not be treated here.

■ 12.6 PARTITIONING THE SUM OF SQUARES

The previous sections of this chapter have discussed various topics associated with simple linear regression and correlation. The subjects of regression, correlation, and analysis of variance (which was presented in Chapter 10) are all related to one another. Before we show this relationship, however, we must discuss how the variation in a regression problem's dependent variable, Y, can be partitioned into two parts.

For a single variable Y, the variation in Y is measured by the sum of squares, or the sum of squared deviations of the Y values from their mean, all of which can be considered as being due to random or unexplained variation; hence, the estimated variance of Y is based on the total sum of squares. In the regression situation, however, some of the observed variation among the sample Y's is associated with the relationship between Y and X. Figure 12.8 shows one observed point (X_i, Y_i) and the point on the fitted line whose coordinates are the means $(\overline{X}, \overline{Y})$. We see that the deviation $Y_i - \overline{Y}$ from the mean is the sum of the deviation of Y_i from the corresponding predicted value $\hat{Y}_i$ and the deviation of $\hat{Y}_i$ from $\overline{Y}$; that is,

$$(Y_i - \overline{Y}) = (Y_i - \hat{Y}_i) + (\hat{Y}_i - \overline{Y}) \tag{12.30}$$

The second element, the deviation of $\hat{Y}_i$ from $\overline{Y}$, is associated with the relationship between Y and X, so that this much of the deviation of Y_i from the mean may be said to be accounted for by, or due

---

■The material in this section is not required in the sequel.

FIGURE 12.8   Subdivision of $Y_i - \overline{Y}$

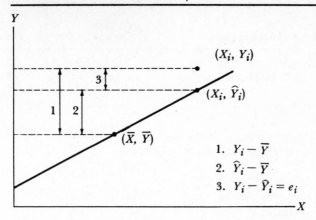

1. $Y_i - \overline{Y}$
2. $\hat{Y}_i - \overline{Y}$
3. $Y_i - \hat{Y}_i = e_i$

to, the regression of Y on X. If we square and add both sides of Equation (12.30) and sum over all values of Y_i, we find that*

$$\Sigma(Y_i - \overline{Y})^2 = \Sigma(Y_i - \hat{Y}_i)^2 + \Sigma(\hat{Y}_i - \overline{Y})^2 \qquad (12.31)$$

We have partitioned the total sum of squares for Y into two parts. One part, $\Sigma(\hat{Y}_i - \overline{Y})^2$, is associated with the regression; the remainder, $\Sigma(Y_i - \hat{Y}_i)^2$, with the random or unexplained variation in the data. These quantities may be found more simply by using the following formulas:

$$\Sigma(\hat{Y}_i - \overline{Y})^2 = b\Sigma xy$$
$$\Sigma(Y_i - \hat{Y}_i)^2 = \Sigma y^2 - b\Sigma xy \qquad (12.32)$$

The partitioning we have discussed is conveniently summarized in Table 12.1. We note that the sum of squares for regression has 1 degree of freedom, the sum of squares of deviations has $n - 2$ degrees of freedom, and the total sum of squares has, as usual, $n - 1$ degrees of freedom. The degrees of freedom, as well as the sum of squares, are partitioned.

For the gas station example of Section 12.2, the total variation of the Y's about their mean $\overline{Y}$ is measured by the total sum of squares with degrees of freedom equal to $n - 1$, or 6:

$$\Sigma y^2 = 102.00$$

The portion of this variation that is accounted for by the linear relation

---

*One would expect the right side of Equation (12.31) to have a term of the form $2\Sigma(Y_i - \hat{Y}_i)(\hat{Y}_i - \overline{Y})$, but it can be shown that this sum is zero.

TABLE 12.1   Partitioned Sum of Squares in Regression

SOURCE OF VARIATION	DEGREES OF FREEDOM	SUM OF SQUARES
Due to regression	1	$b\Sigma xy$
Deviations from regression	$n - 2$	$\Sigma y^2 - b\Sigma xy$
Totals	$n - 1$	Σy^2

between the variables is the sum of squares with 1 degree of freedom associated with, or due to, regression:

$$b\Sigma xy = (1.89)(53) = 100.17$$

The remaining unexplained, or random, variation is the variation of the observed Y's about the estimated line. It is measured by the sum of squares of deviations about the regression line. This sum of squares has $n - 2$, or 5, degrees of freedom and is found by subtraction,

$$\Sigma y^2 - b\Sigma xy = 102.00 - 100.17 = 1.83$$

The summary table is as follows:

SOURCE	df	SUM OF SQUARES
Due to regression	1	100.17
Deviations from regression	5	1.83
Totals	6	102.00

We are now prepared to use the information presented in this section to show the relationship between regression, correlation, and analysis of variance. That relationship is shown in the next section together with a new example which is intended to review and summarize the material covered to this point in the chapter.

■ 12.7   THE RELATIONSHIP BETWEEN REGRESSION, CORRELATION, AND ANALYSIS OF VARIANCE

In Chapter 10, we discussed the topic of analysis of variance. That topic is closely related to the regression and correlation problems we have discussed in this chapter. The purpose of this section is to show that relationship and to review some of the concepts covered in previous sections by working through a completely new illustration problem.

■The material in this section is not required in the sequel.

In order to show the relationship between regression, correlation, and analysis of variance, we need an example in which we find the regression equation $\hat{Y} = a + bX$, the sample standard deviation of the regression line coefficient, s_b, the correlation coefficient, r, and the table in which the regression sum of squares is partitioned. Let us consider the following example and use it to review and reinforce the concepts discussed in previous sections of this chapter.

EXAMPLE 1

A law firm has noticed that its photocopying costs seem to fluctuate rather dramatically from one week to the next. The firm uses a neighborhood photocopy service to copy its legal briefs. The office manager for the firm has indicated to the partners the opinion that the fluctuation in costs is due primarily to the changing work load, which is reflected in the number of cases the firm is preparing for trial. To prove his point, the office manager randomly selected 5 weeks and showed the number of cases in preparation for trial in those weeks (this is variable X) and the photocopying costs in those weeks (this is variable Y expressed in hundreds of dollars). These figures are given in the following table, and they are plotted in Figure 12.9.

X	0	2	7	9	12
Y	1	3	7	6	8

Find the sample regression equation, the standard deviation of the regression coefficient, s_b, the correlation coefficient, r, and the partitioned sum of squares for regression.

The following table shows the calculations that are necessary to find the deviations, squared deviations, and products of deviations needed for the calculations done in a regression and correlation problem.

X	Y	x	y	xy	x^2	y^2
0	1	−6	−4	24	36	16
2	3	−4	−2	8	16	4
7	7	1	2	2	1	4
9	6	3	1	3	9	1
12	8	6	3	18	36	9
30	25	0	0	55	98	34

$$\bar{X} = 30/5 = 6$$

$$\bar{Y} = 25/5 = 5$$

$$\Sigma y^2 = 34$$
$$\Sigma x^2 = 98$$
$$\Sigma xy = 55$$

The equation for the regression line relating X and Y can be found using Formulas (12.7) and (12.8):

FIGURE 12.9 Photocopying costs versus cases in preparation for trial

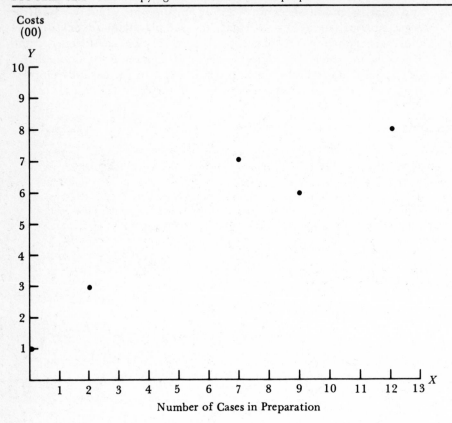

Number of Cases in Preparation

$$b = \Sigma xy / \Sigma x^2 = 55/98 = .56$$

$$a = \overline{Y} - b\overline{X} = 5 - (.56)(6) = 1.64$$

Thus, the regression equation is

$$\hat{Y} = 1.64 + .56X.$$

This equation suggests that the firm's photocopying charges run about $164 per week even when no cases are being prepared. However, weekly costs appear to rise at a rate of $56 per case being prepared.

The standard deviation of the sample regression line's slope, s_b, can be found using Formulas (12.10) and (12.14). This value is

$$s_b = \frac{1.033}{\sqrt{98}} = .104$$

The correlation coefficient can also be found. For this value we could employ Formula (12.11), Formula (12.12), or Formula (12.28). If we use Formula (12.12), we get

$$r^2 = (.56)(55)/34 = .91$$

$$r = .95$$

In Section 12.6 we found that the sum of squared deviations for the Y values (the costs) can be separated into two parts: the sum of squares due to regression and the sum of squares from regression. Figure 12.10 shows the data from the example, the regression line calculated above, and the third week's total deviation from the mean separated into the variation due to regression and the deviation from

FIGURE 12.10 Regression line and partitioned variation at $Y = \$700$

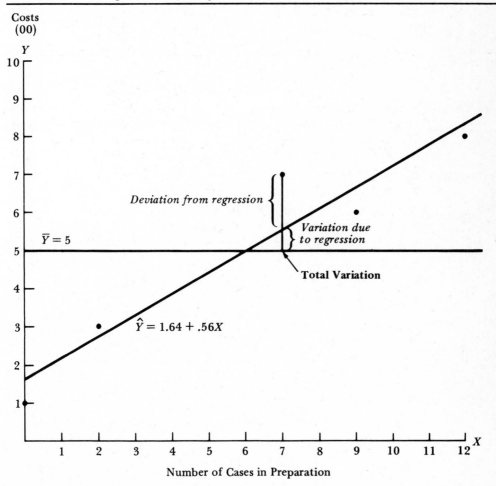

regression. The partitioning of the sum of squares in a regression problem can be presented in a table that has the format of Table 12.1. That table is reproduced here for reference. This table looks something like an analysis-of-variance table. It has one column for the degrees of freedom in each sum of squares. It also has a sum of squared deviations column. However, it does not have a column for the mean squares, which are the sum of squares divided by the degrees of freedom: $MS = SS/df$.

TABLE 12.1   Partitioned Sum of Squares in Regression

SOURCE OF VARIATION	DEGREES OF FREEDOM	SUM OF SQUARES
Due to regression	1	$b\Sigma xy$
Deviations from regression	$n - 2$	$\Sigma y^2 - b\Sigma xy$
Totals	$n - 1$	Σy^2

Table 12.2 presents the regression table for the data in our example of photocopying costs. However, this table has an MS column added, and the MS figures were obtained through division of the SS figures by the degrees of freedom associated with the sums of squares.

For the data in our example we have now obtained the figures we were after:

1.   The sample regression line, $\hat{Y} = 1.64 + .56X$,
2.   The sample standard deviation of the slope, $s_b = .104$,
3.   The sample correlation coefficient, $r = .95$, and
4.   The table of partitioned sums of squares and the mean squares.

With these figures we are prepared to demonstrate the relationship between regression, correlation, and analysis of variance.

We begin by suggesting a null hypothesis: There is no relationship between the costs of photocopying in this law firm and the number of cases being prepared for trial. That is, the relationship we see in

TABLE 12.2   Partitioned Sum of Squares in Photocopying Example

SOURCE OF VARIATION	df	SS	MS
Due to regression	1	30.80	30.80
Deviations from regression	3	3.20	1.07
Totals	4	34.00	

this sample of five randomly selected weeks is purely random variation, and in the population of all weekly figures there would be no relationship between these two variables. This hypothesis can be tested in three mathematically equivalent ways. The method we choose depends on the way we view the problem. If we view it as a regression problem in which we wish to predict photocopying costs from a set, or given, number of cases, then we can test the hypothesis using the regression results. However, if we view the number of cases, X, and the photocopying costs, Y, as two jointly varying random variables whose strength of relationship we wish to measure, then we can test the null hypothesis using the correlation results or analysis of variance on the mean squares in the regression table. All three approaches produce exactly the same conclusions. These three approaches to test the hypothesis of no relationship are discussed in the following sections.

12.7.1 Testing the Hypothesis That $B = 0$

We begin by testing the null hypothesis using the regression results. If the hypothesis just posed is true, then the regression equation's slope in the population is $B = 0$. That is, we should test the hypothesis:

$$H_0: B = 0$$

$$H_a: B \neq 0$$

We will use an α value of .05 and assume for this demonstration that the alternative hypothesis is a "not equal" type. This means that we can use Equation (12.16) and reject the null hypothesis if the t value we calculate exceeds the $t_{.025,3} = 3.182$ value we obtain from Table VI of the Appendix. Our data give:

$$t = b/s_b = .56/.104 = 5.38$$

Thus we reject the hypothesis since 5.2 is much larger than 3.182.

12.7.2 Testing the Hypothesis That $\rho = 0$

If the problem at hand is a correlation problem rather than a regression problem, then we have a second way of testing the hypothesis that there is no relationship between two variables. We use the correlation results and test the hypothesis that rho, the population coefficient of correlation, is zero. That is,

$$H_0: \rho = 0$$

$$H_a: \rho \neq 0$$

We will use an α value of .05 again. But to test this hypothesis we must use Formula (12.29). We find that

$$t = r \sqrt{\frac{n-2}{1-r^2}} = .95 \sqrt{\frac{5-2}{1-(.95)^2}} = 5.3$$

The value of 5.3 exceeds $t_{.025,3} = 3.182$ so we reject the hypothesis that there is no relationship between photocopying costs and the number of cases being prepared for trial. It is important to note that the t value of 5.38 found testing H_0: $B = 0$ is very close to the t value of 5.3 found testing H_0: $\rho = 0$. In fact, these two figures are exactly the same except for rounding errors. Thus, these two hypothesis tests are mathematically equivalent, but the second test is the relevant one when we view both X and Y as random variables.

12.7.3   Testing the Hypothesis Using Mean Squares

A third way of testing the hypothesis that two variables have no relationship to one another is to examine the mean square figures from the regression summary table. Table 12.2 showed that the mean square due to regression was 30.80 and had one degree of freedom while the mean square due to deviations from regression was 1.07 and had three degrees of freedom. In Chapter 10, we found that we could test the significance of two mean squares by taking their ratio and comparing that to the F distribution. In the regression problem, the situation where there is no relationship between the variables being considered is characterized by a regression line of $\hat{Y} = \bar{Y}$. In that case, the mean square due to regression would be zero since the sum of squares of the regression line values about the mean would be zero due to the fact that the regression line and the mean line are the same line. The larger the mean square due to regression, the more likely there is to be a significant relationship between the two variables. The ratio of the mean square due to regression to the mean square of the deviations from regression is an F-distributed variable under the hypothesis of no relationship, if both X and Y are random variables. Thus, the F value computed from our data is

$$F = 30.80/1.07 = 28.79$$

This value exceeds the Table VIII F value of $F_{.05,1,3} = 10.128$ so we reject the hypothesis that the variables have no relationship.

In Chapter 10 we learned that $t_{\alpha/2,\nu}^2 = F_{\alpha,1,\nu}$. The t value we used in the two previous hypothesis tests was $t_{.025,3} = 3.182$. When we square this figure, we obtain $(3.182)^2 = 10.125$, which is very close to the $F_{.05,1,3}$ value we found in the F table. Also, the calculated t

value of 5.3 obtained in the previous hypothesis test can be squared to give $(5.3)^2 = 28.09$, which is approximately the same, except for rounding errors, as the calculated F value previously obtained. Thus, the analysis-of-variance procedure yields exactly the same mathematical results as the hypothesis tests that $B = 0$ and $\rho = 0$.

The previous paragraphs demonstrate that regression, correlation, and analysis of variance are closely related. In order to test the hypothesis that two variables are unrelated to one another, one need only perform one of the following three tests:

1. Calculate a t value for the sample regression coefficient to test $H_0: B = 0$.
2. Calculate a t value for the sample correlation coefficient to test $H_0: \rho = 0$.
3. Calculate an F value for the ratio of the mean squares in the regression table to test H_0:MS due to regression is zero.

The mathematical results produced by each of these tests are identical. Which test is used depends on the way in which we view the variables and, to some extent, the preference of the person performing the test.

12.8 REGRESSION, CORRELATION, AND COMPUTERS

It should be fairly obvious that the calculations of a, b, $s_{Y|X}$, s_b, s_a, $s_{\hat{Y}_0}$, r^2, and the confidence intervals for a regression and correlation problem can become very tedious. This is especially true if n, the number of pairs of X and Y, is large. Thus, computer programs have been prepared which will aid the user in problems of this type. These programs, or packages as they are sometimes called, are stored in some computers' program libraries. The person desiring to use one of these programs need only:

1. Give his or her computer the name or code of the regression and correlation program.
2. Tell the computer how many pairs of X and Y values are in the data set.
3. Supply the computer with the X, Y pairs.

This information is supplied to the computer most often on punched tabulating cards or through teletype terminals.

Once the computer has been given this information, it will automatically execute its regression and correlation program using the information that has been supplied. The program followed by the computer will calculate and print out a, b, r^2, and all the standard deviations

TABLE 12.3

FORMULAS	COMMENTS
1. $b = \dfrac{\Sigma xy}{\Sigma x^2}$ and $a = \bar{Y} - b\bar{X}$	**a.** b and a are the slope and intercept of the least-squares regression line. **b.** The x and y values are deviations from their means: $x_i = X_i - \bar{X}$ and $y_i = Y_i - \bar{Y}$.
2. $s^2_{Y\mid X} = \dfrac{1}{n-2}\Sigma(Y_i - \hat{Y}_i)^2$ $= \dfrac{1}{n-2}[\Sigma y^2 - b\Sigma xy]$	**a.** This figure measures the scatter of the Y_i values around the regression line. The s^2_Y figure measures the scatter of the Y_i values around the mean, $\bar{Y}$. **b.** Formulas (12.10a) to (12.10c) are alternative forms of this formula.
3. $r^2 = 1 - \dfrac{\Sigma(Y_i - \hat{Y}_i)^2}{\Sigma(Y_i - \bar{Y})^2}$ or $r^2 = \dfrac{b\Sigma xy}{\Sigma y^2}$	**a.** r^2 varies between 0 and 1 and measures the relative strength of the relationship in X and Y. **b.** r is the coefficient of correlation and has the same algebraic sign as the slope, b.
4. $s_b = \dfrac{s_{Y\mid X}}{\sqrt{\Sigma x^2}}$	**a.** s_b measures the variation of the sample regression slope, b, around the population regression slope, B, under repeated sampling. **b.** s_b is used in confidence intervals for B and hypothesis tests about B. See Formulas (12.13) and (12.15).
5. $s_{\hat{Y}_0} = s_{Y\mid X}\sqrt{\dfrac{1}{n} + \dfrac{(X_0 - \bar{X})^2}{\Sigma x^2}}$	**a.** This figure estimates the variation of the $\hat{Y}_i$ values around the population regression line at the point $X = X_0$. **b.** The $s_{\hat{Y}_0}$ figure is used in construction of confidence intervals or hypothesis tests concerning the mean value of Y that might be associated with $X = X_0$.
6. $s_{XY} = \widehat{\text{Cov}}(X, Y) = \dfrac{\Sigma xy}{n-1}$	**a.** The covariance can be positive or negative. **b.** Covariance is an intermediate figure on the way to finding the correlation coefficient.

7. $r = \dfrac{s_{XY}}{\sqrt{s_X^2 s_Y^2}} = \dfrac{\Sigma xy}{\sqrt{\Sigma x^2 \, \Sigma y^2}}$

a. This formula is the equivalent of the formula for r in number 3 above.

b. Although the numerical computation is the same, the interpretation of r here is a measure of *joint* variation between X and Y when they are both random variables.

discussed in this chapter. Some of the programs will supply other information such as the mean values of X and Y and the values of the sums of squares.

The important point to note is that a person needs to know very little about computers and computer programming to be able to execute one of these regression and correlation programs. Also, the computer can merely print out the calculations it makes. The interpretation of the resulting information is still the responsibility of the program user.

12.9 SUMMARY

Numerous formulas have been presented in this chapter. Some of the more important ones are summarized in Table 12.3.

It should be noted that the square root of this table's Formula 3 is mathematically equivalent to Formula 7. But the interpretation and uses of r^2 or r in a regression problem are somewhat different than the interpretation of Formula 7's r value in a correlation problem. The values of r^2 and r computed from Formula 3 in a regression problem are used to measure the goodness of fit of the data points around a regression line. In a correlation problem, however, a regression line is usually not even constructed. Thus, the r value computed from Formula 7 in a correlation problem is used to measure the strength of the joint relationship between two variables.

This difference can be illustrated with an example. The attitudes of a company's employees toward their supervision and their compensation were measured using two questionnaires. The correlation coefficient, r, between the employees' scores on the two questionnaires could be *computed* as the square root of Formula 3 *or* by Formula 7. The mathematical result would be the same. However, the interpretation would be the one reserved for Formula 7: the r measures the strength of the joint relationship between attitude toward supervision and attitude toward compensation in this employee group. The interpretation usually assigned to Formula 3 would not make sense since a regression equation would not likely be constructed to "predict" a person's score on one attitude measure on the basis of their score on the other. Thus, the goodness-of-fit interpretation of r would not be logical.

PROBLEMS

1. Given the following pairs of values,
   a. Find the regression equation, $\hat{Y} = a + bX$.
   b. Find the predicted value of Y given $X = 5$.

X	2	4	6	8	10
Y	3	1	7	5	9

2. Given the following pairs of values,
   a. Find the equation for the regression of Y on X.
   b. Find the predicted value of Y given $X = 6$.

X	1	5	9	13	17
Y	7	6	9	8	10

3. Given the following pairs of values,
   a. Find the equation for the regression of Y on X.
   b. Find the predicted value of Y given $X = 8$.

X	0	1	2	3	4	5	6
Y	1	2	3	5	8	11	12

4. Using the data given for Problem 3, find the standard error of estimate, $s_{Y|X}$.

5. Using the data given for Problem 3, find the coefficient of determination, r^2. Explain what this figure means.

6. Using the data given for Problem 3, find:
   a. The standard deviation of the regression line's slope, s_b.
   b. An interval within which you can be 95% confident that B, the population regression line's slope, will lie.

7. Using the data given for Problem 3, find:
   a. The standard deviation of the regression line's intercept, s_a.
   b. An interval within which you can be 90% sure that A, the population regression line's intercept, will lie.

8. Using the data given in Problem 3, find:
   a. $s_{\hat{Y}_0}$ when $X_0 = 8$.
   b. The 95% confidence interval for the value of $\mu_{Y|X}$ when $X_0 = 8$.

9. List the assumptions you needed to make in order to construct the confidence intervals in Problems **6**, **7**, and **8**.

10. A company has figured the "standard number of workers" that should be assigned to a work crew on the production line. In an experiment

to verify the validity of the "standard," the size of the work crew was varied from the standard. In the following figures, X represents the number of people above or below standard size of the crew. Crew sizes from three people below standard ($X = -3$) to three people above standard ($X = 3$) were used. Y represents the number of defects produced in a shift with a crew of the corresponding size, X.

X	-3	-2	-1	0	1	2	3
Y	36	33	24	15	9	6	3

   a. Find the regression equation $\hat{Y} = a + bX$.
   b. Find the predicted value of Y given $X = -1$.
   c. Explain the meaning of $\hat{Y}$ for $X = 0$.
   d. Explain the meaning of the slope, b.

11. Using the data given for Problem **10**, find:
   a. The standard error of estimate, $s_{\eta x}$.
   b. The coefficient of determination.

12. Using the data given for Problem **10**,
   a. Find s_a.
   b. Test the hypothesis that $A = 20$ against the alternative hypothesis that A is less than 20. Let $\alpha = .05$.

13. Using the data given for Problem **10**,
   a. Find s_b.
   b. Test the hypothesis that $B = 0$ against the alternative hypothesis that $B \neq 0$. Let $\alpha = .10$.

14. Using the data given for Problem **10**, assume it is proposed that for tomorrow's shift the work crew should operate with one more person than "standard." Find a 95% confidence interval for the mean number of defects that will be produced by the crew on tomorrow's shift.

15. Find the equation for the regression of Y on X.

X	12	13	14	15	16	17	18
Y	-8	-6	-4	0	6	12	14

16. Find the equation for the regression of Y on X.

X	-10	-6	-2	2	6	10	14	16	20	23
Y	-8	-5	-4	-3	0	4	7	8	11	10

17. In a regression problem: $n = 30$, $\Sigma X_i = 15$, $\Sigma Y_i = 30$, $\Sigma x_i y_i = -30$, $\Sigma x_i^2 = 10$, and $\Sigma y_i^2 = 160$.
   a. Find the regression line, $\hat{Y} = a + bX$.
   b. Estimate the variance, $\sigma_{Y|X}^2$.
   c. Test $H_0: B = 0$ against $H_a: B \neq 0$. Let $\alpha = .05$.

18. In a regression analysis: $n = 18$, $\overline{X} = 6$, $\overline{Y} = 20$, $\Sigma x_i^2 = 100$, $\Sigma y_i^2 = 400$, and $\Sigma x_i y_i = -120$.
   a. Find the equation for the regression of Y on X.

    b.   Find the estimated variance.

    c.   Test the hypothesis $B = 0$ against $H_a: B \neq 0$ using $\alpha = .01$.

    d.   Find and interpret the coefficient of determination, r^2.

19.   In a regression analysis: $n = 25$, $\Sigma X_i = 75$, $\Sigma Y_i = 50$, $\Sigma X_i^2 = 625$, $\Sigma X_i Y_i = 30$, and $\Sigma Y_i^2 = 228$.

    a.   Find the regression equation.

    b.   Find $s_{Y|X}^2$, s_a^2, and s_b^2.

    c.   Test that $B = 0$. Let $\alpha = .01$. Use $H_a: B \neq 0$.

    d.   Find a 95% confidence interval for A.

20.   In a regression analysis: $n = 38$, $\overline{Y} = 20$, $\overline{X} = 7$, $\Sigma y_i^2 = 900$, $\Sigma x_i^2 = 60$, and $\Sigma x_i y_i = 180$.

    a.   Find the regression equation.

    b.   Test the hypothesis that $A = 0$. Let $\alpha = .02$. Use $H_a: A \neq 0$.

    c.   Find a 90% confidence interval for B.

    d.   What fraction of the squared error of prediction in Y can be eliminated using the regression line as a predictor rather than $\overline{Y}$?

21.   Given that $n = 41$, $\overline{Y} = 10$, $\overline{X} = 12$, $\Sigma x_i^2 = 400$, $\Sigma x_i y_i = 100$, and $\Sigma y_i^2 = 64$, find:

    a.   The predicted value of Y given $X = 8$.

    b.   A 99% prediction interval for a mean observation on Y given $X = 8$.

22.   A sample of 32 graduates with B.S. degrees in statistics reported their starting salaries in their first professional positions. The estimated average annual starting salary by year is given by

$$Y = 7130 + 205(X - 1960)$$

    a.   Does a *linear* equation seem valid as a predictor of starting salary?

    b.   How do you interpret the various components of this equation?

    c.   What is the estimated mean annual starting salary for the year 1981?

23.   In the equation $\hat{Y} = a + bX$, where X is pounds and Y is dollars, what units are associated with a? with b? with r^2?

24.   In the equation $Y = a + bX$, Y is the amount of production material used expressed in terms of square feet. The X variable is the number of hours the production line is run. What are the units associated with a? with b? with $s_{Y|X}$?

25.   In the equation $Y = 120 + .32X$, Y is the weekly sales of a store in thousands of dollars, and X is the number of hours the store is open during the week.

    a.   What are the units of measure on a and b?

    b.   If the store stays open 60 hours next week, what is its predicted sales level?

    c.   If the store doesn't open at all next week, what would the equation predict for its sales level?

    d.   If the store doesn't open at all next week, what would you predict for its sales level?

    e.   Why is the answer in Part **c** not valid?

26.   The following are the average incomes per acre, Y, produced by a commercial system for taking game animals over a ten-year period.

a. Find the regression equation predicting income per acre from time in years.
b. Find a 99% confidence interval for B.
c. What fraction of the variation in income is accounted for by the relationship?
d. What is the predicted income per acre for this year? Why is this prediction likely to involve a great deal of error?

YEAR	1956	1957	1958	1959	1960	1961	1962	1963	1964	1965
INCOME PER ACRE	.51	.59	.64	.74	.78	.89	1.01	1.07	1.10	1.18

27. A newly developed low-pressure snow tire has been tested to see how it wears under normal, dry-weather conditions. Twenty of the new tires were mounted on the right front wheels of 20 standard passenger cars. These cars were then driven at high speeds on a dry test track for varying lengths of time. The tread wear in millimeters was then recorded for each tire.
a. From the accompanying data, estimate the relationship between hours driven, X, and tread wear, Y.
b. Find a 95% confidence interval for the slope of the line.
c. Test the hypothesis that the line passes through the origin. That is, test $H_0: A = 0$ against $H_a: A \neq 0$, letting the chance of a Type I error be .05.

HOURS	WEAR	HOURS	WEAR
13	.1	62	.4
25	.2	105	.7
27	.2	88	.6
46	.3	63	.4
18	.1	77	.5
31	.2	109	.7
46	.3	117	.8
57	.4	35	.2
75	.5	98	.6
87	.6	121	.8

28. The following data show the sales produced (in thousands of dollars) by five new salespeople in their second, fourth, sixth, and eighth months on the job. Note that there are 20 observations, each value of X (time) occurring with 5 different Y's (sales figures). Thus, the figure in the lower right corner of the table represents an (X_i, Y_i) point of (8, 26.3).

X = TIME (MONTHS)	2	4	6	8
	21.3	23.2	25.5	25.9
	21.7	23.4	23.6	25.2
Y = SALES ($000)	21.4	23.3	24.7	27.6
	22.1	23.7	26.0	27.1
	20.7	23.0	24.3	26.3

a. Find the regression equation.
b. Test the hypothesis that $B = 1.0$. Use $H_a: B < 1.0$ and let $\alpha = .025$.
c. What fraction of the variation in Y is accounted for by time?
d. Estimate the mean sales given that $X = 5$.
e. Find a 95% confidence interval for the mean sales, given $X = 5$.

29. The following data are the ultimate loads resulting from tests of joints made with different sizes of common nails. Note that the first row indicates (X_i, Y_i) points of $(30, 11.03)$, $(50, 11.97)$, and $(60, 13.69)$. Each following row represents similar pairs, where the first element of the pair is 30, 50, or 60.

NAIL SIZE	30d	50d	60d
	11.03	11.97	13.69
	10.64	13.63	14.82
ULTIMATE LOAD	10.48	13.56	15.23
	10.02	15.20	15.45
	10.31	14.84	16.28

a. Find the equation for the regression of ultimate load on nail size.
b. Find a 98% confidence interval for the slope of the line.
c. Is b significantly different from zero? Use $H_a: B \neq 0$ and $\alpha = .01$.
d. Find a 95% confidence interval for the mean ultimate load of joints of this type made with 30d nails.

30. The following data show the gas mileage obtained in miles per gallon, Y, achieved by a compact car in driving at a constant speed for 3 hours with various amounts, X, of a gasoline additive mixed in a full tank of gas.

Y	X	Y	X
26.8	45	28.1	54
28.1	59	29.2	63
28.5	66	26.9	48
27.0	47	28.6	55
28.8	61	31.0	67
29.7	68	26.4	44
27.5	49	27.7	56
29.0	57	28.2	62
30.2	67	26.7	50
26.3	43	30.8	70

a. Plot the points on graph paper.
b. Find and plot the point $(\bar{X}, \bar{Y})$.
c. Find a and b and, hence, the regression equation $\hat{Y} = a + bX$. Draw this line on the graph.
d. Find the deviation of each Y from the corresponding $\hat{Y}$. Square and sum these deviations.

e. Find the sum of squares for deviations from regression by the formula

$$\Sigma(Y_i - \hat{Y}_i)^2 = \Sigma y_i^2 - b\Sigma x_i y_i$$

How do you account for the difference between this figure and that obtained in Part **d**?

f. Construct a summary table showing the partitioning of the sum of squares and degrees of freedom.

g. Find the estimated variance $s_{Y|X}^2$ and the variance and standard deviation of b.

h. Find the 95% confidence interval for B.

i. Test the hypothesis that $B = 0$ against $H_a: B \neq 0$ using $\alpha = .05$.

j. Test the hypothesis that $\rho = 0$ against $H_a: \rho \neq 0$ using $\alpha = .05$.

k. Test the hypothesis that there is no correlation between mileage and the amount of additive using the data developed in Part **f**. Use $\alpha = .05$.

l. Show that the results in Parts **i**, **j**, and **k** are mathematically identical, except for rounding errors.

31. The following table gives annual salary (in thousands of dollars), the number of years with the company, and the proportion of sales quota met last year for 16 district sales managers for a national office equipment manufacturer.

SALARY	YEARS	QUOTA	SALARY	YEARS	QUOTA
33.0	23	.90	30.0	9	.98
34.1	12	.92	33.1	10	1.90
21.0	9	.89	38.1	17	.88
29.5	13	.59	46.6	17	.97
38.4	12	1.24	21.1	8	.78
25.3	10	.57	38.9	15	.94
21.4	8	.77	38.7	14	.80
47.7	26	.91	40.0	15	.99

a. Find the equation for the regression of salary on years of employment.

b. Find the equation for the regression of salary on proportion of quota met.

c. Is the regression of salary on proportion of quota met significant; i.e., is B significantly different from zero? Use $H_a: B \neq 0$ and $\alpha = .01$.

d. Which variable, years of employment or proportion of quota met, is the better predictor of salary? (*Hint:* Compare the coefficients of determination.)

32. The following data represent the number of hours since rush hour and the carbon monoxide counts (in parts per million) on an August day in the heart of a large western city.

a. Fit the regression of CO count on time.

b. Find a 98% confidence interval for B.

c. Find and interpret the coefficient of determination.

HOURS	CO COUNT
1.0	41.3
1.5	34.0
2.0	32.5
2.5	35.2
3.0	28.8
3.5	25.6
4.0	30.1
4.5	25.7
5.0	22.6
5.5	20.3
6.0	18.8
6.5	17.1

33. Given the following sets of quantities, find the sample correlation coefficient. Test the hypothesis that $\rho = 0$. Let $\alpha = .05$. Use $H_a: \rho \neq 0$.
    a. $n = 11$, $\Sigma y^2 = 400$, $\Sigma xy = 400$, $\Sigma x^2 = 625$
    b. $n = 18$, $\Sigma y^2 = 100$, $\Sigma xy = 36$, $\Sigma x^2 = 36$

34. Given the following information, find the sample correlation coefficient. Test the hypothesis that $\rho = 0$. Let $\alpha = .01$. Use $H_a: \rho \neq 0$.
    a. $n = 29$, $\Sigma y^2 = 64$, $\Sigma xy = 40$, $\Sigma x^2 = 100$
    b. $n = 18$, $\Sigma y^2 = 25$, $\Sigma xy = 54$, $\Sigma x^2 = 144$

35. Find the sample correlation coefficient and test $H_0: \rho = 0$ for the data of
    a. Problem 18.   b. Problem 19.
    Use $H_a: \rho \neq 0$ and $\alpha = .05$.

36. Records were kept on the scores received by 14 job applicants on a manual dexterity test and their production output after one week on the job.
    a. Use the accompanying data to estimate the correlation between the test scores and production output.
    b. Is there a significant correlation?

SCORE	OUTPUT	SCORE	OUTPUT
124	120	66	84
84	103	31	30
13	16	43	62
13	20	19	26
48	86	117	121
61	36	50	93
112	153	72	83

37. The executives at a national weight-loss clinic feel that there is a relationship between the weight of a person entering their program and the number of pounds lost. To confirm this suspicion, they took a sample from their records. The sample data are as follows:

X = BEGINNING WEIGHT	205	165	289	154	142	306	261	177
Y = WEIGHT LOSS	25	15	36	12	15	146	73	50

The regression information for this relationship is:

$$\hat{Y} = -67.78 + .54X, \quad r^2 = .58, \quad s_{y|x} = 31.60, \quad s_b = .186,$$

$$\Sigma x^2 = 28,911.87, \quad \Sigma y^2 = 14,362.00, \quad \Sigma xy = 15,557.48.$$

a. Test the hypothesis that there is no relationship between beginning weight and weight loss using the t test on the slope of the regression line. Let $\alpha = .05$.

b. Repeat the test in Part a using the test on r, the correlation coefficient.

c. Repeat the test in Part a using analysis of variance on the partitioned sums of squares for regression.

d. Show that the results are the same for the three previous tests if you compare the t and F values used.

38. Seven executives working for the same firm have expense accounts. The personnel people in the firm feel that there is a relationship between the executives' annual salaries and the amount they claim in expenses each year. Thus, they ran the following regression analysis relating the variables:

X = SALARY (000)	25	30	35	40	40	38	22
Y = EXPENSES (000)	.9	1.2	2.2	1.1	.8	1.6	.6

The regression results gave the following:

$$\hat{Y} = .227 + .030X, \quad r = .398, \quad s_{y|x} = .547, \quad s_b = .0306,$$

$$\Sigma y^2 = 1.78, \quad \Sigma x^2 = 320.86, \quad \Sigma xy = 9.63.$$

a. Test the hypothesis that there is no relationship between annual salary and expenses claimed using the t test on the regression coefficient b. Use $\alpha = .05$.

b. Repeat the test of Part a using the t test on r.

c. Repeat the test of Part a using analysis of variance on the table of partitioned sums of squares.

d. Show that the mathematical results are the same for all three approaches by comparing the t and F values used.

SAMPLE DATA SET QUESTIONS:

Refer to the 113 applicants for credit listed in the Sample Data Set Appendix of this book.

a. Find the regression line that would allow us to predict JOBINC for the group granted credit as a linear function of AGE.

b. Use the results obtained in the previous question to test the hypothesis $H_0: B = 0$ against the alternative hypothesis $H_a: B \neq 0$ using $\alpha = .10$.

c. Find the coefficient of correlation between TOTBAL and TOTPAY for the applicants who were denied credit.

REFERENCES

12.1   David, F. N. *Tables of the Correlation Coefficient.* Cambridge University Press, Cambridge, 1954.

12.2   Huff, Darrell. *How to Lie with Statistics.* Norton, New York, 1965. Chapter 8.

12.3   Kendall, M. G., and A. Stuart. *The Advanced Theory of Statistics,* 2nd edition. Griffin, London, 1964.

12.4   Mosteller, Frederick, Robert E. K. Rourke, and George B. Thomas, Jr. *Probability with Statistical Applications,* 2nd edition. Addison-Wesley, Reading, Massachusetts, 1970. Chapter 11.

12.5   Ostle Bernard. *Statistics in Research,* 2nd edition. Iowa State University Press, Ames, 1963. Chapters 8 and 9.

12.6   Richmond, Samuel B. *Statistical Analysis,* 2nd edition. Ronald Press, New York, 1964. Chapter 19.

12.7   Snedecor, George W., and William G. Cochran. *Statistical Methods,* 6th edition. Iowa State University Press, Ames, 1967. Chapters 6 and 7.

MULTIPLE REGRESSION

In many business and economics problems, we are interested in the relationship between one variable, say Y, and two or more other variables, say X_1 and X_2. For example, the amount of gas pumped at several service stations might be related to the location of the station as well as the number of pumps. If we let X_1 represent the number of pumps at each of the gas stations in our sample and let X_2 represent daily traffic count past each of these stations, then these two variables can be used together to explain or predict the variations in amount of gas pumped at each station. A logical extension of the simple linear Equation (12.4) would be

$$\mu_{Y|X_1, X_2} = A + B_1 X_1 + B_2 X_2 \qquad (13.1)$$

From only a sample of stations we would obtain only estimates of A, B_1, and B_2. The estimated prediction equation is

$$\hat{Y} = a + b_1 X_1 + b_2 X_2 \tag{13.2}$$

This equation, like the corresponding Equation (12.5), is a linear equation. Unlike that equation it cannot be graphed in two dimensions. Three dimensions would be needed, one for each of the three variables, Y, X_1, and X_2. The three-dimensional scatter diagram would appear as a series of points scattered around a plane rather than a line, the plane being all the points that satisfy the equation. Figure 13.1 shows an example of such a plane.

Any time a regression equation uses two or more X values in explaining or predicting variation in the dependent Y variable, the regression equation is called a *multiple regression* equation. This is because there are *multiple* independent variables. Multiple regression equations are used any time we feel that prediction of the movement in our dependent variable can be made more accurate by using more than one associated variable. In the case of the gas station example, it seems logical that by using traffic count past a station, X_2, in addition to the number of pumps operated by the station, X_1, we would be able to predict better what volume of gasoline the station will pump. Likewise, it would probably be easier to predict the price of a condominium using the value of the land on which it is built in addition to the area of the rooms.

This chapter covers three major topics in multiple regression. Sections 13.1–13.5 present the mathematical procedures used to develop and evaluate multiple regression equations. These procedures, however, are rather tedious to implement in practical problems. Thus, Sections 13.6 and 13.7 demonstrate how one popular multiple regression computer package is used to solve and interpret realistic multiple regression problems. Although the reader may not have access to this particular package, he or she will find that the computer output presented in this chapter is very similar to that produced by most major multiple regression packages. Finally, Section 13.8 discusses some special problems of which the practitioner of multiple regression must be wary.

13.1 FINDING THE COEFFICIENTS FOR A MULTIPLE REGRESSION EQUATION

In this chapter, as in Chapter 12, we will assume that the regression line desired in each case has the linear form of Equation (13.1). That is, we will deal only with the case in which the independent variables, the X's, in the equation have an additive impact on Y. In this section

FIGURE 13.1   Example of regression plane with one dependent variable and two independent variables

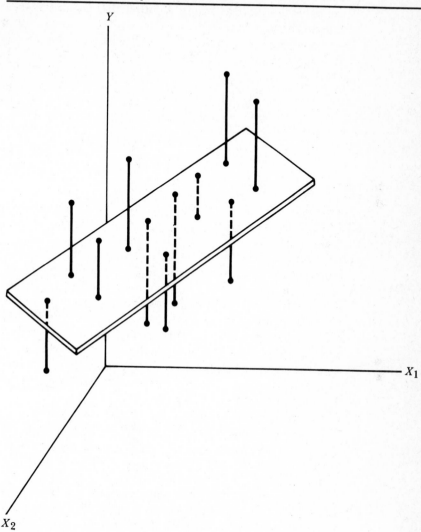

we will deal with the problem of how we use sample data to make estimates of the regression equation's coefficients.

Let us continue with the real estate example presented in Chapter 12. A real estate investment firm is interested in estimating the price of condominiums by using a sample of five recently sold condominium units in a particular city. Let us assume that, in addition to the data presented in Section 12.2, the firm also knows the land costs that can be assigned to each of the condominium units.* The land costs would likely reflect the desirability of the condominium units' location; and the more desirable locations, all other things being equal, would likely contain the higher-priced units.

Consider the following data, which are the data presented in Section 12.2 with the land costs added.

Y = SALES PRICE (000)	X₁ = AREA SQUARE FEET (00)	X₂ = LAND COSTS (000)
36	9	8
80	15	7
44	10	9
55	11	10
35	10	6

This table of numbers indicates that the first condominium sold for $36,000, contained 900 square feet of area, and was built on land that cost $8000 per unit.

Our objective is to find the values of a, b_1, and b_2, which can be used as coefficients in the regression equation represented by Equation (13.2). The value of a is the intercept of the regression plane, and it forms the beginning point for the regression equation. It has little physical meaning in this problem. The value of b_1, however, is the average price change (in thousands of dollars) associated with a one-unit change (100 square feet) in X_1, assuming X_2 is held constant. Likewise, b_2 is the average change in price (in thousands of dollars) associated with a one-unit change (one thousand dollars) in land costs, assuming X_1 is held constant.

In order to determine numerical values for a, b_1, and b_2, we again use the least squares method introduced in Section 12.2. However, in this case we find the values of a, b_1, and b_2 which minimize the vertical distance between the (Y, X_1, X_2) points in the three-dimensional space and the plane determined by the equation $\hat{Y} = a + b_1 X_1 + b_2 X_2$. The coefficients of this equation, with Y as the dependent variable and X_1 and X_2 as the independent variables, can be found by solving a system of three simultaneous linear equations in which a, b_1, and b_2 are the unknowns. These equations are sometimes referred to as the *normal equations* and are as follows:

$$\Sigma Y = \quad an + b_1 \Sigma X_1 \quad + b_2 \Sigma X_2$$
$$\Sigma X_1 Y = a \Sigma X_1 + b_1 \Sigma X_1^2 \quad + b_2 \Sigma X_1 X_2 \qquad (13.3)$$
$$\Sigma X_2 Y = a \Sigma X_2 + b_1 \Sigma X_1 X_2 + b_2 \Sigma X_2^2$$

---

*This would be found by taking the total cost of the land and dividing by the number of units in the condominium complex.

where n is the number of (Y, X_1, X_2) points in the sample. Standard techniques for solving simultaneous equations can be employed to find the values of a, b_1, and b_2 which satisfy the three equations in (13.3).

EXAMPLE 1

Problem: Find the multiple linear regression equation for the data relating condominium sales prices to area and land costs.

Solution: The data necessary to develop the three simultaneous equations in (13.3) are found in the following table.

Y	X_1	X_2	$X_1 Y$	$X_2 Y$	$X_1 X_2$	X_1^2	X_2^2
36	9	8	324	288	72	81	64
80	15	7	1200	560	105	225	49
44	10	9	440	396	90	100	81
55	11	10	605	550	110	121	100
35	10	6	350	210	60	100	36
250	55	40	2919	2004	437	627	330
↓	↓	↓	↓	↓	↓	↓	↓
ΣY	ΣX_1	ΣX_2	$\Sigma X_1 Y$	$\Sigma X_2 Y$	$\Sigma X_1 X_2$	ΣX_1^2	ΣX_2^2

$$n = 5$$

In a realistic problem, of course, far more than five condominiums would be studied. Using these figures, we can form the three equations in Expression (13.3):

$$5a + 55b_1 + 40b_2 = 250$$

$$55a + 627b_1 + 437b_2 = 2919$$

$$40a + 437b_1 + 330b_2 = 2004$$

The solution* of these three equations in three unknowns gives the following values:

$$a = -61.29 \qquad b_1 = 8.07 \qquad b_2 = 2.82$$

This indicates that the multiple linear regression equation is

$$\hat{Y} = -61.29 + 8.07\,X_1 + 2.82\,X_2$$

The intercept value of $a = -61.29$ indicates nothing in a physical meaning for this problem (except for the nonsensical meaning that a condominium with no area, $X_1 = 0$, and built on land that cost nothing, $X_2 = 0$, would sell for $-\$61,290$). The 8.07 coefficient on the X_1 term indicates that an additional 100 square feet in area is associated with an additional \$8070 in price, assuming X_2 is held constant. The 2.82 coefficient on the X_2 term indicates that an additional \$1000 in land cost is associated with an additional \$2820 in price, assuming X_1 is held constant.

---

*Any elementary algebra book contains explanations of several methods for solving simultaneous equations.

The multiple linear regression equation developed using the three equations in (13.3) involves one dependent variable, Y, which is predicted using two independent variables, X_1 and X_2. If more than two independent variables are to be used in the regression equation, then Expression (13.3) must be expanded. For instance, if the least squares regression equation relating the dependent variable, Y, to three independent variables, X_1, X_2, and X_3, then four coefficients are required for the least squares equation $\hat{Y} = a + b_1 X_1 + b_2 X_2 + b_3 X_3$. To find the four values a, b_1, b_2, and b_3, one must solve the following four simultaneous normal equations.

$$
\begin{aligned}
\Sigma Y &= an + b_1 \Sigma X_1 + b_2 \Sigma X_2 + b_3 \Sigma X_3 \\
\Sigma X_1 Y &= a \Sigma X_1 + b_1 \Sigma X_1^2 + b_2 \Sigma X_1 X_2 + b_3 \Sigma X_1 X_3 \\
\Sigma X_2 Y &= a \Sigma X_2 + b_1 \Sigma X_1 X_2 + b_2 \Sigma X_2^2 + b_3 \Sigma X_2 X_3 \\
\Sigma X_3 Y &= a \Sigma X_3 + b_1 \Sigma X_1 X_3 + b_2 \Sigma X_2 X_3 + b_3 \Sigma X_3^2
\end{aligned}
\tag{13.4}
$$

This rather formidable looking set of equations would not usually be solved without the aid of a computer program. However, merely examining the equations in (13.4) can give us insight into the way the general set of equations would be constructed for deriving the least squares coefficients for an equation with any number of variables. The equations in (13.4) follow a pattern and have a symmetry about them. For each additional variable added to the least squares equation, one new row and one new column are added to the normal equations.

In each case the ith new row of equations added for the ith independent variable will be:

$$
\Sigma X_i Y = a \Sigma X_i + b_1 \Sigma X_1 X_i + b_2 \Sigma X_2 X_i + \ldots + b_i \Sigma X_i^2 \tag{13.5}
$$

The new column of expressions added to the end of each equation will be:

$$
\begin{aligned}
&b_i \Sigma X_i \\
&b_i \Sigma X_1 X_i \\
&b_i \Sigma X_2 X_i \\
&\cdots \\
&b_i \Sigma X_i^2
\end{aligned}
\tag{13.6}
$$

The last expression in this column is, of course, the same as the last term in the new row, (13.5).

The terms in a set of normal equations have a symmetry that can be used in setting up the equations. Note that, on the right side of the equals signs, the terms in the ith row and i^{th} column (terms on the diagonal from upper left to lower right) have the form $b_i \Sigma X_i^2$. Also, the summations located in the ith row and j^{th} column are the same as those in the jth row and i^{th} column. The numerical values in Example 1's normal equations exhibit this symmetry. Those readers familiar with computer programming will note how this symmetry can be used to make computer packages like the one described in Section 13.7 operate efficiently.

The examples in the next four sections involve only two independent variables, X_1 and X_2. Thus, their regression equations are developed from Expression (13.3). But all the results discussed in Sections 13.2–13.5 can be extended to problems involving any number of independent variables.

13.2 MEASURING THE ACCURACY OF PREDICTIONS IN MULTIPLE REGRESSION

In Section 12.3 we learned that the accuracy of prediction of a simple regression line could be measured in two ways. First, it was measured by the variance of the errors in prediction, $s_{Y|X}^2$. This figure measures the scatter of the Y_i values around the *simple* regression *line* $\hat{Y} = a + bX$. In *multiple* regression there is a similar measure which measures the scatter of the Y_i values around the regression *plane*. For the case in which two independent variables are used in the regression equation, this figure is found as follows:

$$S_{Y|12}^2 = \frac{1}{n-3} \Sigma (Y_i - \hat{Y}_i)^2 \tag{13.7}$$

The symbol $S_{Y|12}^2$ represents the variance of Y around the regression plane defined by variables X_1 and X_2. In general, the accuracy of predictions for a regression equation with k independent variables, $\hat{Y} = a + b_1 X_1 + b_2 X_2 + \cdots + b_k X_k$, can be expressed as the variance

$$S_{Y|12\dots k}^2 = \frac{1}{n-k-1} \Sigma (Y_i - \hat{Y}_i)^2 \tag{13.8}$$

However, in Section 12.3, we also learned that in regression problems the accuracy of prediction is evaluated in terms of the squared errors. The coefficient of determination, r^2, which we developed to measure this accuracy in simple regression problems has a direct parallel in multiple regression. In general, the *coefficient of multiple determination* is defined as

$$R^2 = 1 - \frac{\Sigma(Y_i - \hat{Y}_i)^2}{\Sigma(Y_i - \overline{Y})^2} \tag{13.9}$$

The same interpretation given to r^2 can be attached to its counterpart, R^2.* The coefficient of multiple determination, R^2, is the proportion of squared error in estimating the Y_i values that can be eliminated using the regression equation $\hat{Y} = a + b_1 X_1 + b_2 X_2 + \cdots + b_k X_k$ as the estimator rather than using $\overline{Y}$ as the estimator. R^2, like r^2, is a unitless number between zero and 1.0. If there is no relationship whatever between the dependent variable, Y, and the independent variables used in the regression equation, $R^2 = 0$. However, if the regression equation can be used to make exact estimates or predictions of the Y_i values, then $R^2 = 1.0$.

EXAMPLE 1

Problem: Use the regression equation relating condominium prices to area and land costs to find $S_{Y|12}^2$ and the coefficient of multiple determination, R^2.

Solution: In Example 1 of the previous section, the regression equation was found to be $\hat{Y} = -61.29 + 8.07 X_1 + 2.82 X_2$. For each of the five condominiums in the sample problem, the areas and land costs were substituted into this equation for the X_1 and X_2 values to obtain the $\hat{Y}_i$ shown in the second column of the following table.

Y_i	$\hat{Y}_i$	$(Y_i - \hat{Y}_i)$	$(Y_i - \hat{Y}_i)^2$
36	33.90	2.10	4.410
80	79.50	.50	.250
44	44.79	− .79	.624
55	55.68	− .68	.462
35	36.33	−1.33	1.769
			7.515

$$S_{Y|12}^2 = \frac{1}{n - k - 1} \Sigma(Y_i - \hat{Y}_i)^2 = \frac{1}{5 - 2 - 1}(7.515) = 3.758$$

Thus, the variance of the condominium prices around the regression plane is 3.758 (thousand dollars)^2. The square root of this figure is the multiple regression equation's standard error of estimate $S_{Y|12} = \$1.939$ thousand. Thus, if the condominium prices are normally distributed around the regression

---

*In our notation we will use the lowercase $s_{Y|X}$ and r^2 figures when referring to the standard error of estimate and the coefficient of determination in simple, single-independent-variable regression problems. The uppercase $S_{Y|12...k}$ and R^2 will refer to the corresponding measures in multiple regression problems.

plane, about two thirds of the prices should fall within $\pm \$1939$ of the value estimated by the regression equation.

The regression equation is a great improvement over the mean condominium price, $\overline{Y} = \$50,000$, as a predictor of a condominium's price. This is indicated by the R^2 value calculated as follows. From Example 1 in Section 12.2, we know that $\Sigma(Y_i - \overline{Y})^2 = 1382$. The previous table shows that $\Sigma(Y_i - \hat{Y}_i)^2 = 7.515$. When we use these two figures in Formula (13.9) we find:

$$R^2 = 1 - \frac{7.515}{1382} = .995$$

This value of R^2 implies that 99.5% of the squared error, or variation, in estimating condominium prices can be eliminated using the regression equation to predict prices rather than saying each condominium will probably sell for the average price of $\overline{Y} = \$50,000$.

13.3 STATISTICAL INFERENCE IN MULTIPLE REGRESSION

At the beginning of Section 13.2, we indicated that the real estate investment firm was assuming a regression line of the form $\mu_{Y|X_1, X_2} = A + B_1 X_1 + B_2 X_2$. However, since they only had a sample taken from the population of all condominiums for sale in the city of interest, we were only able to find $\hat{Y} = a + b_1 X_1 + b_2 X_2$, the sample regression equation.

In Section 12.4 we discussed how the methods of statistical inference could be used to construct confidence intervals or test hypotheses for the slope, B, of the population's simple regression equation. Similar methods to those described in Section 12.4 can be used to answer the question: What are the slopes, B_1 and B_2, of the population regression equation?

In order to construct confidence intervals or test hypotheses to answer this question, we must be able to make the same three assumptions that were listed in Section 12.4:

1. The X_1 and X_2 values are known, that is, nonrandom.
2. The Y values are normally and independently distributed around the regression plane defined by $\mu_{Y|X_1 X_2} = A + B_1 X_1 + B_2 X_2$.
3. For each pair of (X_1, X_2) values, the variance of Y above and below the plane is the same. That is, the distribution of Y_i values around the regression plane is homoscedastic.

If these three conditions are satisfied, then standard confidence interval procedures can be applied to estimate the values of B_1 or B_2. For instance, the $100(1 - \alpha)\%$ confidence interval for B_1 is

$$L = b_1 - t_{\alpha/2,\nu} s_{b_1}$$

$$R = b_1 + t_{\alpha/2,\nu} s_{b_1}$$

(13.10)

where ν is the number of degrees of freedom associated with the mean square error, $S^2_{Y|12}$. In general, $\nu = n - k - 1$ where k is the number of independent variables used in the regression equation. Thus, $\nu = n - 2 - 1 = n - 3$ in the case where X_1 and X_2 are used to estimate Y.

The value of s_{b_1} is called the standard error of the regression coefficient and its value is

$$s_{b_1} = \frac{S_{Y|12}}{\sqrt{\Sigma x_1^2 (1 - r_{12}^2)}}$$

(13.11)

The value r_{12}^2 is the coefficient of determination between X_1 and X_2. Note that if X_1 and X_2 are completely unrelated to one another, $r_{12}^2 = 0$ and Formula (13.11) reduces to Formula (12.14) in the simple regression case. Note also that Σx_1^2 is the sum of the squared *deviations* of the X_1 values around their mean $\bar{X}_1$. That is, $\Sigma x_1^2 = \Sigma (X_1 - \bar{X})^2$. Similarly,

$$s_{b_2} = \frac{S_{Y|12}}{\sqrt{\Sigma x_2^2 (1 - r_{12}^2)}}$$

(13.12)

If we wish to test the hypothesis $H_0: B_1 = 0$, we proceed just as we did in Formula (12.16) and calculate

$$t = \frac{b_1}{s_{b_1}}$$

(13.13)

Under the hypothesis $H_0: B_1 = 0$, this value is t distributed with $n - k - 1$ degrees of freedom. We must note, however, that Expression (13.13) is valid only if the null hypothesis is that B_1 is zero. If the null hypothesis is that B_1 equals some specific value, say B_0, then the numerator of (13.13) is replaced with $b_1 - B_0$. A parallel t test can, of course, be conducted to test hypotheses about the value of B_2.

EXAMPLE 1

Problem: In the condominium example, find a 95% confidence interval for B_1. Also, test the hypothesis $H_0: B_2 = 0$. Let $\alpha = .05$, and use $H_a: B_2 > 0$.

Solution: In order to construct the confidence interval or test the hypothesis, we must find s_{b_1} and s_{b_2} using Formulas (13.11) and (13.12). Finding these two standard errors will be facilitated by the figures in the following table.

X_1	X_2	x_1	x_2	x_1^2	x_2^2	$x_1 x_2$
9	8	-2	0	4	0	0
15	7	4	-1	16	1	-4
10	9	-1	1	1	1	-1
11	10	0	2	0	4	0
10	6	-1	-2	1	4	2
55	40	0	0	22	10	-3
				$\downarrow$	$\downarrow$	$\downarrow$
$\overline{X}_1 = 11$	$\overline{X}_2 = 8$			Σx_1^2	Σx_2^2	$\Sigma x_1 x_2$

In Example 1 of Section 13.2 we found that $s_{Y|12} = 1.939$. Using Formula (12.28) we can quickly find that r_{12}^2 is

$$r_{12}^2 = \frac{(\Sigma x_1 x_2)^2}{\Sigma x_1^2 \Sigma x_2^2} = \frac{(-3)^2}{(22)(10)} = .04$$

This figure indicates that X_1 and X_2 are almost independent of one another. Now we are prepared to calculate s_{b_1} and s_{b_2} as follows:

$$s_{b_1} = \frac{S_{Y|12}}{\sqrt{\Sigma x_1^2 (1 - r_{12}^2)}} = \frac{1.939}{\sqrt{(22)(1 - .04)}} = .42$$

$$s_{b_2} = \frac{S_{Y|12}}{\sqrt{\Sigma x_2^2 (1 - r_{12}^2)}} = \frac{1.939}{\sqrt{(10)(1 - .04)}} = .63$$

With these preliminary figures we are now prepared to answer the questions posed in this example. The sample regression line was determined in Section 13.1 to be $\hat{Y} = -61.29 + 8.07 X_1 + 2.82 X_2$. Thus, the 95% confidence interval for B_1 is found using Formula (13.10):

$$L = 8.07 - (4.303)(.42) = 6.26$$

$$R = 8.07 + (4.303)(.42) = 9.88$$

Thus, an additional 100 square feet in area is associated with a price increase of between $6260 and $9880, all other things being equal. This statement can be made with 95% confidence. If we test the hypothesis $H_0: B_2 = 0$ against the alternative $H_a: B_2 > 0$, we find the following from an extension of Formula (13.13):

$$t = \frac{b_2}{s_{b_2}} = \frac{2.82}{.63} = 4.48$$

The t value that causes us to reject the null hypothesis with $\alpha = .05$ and $\nu = n - 2 - 1 = 2$ is $t_{.05.2} = 2.920$. Thus, we reject the null hypothesis and conclude that B_2 is greater than zero. In these two calculations, however, our conclusions are valid only to the extent that X_1 and X_2 are independent of one another. Since $r_{12}^2 = .04$, these two variables appear to be nearly independent.

This example introduces a problem which we must discuss. If we wish to construct confidence intervals for both B_1 and B_2 or if we wish to perform two hypothesis tests or a combination of these calculations as we did previously, the two results are not independent of one another, unless X_1 and X_2 are independent of one another. In our example $r_{12}^2 = .04$, which indicates that the two variables are relatively independent. Thus, our prior conclusions appear to be valid.

However, in problems where X_1 and X_2 are highly correlated (as would be indicated by a high r_{12}^2 value), we must be cautious about the interpretation we place on confidence intervals or hypothesis tests about B_1 and B_2. Under these circumstances, a confidence interval for B_1 gives a range of values within which we are confident that B_1 lies, assuming that both X_1 and X_2 are in the regression equation. That is, the confidence interval applies only to the coefficient of X_1 in the population regression equation $\hat{Y} = A + B_1 X_1 + B_2 X_2$. The value of B_1 in the population equation relating only X_1 to Y—that is, in $\hat{Y} = B_1 X_1$—might be rather different.

If X_1 and X_2 are highly correlated variables, then the hypothesis test $H_0: B_2 = 0$ is a test of the assertion that variable X_2 adds no *additional* explanatory power to the regression equation over and above that which X_1 has already provided. Thus, failure to reject this hypothesis would not allow us to conclude that X_2 is unrelated to Y. It would only allow us to conclude that there is not enough evidence to suggest that X_2 is related to the *prediction errors* in Y after X_1's contribution to the predictions is considered. Conversely, rejecting the hypothesis $H_0: B_2 = 0$ implies that X_2 is related to the variation in Y not explained by X_1.

The condition where X_1 and X_2 are highly correlated to one another is called *multicollinearity*. This condition and the problems it presents will be discussed more fully in Section 13.8.

13.4 THE BETA COEFFICIENTS

Our example regression equation is $\hat{Y} = -61.29 + 8.07 X_1 + 2.82 X_2$, and the coefficients on X_1 and X_2 imply the following: a 100-square foot change in condominium area is associated with an $8070 change in price while a $1000 change in land cost is associated with a $2820 change in price. One might wish to ask, however, which variable condominium price is more sensitive to, area or land cost? The $b_1 = 8.07$ and $b_2 = 2.82$ cannot be used to answer this question due to the fact that they are expressed in different units. To be exact, the units on b_1 and b_2 are

$b_1 = 8.07$ thousand dollars per 100 square feet of area

$b_2 = 2.82$ thousand dollars per \$1000 of land cost

One way to overcome this problem of differing units is to convert b_1 and b_2 to *beta coefficients*. These are defined as

$$\beta_1 = b_1 \sqrt{\frac{\Sigma x_1^2}{\Sigma y^2}}$$

$$\beta_2 = b_2 \sqrt{\frac{\Sigma x_2^2}{\Sigma y^2}}$$

(13.14)

The beta coefficients are the regression equation's coefficients expressed in terms of the standard deviations of the variables involved. β_1 is the number of standard deviations which Y changes for a one-standard-deviation change in X_1. β_2 is the number of standard deviations Y will change for a one-standard-deviation change in X_2. In our condominium example, we found $\Sigma y^2 = 1382$ (in Section 12.2). Thus,

$$\beta_1 = (8.07) \sqrt{\frac{22}{1382}} = 1.02$$

$$\beta_2 = (2.82) \sqrt{\frac{10}{1382}} = .24$$

These figures indicate that a one-standard-deviation change in X_1 (area) is associated with a 1.02-standard-deviation change in price. But a one-standard-deviation change in X_2 (land cost) is associated with a change of only .24 standard deviations in price. These beta coefficients are pure numbers which can answer our original question about the sensitivity of price to changes in area and land costs. It appears that, in our sample data, price is far more sensitive to changes in area ($\beta_1 = 1.02$) than it is to changes in land cost ($\beta_2 = .24$). In problems that have several independent variables, the variables can be ranked by the absolute values of their beta coefficients to give the relative sensitivity of Y to each of the X_i's.

13.5   PARTITIONING THE SUM OF SQUARES IN MULTIPLE REGRESSION

In Section 12.6, we showed how the sum of squared deviations of the dependent variable, Y, from its mean, $\overline{Y}$, could be expressed as follows:

$$\Sigma(Y_i - \overline{Y})^2 = \Sigma(Y_i - \hat{Y}_i)^2 + \Sigma(\hat{Y}_i - \overline{Y})^2 \qquad (12.31)$$

In the same section we showed how the partitioned sums of squares could be presented in a summary table such as Table 12.1.

The same type of partitioning of the dependent variable can be accomplished in multiple regression. However, there are no short formulas for finding the separate terms in Equation (12.31). We must calculate the individual $\hat{Y}_i$ values to achieve the partitioning suggested by this equation. The following table shows the calculations of the terms in partitioning of prices in our condominium sales example. Many of these computations have been done in previous examples, but they are repeated here for completeness.

Y_i	$\hat{Y}_i$	$Y_i - \overline{Y}$	$(Y_i - \overline{Y})^2$	$Y_i - \hat{Y}_i$	$(Y_i - \hat{Y}_i)^2$	$\hat{Y}_i - \overline{Y}$	$(\hat{Y}_i - \overline{Y})^2$
36	33.90	-14	196	2.10	4.41	-16.10	259.21
80	79.50	30	900	.50	.25	29.50	870.25
44	44.79	-6	36	$-.79$	.62	-5.21	27.14
55	55.68	5	25	$-.68$	.46	5.68	32.26
35	36.33	-15	225	-1.33	1.77	-13.67	186.87
			1382		7.51		1375.73

In this table we can see that, with the exception of rounding differences, Equation (12.31) holds since $\Sigma(Y_i - \hat{Y}_i)^2 + \Sigma(\hat{Y}_i - \overline{Y})^2 = 7.51 + 1375.73 = 1383.24$ which is very close to the value of $\Sigma(Y_i - \overline{Y})^2 = 1382$.

The multiple regression summary table which corresponds to Table 12.1 is Table 13.1. Note that the number of degrees of freedom for the partitioned portions are k for the sum of squares due to regression and $n - k - 1$ for the sum of squared deviations from regression. The value of n is the total number of data observations (in our example, $n = 5$). The k figure indicates the number of independent variables used in the regression equation. For the condominium example, $k = 2$.

In Section 12.7, we showed how the simple regression summary table, Table 12.2, could be extended to include the mean squares (the

---

TABLE 13.1   Partitioned Sums of Squares in Multiple Regression

SOURCE OF VARIATION	DEGREES OF FREEDOM	SUM OF SQUARES	
Due to Regression	$k = 2$	$\Sigma(\hat{Y}_i - \overline{Y})^2 =$	1375.73
Deviations from Regression	$n - k - 1 = 2$	$\Sigma(Y_i - \hat{Y}_i)^2 =$	7.51
Totals	$n - 1 = 4$	$\Sigma(Y_i - \overline{Y})^2 =$	1383.24

sums of squares divided by their degrees of freedom). These mean square figures were then used to test the *overall significance of the simple regression relationship* by calculating the *F* ratio

$$F = \frac{\text{Mean square due to regression}}{\text{Mean square from regression}} \tag{13.15}$$

If this ratio is large, it indicates that the amount of variation in Y that is explained by the regression equation (as measured by the numerator) is large relative to the amount of variation that is left unexplained by the regression equation (as measured by the denominator).

The overall significance of multiple regression relationship can also be tested using Expression (13.15). Table 13.1 is expanded in Table 13.2 to include the mean square calculations. When these mean square figures are used in Formula (13.15), the resulting F value can be used to test the following hypothesis:

$$H_0: B_1 = B_2 = 0$$

H_a: One or more of the B_i values is not equal to zero.

Thus, the F ratio tests the hypothesis that there is no relationship between any of the X_i variables and the dependent variable, Y. The alternative hypothesis is true if one or more of the X_i variables is related to the dependent variable, Y.

$$F = \frac{687.87}{3.76} = 182.94$$

If we are willing to run a 5 percent chance of rejecting the null hypothesis when it is true, then we compare the previously calculated F value to the table $F_{.05,2,2} = 19.000$. The calculated F value is so much larger than the table F value that we can reject the null hypothesis with confidence and conclude that there is a relationship between either X_1 or X_2, or both, and the dependent variable, Y. That is, either area, land cost, or both, are related to the prices of the condominiums in our sample.

TABLE 13.2   Regression Summary Table for Condominium Example

SOURCE OF VARIATION	DEGREES OF FREEDOM	SUM OF SQUARES	MEAN SQUARE
Due to Regression	2	1375.73	687.87
Deviations from Regression	2	7.51	3.76
Totals	4	1383.24	

13.6   STEPWISE MULTIPLE REGRESSION

In Section 13.1, we discussed how the normal equations are used to obtain the coefficients in the sample regression line $\hat{Y} = a + b_1 X_1 + b_2 X_2$. The three equations in Expression (13.3) were solved *simultaneously* for the three values a, b_1, and b_2. There are times, however, when the person constructing a multiple regression equation does not wish to determine the regression equation between the dependent variable, Y, and a set of independent variables (the X's) in this manner. For instance, a person may have a set of fifteen or twenty independent variables that might be associated with the dependent variable whose value he or she would like to predict. However, it would be cumbersome and impractical to construct a regression equation using all fifteen or twenty variables.

This person may wish to construct a regression equation using a subset of original independent variables in order to obtain an equation of a manageable size. However, this raises the problem of which subset of the X's would do the "best" job in predicting the Y_i values. **Stepwise multiple regression** is one procedure commonly used to address this problem.

In a stepwise multiple regression problem, the regression equation is constructed one variable at a time. That is, first the equation $\hat{Y} = a + b_i X_i$ is constructed using the independent variable X_i that is *most highly correlated* with Y. At the second step, a new variable is brought into the equation to produce $\hat{Y} = a + b_i X_i + b_j X_j$, where X_j is that variable which does the best job of explaining the variation in Y *that has not already been explained by X_i*. That is, X_j is the variable that does the best job in eliminating the prediction errors that were present using the first equation. It should be noted that the a and b_i values of the second equation may be somewhat different than they were in the first equation. In the third step of a stepwise regression problem, a new variable is brought into the equation to form $\hat{Y} = a + b_i X_i + b_j X_j + b_k X_k$. Again, X_k is the variable that does the best job in eliminating the prediction errors of the second equation.

This process is usually continued until the person using the equations feels that there are enough variables in the equation or until the multiple R^2 value is sufficiently high for the uses to which the equation will be put. There are also several complex statistical criteria which can be used to determine a stopping point in the stepwise process. These, however, are beyond the scope of this discussion.

It should be obvious to the reader at this point that a stepwise multiple regression problem would involve an enormous amount of computational effort. For this reason, stepwise regression problems are solved using preprogrammed computer packages. Almost any computer installation has one of these packages available. In the next section, we will present and discuss the output of one of the most popular of these computer programs.

13.7   A COMPUTER EXAMPLE OF STEPWISE REGRESSION

In the last two chapters we have been using an artificially simple example wherein we have constructed regression equations to be used in predicting the sales price of condominiums. An actual study involving the objectives outlined in this example would differ from what has been presented in previous sections in three important aspects:

1. Many more condominium sales would be used in the data set. That is, n would be larger than the value of 5 used in previous examples.
2. More variables would be used to describe each condominium. That is, the number of independent variables, k, would be larger than the one or two used in previous examples.
3. A computer program would be used to make the computations discussed in the previous sections. The meanings of the calculated measures would be the same, but the computer would do the tedious computational work.

In this section we will examine a much more realistic example, using regression to predict condominium sales prices and explain the output of a computer program used to solve it. We will use a data set that has $n = 20$ observations of condominium sales and $k = 5$ independent variables. The data we will use in this expanded example are presented in Table 13.3. The dependent variable that we are interested in predicting is still condominium sales price, as indicated in the first column of the table. The five independent variables used to predict sales price describe each condominium's area, land cost (just as in previous examples), distance from the center of town (in miles), number of bedrooms, and parking features.

The last two independent variables deserve special attention. They are not continuous variables like price, area, land cost, and distance. They are discrete. The number of bedrooms is only an integer and takes on values of 1, 2, 3, or 4. The parking variable is used to indicate whether the condominium unit has a covered parking area. A value of 1 indicates that covered parking is available, and a 0 indicates that it is not. Variables that assume only 0 and 1 as their values are called

TABLE 13.3   Expanded Data for Condominium Pricing Example

i	Y	X_1	X_2	X_3	X_4	X_5
Condominium Number	Sales Price (000)	Unit Area (00)	Land Cost (000)	Distance to City Center (Miles)	Number of Bedrooms	Covered Parking (1 = Yes)
1	36	9	8	4	1	0
2	80	15	7	12	4	1
3	44	10	9	6	2	1
4	55	11	10	0	2	1
5	35	10	6	16	2	0
6	62	12	6	10	3	1
7	42	11	11	20	2	0
8	77	16	12	1	3	1
9	32	7	6	17	2	0
10	46	9	10	6	3	1
11	60	10	11	5	3	1
12	36	10	4	11	2	1
13	100	18	14	6	4	1
14	45	10	5	2	2	0
15	46	11	6	13	3	1
16	35	8	8	3	1	0
17	25	7	4	1	1	0
18	48	12	8	16	3	1
19	40	11	9	8	2	1
20	58	13	10	2	3	1

Data interpretation examples:

Condominium 8 sold for $77,000. It contained 1600 square feet and bore $12,000 in land costs. It was located 1 mile from the center of the city, had 3 bedrooms, and a covered parking area.

Condominiums 1–5 are the same condominiums used in previous examples, but additional information about each unit has been added. For example, the first condominium was located 4 miles from the city center, had 1 bedroom and no covered parking.

dummy variables, or *zero-one variables.* Integer and dummy variables create special problems in interpreting regression results. These will be mentioned in the following discussion.

Since the data presented in Table 13.3 are so voluminous, applying regression techniques would be very tedious without the aid of a computer program. Also, since not all of the five independent variables are likely to add significantly to the accuracy of sales price predictions, it would appear that applying a computerized stepwise regression program to these data would be appropriate.

There are many preprogrammed stepwise regression packages that could be applied to the data in Table 13.3. One of the most commonly available is the stepwise regression program that is part of the Statistical Package for the Social Sciences (SPSS). This program, which was developed at the University of Chicago, requires that the user know very little about computers. However, the user must be able to interpret the computer output. The remainder of this section is devoted to presenting the major outputs of the SPSS stepwise regression program using the data in Table 13.3. The output explained in the following sections is very similar to the output from any multiple regression computer package.

VARIABLE	MEAN	STANDARD DEV
PRICE	50,1000	18,5157
AREA	11,0000	2,8098
LAND	8,2000	2,7261
DISTANCE	7,9500	6,0825
BEDROOMS	2,4000	,8826
PARKING	,6500	,4894

The first piece of information printed by the program is the means and standard deviations of all the variables used in the regression problem. These descriptive statistics reveal that for the enlarged data set, the mean price of the condominiums is $50,100. The large standard deviation of $18,515.70 indicates that there is quite a large variation in the prices. The other means and standard deviations are self-explanatory, but the parking variable mean deserves attention. Since parking is a dummy variable that takes on only 0 (for no covered parking) and 1 (for covered parking), the mean of .65 indicates that 65 percent of the condominiums in the example have covered parking available.

The second major output produced by the SPSS stepwise regression program is the correlation matrix. The following matrix gives the correlation coefficients measuring the degree of association between all possible pairs of variables in this problem. Several things should be noted about this matrix. First, all the figures on the main (upper left to lower right) diagonal equal unity. This is because these values

	PRICE	AREA	LAND	DISTANCE	BEDROOMS	PARKING
PRICE	1,00000	,92466	,63251	-,16959	,83803	,58493
AREA	,92466	1,00000	,58406	-,04927	,78528	,57417
LAND	,63251	,58406	1,00000	-,20568	,42438	,37086
DISTANCE	-,16959	-,04927	-,20568	1,00000	,17059	-,12996
BEDROOMS	,83803	,78528	,42438	,17059	1,00000	,70679
PARKING	,58493	,57417	,37086	-,12996	,70679	1,00000

show the correlations of the variables with themselves, and every variable is perfectly correlated ($r = 1.0$) with itself. Also, the values in the correlation matrix are symmetrical around the main diagonal. That is, the figure in row i and column j is the same as the figure in row j and column i: $r_{ij} = r_{ji}$. This is an obvious result since the correlation of variable X_i with variable X_j should be the same as the correlation of X_j with X_i. The order in which we mention the variables is irrelevant.

The first column (and also the first row) of the correlation matrix shows us that the condominium sales prices are most highly correlated with X_1 = area since r_{YX_1} = .92466. The variable next most highly correlated with price is number of bedrooms, X_4 since r_{YX_4} = .83803. The correlation of price and distance is interesting in that it is negative: r_{YX_3} = −.16959. This indicates that there is a slight tendency for the condominiums located away from the city center to be priced lower than those close to the city.

The next important output of the SPSS stepwise regression program is Step 1 of the regression procedure. In this step, the variable most highly correlated with price is used to obtain a simple regression equation $\hat{Y} = a + bX$. We saw from the correlation matrix that X_1 = area is the most highly correlated with price. Thus, the program prints:

```
DEPENDENT VARIABLE,,      PRICE

VARIABLE(S) ENTERED ON STEP NUMBER  1,,      AREA


MULTIPLE R                ,92466
R SQUARE                  ,85500
ADJUSTED R SQUARE         ,84695
STANDARD ERROR            7,24375


•-- •• -- -- -- •• •• --  VARIABLES IN THE EQUATION  -- •• -- -- -- •• •- -• •-- •

VARIABLE            B           BETA        STD ERROR B          F

AREA            6,09333       ,92466         ,59145          106,139
(CONSTANT)     -16,92667
```

This output indicates that the dependent variable is price and that the first variable entered into the equation is area. For the moment we will ignore the next four figures and go directly to the output entitled VARIABLES IN THE EQUATION. The first two columns indicate that the initial regression equation is

$$\hat{Y} = -16.93 + 6.09\,X_1$$

The column headed by BETA indicates that β_1 = .92466. This means that a one-standard-deviation change in X_1 is associated with a .92466-

standard-deviation change in price. The value of .59145, which is labelled STD ERROR B, is s_{b_1}.

We noted in Equation (13.13) that the hypothesis $H_0: B_1 = 0$ could be tested by calculating

$$t = \frac{b_1}{s_{b_1}}$$

However, in Chapter 10 we also showed that the F and t distributions were related in such a way that $t^2_{\alpha/2,\nu} = F_{\alpha,1,\nu}$ and indicated there were times when the calculation of t and F were equivalent. The SPSS stepwise regression program does not print the t value indicated in Equation (13.13), but it prints the corresponding F. This can be shown using our computer output results as follows:

$$t^2 = \left(\frac{b_1}{s_{b_1}}\right)^2 = \left(\frac{6.09333}{.59145}\right)^2 = (10.302)^2 = 106.139 = F$$

This value of F can be used to test the hypothesis $H_0: B_1 = 0$ against the alternative $H_a: B_1 \neq 0$. If we let $\alpha = .01$, then we can compare the calculated F of 106.139 to the table $F_{.01,1,18} = 8.2854$. The calculated F is so much larger than 8.2854, we can reject the hypothesis that $B_1 = 0$, or equivalently, that there is no correlation between price and area.

Let us now return to the four figures we passed over. These values involve the R, R^2, and $S_{Y|1}$ figures associated with this first equation. (Since this first equation is a *simple* regression equation, the exact notation might better be r, r^2, and $s_{Y|1}$ since the lowercase letters are reserved to indicate simple correlation and regression results.) The MULTIPLE R and R SQUARE figures are computed from Formula (13.9). Thus, we can say that 85.5 percent of the square error of prediction can be eliminated by using the first regression equation to predict sales prices rather than just using $\overline{Y} = \$50,100$ as our predictor.

The computer output designated ADJUSTED R SQUARE is the R^2 figure defined in the footnote of Section 12.3. In multiple regression, this figure can be computed as

$$ADJ\ R^2 = \frac{s_Y^2 - S_{Y|12\ \dots\ k}^2}{s_Y^2}$$

but its advantages and disadvantages as a measure of prediction accuracy involve discussions beyond the scope of this text. Thus, we will ignore ADJ R^2 from now on.

The STANDARD ERROR value is just $S_{Y|1}$. It is the standard deviation of the actual prices around the regression line. If the prices are normally distributed about the line, then approximately two-thirds of the condominiums' prices will lie within $\pm \$7,243.75$ of the regression equation's predicted values.

The next major output produced by the SPSS program is a table similar to Table 13.2 which summarizes the partitioning of the sum of squares of the dependent variable. We can see that this table produces an F value of 106.139, which is the same as the F figure associated with the regression coefficient B_1. This equivalency was explained in Section 12.7.

ANALYSIS OF VARIANCE	DF	SUM OF SQUARES	MEAN SQUARE	F
REGRESSION	1,	5569,306 67	5569,306 67	106,138 9
RESIDUAL	18,	944,49333	52,471 85	

At each step, the SPSS program also presents information concerning the independent variables that are not included in the regression equation yet. For each of the variables not in the equation, we have listed first BETA IN. This value is the β_i that will be associated with each of the X_i values *if* they are chosen to enter the regression equation at the next step.

- - -- -- -- -- --   VARIABLES NOT IN THE EQUATION - -- -- -- -- -- -- --

VARIABLE	BETA IN	PARTIAL	TOLERANCE	F
LAND	,14032	,29912	,65888	1,670
DISTANCE	-,12433	-,32612	,99757	2,023
BEDROOMS	,29194	,47467	,38333	4,944
PARKING	,08059	,17327	,67033	,526

The values in the PARTIAL column are *partial correlation coefficients*. They show the correlation between each of the independent variables, X_i's, and the dependent variable, Y, after the influence of X_1, area, has been removed from them. That is, the PARTIAL correlation figure indicates which variable will be the most effective in improving the regression equation's accuracy if it is entered at the next step. For our example, we can see that the number of bedrooms, X_4, is the variable with the highest PARTIAL value, .47467, and thus the variable that will be entered into the regression equation at the next step. The square of the ith PARTIAL figure also has an interpretation. It is the proportion of the *currently unexplained variance* in Y, sales price, that can be eliminated if the variable X_i is entered in the regression equation at the next step. Since R SQUARE = .85500,* the unexplained

---

*In the SPSS program, the PARTIAL figure is related to the R SQUARE value and not to the ADJUSTED R SQUARE value.

variation in Y at this point is $1.0 - .85500 = .14500$, or 14.5 percent. The square of the PARTIAL correlation figure for number of bedrooms is $(.47467)^2 = .2253$. This means that 22.5 percent of the unexplained 14.5 percent can be eliminated by including X_4 as the next variable in the equation. Thus, the new R SQUARE at the second step should be:

$$\text{New R SQUARE} = .85500 + (.2253)(.14500) = .88767$$

We will see in Step 2 that this is indeed the case.

The TOLERANCE column gives values that indicate the relative independence of the X_i's not in the equation to the independent variable that is in the equation. The TOLERANCE values range between 0 and 1. The TOLERANCE value of .99757 for distance indicates that distance and area, X_1, are almost totally independent. TOLERANCE for a variable not in the regression equation is computed as

$$\text{TOLERANCE(variable } j) = 1 - R^2_{j,(\text{in})} \tag{13.16}$$

That is, the TOLERANCE for variable j is one minus the coefficient of determination between variable j and the variables that are currently "in" the regression equation. In our example, the only variable in the equation at this point is X_1. The TOLERANCE figure for distance, X_3, is

$$\text{TOLERANCE(3)} = 1 - R^2_{3,1}$$

The correlation coefficient* between X_3 and X_1 can be found in the fourth row and second column of the correlation matrix (or in row two and column four, due to the symmetry of the matrix).

$$\text{TOLERANCE(3)} = 1 - (-.04927)^2 = 1 - .00243 = .99757$$

Thus, distance and area seem to be rather independent in our data set.

Finally, the F column for the variables not in the equation indicates $F = (b_i/s_{b_i})^2$ values that would be assigned to each variable *if* that variable were to be introduced into the equation at the next step. We see that number of bedrooms has the highest F value for the variables not in the equation, and in the next step we will see that the F associated with b_4 is 4.944.

The SPSS output associated with Step 2 of the stepwise regression program is as follows. Several things can be noted about the results in the second step.

---

*Since there is only one variable in the equation at this point, $R^2_{3,1}$ could be written as $r^2_{3,1}$ or (due to symmetry) $r^2_{1,3}$.

```
VARIABLE(S) ENTERED ON STEP NUMBER  2 ,,      BEDROOMS


MULTIPLE R              ,94216
R SQUARE                ,88767
ADJUSTED R SQUARE       ,87446
STANDARD ERROR         6,56051


•••••••••••••••••-- VARIABLES IN THE EQUATION -••••••••••••••••••••

VARIABLE              B            BETA      STD ERROR B         F

AREA              4,58261        ,69541       ,86517         28,055
BEDROOMS          6,12456        ,29194      2,75435          4,944
(CONSTANT)       -15,00764


ANALYSIS OF VARIANCE     DF      SUM OF SQUARES        MEAN SQUARE                   F
REGRESSION               2,       5782,11469          2891,05734         67,1709
RESIDUAL                17,        731,68531            43,04031


•••••••••••••• VARIABLES NOT IN THE EQUATION •••••••••••••••

VARIABLE           BETA IN      PARTIAL    TOLERANCE           F

LAND               ,15623       ,37750       ,65581         2,659
DISTANCE          -,20959      -,58772       ,88331         8,443
PARKING           -,04142      -,08734       ,49949          ,123
```

First, the variable number of bedrooms, X_4, was entered into the equation as was predicted in the last step. Also, its BETA and F values are just what the previous step indicated. The R SQUARE figure of .88767 at this step is just what was calculated in Step 1. The inclusion of X_4 in the equation reduced the STANDARD ERROR from \$7243.75 to \$6560.51.

If we compare the first and second equations, we find:

$$\hat{Y} = -16.93 + 6.09X_1$$

and

$$\hat{Y} = -15.01 + 4.58X_1 + 6.12X_4$$

With the introduction of X_4, the constant and b_1 changed values. This is due to the fact that in the new equation the predictive impact of the number of bedrooms in each condominium is now included in the b_4 coefficient. In the previous $b_1 = 6.09$ coefficient, the impact of both area and number of bedrooms was included. In the new $b_1 = 4.58$ coefficient, the impact of the number of bedrooms has been isolated. The new equation indicates that the predicted price of a condominium increases \$4583 for each additional 100 square feet in area, assuming the number of bedrooms is held constant. The price increases \$6125 for each additional bedroom, assuming area is held constant.

The two BETA values indicate that price changes are more sensitive to changes in area ($\beta_1 = .69541$) than to changes in number of bedrooms ($\beta_4 = .29194$).

Now we note that the output at Step 2 gives us three *different* F values to interpret. The two F values associated with b_1 and b_4, the regression coefficients for area and bedrooms, measure the marginal significance of the two variables individually. That is, the F associated with area tests the hypothesis that area does not make a significant contribution to price predictions over and above that made by bedrooms. The F value associated with bedrooms tests the hypothesis that the number of bedrooms does not make a significant marginal contribution to price predictions that area has not already made. The null hypotheses can be stated as:

$$H_0: B_1 = 0 \text{ when } X_4 \text{ is in the equation and held constant}$$

$$H_0: B_4 = 0 \text{ when } X_1 \text{ is in the equation and held constant}$$

The computer results can be used to show that each of these hypotheses is rejected since

$$t^2 = (b_1/s_{b_1})^2 = (4.58261/.86517)^2 = 28.055 = F$$

$$t^2 = (b_4/s_{b_4})^2 = (6.12456/2.75435)^2 = 4.944 = F$$

These calculated F values can be compared to the table F value of $F_{.05,1,17} = 4.4513$. Thus, we can conclude that B_1 and B_4 differ from zero when the other variable is in the equation and held constant.

The F value that appears in the table, summarizing the partitioning of the sum of squares, measures the overall significance of the relationship between the dependent variable, Y, and the independent variables, X_1 and X_4, *combined*. The value of this F can be taken as a measure to test the joint hypothesis $H_0: B_1 = B_4 = 0$ against the alternative H_a: one or more of the independent variables is significantly related to Y. The calculated value of this F is 67.17 and should be compared to $F_{.05,2,17} = 3.5915$. Thus, we can confidently reject the hypothesis that the combined predictive power of X_1 and X_4 is zero.

An analysis of the VARIABLES NOT IN THE EQUATION output shows that the next variable to enter the equation is distance. This is because it has the largest (negative) PARTIAL figure. That is, distance can explain $(-.58772)^2 = .345$ of the variation in price *that has not already been explained* by area and bedrooms. The BETA, TOLERANCE, and F values that will be associated with X_3, distance, at Step 3 are also shown.

The output of Step 3 in the SPSS stepwise regression program follows. The new equation is

$$\hat{Y} = -8.43 + 3.92 X_1 - .64 X_3 + 8.53 X_4$$

Note that the predicted sales price drops approximately \$640 for each mile the condominium is located from the center of the city. Note also that the predicted price is most sensitive to changes in area ($\beta_1 = .59523$), followed by bedrooms ($\beta_4 = .40636$) and distance ($\beta_3 = -.20959$). All of the variables' marginal coefficients are individually different from zero at the 5 percent level of significance since the individual F values are all greater than $F_{.05,1,16} = 4.4940$.

VARIABLE(S) ENTERED ON STEP NUMBER 3,,   DISTANCE

MULTIPLE R	,96253
R SQUARE	,92647
ADJUSTED R SQUARE	,91269
STANDARD ERROR	5,47121

-•-•»----•-»•---- VARIABLES IN THE EQUATION -••---••-••-••-••-•

VARIABLE	B	BETA	STD ERROR B	F
AREA	3,92242	,59523	,75645	26,887
BEDROOMS	8,52510	,40636	2,44107	12,197
DISTANCE	-,63799	-,20959	,21957	8,443
(CONSTANT)	-8,43484			

ANALYSIS OF VARIANCE	DF	SUM OF SQUARES	MEAN SQUARE	F
REGRESSION	3,	6034,85321	2011,61774	67,2013
RESIDUAL	16,	478,94679	29,93417	

---••-••-••-•»-•• VARIABLES NOT IN THE EQUATION -••-••-••-••-•--

VARIABLE	BETA IN	PARTIAL	TOLERANCE	F
LAND	,11045	,32265	,62748	1,743
PARKING	-,16442	-,39924	,43353	2,844

MAXIMUM STEP REACHED

The overall significance of the regression equations prediction is measured by the $F = 67.2014$ in the table partitioning the sum of squares. This value far exceeds $F_{.05,3,16} = 3.2389$, so we conclude that variables X_1, X_3, and X_4 combined are significantly related to price, Y.

This program was terminated at the end of Step 3 since it was instructed (through control statements not discussed here) to stop after including three independent variables in the regression equation. The MAXIMUM STEP REACHED indication at the bottom of the output gives the reason for stopping.

At this point, the variables X_2, land, and X_5, parking, are not in the equation. It also appears that, were they to be included, they would have regression coefficients not significantly different from zero. This is indicated by the F values of 1.743 and 2.844 associated with these two variables. Both of them are below $F_{.05,1,15} = 4.5431$, which is the F value that must be exceeded if we wish to reject the hypothesis H_0: $B_i = 0$ at the next step. Thus, it appears that stopping at Step 3 was a good decision. This does not mean that the land and parking variables do not have any association with price or any predictive power. It means that they have no additional significant association with price *over and above* that which is contained in area, distance, and bedrooms. These three independent variables together can account for 92.647 percent (R SQUARE) of the squared variation in price.

13.8   SPECIAL PROBLEMS IN MULTIPLE REGRESSION

Computer programs like the SPSS Stepwise Regression program make the solution to multiple regression problems rather simple to obtain. However, the ease of obtaining a mathematical solution does not relieve the program user from the responsibility of being cautious in using the results. Several special problems can plague the user if he or she is not careful. Some of the more common problems are discussed briefly in this section.

  Violating the basic assumptions of multiple regression can lead us to incorrect conclusions. The reader will recall that these assumptions are:

1. The independent variables are nonrandom.
2. The dependent variable values are normally and independently distributed around the regression plane.
3. The variance of the dependent variable around the regression plane is constant.

If these assumptions are not met, then we may reach erroneous conclusions.

  One of the best ways to determine if some of the basic assumptions of the regression procedure are met is to graph the prediction errors, or *residuals*, against the independent variables after the regression equation has been constructed. Figure 13.2a shows a situation that suggests that the variance around the regression plane is not constant; it seems to be larger for larger values of the independent variable X_i. Figure 13.2b shows another situation that would cause the user of a multiple regression equation to be concerned. The curvilinear relationship between the residuals and the independent variable X_j suggests that the form of the regression equation might better be polynomial. In this case the persons constructing the regression equation might wish to add a new variable, X_k, that is just the square of X_j. That is, for each observation of a set of variables, $X_k = X_j^2$. There are simple control procedures in the SPSS program and in many other

FIGURE 13.2 (a) Variance of residuals not constant with respect to independent variable X_i. (b) Curvilinear relationship between residuals and independent variable X_j

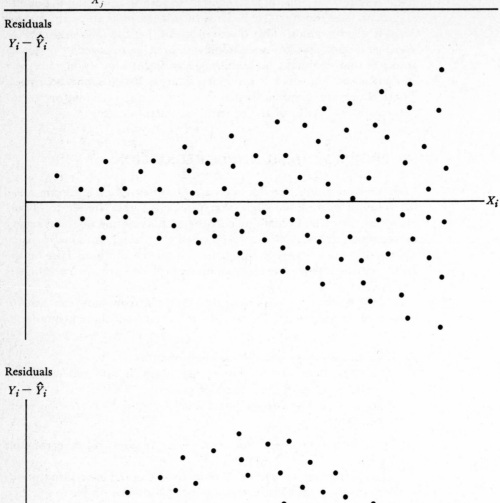

computerized regression packages that allow new independent variables to be created from old ones.

Another special problem encountered by many regression program users is that of *overspecification*. This problem occurs when the program user allows the stepwise program to run "too long." That is, the program goes too many steps and produces a long regression equation containing numerous independent variables. The R^2 measure for such an equation may be quite high, but the increase in the R^2 value produced by the addition of the last several variables may be insignificant, in the statistical sense.

This problem can usually be detected by noting that the coefficients, the b_i values, on several of the independent variables are not significantly different from zero. That is, for several variables, the hypothesis H_0: $B_i = 0$ cannot be rejected. The example in the previous section showed a problem with five independent variables, but only three were included in the final equation. The analysis of the computer output showed that after the third variable had been included, the F values for the remaining two variables not in the equation were not significant at the $\alpha = .05$ level.

The third problem that can create difficulty in multiple regression is that of *multicollinearity*. This problem exists when two or more of the independent variables are highly correlated with one another. For example, we may have a data set in which people's ability to perform a certain type of heavy work is the dependent variable, Y, which we would like to estimate. Two logical independent variables would be the people's heights and weights. However, the heights and weights of people are usually very highly correlated. Thus, this data set would be said to exhibit the characteristic of multicollinearity.

The specific problem introduced by the presence of multicollinearity in the data set involves the reliability of the regression equation's coefficients, the b_i's. This problem can be demonstrated in the two-independent-variable case by the value

$$s_{b_1} = \frac{S_{Y|12}}{\sqrt{\Sigma x_1^2 (1 - r_{12}^2)}}$$

The r_{12}^2 figure in this equation is the coefficient of determination between the two independent variables X_1 and X_2. If these two variables are highly correlated (so that we have multicollinearity), then r_{12}^2 is close to 1.0. This condition causes s_{b_1} to be large. In fact, the use of collinear independent variables can cause s_{b_i} values to be inflated. This, in turn, may cause the regression results user to fail in detecting significant relationships between the dependent variable and some independent variables.

In the example presented in the previous section, we can see the effects of multicollinearity. The first two variables included in the regression equation to predict price were area and number of bedrooms, X_1 and X_4. But the correlation matrix produced in that example shows that $r_{14} = .78528$, which is quite high. This means that area and number of bedrooms are quite highly correlated, which is reasonable since additional bedrooms have additional area. In the initial regression equation produced by Step 1 of the SPSS program, we found that $s_{b_1} = .59145$. But when X_4, number of bedrooms, was introduced in Step 2, we saw s_{b_1} increase to .86517. The increase was due to the multicollinearity of X_1 and X_4. Fortunately, however, the relationship between price and both X_1 and X_4 was so strong that the collinearity did not mask the significance of these two variables.

It is not always possible to detect multicollinearity through examination of the values in the correlation matrix. Such examination will detect only pairwise correlations. We need another measure which will pick up the more complex multicollinear relationships in which one independent variable may be strongly related to a combination of several other independent variables. The TOLERANCE figure explained in the last section measures the collinearity of variables not in the regression equation with those already in the equation at each step. A high TOLERANCE figure suggests that a variable not in the equation is independent of those currently in. In extreme cases of multicollinearity, it may be necessary for those constructing the regression equation to drop one or more of the collinear independent variables in order to obtain regression coefficients that are meaningful and significant. However, there are statistical techniques that can be employed to solve the multicollinearity problem without dropping variables from consideration. Such techniques are beyond the scope of this book.

Another special problem involving regression analysis is that of *extrapolation*. The regression equation constructed from a particular data set is valid for making predictions of the dependent variable, Y, *only if* the independent variables used in the predictions, the X_i, are similar in value to those in the original data set. For instance, in the condominium sales example presented in the last section, the twenty condominiums used to develop the regression equation ranged in area from 700 to 1800 square feet. If we were to attempt to predict the price of a condominium containing 2500 square feet, we would be guilty of extrapolating beyond the range of the independent variable's observed data values. Similarly, if we use any values of the independent variables that are outside the range of the values used in constructing the regression equation, we are extrapolating and cannot place much confidence in our predicted value.

There are situations, however, in which using values of the independent variables that are outside the ranges of the observed data is unavoidable. In these cases, those doing the extrapolation must be careful not to take their estimates of Y too seriously.

The final caution we will give concerns the *assumption of causation* many users of regression inappropriately read into their results. It has been said, and with justification, that among all of the measures treated in this book, the correlation coefficient is the most subject to misinterpretation. One of the main reasons for this is the frequently false assumption that because two variables are related, a change in one *causes* a change in the other. If one has a point to prove, it is extremely easy to succumb to this fallacy and to use a perfectly respectable correlation coefficient to "prove" a cause-and-effect relationship that may not exist.

It has been shown that there is a negative correlation between smoking and grades. If one objects to smoking, one seizes upon this fact as objective evidence that smoking is harmful, that smoking causes low grades. This may be true, but one could also argue that low grades result in increased nervous tension, which, in turn, causes the individual to smoke more. In our gas station example we found that number of pumps and gasoline sales were positively related. But from our regression equation it is impossible to say if more pumps cause higher sales or if higher sales have caused the installation of more pumps or if both variables are related to some other factor, such as location of the station.

This last suggestion illustrates another danger in interpretation of the results of a correlation analysis. Frequently, two variables may appear to be highly correlated when, in fact, they are not directly associated with each other but are both highly correlated with a third variable.

We must also be very careful in our interpretation of the coefficients in a regression equation. We found in the last section that the coefficient on X_4, number of bedrooms, was \$8525.10. This does not mean that one additional bedroom in a condominium will *cause* the price to rise \$8525.10. It merely indicates that there is an *association* between additional bedrooms and price within the data set and that good estimates of price can be obtained if number of bedrooms is weighted by \$8525.10. However, in this example, it makes much more sense to conclude that there is a cause-and-effect relationship between price and number of bedrooms since additional bedrooms require additional materials and labor to be expended during construction. These added costs would naturally force up the price. However, the \$8525.10 figure may not measure the direct effect of those increased costs.

The problems outlined in this section should convince the reader that the application of multiple regression to a problem requires judgment and good sense. Unfortunately, due to the wide availability of easy-to-use computer programs, regression analysis is sometimes carried out in a very mechanical fashion by well-meaning, though uninformed people. On the other hand, when it is applied with care by competent individuals, multiple regression analysis is one of the most powerful tools available to managers and economists today.

PROBLEMS

1. Some of the differences between simple and multiple regression are (you may check as many responses as you want):
   a. Simple regression uses one independent variable and multiple regression uses two or more.
   b. Simple regression equations represent lines, geometrically, and multiple regression equations represent curves.
   c. Uppercase letters like R^2 and $S_{Y|12}$ are used to represent measures of prediction accuracy in multiple regression, while lowercase letters are used in simple regression.
   d. Multiple regression equations usually give more accurate predictions of Y, as long as there are two or more independent variables associated with Y.

2. A regression equation was found to be: $\hat{Y} = 20 + 14X_1 - 7X_2$. Which of the following statements are correct?
   a. A one-unit increase in X_1 causes Y to increase by 14 units.
   b. Y is more highly correlated with X_1 than with X_2 since the coefficient on X_1 is positive.
   c. If the value of X_2 is large enough, it would be possible to obtain negative predictions of Y.
   d. Since the coefficient on X_1 is twice as large as the coefficient on X_2, we can conclude that Y is twice as sensitive to changes in X_1 as it is to changes in X_2.

3. A regression equation was found to be: $\hat{Y} = 1.5 + .2X_1 + 3.1X_2$. The R^2 value for this equation was .95. The X_1 values ranged from -10 to $+15$, and the X_2 values ranged from -20 to $+50$. Which of the following statements are correct?
   a. X_2 is more strongly correlated with Y than is X_1 since its regression coefficient is larger.
   b. When $X_1 = 0$ and $X_2 = 0$, the predicted value of Y is 1.5 and this value has legitimate physical meaning since the data values for X_1 and X_2 span zero.
   c. A very high proportion of the squared error of prediction incurred by using $\bar{Y}$ as the predictor can be eliminated by using the regression equation in making predictions.

4. The closing prices of 6 stocks on the last day of last month seem to be quite highly correlated with their latest reported earnings per share figures and with the percentage of earnings growth they experienced

in the last year. The figures are given in the following table.

Y CLOSING PRICE PER SHARE	X_1 LATEST EARNINGS PER SHARE	X_2 PERCENTAGE EARNINGS GROWTH
$10	$1.50	3
18	2.00	10
22	2.00	8
30	2.50	6
30	3.00	10
40	7.00	−1

Use Equations (13.3) to show that the regression equation is approximately $\hat{Y} = .72 + 5.94X_1 + 1.08X_2$.

5. For the regression equation developed in Problem **4**, give the units and physical interpretation of each coefficient: .72, 5.94, and 1.08.

6. Use Formulas (13.8) and (13.9) to show that $S^2_{Y|12} \cong 35$ and $R^2 \cong .8$ for the regression equation formed in Problem **4**. (Allow for the fact that the answers given previously have been obtained using a good deal of rounding and your answers may differ somewhat.)

7. For the regression equation found in Problem **4**, find an interval within which you can be 95% sure that B_1 lies for all stocks of this type in the population. (*Hint:* $r^2_{12} = .45$.) What do you have to assume about the distribution of stock prices in order to construct this confidence interval?

8. Test the hypothesis $H_0: B_2 = 0$ against $H_a: B_2 > 0$ for the regression equation found in Problem **4**. Let $\alpha = .05$. Note the hint in Problem **7**.

9. For the regression equation found in Problem **4**, find the β coefficients associated with X_1 and X_2. To which variable are the estimates of Y more sensitive, X_1 or X_2?

10. Partition the sum of squares of deviations in stock prices in Problem **4** and construct a table like Table 13.1. Show that the F value computed from this table is approximately 6.6.

11. Use the F value computed in Problem **10** to test the hypothesis that $B_1 = B_2 = 0$ in the population of stock prices. That is, test the hypothesis that there is no relationship between stock prices and either earnings per share or percentage growth in earnings, or both. Let $\alpha = .025$.

12. The tax legislation committee of a certain state's legislature is considering a proposal to give residents of the state a tax rebate. They feel that the rebate would be spent in the state and thus boost the state's economy. The last time the legislature gave the residents of the state a tax rebate was in 1972 (an election year). In order to estimate the impact of a rebate on spending in the state, the chairman of the committee asked a legislative analyst to gather data concerning the ways families spent rebates in 1972. The analyst selected five 1972 rebate recipients (in a realistic problem, many more would be selected) and determined how they spent their rebates using extensive interviews. One of the tables from the report follows.

PERCENTAGE OF REBATE SPENT	FAMILY INCOME (000)	NUMBER OF DEPENDENTS
10	27	1
35	30	4
50	13	4
55	11	5
100	9	6

Construct the regression equation relating these three variables, letting Y = percentage of rebate spent, X_1 = family income (in thousands of dollars), and X_2 = number of dependents in the family, and show that the result is $\hat{Y} = 19 - 1.0X_1 + 12.3X_2$. Interpret the meaning of the constant and two regression coefficients in this equation.

13. Find $S^2_{Y|12}$ and R^2 for the regression equation found in Problem **12**. What does the value of R^2 mean?

14. For the regression equation found in Problem **12**, test the hypothesis $H_0: B_1 = 0$ against the alternative $H_a: B_1 < 0$ using $\alpha = .05$. (*Hint:* $r^2_{12} = .51$.)

15. Find an interval within which you can be 95% sure that B_2 lies for all recipients of 1972 tax rebates in Problem **12**. Note the hint in Problem **14**.

16. Find the β coefficients associated with X_1 and X_2 in the regression equation found in Problem **12**.

17. Find the F value that tests the overall regression relationship for the regression equation in Problem **12**. Conduct the test using $\alpha = .01$. That is, test $H_0: B_1 = B_2 = 0$ using $\alpha = .01$.

18. Discuss the implications your results in Problems **12–17** have for the policies and strategies that might be used by the legislators in proposing the tax rebate.

19. A mining firm has 5 large pumps which pump water from the mines they operate. Management has been concerned lately about the amount of money required to repair malfunctioning pumps. These costs are sums over and above the amounts spent for routine maintenance. In order to get a better idea of the relationship between these costs and other factors connected with the pumps, the operations superintendent ran a regression relating the variables shown in the following table.

Y MEAN MONTHLY REPAIR COSTS OVER THE LAST YEAR	X_1 MEAN WEEKLY HOURS OF OPERATION OVER THE LAST YEAR	X_2 AGE OF THE PUMP AT THE FIRST OF THE YEAR (IN MONTHS)
$643	28	80
613	26	48
494	15	27
250	15	2
400	16	13

Show that the regression equation is $\hat{Y} = 314.7 + .2X_1 + 4.7X_2$. Find the predicted mean monthly repair cost for a pump of this type that averaged 20 hours of operation per week and was 10 months old at the first of the year.

20. Show that $S_{Y|12}^2 \cong 8100$ and $R^2 \cong .84$ for the regression equation found in Problem **19**. What is the meaning of $R^2 = .84$?

21. Upon seeing the results of the regression analysis in Problem **19**, the operations superintendent said, "There certainly isn't a very significant relationship between the variables we used in the regression equation and the repair costs." Verify his statement by:

     a. Testing the hypothesis $H_0: B_1 = 0$ versus $H_a: B_1 > 0$ with $\alpha = .10$. (*Hint:* $r_{12}^2 = .84$. The fact that $R^2 = .84$ also is only coincidence in this problem.)

     b. Testing the hypothesis $H_0: B_2 = 0$ versus $H_a: B_2 > 0$ with $\alpha = .10$.

     c. Testing the hypothesis $H_0: B_1 = B_2 = 0$ using the F test and $\alpha = .05$.

22. Find the β coefficients for X_1 and X_2 in Problem **19**.

23. Discuss the problem of multicollinearity in the data for Problem **19** and the difficulties it introduces.

24. In Section 13.7, the stepwise multiple regression program from the SPSS package was discussed. The data used in the example were presented in Table 13.3. The program was instructed to stop after 3 variables were entered in the regression equation. The computer output that follows shows the final step of that same program when the program was instructed to include all 5 independent variables in the regression equation.

```
VARIABLE(S) ENTERED ON STEP NUMBER  5,,     LAND


MULTIPLE R            ,97284
R SQUARE              ,94641
ADJUSTED R SQUARE     ,92728
STANDARD ERROR       4,99323
```

```
--- ** -- -- ** ** --  VARIABLES IN THE EQUATION -- -- -- -- -- -- -- -- -- -- -- --

VARIABLE               B           BETA      STD ERROR B          F

AREA                3,40373       ,51652       ,75212          20,480
BEDROOMS           11,39093       ,54297      2,74893          17,171
DISTANCE            -,71567      -,23510       ,21884          10,695
PARKING            -6,37248      -,16842      3,55674           3,210
LAND                ,77780       ,11452       ,53070           2,148
(CONSTANT)        -11,22551
```

ANALYSIS OF VARIANCE	DF	SUM OF SQUARES	MEAN SQUARE	F
REGRESSION	5,	6164,74742	1232,94948	49,4518
RESIDUAL	14,	349,05258	24,93233	

```
--- -- ** -- ** --  VARIABLES NOT IN THE EQUATION -- -- -- -- -- ** -- -- -

VARIABLE      BETA IN     PARTIAL     TOLERANCE            F


MAXIMUM STEP REACHED
```

a. Write out the regression equation using all 5 variables.
b. What is the R^2 value calculated using Formula (13.9)?
c. To which of the 5 variables is the predicted price, $\hat{Y}$, most sensitive? Cite a reason for your answer.
d. Upon seeing the results of this regression equation, a condominium developer said, "Why that's ridiculous. Your equation says that having covered parking causes the price of a condominium to drop $6372.48. Your entire equation must be off!" How would you respond?

25. Using the data presented in Problem 24:
a. Test the hypothesis $H_0: B_1 = B_2 = B_3 = B_4 = B_5 = 0$ using $\alpha = .05$.
b. Test the hypothesis $H_0: B_5 = 0$ against $H_a: B_5 \neq 0$ using $\alpha = .05$. (Remember that $X_5 =$ parking.)
c. Test the hypothesis $H_0: B_2 = 0$ against $H_a: B_2 \neq 0$ using $\alpha = .05$.

26. Problem 19 involved a mining firm that was trying to construct a meaningful regression equation associating mean monthly repair costs over the last year, Y, with mean weekly hours of operation of the pumps, X_1, and age of the pump at the first of the year, X_2. That regression equation did not show very significant results. Thus, the management of the company contacted other mine operators who owned similar pumps and expanded their data set using the information on an additional 13 pumps. Also, they added a new variable, X_3, showing the frequency of routine maintenance performed throughout the year. The data for all 18 pumps are presented in the following table.

$Y = RPAIRCST$ Mean Monthly Repair Costs During Last Year	$X_1 = HOURSOP$ Mean Weekly Hours of Operation	$X_2 = AGE$ Age of Pump at First of Year	$X_3 = FREQ$ Frequency of Routine Maintenance
643	28	80	36
613	26	48	12
494	15	27	18
250	15	2	52
400	16	13	26
791	20	70	6
836	31	68	6
124	15	14	39
359	15	29	26
500	15	76	6
611	35	52	13
492	28	61	18
250	19	11	52
612	35	34	39
317	41	16	39
199	19	1	52
216	26	17	52
321	30	35	39

The SPSS stepwise multiple regression program was applied to the data in the table. The preliminary output of the program follows.

a. On the average, how many years old are the pumps?
b. On the average, how much time passes between performances of routine maintenance?

VARIABLE	MEAN	STANDARD DEV
RPAIRCST	446,0000	208,1170
HOURSOP	23,8333	8,3684
AGE	36,3333	26,1826
FREQ	31,1667	16,7200

Note that in the program the labels given the variables were: $Y =$ RPAIRCST, $X_1 =$ HOURSOP, $X_2 =$ AGE, and $X_3 =$ FREQ.

	RPAIRCST	HOURSOP	AGE	FREQ
RPAIRCST	1,00000	,33789	,82155	-,70930
HOURSOP	,33789	1,00000	,24995	-,00526
AGE	,82155	,24995	1,00000	-,72264
FREQ	-,70930	-,00526	-,72264	1,00000

27. Problems **27–31** refer to the successive outputs from the stepwise program.
   a. Which of the four variables has the greatest *relative* dispersion. (*Hint:* Find the coefficient of variation for each variable.)
   b. Which variable is most highly correlated with the dependent variable, Y? What is their r^2 value?
   c. Which two *independent variables* are most strongly correlated with one another? What is their r^2 value?
   d. Which two *independent variables* are least strongly correlated with one another? What is their r^2 value?
28. Step 1 of the stepwise multiple regression output for the data in Problem 27 follows.

DEPENDENT VARIABLE,,   RPAIRCST

VARIABLE(S) ENTERED ON STEP NUMBER 1,,   AGE

MULTIPLE R	,82155
R SQUARE	,67494
ADJUSTED R SQUARE	,65462
STANDARD ERROR	122,30798

---------------- VARIABLES IN THE EQUATION --------------------

VARIABLE	B	BETA	STD ERROR B	F
AGE	6,53020	,82155	1,13297	33,221
(CONSTANT)	208,73591			

   a. Write the regression equation after Step 1 has been performed.
   b. What is R^2 at this point?
   c. Test the significance of b_2. That is, test $H_0: B_2 = 0$ against the alternative $H_a: B_2 \neq 0$ using $\alpha = .01$. (Remember, $X_2 =$ AGE.)

```
ANALYSIS OF VARIANCE      DF       SUM OF SQUARES        MEAN SQUARE              F
REGRESSION                 1,        496968,13189        496968,13189          33,2214
RESIDUAL                  16,        239347,86811        14959,24176
```

```
----------------- VARIABLES NOT IN THE EQUATION -----------------

VARIABLE        BETA IN      PARTIAL    TOLERANCE            F

HOURSGR         ,14138       ,24010      ,93753            ,918
FREQ           -,24199      -,29338      ,47779           1,413
```

    d. Which variable will be the one to enter the regression equation at Step 2? How do you know?

    e. From the previous output, calculate the value of R^2 at the end of Step 2.

29. Step 2 of the stepwise regression output for the data in Problem **27** follows.

```
VARIABLE(S) ENTERED ON STEP NUMBER  2,,      FREQ


MULTIPLE R            ,83840
R SQUARE             ,70292
ADJUSTED R SQUARE    ,66331
STANDARD ERROR     120,76050
```

```
----------------- VARIABLES IN THE EQUATION -----------------

VARIABLE              B           BETA      STD ERROR B          F

AGE              5,140 21       ,64667      1,61834          10,088
FREQ            -3,01209       -,24199      2,53422           1,413
(CONSTANT)     353,11604
```

    a. Write out the regression equation at this point.

    b. What is $S^2_{Y|23}$?

    c. What is R^2?

    d. Test the hypothesis $H_0: B_3 = 0$ against $H_a: B_3 \neq 0$ using $\alpha = .05$. (Remember, $X_3 = $ FREQ.)

    e. What interpretation can be given to the constant term, 353.11604? Explain the meaning of the coefficients on AGE and FREQ.

```
ANALYSIS OF VARIANCE      DF       SUM OF SQUARES        MEAN SQUARE              F
REGRESSION                 2,       517569,52390        258784,76195          17,7455
RESIDUAL                  15,       218746,47610        14583,09841
```

```
----------------- VARIABLES NOT IN THE EQUATION -----------------

VARIABLE        BETA IN      PARTIAL    TOLERANCE            F

HOURSGR         ,20040       ,34357      ,87316           1,874
```

    f. Test the hypothesis $H_0: B_2 = B_3 = 0$ using $\alpha = .05$.

    g. If the program is allowed to run one more step, what will be the F value associated with X_1 in the new equation?

30. The program described in Problems **28** and **29** was not allowed to run through Step 3.

    a. Show that X_1 = HOURSOP was not significantly enough correlated with the unexplained variation in Y to be included in the equation. (*Hint:* Test $H_0 : B_1 = 0$ using $\alpha = .05$ and the F value in the VARIABLES NOT IN THE EQUATION section of Step 2.)

    b. Where should the program be stopped: after Step 1 or after Step 2? Why?

31. Discuss the extent of the collinearity between AGE and FREQ in Problems **28–30**. Which values in the output of Step 2 might be influenced by this collinearity?

32. A company has been experiencing problems recently with turnover among employees in a certain department. The work done in the department is largely unskilled, but the department figures that it must have an employee remain on the job at least one year before the hiring and training costs have been recovered. In order to determine if any of the preemployment information gathered on applicants for jobs in the department is useful in predicting the length of time they will spend on the job, the personnel manager of the company took a random sample of 25 people who recently quit. The manager listed their ages at the date of job application, sex (0 = male, 1 = female), number of weeks worked in the department before quitting, and the score obtained by the applicant in a test administered to determine how well-suited the person is for the job. The data are presented in the following table. The table is followed by the complete SPSS stepwise multiple regression output for this problem. The questions can be answered using information contained in the output.

$Y = JOBTENUR$ Length of Stay on the Job (Weeks)	$X_1 = AGE$ Age at Time of Application (Years)	$X_2 = SEX$ Sex 0 = Male 1 = Female	$X_3 = TEST$ Test Score
45	21	0	67
3	18	0	12
15	34	1	72
85	26	1	19
12	20	1	43
92	42	1	97
1	18	0	60
5	21	0	73
81	51	1	78
10	20	1	50
13	23	1	60
12	22	1	40
30	35	1	68
29	41	1	55
125	30	0	92
432	45	0	99
16	19	1	20
99	26	0	50
582	32	1	89

$Y = JOBTENUR$ Length of Stay on the Job (Weeks)	$X_1 = AGE$ Age at Time of Application (Years)	$X_2 = SEX$ Sex 0 = Male 1 = Female	$X_3 = TEST$ Test Score
376	30	0	91
191	21	1	62
43	20	1	19
23	20	0	29
84	29	1	36
1	19	0	40

VARIABLE	MEAN	STANDARD DEV
JOBTENUR	96,2000	149,1356
AGE	27,3200	9,3885
SEX	,6000	,5000
TEST	56,8400	26,0907

	JOBTENUR	AGE	SEX	TEST
JOBTENUR	1,00000	,39080	-,08270	,56289
AGE	,39080	1,00000	,22368	,62976
SEX	-,08270	,22368	1,00000	-,14245
TEST	,56289	,62976	-,14245	1,00000

VARIABLE(S) ENTERED ON STEP NUMBER   1,,   TEST

MULTIPLE R   ,56289
R SQUARE   ,31685
ADJUSTED R SQUARE   ,28715
STANDARD ERROR   125,91620

●●--●●--●●●●●●●---- VARIABLES IN THE EQUATION -●●●---●●--●●●●●●-●

VARIABLE	B	BETA	STD ERROR B	F
TEST	3,21752	,56289	,98512	10,667
(CONSTANT)	-86,68390			

ANALYSIS OF VARIANCE	DF	SUM OF SQUARES	MEAN SQUARE	F
REGRESSION	1,	169131,56897	169131,56897	10,6674
RESIDUAL	23,	364662,43103	15854,88831	

--●●●●●●--●●-- VARIABLES NOT IN THE EQUATION -●●----●●--●--

VARIABLE	BETA IN	PARTIAL	TOLERANCE	F
AGE	,06018	,05656	,60340	,071
SEX	-,00257	-,00307	,97971	,000

F-LEVEL OR TOLERANCE-LEVEL INSUFFICIENT FOR FURTHER COMPUTATION

a. On the average, how many years are people staying on the job in this department?

b. Explain the meaning of the mean value of .6000 for SEX.

c. Which two *independent variables* are most highly correlated with one another, and what is their r^2 value?

d. Write out the regression equation produced by the SPSS program. Interpret the meaning of the coefficients in the equation.

e. Give all the β coefficients for the variables *in the regression equation* and discuss their meaning.

f. Test the hypothesis $H_0 : B_3 = 0$ against $H_a : B_3 \neq 0$ using $\alpha = .05$. (Remember, $X_3 = $ TEST.)

g. The program stopped with the message *F*-LEVEL OR TOLERANCE-LEVEL INSUFFICIENT FOR FURTHER COMPUTATION. Discuss the possible meaning of this message in terms of the F values listed in the VARIABLES NOT IN THE EQUATION at the time the program stopped.

SAMPLE DATA SET QUESTIONS: *Refer to the 113 applicants for credit listed in the Sample Data Set Appendix of this book.*

a. Use a computer program to obtain the multiple regression equation in which JOBINC is the dependent variable and the independent variables are: SEX, JOBYRS, TOTBAL, and MSTATUS. Use only the applicants who were granted credit and who listed a JOBINC figure.

b. What types of variables are SEX and MSTATUS?

c. Test two separate hypotheses. First test the hypothesis that JOBYRS is unrelated to JOBINC, all other variables being held constant. Then test the hypothesis that TOTBAL is unrelated to JOBINC, all other variables being held constant. Use $H_a : B_i \neq 0$ and $\alpha = .05$ in these two tests.

REFERENCES

13.1 Miller, Robert B., and Dean W. Wichern. *Intermediate Business Statistics.* Holt, Rinehart and Winston, New York, 1977.

13.2 Mendenhall, William, and James E. Reinmuth. *Statistics for Management and Economics,* 2nd edition. Duxbury Press, North Scituate, Mass., 1974. Chapter 12.

13.3 Spurr, William A., and Charles P. Bonini. *Statistical Analysis for Business Decisions.* Irwin, Homewood, Illinois, 1967. Chapter 23.

13.4 Wonnacott, Thomas H., and Ronald J. Wonnacott. *Introductory Statistics for Business and Economics,* 2nd edition. John Wiley & Sons, New York, 1977. Chapter 13.

TIME
SERIES
ANALYSIS

14.1   THE NOTION OF A TIME SERIES

Historical data have been mentioned several times in previous chapters. These data can be used in a number of ways. Historical data are used to estimate probabilities. They are used in regression and correlation problems. They are also used in tests of hypotheses when we want to determine if a particular population currently has the same characteristics it has had in the past.

Much of the historical data available to businessmen and economists have been recorded at regular time intervals. For instance, gross national product figures are easily obtained by the calendar quarter; many companies keep track of revenue and expense figures by the month;

and governmental agencies record their expenses against budget by
the month. The sequence of values recorded over time for any piece
of data forms what is called a **time series.** Thus, a sequence of quarterly
gross national product figures from the first quarter of 1963 through
the last quarter of 1972 would form a time series consisting of 4 ×
10, or 40 values. Monthly sales of a company from January of 1960
through December of 1974 would form a time series of 12 × 15,
or 180 values.

Table 14.1 shows the time series values for monthly employment
in a company for 20 quarters. Sometimes it is difficult to see very
much in a set of numbers, so time series are often presented in graphs
and charts. The data in Table 14.1 are graphed in Figure 14.1. Section
2.3 deals with the graphic presentation of data in histograms and ogives.
Time series data are most often graphed using a *frequency polygon* like
the one in Figure 14.1. The pattern present in the data is quite evident
when the frequency polygon is used. It appears that the general level
of employment was rising slightly over the 20-quarter period, but there
were also marked seasonal fluctuations in employment. The summer
months had especially high employment, while the winter months were
characterized by low employment. This pattern is typical of companies
in the construction or recreation industries.

Observing patterns such as these can be useful to management
in planning for the future. The regular fluctuation suggested by Figure
14.1 can be used in forecasting almost any figure connected to the
company's level of employment. For instance, the company's treasurer
may be interested in budgeting wage and salary figures by quarter
for next year. Thus, an estimate of what the employment level will
be is needed. If all employees are covered by accident insurance, it
would also be helpful to know how many employees the company will
have for purposes of budgeting insurance premium costs. There are
many other tasks in which a forecast of the company employment
level might be useful—from determining how many company uniforms
to order, to planning an expansion of the company cafeteria.

TABLE 14.1   Number of People Employed on Last Day of Quarter

	FIVE YEARS AGO	FOUR YEARS AGO	THREE YEARS AGO	TWO YEARS AGO	LAST YEAR
1st Quarter	150	160	175	180	200
2nd Quarter	175	190	200	220	250
3rd Quarter	450	460	480	495	520
4th Quarter	140	145	160	180	190

FIGURE 14.1   Frequency polygon for quarterly employment figures of Table 14.1

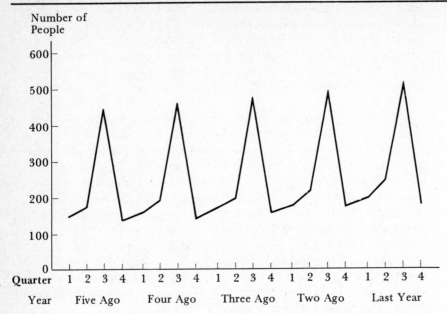

But before one can use the patterns in a set of data as an aid in forecasting, these patterns must be identified and described in a concrete way. For instance, to say that "employment for the company whose data are graphed in Figure 14.1 is rising slightly, is high in the summer, and is low in the winter" is correct but rather vague. It would be more concrete and useful in forecasting to say that "employment seems to be rising at an average rate of about 6 percent per year, and third-quarter employment averages about 2.8 times as much as first-quarter employment." *Time series analysis* consists of breaking down a time series into its component parts and analyzing these parts so that concrete, quantitative statements about the patterns the data reveal can be made. It is these concrete statements that will be useful in forecasting future values of the time series. The sections that follow in this chapter discuss the components of time series and methods for describing each component in a fashion that will be useful for forecasting purposes.

14.2   COMPONENTS OF TIME SERIES

Statisticians usually divide time series into four components: the long-term trend, seasonal variations, cyclical variations, and irregular variations.

FIGURE 14.2 Data with a curved downward trend

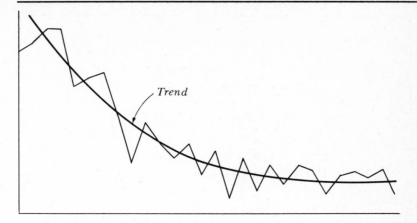

Trend

Trends in time series are the long-term movements of the series that can be characterized by steady or only slightly variable rates of change. Trends can be represented by straight lines (when the data have a steady rate of change) or by smooth curves (when the rate of change is slightly variable). The time series graphed in Figure 14.1 has a slight upward trend which could be described with a straight line. The time series graphed in Figure 14.2 has a downward trend that can be represented with a smooth curve. The basic economic forces that produce long-term trend movements in time series are population changes, inflation, business competition, and technological changes.

It is well known that fuel oil consumption rises in the winter and falls in the summer. But gasoline consumption has an opposite seasonal variation, rising in the summer and falling in the winter. As their name indicates, **seasonal variations** in a time series are those variations that occur rather predictably at a particular time each year. Quite obviously, seasonal variations can be found only in data recorded at intervals of less than a year; quarterly, monthly, or weekly data might well indicate variations of this type. The quarterly time series in Figure 14.1 shows marked seasonal variation in addition to a slight trend.

Cyclical variations are movements in a time series that, like seasonal variations, are recurrent but that, unlike seasonal variations, occur in cycles of longer than one year. Cyclical variations can be found in time series recorded on an annual basis as well as in data recorded at more frequent intervals. The time series in Figure 14.3 is an annual series that shows cyclical variation in the number of housing starts in the United States from 1959 through 1975. Cycles are not constant in amplitude and duration, and this lack of regular pattern makes their future occurrence difficult to predict. Cyclical variations are often

FIGURE 14.3   Private and public housing starts, 1959–1975 (data from the Statistical Abstract of the United States)

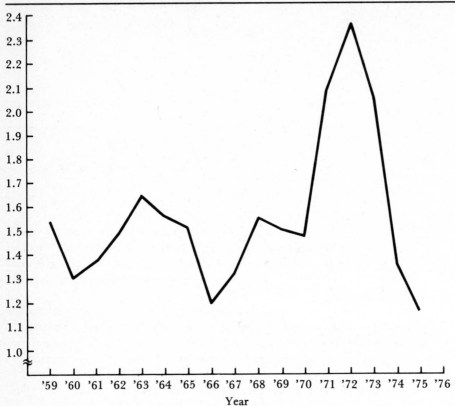

caused by general economic conditions, governmental policy changes, or shifts in consumer taste and purchasing habits.

Irregular variations constitute the class of time series movements that do not fit into the other three categories previously discussed. Essentially, they are the movements left in a time series when trends, seasonal variations, and cyclical variations have been identified. Irregular variations cannot be predicted using historical data, and they are not periodic in nature. They are caused by such factors as changes in the weather, wars, strikes, governmental legislation, and elections. For instance, a particularly harsh winter will cause a variation in crop production and fuel oil usage. A strike in the automotive industry will result in a drop in production. A strike in the steel industry is typically preceded by unusually high production, as steel users stockpile material in anticipation of short supplies. The passage of antipollution legislation in the late 1960s and early 1970s caused a sharp increase in the sales of firms producing air and water purification equipment. All of these unusual changes would be called irregular variations. It is often possible to determine the causes of irregular variations that are large. Figure 14.4 shows the time series of U.S. imports of handguns for private use from 1963 to 1975. The high figures in the 1966–68 period correspond to a period of urban violence and political assassination in this country.

FIGURE 14.4   Imports of handguns for private use (data from the *Statistical Abstract of the United States*)

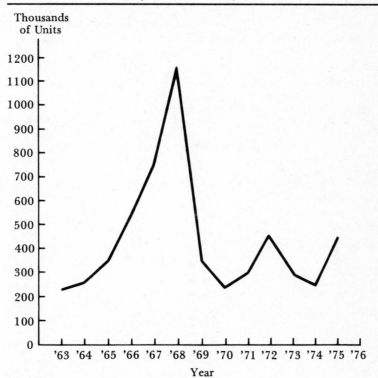

Thousands of Units
Year

14.3 SMOOTHING

Sometimes, the seasonal and irregular variations in time series are so large that it is difficult to determine if trend and cyclical components exist. Large variations can be smoothed from a series using a number of smoothing techniques. Two such techniques which we will discuss are *moving averages* and *exponential smoothing*.

A moving average of the values in a time series is found by averaging two or more consecutive values in the series and letting the computed value replace one of the values averaged. For instance, a three-month average can be constructed using the average of values for January, February, and March as the smoothed figure for February, and the average of data for February, March, and April as the March smoothed figure, and so on. The average is *moving* in the sense that as the average for each new month is calculated, a more recent value of the original time series is brought into the calculation of the average and an earlier one is dropped out.

EXAMPLE 1

Problem: Table 14.2 shows the 24-month sales record for electronic calculators in the business equipment department of the Mammoth Department Store.

TABLE 14.2　Actual and Smoothed Sales of Electronic Calculators at Mammoth Department Store

TIME PERIOD	ACTUAL NUMBER OF UNITS SOLD	FIVE-MONTH MOVING AVERAGE	EXPONENTIALLY SMOOTHED AVERAGE WITH $\alpha = .5$
Two Years Ago			
1　January	21		21.0
2　February	20		20.5
3　March	19	18.4	19.8
4　April	18	17.2	18.9
5　May	14	17.6	16.5
6　June	15	19.4	15.8
7　July	22	20.8	18.9
8　August	28	23.0	23.5
9　September	25	25.0	24.3
10　October	25	24.6	24.7
11　November	25	24.0	24.9
12　December	20	24.0	22.5
Last Year			
13　January	25	23.8	23.8
14　February	25	24.4	24.4
15　March	24	27.6	24.2
16　April	28	29.0	26.1
17　May	36	29.0	31.0
18　June	32	28.8	31.5
19　July	25	27.6	28.3
20　August	23	25.0	25.7
21　September	22	23.0	23.8
22　October	23	23.4	23.4
23　November	22		22.7
24　December	27		24.9

Find the 5-month moving average time series.

Solution: The general formula for finding the 5-month average in month t is:

$$\overline{Y}_t = \frac{Y_{t-2} + Y_{t-1} + Y_t + Y_{t+1} + Y_{t+2}}{5}$$

For instance, the moving average value for November of 2 years ago is:

$$\overline{Y}_{11} = \frac{25 + 25 + 25 + 20 + 25}{5} = 24.0$$

The second column of figures in Table 14.2 shows the moving average values for this series, and the actual and 5-month moving average figures are graphed in Figure 14.5.

FIGURE 14.5   Sales record for electronic calculators in the Mammoth Department Store

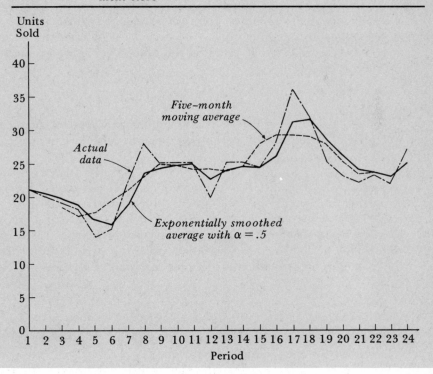

Two important points are demonstrated in Example 1. The first is that a moving average cannot be calculated for the first two and last two periods in the time series. This is because the calculations

for the moving average in January of two years ago would require the sales figures for December and November of three years ago, and these are not given. Similarly, the moving average for November of last year would require the sales figures for both December of last year and January of this year, and this latter figure is not available. In general, the longer the span covered by the moving average, the more periods are lost at the beginning and end of the time series. Thus, a six-month moving average loses three months on each end; a nine-month moving average does not have values for the first four months or the last four months; and so forth.

The second point made by Example 1 is that the smoothed values contain less variation than the original values. This is always true, and the longer the span covered by the moving average, the less variable will be the smoothed series. Thus, a six-month moving average would contain less variation than the five-month series. Of course, if too many periods are used in the calculation of the moving average, the smoothed series becomes nearly a straight line and, rather than being exposed, any trend or cyclical variations that may exist are smoothed out entirely.

A second way of smoothing data is called exponential smoothing. An exponentially smoothed value can be obtained for each time period. Let S_t be the smoothed value for period t. We begin by setting S_1 equal to Y_1, the actual value of the time series in the first period. Then

$$S_2 = \alpha Y_2 + (1 - \alpha) S_1$$

where α is a number between zero and one and is called the *smoothing constant*. For the general period, the smoothed value is

$$S_t = \alpha Y_t + (1 - \alpha) S_{t-1} \qquad 0 \le \alpha \le 1 \qquad \text{(14.1)}$$

Formula (14.1) shows that this period's smoothed value is a weighted average of two figures: the actual value in the time series for this period and the previous smoothed value. But the previous smoothed value is the weighted average of the actual time series value for the previous period and the smoothed value the period before that. If we were to proceed backward in time, we would find that

$$S_t = \alpha Y_t + (1 - \alpha) \alpha Y_{t-1} + (1 - \alpha)^2 \alpha Y_{t-2}$$
$$+ (1 - \alpha)^3 \alpha Y_{t-3} + \ldots + (1 - \alpha)^{t-1} Y_1 \qquad \text{(14.2)}$$

Formula (14.2) shows that S_t is really a weighted sum of all the previous values in the time series back to time $t = 1$. But the values further and further in the past have lighter and lighter weights since α and $1 - \alpha$ are fractional values which become smaller and smaller as we multiply them together and raise them to powers as in the formula.

The statistician smoothing the data can set the value of α anywhere between zero and one, thereby controlling how much weight current and past values in the time series will receive. If α is large, say .95, then S_t will be composed primarily of the current value of the time series Y_t, and only 5 percent of S_t will be dependent on historical values of the series. However, if α is small, say .10, then S_t will be composed primarily of the weighted historical values in the time series and only 10 percent of S_t will be dependent on the current value Y_t. If the statistician wants the smoothed series to follow the actual series quite closely—that is, little "smoothing" is desired—he or she should use a large smoothing constant α. The resulting smoothed series will be very sensitive to changes in the current values of the time series. But if a smoothed series that has most of the random, volatile variation taken out of it is desired, the statistician should choose a small smoothing constant. The resulting series will be quite smooth and reflect historical values of the time series far more than current values.

EXAMPLE 2

Find the exponential average values for the Mammoth Department Store's calculator sales using a smoothing constant of $\alpha = .5$. These values are presented in the third column of numbers in Table 14.2. They are also plotted on the graph in Figure 14.5. The calculations for the first few figures are as follows.

$$S_1 = Y_1 \qquad\qquad\qquad\qquad\qquad\qquad = 21.0$$

$$S_2 = \alpha Y_2 + (1 - \alpha) S_1 = .5(20) + .5(21.0) = 20.5$$

$$S_3 = \alpha Y_3 + (1 - \alpha) S_2 = .5(19) + .5(20.5) = 19.8$$

$$S_4 = \alpha Y_4 + (1 - \alpha) S_3 = .5(18) + .5(19.8) = 18.9$$

etc.

14.4 MEASURING THE COMPONENTS OF A TIME SERIES

When the original data of a time series have been smoothed, the results often uncover evidence that the time series contains elements of trend, seasonal, and/or cyclical variations. The statistician may well be interested in obtaining reliable means of measuring these components to use in a model or equation to forecast future values of the time series.

The classical equation that is used to represent a time series is the following:

$$\text{Actual value} = T \times S \times C \times I \qquad\qquad (14.3)$$

where the trend value T is measured in the same units as the time series, and S, C, and I are indices of seasonal, cyclical, and irregular variation, respectively. If we wish to forecast a future period's value of the time series, the forecast can be found by doing the following:

STEP 1. Project a trend value.

STEP 2. Multiply the trend value by a seasonal index number that measures the typical deviation of the time series from the trend for that season.

STEP 3. Multiply the value in Step 2 by an index that measures the cyclical variation expected for that period.

Since irregular variations are, by their very nature, unpredictable, the forecast is completed after the third step. That is,

$$\text{Forecasted value} = T \times S \times C \qquad (14.4)$$

Thus, it would be beneficial if the forecaster were to have some numerical measures of T, S, and C to use in Equation (14.4). Section 14.5 contains an explanation of how the trend, T, can be measured in a time series. Sections 14.6 and 14.7 show various means of measuring the seasonal variation, S. Section 14.8 discusses the analysis of cycles, C, and the last section in the chapter deals with how the knowledge of the components in a time series can be used in both short- and long-term forecasting.

14.5 TREND ANALYSIS

When the original data or the smoothed data in a time series show evidence of trend, the statistician may well be interested in finding a suitable way of measuring and representing the trend. If the time series shows a steady upward or downward movement, a straight line can be used to represent the trend. If the overall rise or fall occurs at an increasing or decreasing rate, a curved line should be used. In either case, the most common method for finding a trend line is to fit a least-squares line to the data. The procedure is the same as for fitting regression least-squares lines to pairs of observations (see Chapter 12). In the present case, the pairs of points are the values of the time series data paired with the date of the observations.

EXAMPLE 1

Problem: The data in Table 14.3 represent the closing price of a speculative stock over the last 9 quarters of trading. Monthly, weekly, or daily data have not been presented since the stock's price fluctuates so much over short periods

TABLE 14.3 **Closing Prices for Speculative Stock over Nine Quarters**

QUARTER	PRICE OF STOCK ON LAST DAY OF QUARTER
1	$37.75
2	$44.00
3	$46.25
4	$44.00
5	$51.75
6	$49.00
7	$58.50
8	$58.00
9	$70.25

of time. The general rise in price is evident from the quarterly data. Fit a least-squares linear trend line to these data.

Solution: The general form of the linear trend line is

$$\hat{Y} = a + bt \qquad (14.5)$$

where t is a numerical value for the time period measured from the origin of the time series. The selection of the origin for the series is rather arbitrary. For instance, the quarters in Table 14.3 could have been numbered from 0 to 8 instead of 1 to 9. Equations (12.7) and (12.8) can be used to obtain the values for a and b, the coefficients of the least-squares lines. In this case

$$a = 34.26 \quad \text{and} \quad b = 3.36$$

so that the least-squares trend line becomes

$$\hat{Y} = 34.26 + 3.36t$$

These coefficients can be interpreted as meaning that in time period $t = 0$, the trend value of the stock would be $34.26 per share and that over the 9-quarter period the price of the stock rose an average of $3.36 per share per quarter. The actual values of the time series and the least-squares trend line are plotted in Figure 14.6.

This example illustrates two important points about the least-squares line. The first is that the least-squares method of fitting a line to a set of points gives heavy weight to the larger fluctuations. This is because large deviations result in *very large squared deviations,* which strongly influence the values of a and b calculated by Equations (12.7) and (12.8). Thus, a few large fluctuations like the rapid price rises in quarters seven and nine can unduly influence the slope of the least-squares line. Note that in Figure 14.6 the seventh- and ninth-quarter price

FIGURE 14.6   Actual values and least-squares trend line for closing prices of a speculative stock

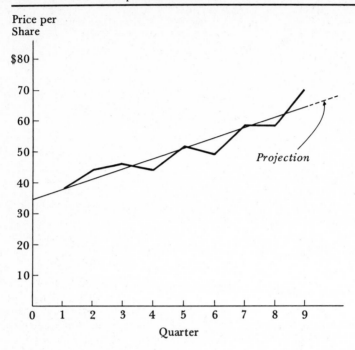

rises have pulled the line upward so that the price fluctuations in the first six quarters are not *around* the line but near or below it. The second point illustrated by Example 1 concerns the shape of the trend line. There is a slight indication that the general rate of price increase for the stock is increasing. This means that a trend curve might better describe the movement of this time series. A least-squares curve could be fitted to the time series using the multiple regression techniques discussed in Chapter 13.

EXAMPLE 2

Problem: Fit a polynomial trend line of the form $\hat{Y} = a + b_1 t + b_2 t^2$ to the data in Table 14.3.

Solution: In order to perform the multivariate regression required for the solution to this problem, we must compute the second independent variable needed in the equation, t^2. The following table shows the data from Table 14.3 rearranged in the format used for the problems in Chapter 13. Note that the third column is just the square of the second.

Y CLOSING PRICES OF STOCK	t QUARTER	t^2 (QUARTER)^2
37.75	1	1
44.00	2	4
46.25	3	9
44.00	4	16
51.75	5	25
49.00	6	36
58.50	7	49
58.00	8	64
70.25	9	81

With the data in this form, we could form Equations (13.3) and solve for a, b_1, and b_2 just as though we were finding the regression equation coefficients for any two-independent-variable equation $\hat{Y} = a + b_1 X_1 + b_2 X_2$. The only difference here is that $X_1 = t$ and $X_2 = t^2$. However, a computer program similar to the one discussed in Section 13.7 was used to find the coefficients for this equation. Its results gave

$$\hat{Y} = 39.90 + .28t + .31t^2$$

This equation does a slightly better job in fitting the data points than does the one produced in Example 1. This is shown by the fact that R^2 for this equation is .91, and the r^2 for the simple regression equation of Example 1 is .87. In this particular example, the indications of curved trend are so slight that it would probably be best to wait a few more quarters before concluding that the trend is not linear.

14.6 SEASONAL INDICES

Next we turn our attention to measuring the seasonal variation in a time series, S, using seasonal indices. Seasonal indices are numbers which vary from a base of 100. When a month (or week or quarter or other seasonal period) is assigned an index value of 100, this indicates that that month has no seasonal variation. Two methods of obtaining seasonal indices are presented here. The first involves finding a seasonal index that compares the seasonal mean value to a grand mean value. This method is most appropriate for time series that do not have strong trends or cyclical variations. The second method involves comparing each season's actual value to a yearly moving average to obtain an index. The indices arrived at are then averaged over all the time periods in the series. This method is much more widely used since it can give meaningful seasonal indices for data with strong trends and cyclical variations.

The examples in this section consist of problems in which seasonal indices are desired for each of the four quarters in a year. If indices were to be calculated for each month of the year, techniques employed would be the same. We obtain quarterly indices here merely to demonstrate the method without using excessive amounts of space.

EXAMPLE 1

Problem: Table 14.4 presents 10 years of quarterly data for the sales of XB-7, a product used in combating certain types of lawn larvae. It is recommended by the manufacturer that XB-7 be applied in the spring and fall. Find the seasonal sales indices for the four quarters of the year.

Solution: The indices have been found in Table 14.4 using the first method of calculating seasonal indices. The mean sales value for each quarter is taken over the 10-year period and then is expressed as a percentage of the grand mean over all quarters in the 10-year period. The grand mean is just the mean of the four quarterly means. Thus, the seasonal index for the first quarter is the mean for the quarter divided by the grand mean times 100:

$$\frac{331.8}{411.48} \times 100 = 80.7$$

TABLE 14.4   Calculation of Quarterly Sales Index for Product XB-7
(Figures in tons)

YEAR	FIRST QUARTER	SECOND QUARTER	THIRD QUARTER	FOURTH QUARTER
1	257	288	263	311
2	291	368	341	408
3	319	485	325	381
4	305	364	336	383
5	332	435	410	449
6	368	520	415	444
7	332	464	405	468
8	351	440	411	668
9	355	504	449	527
10	408	490	740	649
Total	3318	4358	4095	4688
Grand total				16,459
Quarterly mean[a]	331.8	435.8	409.5	468.8
Grand mean[b]				411.48
Quarterly index[c]	80.7	105.9	99.5	113.9
Sum of indices				400.0

[a] The total for the quarter divided by 10.
[b] (Sum of quarterly means)/4, or 16,459/40.
[c] (Quarterly mean/grand mean) × 100.

Thus, on the average, the first quarter's sales are only about 80.7% of the average quarter's sales. On the other hand, the fourth quarter's sales are about 13.9% above the average quarter's sales. Since the seasonal indices vary from a base of 100, the four quarterly indices add up to 400. When monthly indices are calculated, their total should be 1200, averaging a value of 100 for each month.

EXAMPLE 2

Problem: Table 14.5 presents in somewhat different form the 10 years of quarterly data for the sales of product XB-7. In this example, seasonal indices are to be computed using the method of comparing actual sales values to a four-quarter moving average.

Solution: There are 5 steps in this process. The first 3 are shown in Table 14.5 and the fourth and fifth in Table 14.6.

STEP 1. *Determine a 4-quarter moving average.* This average is found by summing four adjacent values and dividing the total by 4. Now, to which quarter should this moving average be assigned? In Example 1 of Section 14.3, where a 5-month moving average was calculated, this question did not arise. This was because with an odd number of periods (5) involved in the moving average, each of the moving average values could be assigned to the middle (third) period of the 5-period unit. However, when the number of periods is even (like 4), the moving average cannot be assigned to a "middle" period since one does not exist. The "middle" of 4 quarters is between the second and third quarter. Thus, the moving averages in the second column of Table 14.5 are centered *between* quarters. In order to find a moving average that can be assigned to a particular quarter, it is necessary to perform Step 2.

STEP 2. *Form a 2-quarter moving average from the 4-quarter moving average column.* This figure, column 3 of the table, is now centered *on* a quarter instead of *between* quarters. This figure is also a 4-quarter moving average that is centered on a particular quarter. For instance, the first moving average is centered on the third quarter of the first year. It is actually a weighted average of quarterly sales data with the following weights:

YEAR	QUARTER	WEIGHT
1	First	1
1	Second	2
1	Third	2
1	Fourth	2
2	First	1
	Total Weights	8

Similarly, each quarter's moving average consists of data from 5 quarters where the first and last quarter are weighted once and the intervening quarters are weighted twice.

TABLE 14.5   Calculations for Steps 1–3 in Finding Seasonal Sales Indices of Product XB-7 (Amounts in tons)

YEAR & QUARTER	COLUMN 1 Actual Data	COLUMN 2 Four-Quarter Moving Average of Column 1	COLUMN 3 Two-Quarter Moving Average of Column 2	COLUMN 4 Actual Data as Percent of Column 3
1—First	257			
1—Second	288	—	—	—
1—Third	263	279.75	284.0	92.6
1—Fourth	311	288.25	298.2	104.3
2—First	291	308.25	318.0	91.5
2—Second	368	327.75	339.9	108.3
2—Third	341	352.00	355.5	95.9
2—Fourth	408	359.00	373.6	109.2
3—First	319	388.25	386.2	82.6
3—Second	485	384.25	380.9	127.3
3—Third	325	377.50	375.8	86.5
3—Fourth	381	374.00	358.9	106.2
4—First	305	343.75	345.1	88.4
4—Second	364	346.50	346.8	105.0
4—Third	336	347.00	350.4	95.9
4—Fourth	383	353.75	362.6	105.6
5—First	332	371.50	380.8	87.2
5—Second	435	390.00	398.2	109.2
5—Third	410	406.50	411.0	99.8
5—Fourth	449	415.50	426.1	105.4
6—First	368	436.75	437.4	84.1
6—Second	520	438.00	437.4	118.9
6—Third	415	436.75	432.2	96.0
6—Fourth	444	427.75	420.8	105.5
7—First	332	413.75	412.5	80.5
7—Second	464	411.25	414.2	112.0
7—Third	405	417.25	419.6	96.5
7—Fourth	468	422.00	419.0	111.7
8—First	351	416.00	416.8	84.2
8—Second	440	417.50	442.5	99.4
8—Third	411	467.50	468.0	87.8
8—Fourth	668	468.50	476.5	140.2
9—First	355	484.50	489.2	72.6
9—Second	504	494.00	476.4	105.8
9—Third	449	458.75	465.4	96.5
9—Fourth	527	472.00	470.2	112.1
10—First	408	468.50	504.9	80.8
10—Second	490	541.25	556.5	88.1
10—Third	740	571.75	—	—
10—Fourth	649	—	—	—

STEP 3. *Express each quarter's actual figure as a percentage of its moving average.* The fourth column of Table 14.5 contains these percentages.

STEP 4. *For each quarter (first, second, third, and fourth) find the mean percentage*

TABLE 14.6 Calculations for Steps 4 and 5 in Finding Seasonal
            Sales Indices of Product XB-7

| | | QUARTER: | | |
YEAR	First	Second	Third	Fourth
1	—	—	92.6	104.3
2	91.5	108.3	95.9	109.2
3	82.6	127.3	86.5	106.2
4	88.4	105.0	95.9	105.6
5	87.2	109.2	99.8	105.4
6	84.1	118.9	96.0	105.5
7	80.5	112.0	96.5	111.7
8	84.2	99.4	87.8	140.2
9	72.6	105.8	96.5	112.1
10	80.8	88.1	—	—
Mean percentage	83.5	108.2	94.1	111.1
Total[a]				396.9
Seasonal index	84.2	109.0	94.8	112.0
Total[b]				400.0

[a]Total of percentages.
[b]Total of seasonal indices.

value from step 4 over all the years. This is done in Table 14.6. For
instance, the mean percentage value for the first quarter is 83.5. This
indicates that on the average, the sales figure for the first quarter
was 83.5% of the four-quarter moving average centered on the first
quarter. But these mean percentages cannot be used as seasonal indices
themselves. This is because seasonal indices vary from a base of 100.
The four mean percentages do not add up to 400—they add to 396.9.
Thus, a final calculation is needed to obtain a seasonal index.

STEP 5. *Adjust the mean percentages so they add to a total of 400.* This is accomplished
by multiplying each mean percentage figure by an adjustment factor
of 400/396.9, or 1.008. The final result is a new percentage figure
which varies from a base of 100 and represents the seasonal index.

The steps outlined in Example 2 are very similar to those that
would be followed in calculating monthly seasonal indices. The minor
modifications that would be made can be seen by comparing the steps
of the example with the following procedure:

STEP 1. Determine a twelve-month moving average.

STEP 2. Obtain a two-month moving average of the figures found
in Step 1. This is a modified twelve-month moving average.

STEP 3. Express each month's actual figure as a percentage of the
moving average in Step 3.

STEP 4. For each month (January, February, . . . , December) find the mean percentage value from Step 3 over all years.

STEP 5.   Adjust the mean percentages so they total 1200.

The seasonal indices calculated in Examples 1 and 2 do not differ a great deal from one another. They both show that sales are high in the spring and fall and low in the winter and summer. However, the indices developed in Example 2 are preferable to those from Example 1. This is due to the fact that the data over the ten-year period used in the examples show a rather strong trend, and the second of our two methods for developing seasonal indices is most appropriate for data with strong trend.

The difficulty of calculating seasonal indices by the second method is usually not a serious factor since, in most problems where much data is involved, computer programs are written to perform the calculations.

14.7   FINDING SEASONAL MEASURES USING MULTIPLE REGRESSION

In the last section, we found that the second, more complex method of finding seasonal indices is the appropriate method for data containing a strong trend. However, it is possible to combine our knowledge of regression and time series in a method that both accounts for trends in the data *and* requires less work than the second method in Section 14.6. One need only have access to a preprogrammed multiple regression package and a computer.

The method proceeds as follows. First, we must select one season as the "base" season. If the data are quarterly, we may wish to select the first calendar quarter, winter, as the base season. All other seasonal indices will be measured with respect to this base season. There are no absolute rules governing the selection of the base season.

Next, the time series data are recorded over the historical time period, using dummy variables to represent those seasons other than the base season. For example, one year of quarterly sales data for a company might be $50, $125, $105, and $60 thousand. If winter were selected as the base season, then these data could be recorded with the use of dummy variables as follows.

Y SALES	t TIME	S_2 SPRING	S_3 SUMMER	S_4 FALL
50	1	0	0	0
125	2	1	0	0
105	3	0	1	0
60	4	0	0	1

Note that the seasonal variables—spring, summer, and fall—are all zero in the time periods that do not correspond to their season. This notation allows us to represent all four seasons with only three variables. For instance, the winter period is represented by three zeroes in the seasonal variables, (0, 0, 0). Spring is represented by (1, 0, 0), summer by (0, 1, 0), and fall by (0, 0, 1).

With the data represented in this fashion, it is possible to construct a regression equation of the form

$$\hat{Y} = a + b_1 t + b_2 S_2 + b_3 S_3 + b_4 S_4 \tag{14.6}$$

where $\hat{Y}$ represents the predicted sales figure, t is the time period, and the S_i values are the dummy variables representing spring, summer, and fall. To obtain a forecast using this regression equation, we must determine the time period t and the season S_i for which the forecast is desired. If we desire a forecast of sales for the summer quarter following the last data point, then $t = 7$ and the seasonal indications are (0, 1, 0). The forecast would then be

$$\hat{Y} = a + b_1 (7) + b_2 (0) + b_3 (1) + b_4 (0)$$
$$= a + b_1 (7) + b_3$$

In this result, we can see that the forecast consists of the trend value $a + b_1 (7)$ plus the amount b_3. This b_3 is the seasonal adjustment figure for summer sales. The same line of reasoning shows that b_2 is the seasonal adjustment figure for spring and that b_4 gives this adjustment for fall. Since winter is the base quarter, no seasonal adjustment figure is added to the trend. In realistic problems, of course, more than four data points are used to obtain the regression equation.

EXAMPLE 1

Problem: The figures in the following table represent the quarterly fuel consumption of a small trucking firm over the past 5 years. Find the seasonal adjustments for fuel consumption by multiple regression using dummy seasonal variables. Predict fuel consumption for fall of this year.

	WINTER	SPRING	SUMMER	FALL
5 Years Ago	80	100	150	105
4 Years Ago	95	115	170	105
3 Years Ago	110	150	190	145
2 Years Ago	115	175	210	180
Last Year	120	185	220	190

(Thousands of Gallons)

Solution: The first step in the solution is to select a base season. For convenience we will use winter as that base. Winter seems to be the lowest fuel consumption season and is perhaps appropriate as a "base." Next, we record the data over the last 5 years using dummy variables to represent the spring, summer, and fall seasons. Table 14.7 shows how these data are recorded. When those data were entered into a multiple regression program like the SPSS program described in Chapter 13, the following regression equation resulted:

$$\hat{Y} = 61.5 + 4.7t + 36.3S_2 + 74.6S_3 + 26.8S_4$$

This regression line indicates that the forecast would be $\hat{Y} = 61.5 + 4.7(0) = 61.5$ at time $t = 0$. That is, the constant term can be interpreted as the equation's estimate of the fuel consumption trend for fall of 6 years ago, before seasonal adjustments are made. The 4.7 figure represents the trend and suggests that, all other things being constant, the fuel consumption has been rising at a rate of 4700 gallons per quarter. The seasonal adjustment figures are 36.3, 74.6, and 26.8 for spring, summer, and fall, respectively. There is no seasonal adjustment for winter. This means that winter forecasts are equal to the trend. However, spring's consumption is 36,300 gallons above the trend, summer's is 74,600 above trend, and fall's is 26,800 above trend.

The forecast for fall of this year would be made using $t = 24$ and a set of dummy variables for fall, (0, 0, 1). Thus,

TABLE 14.7   Data for Fuel Consumption Example with Time and Dummy Variables Included

YEAR	QUARTER	t	Y ACTUAL DATA	S_2 SPRING	S_3 SUMMER	S_4 FALL	$\hat{Y}$ PREDICTED VALUES
Five Ago	Winter	1	80	0	0	0	66.2
	Spring	2	100	1	0	0	107.2
	Summer	3	150	0	1	0	150.2
	Fall	4	105	0	0	1	107.2
Four Ago	Winter	5	95	0	0	0	85.1
	Spring	6	115	1	0	0	126.1
	Summer	7	170	0	1	0	169.1
	Fall	8	105	0	0	1	126.1
Three Ago	Winter	9	110	0	0	0	104.0
	Spring	10	150	1	0	0	145.0
	Summer	11	190	0	1	0	188.0
	Fall	12	145	0	0	1	145.0
Two Ago	Winter	13	115	0	0	0	122.9
	Spring	14	175	1	0	0	163.9
	Summer	15	210	0	1	0	206.9
	Fall	16	180	0	0	1	163.9
Last Year	Winter	17	120	0	0	0	141.8
	Spring	18	185	1	0	0	182.8
	Summer	19	220	0	1	0	225.8
	Fall	20	190	0	0	1	182.8

$$\hat{Y} = 61.5 + 4.7(24) + 36.3(0) + 74.6(0) + 26.8(1)$$

$$= 61.5 + 112.8 + 26.8$$

$$= 201.1 \qquad \text{or } 201,100 \text{ gallons}$$

There is some uncertainty about this prediction, however, since we are guilty of extrapolating our forecast beyond the range of the independent variable, t. That is, $t = 24$ is beyond the range of t values used to develop the regression equation. Unfortunately, in forecasting using a time series regression equation, there is no way to avoid such extrapolations. The predicted consumption values for the five-year period are shown in the last column of Table 14.7. These figures will be used in the discussion for Section 14.8. It should be noted that the R^2 figure for the regression equation developed in this example is .94. This indicates that the regression equation does a rather good job of fitting the predicted to the actual sales values.

In Section 14.4 we mentioned that Equation (14.3) represents the classical time series model where

$$\text{Actual value} = T \times S \times C \times I$$

This model is sometimes referred to as the *multiplicative model* since all the components are multiplied together. This name also helps to distinguish the model from the *additive model* which represents a time series value as follows:

$$\text{Actual value} = T + S + C + I \qquad (14.7)$$

When we use the regression Equation (14.6) to forecast a future value of a time series, we are using a short form of this additive model. This is because use of the dummy variables to represent the seasons always results in a forecast of the form

$$\hat{Y} = T + S_i \qquad (14.8)$$

where S_i is the amount to be *added* to the trend for the ith season. An even better forecast could be obtained if we had some measure of the likely cyclical and irregular variations that could be added to Equation (14.8) to predict the time series value more accurately.

One may wonder which method of representing time series values is more appropriate

$$\text{Actual value} = T \times S \times C \times I$$

or

$$\text{Actual value} = T + S + C + I$$

The answer, of course, depends on the nature of the time series involved. On the whole, however, the first method seems to be preferred by most statisticians.

To understand this preference we can examine the forecast of fuel consumption produced in Example 1 of this section. That forecast consisted of the trend value $T = 61.5 + 4.7(24) = 174.3$ thousand gallons plus 26.8 thousand gallons for the fall seasonal figure. Thus, the total forecast was 201.1 thousand gallons. One might say that the seasonal variation for fall is $(26.8/174.3) \times 100 = 15.4$ percent above the trend. One might also say that the seasonal variation for fall is 26.8 thousand gallons above the trend. For this prediction, either statement is correct and they both say the same thing and result in the same forecast.

But for a future fall forecast, say next year, the two views of seasonal variation (multiplicative and additive) give different forecasts. In that future year, the additive model requires that we still add 26.8 thousand gallons to the trend figure to obtain our forecast. The multiplicative approach would involve raising the forecast by a figure similar to the 15.4 percent. If there is a strong upward trend in a time series, the 15.4 percent will be quite a bit larger than 26.8 thousand gallons.

The basic difference between the multiplicative and additive models is that the first assumes that seasonal variations produce the same *percentage* change each time a season arrives while the second assumes that the seasonal variations produce the same *amount* of change. Most analysts agree that typical time series behave in the former manner— seasonal variations seem to be constant percentages through the years rather than constant amounts. Thus, the multiplicative model is more generally used in forecasting except in short-term forecasting situations or where the time series exhibits little trend. In these cases, the percentage change and the additive change are very nearly the same.

14.8   CYCLICAL VARIATION

Cyclical variations are movements in a time series that recur over periods of longer than a year. These variations have no particular patterns of the sort we see in seasonal variations. Since cycles differ from one another in their amplitudes and durations, it is not reasonable to develop indices to measure their "average" strength or amplitude. It is possible, however, to analyze historical data and identify cyclical variations. One method for doing this is outlined here.

Throughout this analysis of cyclical variation, we will represent the time series values using the multiplicative model:

$$\text{Actual value} = T \times S \times C \times I$$

where the trend value T is measured in the same units as the time series, and S, C, and I are indices of seasonal, cyclical, and irregular variation, respectively. If the data for the time series are annual, then there can be no seasonal variation in the time series and

$$\text{Actual value} = T \times C \times I \qquad (14.9)$$

Generally, moving averages are considered to contain only trend and cyclical components. If the moving average is a twelve-month moving average, it does not contain seasonal variations and the irregular variations have been averaged out. Thus

$$\text{Twelve-month moving average} = T \times C \qquad (14.10)$$

For quarterly data,

$$\text{Four-quarter moving average} = T \times C \qquad (14.11)$$

A twelve-month moving average can be obtained for each month in a time series (except the first six and last six), or a four-quarter moving average can be obtained for each quarter (except the first two and last two), using the method illustrated in column 3 of Table 14.5. If a least-squares trend equation has been constructed for the time series, then a trend value can also be obtained for each month or quarter. If the trend value is divided into the moving average, the resulting ratio is a measure of the cyclical component for that period. That is,

$$\frac{\text{Moving average}}{\text{Trend}} = \frac{T \times C}{T} = C \qquad (14.12)$$

This ratio is usually expressed as a percentage. A value of $C = 100$ would indicate that there is no cyclical variation in the time series for that period. If this percentage is determined for each period in the time series, its movements can serve to indicate the cyclical variations of the series.

There is another method of arriving at a measure of cyclical variation. It involves dividing each value in the time series by the trend. If the actual time series data are annual, then

$$\frac{\text{Actual value}}{\text{Trend value}} = \frac{T \times C \times I}{T} = C \times I \qquad (14.13)$$

If the actual data are monthly or quarterly, then the actual value can be divided by the trend value adjusted for seasonal variation. For instance, the trend value may be 60 for a particular month, and the seasonal index for that month may be 105. Then the trend value adjusted for seasonal variation would be 60 × 1.05, or 63. When the actual figures are divided by this trend adjusted for seasonal variation, the result is

$$\frac{\text{Actual value}}{\text{Adjusted trend}} = \frac{T \times S \times C \times I}{T \times S} = C \times I \qquad (14.14)$$

The results in Equations (14.13) and (14.14) are both proportions that are usually converted to percentages. Both these proportions contain irregular variation in addition to cyclical variation. A moving average of these ratios can be taken to average out the irregular variations. The final result, a set of percentages for each period in the time series, offers a measure of the cyclical variation in the series.

These percentage variations are not very useful in forecasting changes in time series values since cycles are by their nature somewhat irregular in duration and amplitude. However, the percentages can be used to identify turning points in cycles for historical time series. They can also be used to adjust historical data to remove cyclical components in much the same way as seasonal indices are used to adjust historical data values to deseasonalize them. (Methods of deseasonalization are discussed in Section 14.9.)

EXAMPLE 1

Problem: In Section 14.6, data on sales of a lawn treatment product, XB-7, were used to develop seasonal indices. The second method yielded quarterly seasonal indices of 84.2, 109.0, 94.8, and 112.0. The simple linear trend line for the 10 years of data used in that example is

$$\hat{Y} = 278.9 + 6.46t$$

Use this trend line and the seasonal indices to remove the trend and seasonal variation from the last year of sales data.

Solution: As a first step in the solution, we must use the trend line to obtain trend values for the last 4 quarters:

$$\hat{Y}_{37} = 278.9 + (6.46)(37) = 517.9$$

$$\hat{Y}_{38} = 278.9 + (6.46)(38) = 524.4$$

$$\hat{Y}_{39} = 278.9 + (6.46)(39) = 530.8$$

$$\hat{Y}_{40} = 278.9 + (6.46)(40) = 537.3$$

Next, we adjust these trend values for seasonal variation using the seasonal indices:

$$(517.9)(0.842) = 436.1$$

$$(524.4)(1.090) = 571.6$$

$$(530.8)(0.948) = 503.2$$

$$(537.3)(1.120) = 601.8$$

Note that in these calculations the seasonal indices were expressed as proportions, not as percentages. The resulting figures contain trend and seasonal variation, $T \times S$. If we divide these figures into the actual values $= T \times S \times C \times I$, we obtain just $C \times I$, which measures the cyclical and irregular components of the time series.

$$(408/436.1) \times 100 = 93.6$$

$$(490/571.6) \times 100 = 85.7$$

$$(740/503.2) \times 100 = 147.1$$

$$(649/601.8) \times 100 = 107.8$$

These four figures contain only cyclical and irregular variation. Since the third quarter's figure of 147.1 is so much larger than the others, it appears that there was some irregular influence in the sales during that period (perhaps a large promotion of the product) which may even have carried over into the fourth quarter.

14.9 USING TIME SERIES RESULTS AND FORECASTING

In the previous sections of this chapter, we have discussed methods of analyzing historical time series data in order to separate the series into its components of trend, seasonal, cyclical, and irregular variation. While the analysis of a time series to identify its components is interesting, the work involved can be justified only if the results can somehow be used. This section explains some of the ways in which results from time series analysis can be used in adjusting data and in forecasting. Forecasting using the additive time series model was described in Section 14.7. This section is limited to uses of time series results obtained from analysis of the multiplicative model.

One of the most common uses of seasonal indices is adjustment of actual data to account for seasonal variation. For instance, the company that sells product XB-7, which was described in Examples 1 and 2 of Section 14.6, might have a goal of selling 2350 tons of the product next year. This averages out to 587.5 tons per quarter. If they sell 500 tons the first quarter, it looks at first glance as though they are below their target sales level. However, due to seasonal variations in sales, the first quarter is always slow. If we use the seasonal index found in Example 2 for the first quarter, we know from historical data that the first quarter represents about 84.2 percent of the average quarterly sales for the year. Taking this seasonal variation into account, we would expect sales of only .842 × 587.5, or 494.7 tons, in the first quarter.

In the same way, the .842 figure can be used to adjust the actual sales figure to a deseasonalized value. That is, if seasonal variation is taken into account, the actual sales figure of 500 tons is equivalent to an adjusted sales value of 500/.842, or 593.8 tons. This figure can then be compared to the average quarterly sales target of 587.5 tons needed to attain the yearly goal. Either way the result is expressed, the company seems to be ahead of schedule in meeting its goal.

Most of the economic time series statistics we see in publications read by the general public are "seasonally adjusted." This means that the actual figure has been deseasonalized by dividing it by the appropriate seasonal index expressed as a proportion rather than a percentage. The deseasonalized figures show trends and cyclical variations much more clearly. The irregular variations are still present, however, as can be seen in Equation (14.15).

$$\frac{T \times S \times C \times I}{S} = T \times C \times I \qquad (14.15)$$

The results of time series analysis can be used in several different forecasting situations. Some of these are presented here in the form of examples. The first example demonstrates how a yearly forecast can be broken down into forecasts for each season of the year using seasonal indices. The second example demonstrates how a forecast can be developed for a particular month or quarter using a trend equation and a seasonal index. The third example shows how the next period's (month or quarter) forecast can be obtained using only this period's data and seasonal indices.

EXAMPLE 1

Problem: The Huntsman Specialty Building Products Company would like to develop a forecast of monthly sales for the coming year. An industry sales forecast financed by the industry-wide trade association and Huntsman's knowledge of their historical percentage of industry sales have enabled the company to determine that its annual sales for the coming year will be about $15 million. Use the following monthly sales indices to break down the annual forecast into monthly forecasts.

MONTH	INDEX	MONTH	INDEX
January	86	April	100
February	88	May	104
March	92	June	114

MONTH	INDEX	MONTH	INDEX
July	119	October	98
August	106	November	97
September	100	December	96

Solution: Since the total of all the seasonal indices is 1200, the monthly forecasts can be expressed as

$$\frac{\text{Seasonal index}}{1200} \times \text{Annual forecast}$$

Thus, the forecast figures are calculated as follows:

$$\text{January: } \frac{86}{1200} \times \$15 \text{ million} = \$1.075 \text{ million}$$

$$\text{February: } \frac{88}{1200} \times \$15 \text{ million} = \$1.100 \text{ million}$$

etc. The entire set of 12 monthly forecasts is as follows:

January	$1.075 million	July	$ 1.488 million
February	1.100 million	August	1.325 million
March	1.150 million	September	1.250 million
April	1.250 million	October	1.225 million
May	1.300 million	November	1.212 million
June	1.425 million	December	1.200 million
		Total	$15.000 million

EXAMPLE 2

In January of 1979 the executive assistant to the governor of a certain state was given the task of forecasting state revenues from sales taxes during each quarter of the year 1980. In order to do this, the assistant used a quarterly trend equation and seasonal indices supplied by the state treasurer's office. The trend equation and indices had been developed using sales tax collection data from the past 24 years. The data were adjusted to eliminate differences in the state sales tax rate over the 24-year period. The trend equation was

$$\hat{Y} = \$151 + \$3.1t$$

where all figures are in millions of dollars and $t = 1$ occurred in the first quarter of 1955. Since the first quarter of 1955 was the period in which $t = 1$, the first quarter of 1980 was $t = 101$, the second quarter was $t = 102$, and so forth. The quarterly sales tax collection indices were

First quarter	80
Second quarter	95
Third quarter	104
Fourth quarter	121

The governor's assistant proceeded to develop the forecast using the formula

$$\text{Forecast} = T \times \frac{S}{100} \qquad (14.16)$$

Recall that forecasts of cyclical and irregular variations cannot be developed from historical data using techniques presented in this chapter since they do not follow predictable patterns. Thus, the projection was based solely on the trend and seasonal data. Any deviations from the forecast would then be due mainly to business-cycle fluctuations and irregular variations. Combining the forecasting Formula (14.16) with the trend equation, the assistant arrived at a general working formula

$$\hat{Y}_{q,yr} = (151 + 3.1t)\frac{S}{100}$$

where the first subscript on the $\hat{Y}$ value is the quarter and the second is the year. The forecasted revenues by quarter were:

$$\hat{Y}_{1,80} = [151 + 3.1(101)](80/100) \quad = 464.1 \times .80 = \quad 371.3$$

$$\hat{Y}_{2,80} = [151 + 3.1(102)](95/100) \quad = 467.2 \times .95 = \quad 443.8$$

$$\hat{Y}_{3,80} = [151 + 3.1(103)](104/100) = 470.3 \times 1.04 = \quad 489.1$$

$$\hat{Y}_{4,80} = [151 + 3.1(104)](121/100) = 473.4 \times 1.21 = \quad 572.8$$

$$\text{Total} \quad 1877.0$$

The sum of the quarterly forecasts gave an annual forecast of $1.877 billion in sales tax revenues for the year 1980.

The method of forecasting demonstrated in Example 2 consists of using the trend line and seasonal indices. This method could be used to make projections to the year 2000 and beyond. It should be remembered, however, that projections into the future can become very unreliable past one or two years. Projections five to ten years in the future can be made with confidence only for those time series proven to be highly stable and regular in the past.

EXAMPLE 3

The purchasing department of the Sky Kitchen Catering Service is planning its ordering for the current month. Sky Kitchen prepares meals for airlines to serve to passengers on flights departing from a city in the Midwest. They have found that the number of meals they are asked to prepare each month follows the same seasonal pattern as airline traffic. Thus, they can use the local airport's airline traffic seasonal indices to help them project the number of meals they must prepare. Last month they prepared 9232 meals. The purchasing agent would like to know how many meals to plan for this month.

The last month was June with a seasonal airline traffic index of 118, and the current month is July with a seasonal index of 124. The projection for this month's meals can be found using the following relationship:

$$\frac{\text{Meals prepared in July}}{\text{Meals prepared in June}} = \frac{\text{Seasonal index for July}}{\text{Seasonal index for June}}$$

That is, the ratio of actual meals prepared in July to those prepared in June should be about the same as the ratio of the airline traffic seasonal index for July to the index for June. All of the values in the formula are known but one—the meals to be prepared in July. If we let this value be the unknown X, then we have:

$$\frac{X}{9232} = \frac{124}{118}$$

or

$$X = 9232 \times \frac{124}{118} = 9701$$

The purchasing agent can plan for about 9700 meals.

The method of forecasting described in Example 3 can be used to obtain only short-range forecasts. The number of meals the catering service should plan on serving in August could also be calculated using the ratio of the August index to the June index. In fact, projections could be made through the end of the year. However, projections past 60 or 90 days using the method may be in error. A 60-day projection may be needed for August if the final figures on the number of meals served in July are not available on the first of August. This is a common situation, where the last period's data are not available in time to use in the next period's forecast. The forecast must be based on data from two periods back. In Example 3 the August forecast might be

$$X = \text{Meals prepared in June} \times \frac{\text{Seasonal index for August}}{\text{Seasonal index for June}}$$

But once again, the greater the time between the present and the forecast period, the less confidence we have in the forecast.

PROBLEMS

1. Consider a time series whose first value was recorded in January of 1930. The last period for which there are records is September of 1962.
    a. How many full months of data are available?
    b. How many full quarters of data are available?
    c. How many full years of data are available?
2. A certain company started keeping weekly sales records on Monday in the first week of July of 1964. How many weeks of data will they have collected by next Friday?
3. Give an example of a time series you think would have:
    a. A moderately increasing linear trend.
    b. A decreasing linear trend.
    c. A curvilinear trend.
4. Give an example of a time series you think would have:
    a. High seasonal indices in the first half of the year and low seasonal indices in the second half of the year.
    b. High seasonal indices in winter and summer and low seasonal indices in the spring and fall.
5. Give an example of a time series that would show a large irregular variation (up or down) like the one in Figure 14.4. What is the cause of this variation?
6. The following figures show the percentage of new people hired each year in a company, broken down by the quarter in which the new employees actually joined the company.

	1ST QUARTER	2ND QUARTER	3RD QUARTER	4TH QUARTER
3 Years Ago	24.7	26.5	22.8	26.0
2 Years Ago	24.5	25.7	25.0	24.8
1 Year Ago	26.6	24.0	26.6	22.8

    a. Construct a frequency polygon from these data.
    b. Does there seem to be any seasonal variation in the percentage of new people joining the company?
    c. There does not seem to be any trend in these data. However, the personnel director of the company says, "We are hiring more people every year." Why does the trend in hiring fail to show up?
7. The following figures indicate the number of mergers that took place in a certain industry over a 19-year period.
    a. Plot these data on a frequency polygon.
    b. What type of trend (linear or curved) might best be fit to this time series?
    c. Is there evidence of seasonal variation in this series?

YEAR	MERGERS	YEAR	MERGERS
1960	23	1970	125
1961	23	1971	140

YEAR	MERGERS	YEAR	MERGERS
1962	31	1972	160
1963	23	1973	150
1964	32	1974	165
1965	32	1975	192
1966	42	1976	210
1967	64	1977	250
1968	47	1978	300
1969	96		

8. When a 4-month moving average is found for a time series, how many months do not have averages associated with them:
   a. At the first of the time series?
   b. At the end of the time series?

9. When a 12-month moving average is found for a time series that begins in January of one year and ends in December of another year, which month has:
   a. The first moving average value?
   b. The last moving average value?

10. Find the 3-year moving average values for the merger time series in Problem 7.

11. Find a 4-year moving average series for the merger data in Problem 7. Center the average *on* the years.

12. Find the exponentially smoothed series for the last 10 years in Problem 7:
   a. Using $\alpha = .2$.
   b. Using $\alpha = .8$.

13. Judging from your results in Problem 12, which value of α should be used if the person smoothing the merger data wants the smoothed series to be highly responsive to sudden changes in the merger rate?

14. Fit a least-squares trend line to the merger data in Problem 7. Let $t = 1$ for 1960.

15. What would an exponentially smoothed time series look like if the value $\alpha = 1.0$ were used?

16. The following time series show the number of firms in a certain industry over a 25-year period.

YEAR	FIRMS	YEAR	FIRMS	YEAR	FIRMS
1955	437	1964	906	1972	1071
1956	467	1965	941	1973	1067
1957	526	1966	968	1974	1049
1958	683	1967	981	1975	1098
1959	739	1968	1011	1976	1114
1960	772	1969	1051	1977	1151
1961	804	1970	1056	1978	1163
1962	841	1971	1063	1979	1202
1963	873				

   a. Find the 5-year moving averages for this series.

b. Find the exponentially smoothed averages for this series with $\alpha = .6$.

c. Plot the two averages on the same piece of graph paper.

17. The following quarterly data show the number of appliances (in thousands) returned to a particular manufacturer for warranty service over the past 5 years.

	1ST QUARTER	2ND QUARTER	3RD QUARTER	4TH QUARTER
5 Years Ago	1.2	.8	.6	1.1
4 Years Ago	1.7	1.2	1.0	1.5
3 Years Ago	3.1	3.5	3.5	3.2
2 Years Ago	2.6	2.2	1.9	2.5
1 Year Ago	2.9	2.5	2.2	3.0

a. Plot a frequency histogram of this time series.

b. Find the equation of the least-squares linear trend line that fits this time series. Let $t = 1$ be first quarter 5 years ago.

c. What would the trend line value be for the second quarter of the current year, i.e., two periods beyond the end of the actual data?

18. Fit a trend line to the time series in Problem **16** visually. Then estimate the coefficients a and b for this line.

19. Determine the quarterly seasonal indices for the warranty service time series in Problem **17** using the method described in Example 1 of Section 14.6. Then find values for the same indices using the method described in Example 2 of Section 14.6. If you ignore the differences in the calculations required, which of the two methods of finding seasonal indices is preferable for calculating seasonal indices for this series? Why?

20. Use the method described in Section 14.7 to obtain seasonal figures for the data given in Problem **17**. (*Hint:* This method requires a least squares trend line. A multiple regression computer program must be applied to the data using $t = 1$ in the first quarter 5 years ago and using 3 dummy variables. Let the first quarter be the base quarter.)

21. Compare the data 3 years ago in Problem **17** to the other years. Give a one- or two-sentence discussion of how that year of data might affect the trend line constructed in Problem **20**.

22. The following time series shows the number of building permits issued by month for the largest county in a large western state.

	NUMBER OF PERMITS IN:		
MONTH	1977	1978	1979
January	1015	664	1106
February	901	743	1022
March	1319	1147	1679
April	1590	1284	2011
May	1555	1250	1985

MONTH	NUMBER OF PERMITS IN:		
	1977	1978	1979
June	1473	1352	1938
July	1252	1408	1943
August	1249	1287	2045
September	1293	1309	1738
October	1234	1409	1797
November	946	1269	1722
December	841	1214	1496

a. Use the method of Example 1 in Section 14.6 to find monthly seasonal construction indices.
b. Use the method of Example 2 in Section 14.6 to find monthly seasonal construction indices.
c. Discuss the advantages of the first method over the second method for this time series.

23. Use the method described in Section 14.7 to find seasonal indices for the data given in Problem **22**. (*Hint:* A computer program which performs multiple regressions should be used along with 11 dummy variables. Let January be the base.)

24. The values in the following table show the percentage of students dropping a particular class over the last 3 years at a school using the 3-quarter academic year.

	FALL	WINTER	SPRING
Two Years Ago	5	4	8
Last Year	6	4	10
This year	9	8	12

a. Use the method of Example 1 in Section 14.6 to find the indices of drops for the academic quarters.
b. Use the method of Example 2 in Section 14.6 to find the indices of drops for the academic quarters. (*Note:* Step 2 in the process can be skipped. This is because Step 2 is only used to center the moving average on a period in the data set. Since this data set has only 3 periods per year, the 3-quarter moving average computed in Step 1 will already be centered.)

25. A metropolitan bus company has supplied the following data showing the number of accidents involving their buses over the past 5 years.

	WINTER	SPRING	SUMMER	FALL
5 years Ago	15	10	9	12
4 Years Ago	15	14	11	15
3 Years Ago	22	21	12	19
2 Years Ago	45	22	14	20
Last Year	32	25	18	24

a. Use the method of Example 1 in Section 14.6 to find the 4 seasonal indices for accidents.

b. Use the method described in Section 14.7 to find the 4 seasonal figures. (*Hint:* Use winter of 5 years ago as $t = 1$ and let winter be the base quarter.)

26. Use the method described in Example 2 of Section 14.6 to find the 4 seasonal indices for the data in Problem **25**.

27. In Problem **25** note the value for winter, 2 years ago. Does it appear that this value is out of line with the others due to cyclical or irregular variation? What might have caused this unusually high value?

28. The trend line for the warranty data in Problem **17** is $\hat{Y} = 1.0 + .1t$. Use this trend line and the indices found in the first part of Problem **19** to express the last 2 years of original data as $(C \times I) \times 100$. Discuss why the first method of finding seasonal indices in Problem **19** may be appropriate.

29. For the data in Problem **24**, the trend line is $\hat{Y} = 3.50 + .77t$, where $t = 1$ in the fall quarter, 2 years ago. Use the seasonal indices found in Part **b** of Problem **24** and this trend line to remove the trend and seasonal variation in the data and express the results as $(C \times I) \times 100$. Why would it be impossible for these data to show marked cyclical variation?

30. If actual billings for the investment consulting firm of Johnson, Pratt & Stewart Associates were $79,852 in March, and if the March seasonal index for this firm's billings is 107, then what is the seasonally adjusted March billing figure? What would be the expected annual billings based on the March figure?

31. Assume that the U.S. Department of Commerce reports seasonally adjusted personal income for the first quarter of the year to be $412.5 billion. If the quarterly seasonal index of personal income is 96, what was the actual personal income figure for the quarter?

32. The following time series represent the number of patients received in a hospital emergency room. The seasonal indices for each quarter are also given. Find the seasonally adjusted figures for the time series. Will these seasonal indices tell the emergency room manager how much staffing and what supplies to order for each quarter?

	1ST QUARTER	2ND QUARTER	3RD QUARTER	4TH QUARTER
Patient Visits	16,440	12,295	10,667	13,379
Seasonal Index	118	75	90	117

33. If the manager of the emergency room in Problem **32** has forecasted 60,000 patient visits to the emergency room next year, how should he break down his forecast by quarter?

34. The Peterson Products Company has determined the following seasonal indices for their expenditures on materials for production:

MONTH	INDEX	MONTH	INDEX
January	50	July	90
February	90	August	130
March	140	September	170
April	65	October	140
May	105	November	90
June	90	December	40

    a. Do these indices vary from a base of 100? How can you tell?

    b. If the total annual expenditure for materials at Peterson Products next year is expected to be $2.4 million, find the figures that would be budgeted for each month.

35. If the Peterson Products Company of Problem **34** spends $122,000 for materials in January and $197,500 in February next year:

    a. What would be their two seasonally adjusted figures?

    b. What would be their expected annual material expenditure based on these first two months?

36. Sales of the Jenkins Leasing Company have a trend line that is represented by the equation $\hat{Y}_t = \$55,000 + \$1250t$, where $t = 1$ in the first quarter of 1960. The four quarterly seasonal indices of sales are: 107, 100, 82, and 111.

    a. Approximately how much have sales been increasing each quarter in the past?

    b. What was the approximate level of sales in the first quarter of 1960?

    c. Forecast quarterly sales for the year 1982.

37. The dollar volume of small-business loans made by the Midstate Valley Bank and Trust Company has followed a trend line of $\hat{Y}_t = \$152,000 + \$950t$, where the time period involved is months and $t = 1$ in July of 1965. Since most of these small-business loans are used to finance inventories in the later half of the year, the seasonal indices on these loans are larger in the second half than in the first half. The monthly seasonal indices are as follows:

January	56	July	100
February	64	August	110
March	62	September	137
April	99	October	167
May	83	November	191
June	67	December	64

Total 1200

    a. What was the approximate dollar volume of loans in July of 1965?

    b. Forecast the loan volume for March of the year 2000.

    c. Forecast the loan volume for May of 1988.

    d. Which of the two forecasts above is likely to be more accurate? Why?

    e. Forecast the monthly loan volumes for the year 1982.

38. Late Central Airlines generated 172,889 passenger miles last month. Their seasonal index for passenger miles for last month is 110, but the index for this month is only 92. Using these pieces of information, forecast this month's figure for number of passenger miles flown.

39. If Late Central (Problem **38**) has a seasonal index of passenger miles of 98 for next month, what would be the forecast for next month's figure?

40. On February 1 the president of Midstate Valley Bank and Trust Company (Problem **37**) received a report showing that the bank had loaned $238,560.00 to small businessmen in January. Use this figure and the January and February seasonal indices to forecast the loan volume for February.

41. The quarterly indices for housing starts in a certain state are: winter—60, spring—170, summer—100, and fall—70.

   a. If the state's Department of Housing expects 140 thousand housing starts this year, how should they divide their annual forecast between the four quarters?

   b. Assume that during the winter of this year the state issued permits for 20,000 housing starts. What is the seasonally adjusted figure?

   c. On the basis of winter's 20,000 housing starts, how should the Department of Housing revise its annual forecast? Give a new numerical forecast for the year.

42. Assume the Department of Housing mentioned in Problem **41** developed a least squares regression equation relating housing starts, Y, to time, t. That equation has the form:

$$\hat{Y} = 10,000 + 500t + 12.5t^2$$

where the winter of 1970 is the quarter in which $t = 1$.

   a. What would the Department of Housing forecast for the spring of 1983? (Recall the note of caution in Section 13.8 about the accuracy of predictions using extrapolation, however.)

   b. Which of the following statements best describes the way in which housing starts have been changing over time?

     (i)   They have been growing at a declining rate.
     (ii)   They have been falling at an increasing rate.
     (iii) They have been growing at an increasing rate.
     (iv) They have been growing at a constant rate.
     (v)   They have not been growing or declining.

SAMPLE DATA SET QUESTIONS:

Refer to the 113 applicants for credit listed in the Sample Data Set Appendix of this book.

   a. Are any of the variables listed in that sample data set "time series" variables?

   b. List some variables which the department store's credit office would probably track on a time series basis.

   c. In which months would the seasonal indices for "number of credit applications made" likely be high for this department store?

REFERENCES

14.1 Lapin, Lawrence L. *Statistics for Modern Business Decisions.* Harcourt Brace Jovanovich, New York, 1973. Chapter 15.

14.2 McElroy, Elam E. *Applied Business Statistics.* Holden-Day, San Francisco, 1971. Chapters 11–14.

14.3 Mendenhall, William, and James E. Reinmuth. *Statistics for Management and Economics,* 2nd Edition. Duxbury Press, Belmont, California, 1974. Chapter 14.

INDEX
NUMBERS

The concept of an index number was introduced in Chapter 14. Seasonal indices were developed there to express the variation of time series values in a season (week, month, or quarter) from some base value (usually the mean weekly, monthly, or quarterly value) for the year. Thus, if the month of March has a seasonal index of 110, March figures of the time series are usually 10 percent above the figures for the average month of the year.

There are other types of indices that show changes in a time series over time. These index numbers express the variation of time series values from some base value too, but in this case the base value is not an average of values over a year but instead some fixed value in the time series. For instance, if sales of a company in 1979 were

20 percent above what they were in 1975, an index of the time series for 1979 sales would be 120. The base figure is the fixed value of the 1975 sales. Thus, index numbers as they are used in this chapter measure variation of a time series from a value of that time series at some fixed point in time.

15.1 THE SIMPLE AGGREGATE INDEX

Table 15.1 shows the total electrical power generation in the United States for the years 1968, 1971, and 1975 in billions of kilowatt-hours. The ratio of the 1971 figure to that for 1968 is 1614/1329, or 1.214; multiplication by 100 gives 121.4, the second entry in the second row. We can say that in 1971 electrical power generation was 121.4 percent of what it was in 1968, or that generation was up 21.4 percent. The other two entries in the second row, 100.0 and 144.2, are computed in like manner: $100 \times 1329/1329$ and $100 \times 1916/1329$, respectively. The 1975 generation of power was 44.2 percent above the 1968 generation.

An index number is a ratio, multiplied by 100 to put it in the form of a percentage. The indices in the second row of Table 15.1 all have for their denominator 1329, the generation figure for 1968. In computing these index numbers, we have adopted this *base year* of 1968 as a standard of comparison. If we were to divide the four production figures by 1614, the figure for 1971, we would obtain the power generation indices with 1971 as the base year. These indices are given in the last row of the table. We see, for example, that power generation in 1968 was 82.3 percent of what it became in 1971.

An index number series gives a clear picture of trend. The theory and computation of an index number are ordinarily more complicated than in our simple example, because an index number ordinarily

TABLE 15.1   U.S. Electric Power Generation

	YEAR:		
	1968	*1971*	*1975*
Electric power in billions of kilowatt-hours	1329	1614	1916
Generation indices for:			
Base year 1968	100.0	121.4	144.2
Base year 1971	82.3	100.0	118.7

Source of data: *Statistical Abstract of the United States.*

represents the total combined effect of changes in a number of varying quantities. The best-known index is the U.S. Bureau of Labor Statistics Consumer Price Index, or CPI, usually called the cost-of-living index. The CPI is a statistical measure of changes in prices of goods and services bought by urban residents. Its importance lies in part in the fact that union contracts often contain cost-of-living escalation clauses. Part of its importance lies in the fact that governmental fiscal and economic policies are influenced by its movements. And finally, changes in the CPI have a great impact on the psychology of consumers in this country.

The prices of hundreds of items go into the CPI. An extensive discussion of the CPI is given later in this chapter, but to make its structure clear, we shall construct a simple illustrative example involving the four items whose prices are listed in Table 15.2. We take 1960 as the base year, and the problem is to construct a sensible index for 1970 and 1980 compared with 1960. The first procedure that comes to mind is simply to add the prices for a year and divide each sum by $4.56, the sum for 1960. This gives the *simple aggregate* indices

$$100 \times \frac{4.56}{4.56} = 100.0 \qquad 100 \times \frac{4.53}{4.56} = 99.3 \qquad 100 \times \frac{6.71}{4.56} = 147.1$$

In other words, if P_{0i}, P_{1i}, and P_{2i} represent the respective prices of item i ($i = 1, 2, 3, 4$) for each of the three years, we compute

$$100 \times \frac{\Sigma_i P_{0i}}{\Sigma_i P_{0i}} = 100 \qquad 100 \times \frac{\Sigma_i P_{1i}}{\Sigma_i P_{0i}} \qquad 100 \times \frac{\Sigma_i P_{2i}}{\Sigma_i P_{0i}} \qquad (15.1)$$

But the index of 99.3 for 1970 makes it appear as if prices dropped from 1960 to 1970, a conclusion that defies reason for those who lived through that period of time! A glance at the table shows that although

TABLE 15.2   A Simple Aggregate Price Index

ITEM	1960 PRICE P_{0i}	1970 PRICE P_{1i}	1980 PRICE P_{2i}
1. Milk (dollars/qt.)	$ .26	$ .31	$ .56
2. Steak (dollars/lb.)	1.05	1.17	2.15
3. Butter (dollars/lb.)	.75	.85	1.40
4. Pepper (dollars/lb.)	2.50	2.20	2.60
Total	$4.56	$4.53	$6.71
Price indices computed by Formulas (15.1)	100.0	99.3	147.1

the prices of milk, steak, and butter rose considerably in that period, this increase was more than offset in our computation by the decrease in the price of pepper, which costs a great deal per pound. Since pepper is surely a small part of the average budget, we might get a better idea of food cost changes during the 1960s if we throw out the figures for pepper. This leads to new indices:

	1960	1970	1980
Total (pepper omitted)	$ 2.06	$ 2.33	$ 4.11
Price index	100.00	113.11	199.51

Now the indices reflect an increase in prices. Acting on the observation that pepper played a disproportionate role in the first aggregate index seems to have put our computed figures more in line with what we know to have been the actual situation. But simply throwing out an item that does not seem important in the determination of an index is hardly a satisfactory way to determine the value of the index. Two people constructing an index number to measure the same basic price changes might disagree on which items should be thrown out. Or they might fail to delete from their working data an item that distorts the calculations in a less obvious way than the pepper in our example.

Considerations such as these suggest quite clearly that the prices in our data must somehow be weighted according to their importance in the food budget. The question of the units for which the prices are listed also suggests that prices should be weighted. The milk price is given in dollars per quart, but why not use gallons instead? Why is a quart of milk rather than a gallon comparable to a pound, the unit for the other three items? If milk cost is expressed in dollars per gallon, it becomes four times what it was and exerts proportionally greater influence in the computation. Table 15.3 shows how the simple aggregate price index would change if milk costs were expressed in

TABLE 15.3 A Simple Aggregate Price Index Using Different Purchasing Units

ITEM	1960 PRICE P_{0i}	1970 PRICE P_{1i}	1980 PRICE P_{2i}
1. Milk (dollars/gal.)	$1.04	$1.24	$2.24
2. Steak (dollars/lb.)	1.05	1.17	2.15
3. Butter (dollars/lb.)	.75	.85	1.40
Total	$2.84	$3.26	$5.79
Price indices computed by Formulas (15.1)	100.0	114.8	203.9

dollars per gallon and pepper were left out of the computations. Note that when milk prices are expressed in dollars per gallon, the 1970 prices are 14.8 percent above the 1960 prices, but when they are expressed in dollars per quart, 1970 prices are only 13.1 percent above 1960 prices. One might wonder which of these percentage increases is correct. The answer is that neither truly reflects the percentage price rise in food costs. The only thing that can be said is:

1. One quart of milk, a pound of steak, and a pound of butter cost 13.1 percent more in 1970 than they did in 1960, and
2. One gallon of milk, a pound of steak, and a pound of butter cost 14.8 percent more in 1970 than they did in 1960.

By changing the units of measure, a statistician could construct a simple aggregate index to express almost any percentage price change desired.

In general, when an index number is constructed to reflect the combined effect of changes in a number of varying quantities, the simple aggregate index should not be used. A simple aggregate price index can be used legitimately only under the special conditions that all the prices are expressed in the same units (like dollars per pound) and equal amounts of each item are purchased.

15.2 PRICE INDICES WITH QUANTITY WEIGHTS

It was shown in the last section that the index for changes in prices over a period of time depends on the quantities of the items that are purchased. Thus, it seems reasonable to weight changes in the prices of goods by the quantities of goods purchased in order to reflect the relative importance of each good or item in the total expenditures. This section presents various weighting schemes designed to accomplish this end.

Consider Table 15.4. This table shows the quantities for each of four items purchased by a typical family in 1960, 1970, and 1980.

TABLE 15.4   Market Baskets for 1960, 1970, and 1980

ITEM	1960 q_{0i}	1970 q_{1i}	1980 q_{2i}
1. Milk	728 qt.	735 qt.	737 qt.
2. Steak	312 lb.	320 lb.	350 lb.
3. Butter	55 lb.	56 lb.	56 lb.
4. Pepper	.3 lb.	.3 lb.	.3 lb.

The four food items, together with the quantities in which they were purchased, are called a "market basket." The market baskets are different for each of the years since different quantities of goods were purchased in each year. For any item i, the 1960 price was p_{0i} and the 1960 quantity was q_{0i}. Hence, $p_{0i}q_{0i}$ was the total amount spent on item i in the base year. For example, 728 quarts of milk at $.26 per quart (from Table 15.2) comes to $.26 \times 728$, or $189.28, for 1960 expenditures for milk. Table 15.5 shows the amounts spent for each product in each year. In 1970 the typical family bought 735 quarts of milk (from Table 15.4) at $.31 per quart (from Table 15.2) and thus spent $227.85 on milk in 1970 (milk expenditure figure in Table 15.5). The table also shows the total spent on each year's market basket of goods.

A set of indices known as **weighted aggregate indices** can be constructed from these total expenditure figures. Each weighted aggregate index number is a ratio of the total expenditure on the market basket for a particular year to the total expenditure for a base year; that is,

$$100 \times \frac{\Sigma_i p_{0i} q_{0i}}{\Sigma_i p_{0i} q_{0i}} = 100, \qquad 100 \times \frac{\Sigma_i p_{1i} q_{1i}}{\Sigma_i p_{0i} q_{0i}}, \qquad 100 \times \frac{\Sigma_i p_{2i} q_{2i}}{\Sigma_i p_{0i} q_{0i}} \qquad \textbf{(15.2)}$$

The last row in Table 15.5 consists of weighted aggregate indices reflecting the change in the total amount of money spent on the four food items over the period 1960–1980. Since the quantities of items purchased changed, these indices reflect changes in both price and amount purchased. It would be appropriate, then, to say that the *food bill* for these four items rose approximately 122.7 percent from 1960 to 1980. However, it would *not* be appropriate to say that *prices* for these four items rose 122.7 percent, since the percentage increase reflects some increase in amount of goods purchased, too. Thus, the 122.7 percent figure reflects changes in price and standard of living.

TABLE 15.5   Total Expenditures for Yearly Market Baskets and Weighted Aggregate Indices

ITEM	1960 COST $p_{0i} q_{0i}$	1970 COST $p_{1i} q_{1i}$	1980 COST $p_{2i} q_{2i}$
1. Milk	$189.28	$227.85	$ 412.72
2. Steak	327.60	374.40	752.50
3. Butter	41.25	47.60	78.40
4. Pepper	.75	.66	.78
Total	$558.88	$650.51	$1244.40
Weighted aggregate indices computed by Formulas (15.2)	100.0	116.4	222.7

What do we do if we desire to construct index numbers which reflect only changes in price over a period of time? We must hold the market basket constant in some way. There are two ways to do this. First, we might assume that the amount of each item purchased each year by a typical family does not change from the base-year amount. Thus, we could assume that the typical family purchased 728 quarts of milk, 312 pounds of steak, 55 pounds of butter, and .3 pound of pepper in each of the three years for which we are constructing indices. Then we could use these base-year quantities to weight the prices in each of the following years. The sums of the expenditures in each of the years would then reflect changes in *price only* over the period with which we are working.

This weighting has been done in Table 15.6. The 1960 cost column in this table is just the same as the one in Table 15.5. The cost columns for 1970 and 1980, however, represent new figures. This is because they were calculated assuming that in 1970 and 1980 the typical family purchased the same amount of each item as they purchased in 1960. Thus, the total expenditure for milk in 1970 was computed to be $.31 per quart (the 1970 price) times 728 quarts (the 1960 quantity), or $225.68. Notice that pepper plays a negligible role, as it should. The 1970 pepper expenditure was down 9 cents from 1960, but this hardly affects the total picture. The indices in the last row of Table 15.6 were found by dividing the total expenditure in 1960 into each of the other years' total expenditure figures. The resulting indices, called the *Laspeyres index numbers,* measure the change in cost for a fixed buying pattern. The Laspeyres index number for the nth period is

$$L_{n:0} = 100 \times \frac{\Sigma_i p_{ni} q_{0i}}{\Sigma_i p_{0i} q_{0i}} \qquad (15.3)$$

TABLE 15.6   Price Index with Base-Year Quantity Weights
(The Laspeyres index number)

ITEM	1960 QUANTITY q_{0i}		1960 COST $p_{0i} q_{0i}$	1970 COST $p_{1i} q_{0i}$	1980 COST $p_{2i} q_{0i}$
1. Milk	728 qt.		$189.28	$225.68	$ 407.68
2. Steak	312 lb.		327.60	365.04	670.80
3. Butter	55 lb.		41.25	46.75	77.00
4. Pepper	.3 lb.		.75	.66	.78
		Total	$558.88	$638.13	$1156.26
Price indices computed by Equation (15.3)			100.0	114.2	206.9

The Laspeyres index has the algebraically equivalent form

$$L_{n:0} = 100 \times \frac{\Sigma_i p_{0i} q_{0i} (p_{ni}/p_{0i})}{\Sigma_i p_{0i} q_{0i}}$$

In other words, $L_{n:0}$ is 100 times the weighted average [see Equation (3.5)] of the price ratios p_{ni}/p_{0i} for the various items, the weights being the amounts $p_{0i} q_{0i}$ spent in the base period.

The actual CPI is computed as in our example, although it involves not 4 food items but around 400 items, including food, transportation, medical costs, and so on. The general description of a price index is this: Let the index i refer to the item; i ranges from 1 to 4 in our example and from 1 to about 400 for the real CPI. Let the index n refer to the time period; n ranges over 0, 1, and 2 in our example, but any number of times can be considered and the period could be a month, say, rather than ten years. The price of item i in period n is p_{ni}, and the quantity is q_{ni}. The amount spent on item i in the base period is $p_{0i} q_{0i}$, for a grand total of $\Sigma_i p_{0i} q_{0i}$. If in period n the quantities were q_{0i} as in the base period, the total for item i would be $p_{ni} q_{0i}$ and the grand total expended would be $\Sigma_i p_{ni} q_{0i}$.

It is not mandatory that we use the base-year quantities as weights for the prices. We could just as well have used the 1970 or the 1980 quantities. However, there is one index which uses the nth-period quantities as weights for both the base-period and the nth-period prices. This is the *Paasche index number,* defined by

$$P_{n:0} = 100 \times \frac{\Sigma_i p_{ni} q_{ni}}{\Sigma_i p_{0i} q_{ni}} = 100 \times \frac{\Sigma_i p_{0i} q_{ni} (p_{ni}/p_{0i})}{\Sigma_i p_{0i} q_{ni}} \qquad (15.4)$$

The prices in both the numerator and denominator of the Paasche index number are weighted by q_{ni}. Thus, from one period to the next, a new $P_{n:0}$ index is calculated; and new weights, new q_{ni} values, are needed for each period. The Laspeyres index is more convenient to use on a continuing basis, because the weights remain fixed from one period to the next. For this reason, the Laspeyres index is used far more often than the Paasche index for measuring price changes.

There is one final index of price change which has been proposed over the years. It is called *Fisher's ideal index number,* "ideal" referring to its several subtle mathematical properties, which will not be discussed here. The Fisher index number is found by calculating the Laspeyres and the Paasche indices first and then taking the square root of the product of these numbers. That is,

$$F_{n:0} = \sqrt{L_{n:0} \times P_{n:0}} \qquad (15.5)$$

Since Fisher's index involves the Paasche index, it shares that index's disadvantage of requiring a new set of quantity weights q_{ni} for each period. The ideal index number may have many ideal mathematical properties, but it is not so ideal from a practical point of view.

Index numbers can be used to measure more than just price changes. Sometimes it is desirable to measure physical-volume changes. Index numbers that show changes in physical volume often measure production output. A weighted aggregate method can be used to construct an index to show the change from period 0 to period n:

$$I_{n:0} = \frac{\Sigma_i p_{0i} q_{ni}}{\Sigma_i p_{0i} q_{0i}} \times 100 \qquad (15.6)$$

The quantities in Formula (15.6) are weighted by the prices in the base period.

Table 15.7 shows the calculation of a physical-volume index for the sales output of three items carried by a hardware store. The sales output in 1979 is compared to the sales output in the 1957–59 period in the index $I_{1973:0} = 204$, where in base period 0 (the 1957–59 period) the value of the index is 100. The index was calculated using Formula (15.6) and shows only the change in physical volume from the base period to 1979 since the base-period prices were used as weights for the 1979 quantities. It should be noted that the prices and quantities used for the three-year base period were the mean prices and mean quantities during those years. The value columns were found by multiplying the base-period mean prices by the corresponding quantities in the base period and in 1979. Thus, the figure of $2805.60 for the 1979 value of hammers is the revenue that would have been generated in 1979 by selling the 1979 quantity of 668 hammers at 1957–59 prices. The total value of $8957.00 is the value of the 1979 sales valued at the 1957–59 prices. The index of 100 × ($8957.00/$4390.60) = 204 shows that physical volume of sales rose 104 percent from 1957–59 to 1979.

The reader may be confused at this point since so many ways of calculating price indices are shown. He or she may ask: But what is the *true* price increase? The correct response to that question is another question: The *true* price increase of *what?* Each of the price indices discussed in this section is a *true* measure of price change. The indices merely measure changes for different quantities of goods. The simple aggregate index measures the true price change of one unit of each good. The weighted aggregate index measures the true change in the cost of living from one period to the next. But this index is not a pure price index in that it reflects both price and quantity changes. The Laspeyres index measures only price changes. It measures the same set of items purchased in the given year as were really purchased

TABLE 15.7   Weighted Aggregate Physical-Volume Index

BASE PERIOD 1957–59:

ITEM	Mean Price Per Unit p_{0i}	Mean Sales Per Year q_{0i}	1979 QUANTITIES q_{ni}	BASE-PERIOD VALUES $p_{0i}q_{0i}$	1979 VALUES $p_{0i}q_{ni}$
1. Hammers	$4.20	320	668	$1344.00	$2805.60
2. Saws	8.60	292	591	2511.20	5082.60
3. Nails	.20/lb.	2677 lbs.	5344 lbs.	535.40	1068.80
Total				$4390.60	$8957.00
				100.0	204.0

Physical-volume index computed by Formula (15.6)

in the base year. And the Paasche index number measures only price changes, but it assumes that the purchaser bought in the base year exactly what was bought in the given year, and this is seldom true. If the items in the market basket do not change, and if the quantities purchased do not change, then the weighted aggregate, Laspeyres, and Paasche indices all give the same value.

15.3   SELECTING A BASE PERIOD

Index number series are used primarily for comparative purposes. When the Consumer Price Index hit 120 in 1971, people concluded that prices had risen 20 percent compared to the base-period prices for the CPI, which are the mean prices for 1967. But, one might wonder, why is the 1967 period used as a base period? Why not 1966 or 1970–72? There are several criteria which should be considered when a base period is chosen for an index number series.

The first criterion is the time criterion: The base period should be fairly recent. Part of the justification for this criterion rests on the reasoning that an index number should help people quantitatively compare present conditions to past conditions. If the comparison is to be meaningful, the past (base period) should be recent enough that the person making the comparison can remember what conditions were like. It does not mean much to the average person if he is told that prices now are 2000 percent above what they were in the Middle Ages; since he has never experienced a medieval standard of living, that figure is not meaningful. However, if he is told that prices are now 8 percent above what they were last year, he can remember last year and knows that he has not changed his buying habits much since that time. The comparison tells him quite plainly that his standard of living this year is costing him 8 percent more than it did last year.

A second criterion for selecting a base period is that it should in some sense be a period of normal activity for the series whose index is sought. If one were to construct an index of imports of handguns for private use into the United States, one would not likely use the years 1967–68 for the base period. Figure 14.4 shows why this is true. The civil unrest the country experienced during these two years was cause for extremely high imports of handguns. Thus, the figures in 1967 and 1968 could hardly be considered "normal." One way to arrive at a normal base period is to agglomerate several periods representing different stages in a business cycle. Many government indices were formerly based on the period 1957–59. The base-year values for these indices were found by averaging the figures in the three years from 1957 to 1959, a period that included the business boom of 1957, the recession of 1958, and the economic recovery of 1959. The single year 1967 is now the base period for most general-purpose indices prepared by governmental agencies. This base, adopted in March of 1970, was chosen in part because 1967 was the last year of a period of relatively stable prices (1959–1967) prior to the price surges that occurred in 1968 and accelerated in the 1970s.

A third criterion in selecting a base period is that of comparability. It is often necessary to compare one index to another. In order for comparisons to be valid, the indices should have the same base period. One company might find that an index of its raw materials costs is 105. Another might find that its index of costs is 145. It would not be safe to conclude that costs have risen higher for the second company, however, without knowing the base from which the measurements were made. If the first company were using a 1977–78 base period and the other were using a 1957–59 base period, then the indices are not comparable. One says that costs have risen 5 percent from the 1977–78 period while the other says that costs have risen 45 percent from the 1957–59 period. Most governmental agencies that publish indices cooperate and use the same base period, currently 1967.

A final criterion for selecting a base period is that of the availability of data. The base period should be a period for which accurate and complete data are available. Thus, base periods back in the seventeenth century are likely to be unsatisfactory not only for having little relevance to us because they are not recent, but also for lacking accurate data with which to calculate the base values of the index. It might be best to use census years as base years, since for these years complete data, rather than sample data, are available.

15.4 THE CONSUMER PRICE INDEX

The United States government publishes hundreds of indices, but the most important are the Consumer Price Indices, the CPI's. The CPI was mentioned in Section 15.1 as though there were only one Consumer Price Index. Actually, there are currently three being published on a regular basis. They will be discussed in detail in this section due to their importance and impact on our economy.

The first Consumer Price Index was published in 1919 by the Bureau of Labor Statistics. Its purpose was to help set new inflation-adjusted wage levels for workers in shipbuilding yards. Since that time, the CPI has become this country's principal measure of price change and is used by many government agencies and all segments of the economy to measure inflation rates. To accommodate the changing use of the CPI, it has been revised several times: in 1940, 1946, 1953,

1964, and 1978. Currently, three indices are published. The first is the Consumer Price Index for Urban Wage Earners and Clerical Workers. This index is the old version of the CPI which grew out of the revisions from 1919 through 1964. In the following discussions it will be referred to as the "old CPI" for convenience. As its name implies, the old CPI is intended for measuring changes in the prices of items purchased by urban wage earners and clerical workers. However, this is a rather limited segment of the work force. It accounts for only about 35 to 40 percent of the nation's population.

For this reason, in 1970 the Bureau of Labor Statistics determined that it would develop a new consumer price index called the Consumer Price Index for All Urban Consumers. In the following discussion, this index will be referred to as the CPI-U, where the "U" indicates that it applies to all urban consumers. This new index covers a much larger group in the population—approximately 80 percent of all people in the country. It was first published in mid-1978.

The third index also was published for the first time in 1978. It is called the Revised Consumer Price Index for Urban Wage Earners and Clerical Workers. It is often referred to as the CPI-W, where the "W" stands for workers. This index is very similar to the old CPI, but it uses different quantity weights. These three indices are compared in the following discussion.

The old CPI is published monthly by the U.S. Bureau of Labor Statistics and appears in the *Monthly Labor Review*. The list of items priced consists of over 400 goods and services purchased by families of urban wage earners and clerical workers and by single persons living alone. The price changes from the base period of 1967 are weighted by the quantities in the "market basket" of over 400 goods purchased by these consumers in the 1960–61 period. Separate indices are published for different categories of the market basket. There are indices for food, housing, transportation, medical care costs, and many others. Within each of these major categories, there are indices for minor items. For instance, within the major category of food there are indices for cereals and bakery products, meat, and dairy products.

The price data for the old CPI are gathered in 56 urban areas. The *Monthly Labor Review* publishes CPI figures for 24 large cities in the United States, but indices for all 56 separate areas are not published.

The CPI-U differs from the old CPI in several ways. First, the quantity weights used were obtained from a 1972–73 survey of consumer purchasing patterns. Thus, the CPI-U has more up-to-date quantity weights. Several items not priced in the old CPI are included in the product and service list for the new CPI-U. Some of the new items include: calculators, crockpots, joggers' warm-up suits, women's tennis outfits, and auto rentals. The new CPI-U has different weights assigned to the major purchase categories. These new weights reflect the change in the amounts of items purchased by U.S. consumers between the 1960–61 and 1972–73 periods. Table 15.8 shows the relative weights assigned to the major spending categories for the old CPI and the CPI-U.

The new CPI-U is constructed from data sampled in 85 urban areas rather than the 56 used by the old CPI. Also, the CPI-U uses a new method of pricing items. Under the old CPI computation techniques, about 400 very specific items are priced each month. Under the CPI-U technique, the items priced are somewhat more broadly defined. For instance, an item priced from the old CPI list might be "fees for piano and organ lessons," but in the CPI-U list this item would be part of a larger group called "fees for lessons—piano, swimming, golf, tennis, etc."

Both the old CPI and the CPI-U have a base year of 1967 for prices. However, the current data for the CPI-U will be gathered at many more points of purchase.

The CPI-W was introduced at the same time as the CPI-U. However, like the old CPI it covers only urban wage earners and clerical workers—about 35 to 40 percent of the population. It differs from the other two indices in the following ways. First, its quantity weights were taken from the 1972–73 spending pattern survey—but only the weights for urban wage earners and clerical workers were used to obtain these weights. The weights in the CPI-U came from *all* urban consumer purchases. The weights in the old CPI came from the 1960–61 survey of wage earners and clerical workers. In all other respects, however, the CPI-W is like the CPI-U. Table 15.9 summarizes the similarities and differences in the three indices.

TABLE 15.8   Relative Weights Used in Major
Price Index Categories

| | *WEIGHTS:* | |
CATEGORY	*Old CPI*	*CPI-U*
Housing	35.5	43.9
Food and Beverages	26.2	18.8
Transportation	13.3	18.0
Apparel	9.0	5.8
Medical Care	6.9	5.0
Entertainment	3.7	4.1
Other	5.4	4.4
Total	100.0	100.0

TABLE 15.9   Three Consumer Price Indices Compared

CATEGORY	OLD CPI	CPI-U	CPI-W
Number of Urban Areas Sampled	56	85	85
Base Year	1967	1967	1967
Percent of Population Covered	35–40	80	35–40
Weights Taken from Consumer Expenditure Survey Done in	1960–61	1972–73	1972–73
Published in	All three in *Monthly Labor Review*		
Price Gathering Means	All three use personal visits, telephone calls, and occasional mail surveys.		

The fact that there are now three consumer price indices complicates matters for the user. Now one must be more careful than ever in selecting and applying an index. These three indicators measure three different price changes.

1. The old CPI measures the changes in prices from 1967 of goods and services that were typically purchased in the 1960–61 period by urban wage earners and clerical workers only.

2. The CPI-U measures the changes in prices from 1967 of goods and services that were typically purchased in the 1972–73 period by *all* urban consumers.

3. The CPI-W measures the changes in prices from 1967 of goods and services that were typically purchased in the 1972–73 period by urban wage earners and clerical workers only.

Originally, the Bureau of Labor Statistics intended to drop publication of the old CPI when the two new indices were introduced. This action was resisted by many national labor leaders, who convinced the Bureau to retain the old CPI—at least until some experience with the new indices has been gained.

The reason for the labor movement's concern is that over 50 percent of the United States population have a portion of their income tied to the old CPI in some manner. It may take several years before contracts and government programs in which payments depend on the old CPI can be rewritten to include use of the new indices for measuring inflation.

It cost the U.S. government more than $50 million to produce the 1978 revised indices. One might question why so much money would be spent on this work. The reason is that the Bureau of Labor Statistics is concerned that it produce price indices that accurately reflect inflation rates. The money seems well worth it if the new indices prove over the years that they do indeed better reflect true price changes. Under the old CPI, an error of one-tenth of 1 percent in calculation would result in the misallocation of over $100 million in annual wages and payments. Thus, it appears that the cost of inaccuracy in measuring consumer prices is even higher than the considerable cost of accuracy.

One of the limitations of any price index is that it does not reflect changes in the quality of a product. For instance, a standard automobile with no options might have cost $2500 in 1960; the same model of car with no options might have cost $8000 in 1978. It would be fallacious to conclude that the price index for this model of automobile had gone up to 320 based on the 1960 price at 100. The problem is that the 1978 automobile was not the same car as the 1960 version. The later model would have as standard features safety equipment, pollution control equipment, and luxury items not included in the 1960 model. These differences in quality affect the price of the automobile so that the $5500 price rise cannot be attributed solely to inflation. In order to eliminate as far as possible the differences in quality of products purchased from period to period, the items priced for the CPI are specified in enough detail to ensure that a similar item is priced for each period.

Another important general-purpose index is the Wholesale Price Index, WPI. Dating back to 1890, this index is the oldest continuous time series published by the U.S. Bureau of Labor Statistics. The WPI measures changes in prices of 2450 goods and commodities sold in the primary markets of the United States. Like the CPI, it is published monthly in the *Monthly Labor Review,* where it is broken down into numerous categories. The Wholesale Price Index uses 1967 prices for the base year, and the weights used are the net shipments of goods derived from the industrial censuses of 1963. This index is watched very carefully by businessmen and economists, since rises in wholesale prices usually precede by a month or two rises in retail prices.

15.5   USES OF INDICES

One of the most common uses of general-purpose price indices is the adjusting of dollar values, often called "deflating" or reducing to "constant dollars." Suppose that a manufacturer sold $1,400,000 worth of goods in 1967 and $2,600,000 worth of goods in 1979. His sales volume has increased to $100 \times (2,600,000/1,400,000)$, or 186 percent of the 1967 figure. However, part of that increase is due to inflation. A quick way to adjust for the effect of inflation would be to reduce the 1979 sales figure to 1967 dollars. This could be done with the

1979 WPI, since the company involved is a manufacturer. The calculation is as follows:

YEAR	WPI	ACTUAL SALES (IN MILLIONS)	CONSTANT-DOLLAR SALES (IN MILLIONS)
1967	100.0	$1.4	$1.4/1.000 = $1.40
1979	173.9	$2.6	$2.6/1.739 = $1.50

The 1979 sales would have amounted to only about $1,500,000 at the 1967 prices. In this example the composite WPI was used, but in an actual example the manufacturer could use the subindex within the WPI group which pertained to his industry. If he had his own detailed price and sales-quantity data convenient, he could compute a constant-dollar sales figure more accurately using Formula (15.6) in Section 15.2. The deflating method demonstrated in our example results in only approximate constant-dollar figures, since the items whose prices are being deflated might not have experienced the average price changes reflected in the CPIs or the WPI.

The relevant Consumer Price Index is often used to deflate wage and income figures. Suppose that a typical urban wage earner had a salary of $6780 per year in 1967. By 1972, the worker was earning $8200 per year, but was unhappy. To see the reason for the worker's dissatisfaction we must express the new wage in terms of its purchasing power:

YEAR	CPI	WAGE	CONSTANT-DOLLAR WAGE
1967	100	$6780	$6780/1.00 = $6780
1972	126	$8200	$8200/1.26 = $6508

In constant-dollar terms, our wage earner could purchase only $6508 worth of 1967 goods with the 1972 salary of $8200. That is, the wage earner had actually lost ground in terms of salary buying power.

The reciprocal of a CPI is used to express the "purchasing power" of the dollar. In the preceding example, a 1972 dollar would purchase only $1 × (100/126), or $.79 worth, of 1967 goods. Another way of stating this is to say that $1 in 1972 purchased only 79 percent of what it would purchase in 1967.

The most well-known economic series in the United States is the Gross National Product, GNP. It is usually deflated by an appropriate price index and expressed in terms of constant dollars. This means that changes in GNP reflect changes in national economic output rather than just price. The changes in constant-dollar GNP are called "real"

GNP changes to distinguish them from changes caused by both price and output changes.

The Dow-Jones Industrial Average of stock prices is an index number of sorts. It is better called an "indicator" than an index number. It does not measure changes from some base period, but is supposed to be the mean price of 30 specific industrial stocks. However, its value is not found by adding the prices of these stocks together and dividing by 30. This is because over a period of time the 30 stocks in the industrial average have produced dividends and split. Also, there have been mergers of one company's stock with another; and periodically some stocks are dropped from the list of the 30 industrials and new ones are added. The last change in the makeup of the list of 30 industrials was made in June of 1979, the first such change in 20 years. In order to account for these changes, adjustments have been made in the denominator used with this average.

Suppose, for instance, that we desire to construct a price indicator like the Dow-Jones Industrial Average, but we include only three stocks. On the first day that we publish our indicator, the prices of the stocks are $15, $20, and $25 per share. Then our mean price would be

$$\frac{\$15 + \$20 + \$25}{3} = \$20$$

On the second day the second stock splits two for one. The price on the second stock should fall to somewhere in the neighborhood of $10 per share. If we ignored the split, we would find that with no real change in stock prices, our indicator would change. It would now be

$$\frac{\$15 + \$10 + \$25}{3} = \$16.67$$

This would not be proper since the price indicator should not change if the prices do not change. The drop in the average was due only to a stock split. Thus, the denominator of our average must be adjusted to a new value D such that at the same prices for the stocks ($15, $10 on the split stock, and $25) we still have an indicator average of $20. Thus, we seek a D value such that

$$\frac{\$15 + \$10 + \$25}{D} = \$20$$

The D value that satisfies this equation is 2.5. Now if the split stock's price rises to $11 per share after the split while the other prices remain unchanged, the new value of the indicator is

$$\frac{\$15 + \$11 + \$25}{2.5} = \$20.40$$

Over the years the many stock splits, stock dividends, and mergers that have occurred among the Dow-Jones 30 industrials have caused the denominator to drop far below the original value of 30. The value is now less than 1.5, and it drops further each year.

The Dow-Jones Industrial Average is as much a psychological indicator as a true index of general stock price movement on the New York Stock Exchange. The 30 stocks making up this average are far from typical or average stocks on the exchange. Sometimes their price movements differ considerably from the movements in the general price level of stocks. When the Dow-Jones Industrial Average closed over the 1000 mark for the first time in the fall of 1972, many financial analysts correctly suggested that the stock market in general was not setting any new records. It was only the 30 industrials and a relatively few other stocks that were reaching new highs; most of the rest of the stocks were still not experiencing significant price rises. This lack of representativeness is one reason that the New York Stock Exchange Index was developed in the late 1960s as an average price of *all* stocks on the New York Stock Exchange.

TABLE 15.10   Summary of Formulas

NAME	FORMULA	PRICE CHANGE FROM PERIOD 0 TO PERIOD n FOR
Simple Aggregate	$100 \times \dfrac{\Sigma_i P_{ni}}{\Sigma_i P_{0i}}$	One unit of each item in the market basket.
Weighted Aggregate	$100 \times \dfrac{\Sigma_i p_{ni} q_{ni}}{\Sigma_i p_{0i} q_{0i}}$	Total amount spent on all items in the market basket.
Laspeyres	$100 \times \dfrac{\Sigma_i p_{ni} q_{0i}}{\Sigma_i p_{0i} q_{0i}}$	The market basket of goods purchased in the base period.
Paasche	$100 \times \dfrac{\Sigma_i p_{ni} q_{ni}}{\Sigma_i p_{0i} q_{ni}}$	The market basket of goods purchased in period n.
Fisher's Ideal Index	$\sqrt{L_{n:0} \times P_{n:0}}$	A mixed market basket between the periods.

15.6   SUMMARY

Several price index formulas have been presented in this chapter. Each index measures price changes in a slightly different way. Table 15.10 summarizes the formulas and what they measure.

PROBLEMS

1. For the accompanying retail sales volumes (sales in thousands of dollars) find the sales indices:

YEAR	1970	1972	1974	1976	1978
SALES	$64,978	$65,810	$68,352	$73,409	$78,992

   a. With 1970 as the base year.
   b. With 1978 as the base year.

2. An alloy is made up of 25% metal X, 35% metal Y, and 40% metal Z. During the last year, prices of the metals have changed:

	PRICES PER POUND:	
	Last Year	This Year
Metal X	$1.19	$2.50
Metal Y	6.72	6.89
Metal Z	3.50	2.90

   a. Find the simple aggregate index for the cost of the alloy.
   b. Find the weighted aggregate index for the cost of the alloy.
   c. How can you explain the values of these index numbers in view of the fact that only two out of the three constituent metal prices rose?

3. Find the simple aggregate index number of the following data:

ITEM	1967:		1980:	
	Unit Price	Units Sold	Unit Price	Units Sold
A	$ 1.50	350	$ 3.00	600
B	15.00	100	18.50	125
C	8.50	200	15.00	300

4. Use the data in Problem 3 and find the weighted aggregate index number.
5. Use the data in Problem 3 and find the Laspeyres index number.
6. Use the data in Problem 3 and find the Paasche index number.
7. Use the data in Problem 3 and find Fisher's ideal index number.
8. A company buys three major units from subcontractors. The company then assembles the three units and markets the total product. The prices

charged by the subcontractors have changed over the past few years. The price and quantity data are as follows.

| | FIVE YEARS AGO: | | THIS YEAR: | |
ITEM	Unit Price	Units Ordered	Unit Price	Units Ordered
A	$15.00	200	$20.00	250
B	8.00	150	9.00	100
C	12.00	300	16.00	500

a. Find a price index reflecting the change in price of buying one of each item. What is the name of this index?

b. Find and name the index reflecting the change in price of the components bought at the level of this year's purchase amounts.

c. Find and name the price index reflecting the change in the total amount of money spent for subcontracted parts over the last 5 years.

d. Find and name the price index reflecting the changes in price of the components bought at the level of usage experienced 5 years ago.

9. A coal mining company is currently enjoying the boom in both price and demand for its product. It produces two types of coal: anthracite and bituminous. The following figures show the price per ton and demand (in thousands of tons) in three different years.

| | 1970: | | 1975: | | 1980: | |
TYPE	Price	Quantity	Price	Quantity	Price	Quantity
Anthracite	$15	500	$23	1000	$42	2000
Bituminous	$12	1000	$18	2000	$34	3000

a. Find the simple aggregate index between 1970 and 1975.

b. Find the weighted aggregate index between 1970 and 1980.

10. Use the data in Problem 9 to find the Laspeyres index number between:
a. 1970 and 1975.
b. 1970 and 1980.

11. Use the data in Problem 9 to find the Paasche index number between:
a. 1970 and 1975.
b. 1970 and 1980.

12. For the following data, find the Laspeyres retail price index for each year, using 1967 as the base.

Retail Price of Selected Electrical Appliances

| | AVERAGE UNIT PRICE: | | | THOUSANDS OF |
APPLIANCE	1967	1970	1978	UNITS SOLD, 1967
A	$295	$305	$415	6530
B	334	340	390	2050
C	250	261	360	1500
D	43	44	62	950

13. Which of the following index numbers could be found using the data in Problem **12**?
    a. Simple aggregate index.
    b. Weighted aggregate index.
    c. Laspeyres index.
    d. Paasche index.
    e. Fisher's ideal index.
    f. Physical-volume index.

14. Using the data in Problem **3**, find a physical-volume index for 1980 weighting quantities with 1967 base-year prices.

15. Using the data in Problem **8**, find the physical-volume index for this year, weighting quantities by prices 5 years ago.

16. Using the data in Problem **9**, find the physical-volume index for 1980. Use 1970 prices as weights and 1970 as the base year.

17. Using the data in Problem **9**, find the physical-volume index for 1975. Use 1970 prices as weights and 1970 as the base year.

18. In 1963 the CPI was 106.7; in 1964 it was 108.1. If a worker in 1963 earned $2.54 per hour and in 1964 earned $2.60 per hour, did her purchasing power increase or decrease? By how much?

19. If you earned $9842 in 1970 when the CPI in your city was 110 and you earned $10,992 in 1974 when your city's CPI hit 130, express your 1974 purchasing power as a percentage of your 1970 purchasing power.

20. Assume the CPI for your city had the following values:

YEAR	1970	1972	1974	1976	1978
VALUE	110	128	148	164	182

    a. Find the purchasing power of the dollar in each of these years as a proportion of the 1967 dollar.
    b. Explain the meaning of these figures. (*Hint:* What percentage of 1967 goods could you buy in these years?)

21. A company paid out wages of $10,468,500 in 1967 and $17,450,000 for the same number of workers in fiscal 1974, when the CPI was 158.
    a. Did the workers receive a "real" wage increase?
    b. What was the percentage change in the real wage?
    c. What were 1974's wages in 1967 constant dollars?
    d. What were 1967's wages in 1974 constant dollars?

22. Why would an index series of transportation costs based in the year 1840 be useless today?

23. Go to the latest issue of the *Monthly Labor Review* and find the most current price index for:
    a. All consumer goods and services, CPI-U.
    b. Ice cream, old CPI.
    c. Transportation, CPI-W.
    d. Housing, old CPI.

24. Give examples of two goods or commodities for which:
    a. Quality has changed significantly in the past 25 years.
    b. Quality has remained unchanged for the past 25 years.

25. Suppose you have a stock market indicator made up of five common stocks selling at the prices per share indicated in the accompanying table.

STOCK	PRICE PER SHARE
1	$101.00
2	82.75
3	44.50
4	27.75
5	18.50

   a. Find the market average.

   b. Suppose that stock 1 split two for one and stock 4 split three for one. Find the new denominator D for your average.

   c. After the splits, the prices settled to the values shown in the following table. Find the indicator's new value.

STOCK	PRICE PER SHARE
1	$51.00
2	88.75
3	48.50
4	9.50
5	18.25

26. The price indicator in Problem **25** used weights of 1.0 for each stock when the indicator was first constructed. What other weighting factors might have been used?

27. On Wednesday afternoon, Professor Ralph R. Bradshaw received a long distance telephone call. The caller identified himself as the personnel manager of a large San Francisco company.

> "I understand you're a good statistician, and I need help on a statistical type problem," the caller began. Flattered, Professor Bradshaw offered to do what he could.
>
> "We're transferring an employee from San Francisco to your city, next month, and we need to adjust salary for the cost of living difference between the two locations. I can't find a cost of living index for your city. Are there any local organizations in your area that have an index I might use? You know, the Chamber of Commerce, state industrial promotion board, or utility companies?"
>
> Professor Bradshaw responded that some indices were available but that they would be useless for the purpose the personnel manager proposed. "What do you mean—useless?" asked the caller.

Discuss what Professor Bradshaw might tell his caller.

SAMPLE DATA SET QUESTIONS: *Refer to the 113 applicants for credit listed in the Sample Data Set Appendix of this book.*

a. Is the cost of consumer credit (interest on consumer credit loans) part of the Consumer Price Index?
b. As the Consumer Price Index rises, what should be the impact of this rise on the credit granting policies and procedures of the department store's credit office?

REFERENCES

15.1 *The Consumer Price Index: Concepts and Content Over the Years.* U.S. Department of Labor, Bureau of Labor Statistics, Report 517, 1977.

15.2 *The Consumer Price Index Revision—1978.* U.S. Department of Labor, Bureau of Labor Statistics, 1978.

15.3 Lapin, Lawrence L. *Statistics for Modern Business Decisions.* Harcourt, Brace, Jovanovich, New York, 1973. Chapter 16.

15.4 McElroy, Elam E. *Applied Business Statistics.* Holden-Day, San Francisco, 1971. Chapter 10.

16

DECISION
THEORY

In previous chapters, we have dealt with subjects that involve *decision
making*. For instance, we have discussed the subject of probability with
the aim of *deciding* what the chances are for the occurrence of certain
possible events (such as drawing an ace from a deck of cards). We
have studied the subject of statistical estimation with the aim of *deciding*
what the mean of a population might be (such as the mean daily sales
of a product), given only sample information from which to work.
In both of these cases, and in most of the others we have discussed,
the decisions involved have been only preliminary in nature. That is,
they are decisions that are usually made prior to taking some decisive
action (like betting in a card game or marketing a new product).

Before a decision maker takes a decisive action, he usually considers what the costs or payoffs may be. That is, he considers the economic consequences of his actions. Up until this point we have ignored economic consequences even in hypothesis testing and other types of problems where decisive action was implied. In earlier chapters we have suggested that one might want to reject an hypothesis when sample evidence and the hypothesis are very inconsistent with one another. In dealing with practical decision making, we must learn to use a somewhat different standard. If falsely rejecting an hypothesis is very costly to a decision maker, he might require a great deal of evidence that the hypothesis is false before he rejects it. But if the same error is not significant in terms of its economic consequences, the decision maker might be willing to reject the hypothesis on the basis of very little contradictory evidence. In previous chapters we have dealt only with the probabilities of being wrong in testing hypotheses (the α and β values). In this chapter we will introduce the economic consequences of taking various actions.

The hypothesis-testing problem is only a special case in a general class of problems where a decision maker must choose between two or more courses of action. Examples of the types of choices that have to be made by decision makers every day are:

Accept or reject an hypothesis
Market or not market a new product
Buy machine model A, model B, or model C
Build a new plant in England, Germany, or the United States
Invest in stocks, bonds, or mutual funds
Open a branch office or not open one
Sue a competitor or try out-of-court settlement
Use high-, low-, or moderate-quality materials

The list is endless. But all these decision problems have characteristics in common. The best course of action in each case depends on the economic consequences of the alternative actions *and* the probabilities that these economic consequences will be realized. The objective of this chapter is to present ways in which the economic consequences and the probabilities of realizing them can be combined to determine which of several courses of action is "best" for the decision maker.

16.1   THE PAYOFF TABLE

The first step in determining which is the best of several actions is to define exactly what alternatives the decision maker has and what the economic consequences of choosing these alternatives may be. A **payoff table** can help present this information in a clear way.

A payoff table is a rectangular array of numbers. Usually, it is drawn so that the columns of the array represent each of the alternative actions A_j from which the decision maker must choose. The rows of the array represent events E_i that might take place, or current "states of nature" that will influence the consequences of the decision maker's actions. The numbers in the array at any particular row or column, X_{ij}, represent the economic consequences which will accrue to the decision maker if he takes the action represented by that column, j, and the event represented by that row, i, occurs.

EXAMPLE 1

Consider the following betting game. The player tosses a coin three times or until a head appears, whichever comes first. If a head appears on the first toss, the player receives $1; if the first head appears on the second toss the player receives $2; if it appears on the third toss, the player receives $4; and if no head appears in three tosses, the player must pay $8. Someone presented with the opportunity of playing this game must decide whether or not to play. The payoff table faced by such a person is presented in Table 16.1.

This payoff table has two columns since the person making the decision has two choices—to play or not to play. The table has four rows, each representing an event or state of nature which might take place. The payoffs in the table represent what the decision maker would receive under the various combinations of decisions and events. The negative value indicates that there is to be a pay *out*, not a payoff.

TABLE 16.1   Payoff Table for Betting Game

EVENTS \ ACTIONS	A_1 PLAY	A_2 NOT PLAY
E_1 Head on first toss	1	0
E_2 First head on second toss	2	0
E_3 First head on third toss	4	0
E_4 No head appears	−8	0

EXAMPLE 2

Consider the decision faced by a person who owns a small business. She has been offered $80,000 for this business if she sells this month. However, she is tempted to hold onto the business for a year since the business may be awarded a contract that, according to her best calculations, will yield $30,000 in profits over its one-year life. One year of operation without the contract will not yield any profit since she is operating just at the break-even point. With this in mind, the owner approached the potential buyer of her business and asked if the buyer would consider waiting a year. The purchaser responded positively, but added that the purchase price would be $70,000 in that case and would have to be agreed upon now. The owner faces the payoff situation shown in Table 16.2.

The owner must decide between selling now or holding her business for one year. If she sells now, she will receive $80,000 for sure. Her payoff one year from now is $70,000 plus whatever is realized from the contract bid. (Actually, the figures for payoffs one year away should be discounted.)

TABLE 16.2 Payoff Table for Business Sale
(Figures in thousands of dollars)

EVENTS \ ACTIONS	A_1 SELL NOW	A_2 HOLD A YEAR
E_1 Contract awarded	80	100
E_2 Contract not awarded	80	70

Payoff tables can be used as well in problems where all the consequences are costs. Rather than indicating all the costs as negative payoffs, users of such tables simply enter the costs as positive figures and then keep in mind the fact that the values are costs, not profits. Payoff tables can also be used in problems where the results of an action are not expressed in terms of monetary values. For instance, a payoff table could be constructed showing the production output associated with several different plant layouts (different actions) under various possible demand situations (different events).

The payoff tables discussed in the examples of this section are helpful in laying out a statement of the problem at hand. However, the payoff table by itself does not *usually* indicate which course of action should be taken. It is only under very special circumstances that one can make a decision from the payoff table alone. Consider the payoff problem formulated in Table 16.3. It is apparent from this table that action 3 is the preferred action since its consequences are better than those for the other two actions, *regardless of what event*

TABLE 16.3 Payoff Table with a Dominant Action

EVENTS \ ACTIONS	ACTION 1	ACTION 2	ACTION 3
Event 1	−5	2	8
Event 2	7	−3	9
Event 3	4	6	7

occurs. Action 3 is said to *dominate* the other actions. One action dominates another when its payoffs are equal to or better than the other's regardless of what event occurs. In Examples 1 and 2 of this section, there are no dominant actions. The decision makers must rely on other information to help them make their decisions.

In the examples that have been presented so far in this section the payoffs that appear in the cells of the payoff tables have been given in the problem statements. However, it is often necessary for the decision maker to calculate these payoffs using information about the decision-making situation. The following example presents such a situation.

EXAMPLE 3

Bill Higgins has an opportunity to buy up to 3 carloads of cement (which is used as an ingredient in concrete). The Monrock Company is willing to buy some of the cement from him, but they haven't told him how many carloads they will take. He must make his decision today. Any cement not bought by Monrock can be sold to a broker. Construct Mr. Higgins' payoff table using the following facts:

Cost to Higgins	$65/ton
Tons per carload	120
Price to Monrock	$80/ton
Price to broker if Monrock doesn't buy	$60/ton

Since Higgins must buy in carload lots, his possible actions are: buy 0, buy 1, buy 2, or buy 3 carloads. Monrock may order any number of carloads, but Higgins would be concerned only with their orders up to 3 since this is the maximum number of carloads he could sell them. Thus, the events are: order 0, order 1, order 2, or order 3 or more carloads.

Each cell value in the payoff table can be found by computing total revenue less total cost:

$$\text{Profit} = \text{Total Revenue} - \text{Total Cost}$$

$$= (\text{Revenue from Monrock orders}$$

$$+ \text{Revenue from broker})$$

$$- (\text{Cost of cement ordered})$$

Since all the events and actions are expressed in terms of carloads, the costs and profits must also be so expressed. Thus, the cost of one carload is ($65/ton)(120 tons/carload) = $7800/carload. The revenue from Monrock on a carload is ($80/ton)(120 tons/carload) = $9600/carload. The revenue from the broker for a purchased carload that Monrock does not order is ($60/ton)(120 ton/carload) = $7200/carload. The payoff values for the "order 2" event row could be found as follows:

	ASSUMING MONROCK ORDERS 2 CARLOADS:			
	Buy 0	Buy 1	Buy 2	Buy 3
Revenue from Monrock	0	9600	19200	19200
Revenue from broker	0	0	0	7200
Cost of cement	0	(7800)	(15600)	(23400)
Payoff	0	1800	3600	3000

The payoffs from each row can be computed in the same fashion. The entire table follows, and the reader can verify the values in each cell.

ACTIONS / EVENTS		HIGGINS' ACTIONS:			
		Buy 0	Buy 1	Buy 2	Buy 3
Monrock's Orders	Order 0	0	−600	−1200	−1800
	Order 1	0	1800	1200	600
	Order 2	0	1800	3600	3000
	Order 3+	0	1800	3600	5400

This table has no dominant actions, and it is not obvious how many carloads Mr. Higgins should buy. He needs more information before he can determine his best action. The type of information he needs is discussed in the next section.

16.2   PRIOR PROBABILITIES

A decision maker faced with a payoff table would have no problem choosing an action if he knew what event was going to occur. He would simply examine the row associated with that event and choose the action giving the best payoff in that row. For instance, if the business owner in Example 2 of the preceding section *knew* the contract was not going to be awarded, she would sell her business now. But she does not know what event is going to take place. Thus, she is trying to make her decison under uncertainty. It is this uncertainty that makes the decision-making process difficult.

A decision maker can more sharply state his feelings about the uncertainty he faces if he assigns probabilities to each of the events in his payoff table. For the betting game in Example 1 of the previous section, this is a relatively easy matter involving the probability theory of Chapter 4. The events of Example 1 were, again:

$$E_1 = \text{Head on the first toss}$$

$$E_2 = \text{First head on the second toss}$$

$$E_3 = \text{First head on the third toss}$$

$$E_4 = \text{No head appears}$$

The probability of a head on the first toss of the coin is $P(E_1) = \frac{1}{2}$. If the first head (H) is to appear on the second toss, the first toss must be a tail (T). Using the notation of Chapter 4, $P(E_2) = P(TH)$. Since tosses of a coin are independent, we can multiply the probabilities of a tail and a head together, and

$$P(E_2) = P(TH) = \frac{1}{2} \cdot \frac{1}{2} = \frac{1}{4} = .250$$

In a similar manner

$$P(E_3) = P(TTH) = \frac{1}{2} \cdot \frac{1}{2} \cdot \frac{1}{2} = \frac{1}{8} = .125$$

and

$$P(E_4) = P(TTT) = \frac{1}{2} \cdot \frac{1}{2} \cdot \frac{1}{2} = \frac{1}{8} = .125$$

Now a probability column can be added to the payoff table of Example 1. The new table is presented as Table 16.4.

The task of adding probabilities to the payoff table of Example 2 in the last section is not so simple. The events of Example 2 were, again:

$$E_1 = \text{Contract awarded}$$

$$E_2 = \text{Contract not awarded}$$

In the coin-tossing game, we could calculate the *exact* probabilities of the events from our knowledge of the physical structure of a coin and our knowledge of probabilities. In this new situation we have little to go on except the business owner's subjective feelings about the relative

---

TABLE 16.4   Payoff Table 16.1 with Probabilities

EVENTS *ACTIONS*	PROBABILITIES	A_1 PLAY	A_2 NOT PLAY
E_1 Head on first toss	.500	1	0
E_2 First head on second toss	.250	2	0
E_3 First head on third toss	.125	4	0
E_4 No head appears	.125	−8	0

likelihood of the two events. These feelings would be based on her knowledge of the contract and the other bidders, and her experience in bidding on similar, if not identical, contracts. It would help her in evaluating the two possible actions—sell now versus hold the business for one year—if she could make an explicit statement about the subjective probabilities of the two events. Suppose that after some consideration the business owner states that there is a .20 chance that she will win the contract (and thus an .80 chance that she will not). The payoff table then becomes Table 16.5.

Subjective probability estimates were introduced in Section 4.7. There are some statistical experts who claim that subjective probabilities like those just discussed are meaningless. They feel that subjective probabilities are only guesses and thus very likely to be in error. It is true that people's guesses about the probabilities of various events can be very far from correct. Yet if the person making the guesses is close to the problem and has had a great deal of experience in similar situations, guesses may be quite good even though they may not be exact or correct in all cases. Furthermore, few decision makers can really avoid being influenced in their choice of actions by their subjective feelings concerning the likelihood of each event. If this is true, argue those who favor using subjective probabilities in decision making, isn't it best if the decision maker makes an explicit statement of these feelings in the form of subjective probabilities? He is going to be influenced by his subjective probabilities in any case, so we might as well get them out in the open where we can see them and use them. The material presented in the remainder of this book assumes that subjective probabilities are legitimate and can be entered into the probability columns of payoff tables.

Sometimes it is possible to reevaluate the probabilities entered into a payoff table through some form of experimentation or sampling; the reevaluation process will be discussed in Chapter 17. The probabilities entered into a payoff table before any such experimentation takes place are called **prior probabilities**. They are the probabilities of the various events *prior* to any experimentation or sampling aimed at better determining what event will occur.

TABLE 16.5 Payoff Table 16.2 with Probabilities (Figures in thousands of dollars)

ACTIONS EVENTS	PROBABILITIES	A_1 SELL NOW	A_2 HOLD A YEAR
E_1 Contract awarded	.20	80	100
E_2 Contract not awarded	.80	80	70

16.3   EXPECTED MONETARY VALUE

Entering the probabilities of each event into the payoff table helps to sharpen the statement of the problem facing the decision maker. But it still does not indicate directly which action should be taken. This is because the appropriateness of an action in any problem depends on the combination of the possible payoffs together with the probabilities of realizing those payoffs. Thus, we need a method of combining an action's payoffs with the events' probabilities. This method was presented in Section 5.6, where the expected value of random variables was discussed.

If X is a random variable, the expected value of X, denoted by $E(X)$, is the weighted mean of the possible values X might take. The weights are the probabilities associated with these values. Equation (5.1) is repeated here

$$E(X) = \sum_{r} rP(X = r) \qquad \text{(16.1)}$$

where the sum is over r, an index that, during summation, takes on all the values X can assume.

If the decision maker chooses a particular action A_j, the payoff is a random variable which may assume any of the values X listed in the jth column of the payoff table. If the value X_{ij} is the payoff associated with the ith event's occurrence after the jth action has been taken, then we can define the expected monetary value of A_j to be

$$\text{EMV}_j = \sum_{i=1}^{m} X_{ij} P(E_i) \qquad j = 1, 2, \ldots, n \qquad \text{(16.2)}$$

where the summation is over m events and there are n actions from which to choose. There is an expected monetary value for every action in a payoff table, and this figure is a weighted mean of the payoffs associated with the action. It is a figure combining the probabilities of the events with the payoffs in the table. Thus, the decision maker can choose the action with the highest weighted payoff, the one with the highest expected monetary value.

EXAMPLE 1

Consider the betting game again. The payoff table for this game is presented again as Table 16.6.

The EMV of each of the actions is computed as follows.

TABLE 16.6   Payoff Table for Betting Game

ACTIONS EVENTS	PROBABILITIES	A_1 PLAY	A_2 NOT PLAY
E_1 Head on first toss	.500	1	0
E_2 First head on second toss	.250	2	0
E_3 First head on third toss	.125	4	0
E_4 No head appears	.125	−8	0

$$\text{EMV}_1 = \$1(.500) + \$2(.250) + \$4(.125) - \$8(.125)$$

$$= \$.50 + \$.50 + \$.50 - \$1.00 = \$.50$$

$$\text{EMV}_2 = \$0(.500) + \$0(.250) + \$0(.125) + \$0(.125)$$

$$= \$0$$

The decision maker would play the game (choose A_1) if he desires the action with the highest expected monetary value. Of course, he cannot *expect* a $.50 gain if he plays; he either gains $1, $2, or $4 or loses $8 on each play of the game. The EMV$_1$ value of $.50 means that if the decision maker plays the game many, many times, he will gain $.50 per play on the average. If he were to play the game 1000 times and were to add up his winnings and subtract his losses at the end of that period, his net gain would be about $500, or $.50 per play.

EXAMPLE 2

The calculation of the expected monetary values of the actions confronted by the business owner who wants to sell her business are performed in the same way. Her payoff table is repeated as Table 16.7.

The EMV's of the actions are

$$\text{EMV}_1 = \$80(.20) + \$80(.80) = \$16 + \$64 = \$80 \text{ (thousand)}$$

$$\text{EMV}_2 = \$100(.20) + \$70(.80) = \$20 + \$56 = \$76 \text{ (thousand)}$$

It appears that the best decision based on expected monetary value is to sell the business now. The figure of $80,000 is the actual price the owner will receive for her business; it is not difficult to interpret. But what is the meaning

TABLE 16.7   Payoff Table for Business Sale
(Figures in thousands of dollars)

ACTIONS EVENTS	PROBABILITIES	A_1 SELL NOW	A_2 HOLD A YEAR
E_1 Contract awarded	.20	80	100
E_2 Contract not awarded	.80	80	70

of EMV$_2$ = \$76,000? In a sense it means that if the owner were to face this decision many times in a row and were to choose to hold her business for one year each time, then some of the time she would get \$100,000 and some of the time she would get \$70,000, and her gains each time she made this decision would average out to \$76,000 per sale of the business.

In the previous example it was somewhat ridiculous to suggest that the business owner would face the task of selling her business several times in the future. Her decision is a one-shot decision. Interpreting expected monetary value as the long-run average payoff when there is no long run is somewhat misleading. In one-shot decision problems like those in Example 2, we would do better to view the expected monetary value as an artificial number which shows us the weighted average of the payoffs associated with an action. Then we choose that action which has the highest weighted average of payoffs.

16.4   THE OPPORTUNITY LOSS TABLE

Corresponding to every payoff table there is another table called the *opportunity loss table*. A decision maker might construct an opportunity loss table rather than a payoff table as an aid to choosing the best course of action. However, there is usually less work in constructing a payoff table first and then converting it into an opportunity loss table. Consider the payoff Table 16.8.

To construct an opportunity loss table we proceed as follows: First, we assume that one of the events has already taken place. Then for each of the possible actions we ask ourselves the question: how much better off could we have been if we had chosen one of the other actions open to us? For example, assume that event 1 has occurred. Then assume action 1 was the chosen action and payoff of 40 has been received. We now ask ourselves how much better off we could have been if we had chosen either of the other two actions, action

TABLE 16.8   Hypothetical Payoff Table

EVENTS ACTIONS	PROBABILITIES	ACTION 1	ACTION 2	ACTION 3
Event 1	.45	40	12	−5
Event 2	.30	−10	50	18
Event 3	.25	28	−2	20

2 or action 3. Since their payoffs are 12 and −5 respectively, we could not have put ourselves in a better situation by choosing either of them. Thus, our action 1 has zero opportunity loss associated with it when event 1 occurs; we did not have the opportunity to do better than a payoff of 40. Next we assume that instead of action 1 we had chosen action 2 and that event 1 occurred. Under these conditions our payoff would be 12. We ask ourselves again how much better off we could have been by choosing some other action. If we had chosen action 3, we would have been worse off since the payoff there is −5. But if we had chosen action 1 instead of action 2, we would have gained 40 and thus could have been better off by $40 - 12$, or 28. By choosing action 2 when event 1 occurs we would *lose the opportunity* to gain an additional 28 units of payoff. Thus, the opportunity loss associated with event 1 and action 2 is 28. To obtain the opportunity loss for event 1 and action 3, we proceed in the same way. In the case that event 1 occurs, either action 1 or action 2 would yield a better payoff than the payoff value of −5 associated with action 3. Action 1, with payoff 40, again is the best action, so choosing action 3 when event 1 occurs has an opportunity loss of $40 - (-5)$, or 45. That is, we would be better off by 45 payoff units if action 1 were chosen instead of action 3 when event 1 occurs.

The opportunity loss table is sometimes called the *regret table.* The values in this table contain measures of how badly we might feel if we choose the wrong action. That is, these values measure the feelings of regret we would experience.

In general, the opportunity loss for any action in a particular row of a payoff table is computed as the difference between that action's payoff and the best payoff in that row. If we let X_i^* be the best payoff in the ith row, then the opportunity loss for action j in the face of event i is

$$\mathrm{OL}_{ij} = |X_i^* - X_{ij}| \tag{16.3}$$

We use the absolute values sign in Formula (16.3) for the following reason. If the payoff table being used contains profits as entries, then X_i^* will be the largest value in the ith row and OL_{ij} will be a positive number. But if the payoff table contains costs, then X_i^* is the smallest value in the ith row and OL_{ij} will be negative. Since opportunity loss figures are customarily recorded as positive numbers, the absolute value sign in Formula (16.3) allows us to record only the *difference* between X_i^* and X_{ij} in the cases where the payoff table lists costs.

The computations for the opportunity losses associated with the second and third events of Table 16.8 are as follows.

$$\mathrm{OL}_{21} = 50 - (-10) = 60 \qquad \mathrm{OL}_{22} = 50 - 50 = 0$$

$$OL_{23} = 50 - 18 = 32 \qquad OL_{31} = 28 - 28 = 0$$

$$OL_{32} = 28 - (-2) = 30 \qquad OL_{33} = 28 - 20 = 8$$

Table 16.9 is the entire opportunity loss table corresponding to the payoff Table 16.8.

There are several observations that should be made about Table 16.9. First, it contains all positive or zero values. Even though the entries in the table are considered losses, they are made positive, and the user of the table simply remembers that he or she is dealing with a table of losses. Use of the absolute value sign in Formula (16.3) assures us that we will always have positive entries. Second, there is one zero value in each row of the table. There will always be at least one zero value in each row of an opportunity loss table. The zero value appears at the position occupied by the best payoff value for each row of the corresponding payoff table. Finally, the entries in the opportunity loss table are measures of the cost to the decision maker of not knowing which event is going to occur. If the decision maker knew in advance what event was to take place, he would always choose the best action and would never incur an opportunity loss. The positive opportunity losses in the table, then, are measures of the losses he may incur simply through not knowing in advance what event will occur. They are measures of the cost of the uncertainty he faces.

The calculation of opportunity losses associated with an event in a payoff table of *costs* is demonstrated as follows. Assume that event i has the following costs associated with it.

	COSTS OF:		
	Act 1	Act 2	Act 3
Event i	5000	3500	6500

The opportunity loss associated with the first act can be found using Formula (16.3). In this case $X_i^* = 3500$ since this is the *lowest cost* associated with event i. Thus

TABLE 16.9 Opportunity Loss Table for Payoff Table 16.8

ACTIONS EVENTS	PROBABILITIES	ACTION 1	ACTION 2	ACTION 3
Event 1	.45	0	28	45
Event 2	.30	60	0	32
Event 3	.25	0	30	8

$$OL_{i1} = |3500 - 5000| = 1500$$

$$OL_{i2} = |3500 - 3500| = 0$$

$$OL_{i3} = |3500 - 6500| = 3000$$

That is, the opportunity loss associated with action 1 and event i is 1500. This is because when event i occurs and we have chosen action 1, the cost is 5000. But we could have been 1500 *better off* had we chosen action 2 and incurred a cost of only 3500. The opportunity loss associated with action 2 is zero since action 2 is the lowest-cost act when event i occurs. Finally, action 3 has an opportunity loss of 3000 since we could have saved this much by choosing action 2 at a cost of 3500 rather than action 3 at a cost of 6500.

How does a measure like opportunity loss help determine the best course of action? In the last section we approached the problem of selecting the best action by calculating the expected monetary value of each of the possible actions and choosing the action with the highest EMV. The calculation of EMV for payoff Table 16.8 yields the following results:

$$EMV_1 = 40(.45) - 10(.30) + 28(.25)$$

$$= 18.0 - 3.0 + 7.0 = 22.0$$

$$EMV_2 = 12(.45) + 50(.30) - 2(.25)$$

$$= 5.4 + 15.0 - .5 = 19.9$$

$$EMV_3 = -5(.45) + 18(.30) + 20(.25)$$

$$= -2.25 + 5.4 + 5.0 = 8.15$$

By the criterion of EMV, action 1, with the highest EMV, appears to be the best.

An alternative criterion is suggested by the measures of the cost of our uncertainty about which event will occur. That is, we might want to choose that action which has the minimum expected opportunity loss (EOL) rather than the one with the highest EMV. The expected opportunity losses for each action are calculated in the same way as the EMV values except that we work from the opportunity loss Table 16.9. The calculations are:

$$EOL_1 = 0(.45) + 60(.30) + 0(.25)$$

$$= 0.0 + 18.0 + 0.0 = 18.0$$

$$EOL_2 = 28(.45) + 0(.30) + 30(.25)$$

$$= 12.6 + 0.0 + 7.5 = 20.1$$

$$EOL_3 = 45(.45) + 32(.30) + 8(.25)$$

$$= 20.25 + 9.6 + 2.0 = 31.85$$

It appears that action 1 is also the most desirable action if we choose to minimize the expected opportunity loss, since $EOL_1 = 18.0$ is the lowest of the three figures.

It is no coincidence that action 1 has both the lowest EOL and the highest EMV. *It is always true that the action with the best expected monetary value computed from the payoff table also has the lowest expected opportunity loss computed from the opportunity loss table.*

16.5   EXPECTED VALUE OF PERFECT INFORMATION

The best decision for any problem of the type discussed in this chapter can be found using the payoff table or the opportunity loss table. These tables simply present two views for the same problem. However, there is one advantage to using the opportunity loss table. Since the entries in this table represent the costs of not knowing what event is going to take place, the expected opportunity loss of the best action represents the expected cost of uncertainty for the best action. That is,

$$EOL^* = \text{Expected cost of uncertainty} \qquad (16.4)$$

This relationship is important because it gives an upper limit on how much a decision maker should be willing to spend to eliminate the uncertainty facing him. The most anyone would spend to remove uncertainty is the amount that that uncertainty costs. Thus, the decision maker should be willing to pay an amount equal to or less than the expected cost of uncertainty for a perfect predictor of which event will take place. This amount is called the *expected value of perfect information* (EVPI). So

$$EVPI = EOL^* \qquad (16.5)$$

According to this relationship, the person faced with payoff Table 16.8 should be willing to pay a maximum of 18.0 units for information about which event is going to occur. Actually, the person would pay substantially less than this if the information were only a forecast and not a guaranteed accurate prediction.

EXAMPLE 1

Consider the betting game discussed in Section 16.1. The opportunity loss table for that game is presented in Table 16.10, and the expected opportunity losses are calculated as follows.

TABLE 16.10 Opportunity Loss Table for Betting Game

EVENTS \ ACTIONS	PROBABILITIES	A_1 PLAY	A_2 NOT PLAY
E_1 Head on first toss	.500	0	1
E_2 First head on second toss	.250	0	2
E_3 First head on third toss	.125	0	4
E_4 No head appears	.125	8	0

$$EOL_1 = 0(.500) + 0(.250) + 0(.125) + 8(.125)$$

$$= 0.0 + 0.0 + 0.0 + 1.0 = 1.0$$

$$EOL_2 = 1(.500) + 2(.250) + 4(.125) + 0(.125)$$

$$= 0.5 + 0.5 + 0.5 + 0.0 = 1.5$$

So $EOL^* = 1.0$ and action 1 is the preferred action. Also, $EVPI = EOL^* = 1.0$. This means that a player would be willing to pay up to $1 per play to learn in advance which event was going to occur.

EXAMPLE 2

The business owner who has the opportunity to sell her business now or hold it for one year may construct from payoff Table 16.5 the opportunity loss table presented as Table 16.11. From the tabulated values she may calculate the expected opportunity losses.

$$EOL_1 = 20(.20) + 0(.80) = 4.0 + 0.0 = 4.0$$

$$EOL_2 = 0(.20) + 10(.80) = 0.0 + 8.0 = 8.0$$

So EOL^* is $4000 and the action "sell now" is the best action. But the owner should be willing to spend up to $4000 to gain more information about her chances of winning the contract, since $EOL^* = EVPI$.

TABLE 16.11 Opportunity Loss Table for Business Sale (Figures in thousands of dollars)

EVENTS \ ACTIONS	PROBABILITIES	A_1 SELL NOW	A_2 HOLD A YEAR
E_1 Contract awarded	.20	20	0
E_2 Contract not awarded	.80	0	10

There is a second way of calculating the expected value of perfect information which may help to clarify its meaning. Once again, consider the betting game we have referred to so often. The payoff table for this game is reproduced again as Table 16.12. Let us assume for a

TABLE 16.12   Payoff Table for Betting Game

EVENTS \ ACTIONS	PROBABILITIES	A_1 PLAY	A_2 NOT PLAY
E_1 Head on first toss	.500	1	0
E_2 First head on second toss	.250	2	0
E_3 First head on third toss	.125	4	0
E_4 No head appears	.125	−8	0

moment that the decision maker is offered the chance to play this game many times in a row. If he knew in advance that event 1 were going to occur, that is, if he had a perfect predictor, he would surely play since he would gain $1. Also, if he knew that E_2 or E_3 were going to occur, he would play and receive $2 or $4. But if he knew E_4 were going to occur, he would surely decline the offer to play and end up gaining and losing nothing. Thus, a perfect predictor would allow the decision maker to earn $1 on 50 percent of the plays, $2 on 25 percent of the plays, $4 on 12.5 percent of the plays, and $0 on 12.5 percent of the plays. The average payoff per play using a perfect predictor is called the *expected payoff under certainty* (EPUC). For the betting game this is

$$EPUC = 1(.500) + 2(.250) + 4(.125) + 0(.125)$$

$$= .50 + .50 + .50 + .00 = 1.50$$

Using a perfect predictor, the decision maker could average $1.50 per play. In Section 16.3 we found that the expected monetary value of the decision to play the game every time was $.50. If the decision maker can average $.50 per play without perfect information and $1.50 per play with perfect information, he knows that the per-play value of the perfect information is $1.50 − $.50, or $1.00. That is, the expected value of perfect information is the amount by which the expected payoff under certainty differs from the expected monetary value. Or

$$EVPI = |EPUC − EMV^*| \qquad (16.6)$$

Since according to Expression (16.5) the expected value of perfect information also equals the expected opportunity loss of the best action, we have

$$EVPI = |EPUC − EMV^*| = EOL^* \qquad (16.7)$$

In Example 1 of this section, the EOL of the best action (to play the game) was calculated to be $1, so the relationship checks out.

EXAMPLE 3

In Example 2 of this section, the EVPI for the business owner facing the decision about selling her business was calculated to be $4000. Does this coincide with the value of the EVPI calculated by the alternative method? If the owner had a perfect predictor and knew for sure which event were going to take place, then when event 1 was imminent, she would choose action 2, to hold, and gain 100. This information can be seen from Table 16.7. When event 2 was imminent, she would choose action 1, to sell, and gain 80. These two possibilities are presented in Table 16.13.

TABLE 16.13   Best Payoffs under Certainty for Business Sale

EVENT KNOWN TO BE IMMINENT	BEST ACTION	PAYOFF FOR BEST ACTION
E_1 Contract awarded	A_2 Hold	100
E_2 Contract not awarded	A_1 Sell	80

Since there is only a 20 percent chance that a perfect predictor would forecast E_1 and an 80 percent chance it would forecast E_2, the EPUC is

$$\text{EPUC} = 100(.20) + 80(.80) = 20 + 64 = 84$$

In Section 16.4 the EMV* was found to be $80,000, so by Equation (16.6) the expected value of perfect information is found to be

$$\text{EVPI} = |84 - 80| = 4$$

or, since all figures are in thousands of dollars, $4000.

16.6   UTILITY

In all the examples in this chapter we have assumed that a rational decision maker would choose the action with the highest expected monetary value. That is, we have assumed that expected monetary value is the best decision-making criterion. This seems logical in many instances. However, consider the following situation. Imagine for a moment that a gambler offered you two alternatives. He would pay you $10,000 for sure *or* he would allow you to engage in this game: You toss a coin; if it comes up heads you win $200,000, and if it comes up tails you must pay $160,000. Your payoff table is presented in Table 16.14.

Suppose you decide to approach this situation by figuring the expected monetary value for each of the two actions.

TABLE 16.14   **Payoff Table for Gamble**

EVENTS \ ACTIONS	PROBABILITIES	A_1 SURE THING	A_2 TOSS COIN
E_1 Heads	.5	10,000	200,000
E_2 Tails	.5	10,000	−160,000

$$\text{EMV}_1 = 10,000(.5) + 10,000(.5)$$

$$= 5000 + 5000 = 10,000$$

$$\text{EMV}_2 = 200,000(.5) - 160,000(.5)$$

$$= 100,000 - 80,000 = 20,000$$

The decision to toss the coin has an expected monetary value of $20,000 and thus appears to be better than the decision simply to accept the $10,000. Yet you might justifiably feel that you would rather have the $10,000 straight out than toss the coin and run the risk of having to pay the gambler $160,000. That is, you may prefer a sure $10,000 to a gamble with an EMV of $20,000. This very natural feeling is based on the fact that people make decisions in terms of the **utility** involved in the problem. The reason that many people would take $10,000 for sure instead of tossing the coin is that their utility for the sure $10,000 is higher than their utility for the gambling situation—even though the gambling situation has a higher expected monetary value.

A person's utility for an amount of money, a gamble, or even a physical object is, in a sense, the amount of well-being these things produce in the person. Thus, utility is a very personal concept. One person's utility for an amount of money may be very different from another person's utility for the same amount. A millionaire might easily choose to engage in the gamble of tossing a coin for a $200,000 gain or a $160,000 loss. This is because a millionaire could afford the loss of $160,000 without much difficulty, and the possible gain of $10,000 is peanuts.

Since sheer amount of money does not always measure desirability or utility to a decision maker, the payoff tables discussed in the previous sections should really be filled in with utility figures rather than dollar amounts. If this were done, decision makers could choose the action with the highest expected utility value rather than the highest expected monetary value. The problem with trying to do this is that it is very difficult to measure utility, despite the significant work that has been done in this area (see References 16.3 and 16.6). This chapter will not deal with actual numerical measures of utility. Only cases where monetary value does serve as a good measure or where different decision techniques can be used to express utility will be discussed.

Let us reconsider the gambling situation presented at the beginning of this section and change the scale of the problem. Suppose you are offered the chance to take $.10 for sure *or* toss a coin and receive $2.00 for a head and pay $1.60 for a tail. This is the same problem as presented earlier except that the consequences are one hundred thousand times smaller. The expected value of the sure thing is $.10, and the expected value of tossing the coin is $.20. Many people who would not have tossed the coin when the payoffs were $200,000 and −$160,000 would not hesitate to do so with the smaller payoffs. This is because, in general, for small amounts of money the monetary value is a good measure of utility. In cases involving relatively small dollar amounts the action with the highest expected monetary value is also the one with the highest expected utility.

But what does "relatively small" mean? It means small relative to the total wealth of the person or organization making the decision. Thus, a millionaire would likely be willing to toss the coin for the large stakes while a college student would not. However, the student might change his or her mind once the payoffs had been reduced by a factor of 100,000. Twenty thousand dollars seems a small amount to some very wealthy people just as $.20 seems a small amount to most college students.

Thus, *expected monetary value seems to be a good decision-making criterion when the payoffs involved are small relative to the overall wealth, assets, or budget of the decision maker,* but when the payoffs are large the utility criterion must be considered. How does the decision maker take utility into account in the situation involving relatively large payoffs? He might consider using an alternative decision-making criterion like one of those discussed in the following section.

16.7   DECISION CRITERIA

By using expected monetary value as the basis for making decisions, the decision maker will end up doing well on the average or in the long run. But if someone is very worried about the short run or doesn't believe that the laws of probability hold in his case, then he might do better by abandoning the use of expected monetary value in evaluating alternative courses of action. Several other methods for selecting a course of action are presented in this section. Their use will be demonstrated on payoff Table 16.15.

The **optimist's criterion** is sometimes called the *maximax* criterion. Simply stated, it suggests that the decision maker should select that

TABLE 16.15   Demonstration Payoff Table

EVENTS	PROBABILITIES	A_1	A_2	A_3	A_4
E_1	.45	8	−6	12	−4
E_2	.30	10	12	15	−8
E_3	.25	6	18	−10	20

action with the best payoff. Since 20 is the highest payoff in Table 16.15, the optimistic decision maker would select the action associated with that payoff, A_4. This ignores completely the fact that event 3, which must occur if the decision maker is to realize a payoff of 20, is the least likely of the three events. Even if E_3 had a probability of .001, the optimistic decision maker would make the same choice. The optimist's criterion ignores event probabilities and all payoffs in the table except the best one. Thus, even though A_4 is associated not only with the highest but also with some of the lowest payoffs in the table, the optimist's decision is still A_4.

One can see why this is called the optimist's criterion. A person who is truly optimistic (and only optimistic) *would* ignore the probabilities of events and possible adverse payoffs in order to obtain a large payoff. The reason for the name "maximax criterion" is not so clear. It can be explained in the following way. Suppose the decision maker examined each action separately and chose the maximum payoff associated with each. For the situation represented in Table 16.15, he would find that the four actions had maximum payoffs of 10, 18, 15, and 20, respectively. Then he would choose that action which produced the *maximum* of these *maximums*. By choosing the largest payoff in the table, the maximax decision maker behaves as though he were selecting the action that maximizes the maximum payoff for each action. This two-step maximization process leads to the term *maximax*.

However, this term cannot be applied to the process that would be employed by an optimist approaching a payoff table of costs. The optimist would look at each action and list the *minimum* cost associated with each action and then choose the *minimum* of these minimum costs. Thus, the optimist's criterion would have to be called the *minimin* criterion when applied to payoff tables filled with costs. To avoid this confusion, we will use only one term: the *optimist's criterion*.

Given the obvious drawbacks of the optimist's criterion—put simply, that it ignores everything in the payoff table except the best payoff— one might wonder why anyone would consider it as an aid to decision making. It is true that there are very few circumstances where this method of selecting an action is valid. A manager whose business was in danger of bankruptcy and who needed a large payoff in order to save it from ruin might rationally choose the largest payoff in the table as a guide. For instance, a manager who needs 19 units or more to save a business might use the optimist's criterion on payoff Table 16.15, since only the optimistic decision will give the amount needed to remedy the situation. In this case, that decision seems reasonable.

The **pessimist's criterion** is, in some respects, the opposite of the optimist's criterion. A pessimist assumes that no matter what action is taken, the worst possible event will take place. Thus, the pessimistic decision maker feels that if he takes action 1 in Table 16.15, event 3 is bound to happen since it yields the lowest payoff for A_1, a payoff of 6. And if a pessimist were to take action 2, he feels that event 1 is sure to take place since that would yield the worst possible payoff for A_2, a value of −6. In short, no matter which action he takes, he is bound to end up with the minimum payoff for that action: 6, −6, −10, or −8. But even a pessimist will make the best of a bad situation. Thus, the pessimistic decision maker selects that action which will *maximize* these *minimum* outcomes. This is action 1, with a guaranteed payoff of at least 6. (This two-step process leads some people to call the pessimist's criterion the *maximin* criterion.)

The pessimistic decision maker ignores the probabilities in the payoff table and concentrates only on the adverse outcomes of each action. This method of selecting an action guarantees against large losses and establishes a floor below which the payoff cannot fall. However, it also leads to decisions where large payoffs may not be realized even though they may be quite probable.

The pessimist's decision criterion is appropriate in situations where an organization or person is in financial difficulty and cannot afford a large loss. It may be used to select those actions that ensure against the chances of disastrous results at the expense of passing up possible good results. This is a very conservative approach to decision making and is appropriate only in cases where it is imperative that adverse results be avoided.

The **maximum likelihood criterion** does not ignore probabilities. In fact, the first step in making a maximum likelihood decision is to determine which event is most likely to occur. In payoff Table 16.15, that is event 1, which has probability .45 of occurring. The second step in applying the maximum likelihood criterion involves listing the payoffs associated with that most probable event. For our example, the payoffs associated with E_1 are 8, −6, 12, and −4. The final step is selecting the action with the best payoff among these. Thus, the maximum likelihood decision maker would select action 3, since the payoff of 12 is the maximum value for the payoffs of the most likely event.

This criterion ignores the probabilities and payoffs of all the events in the table except that one event with the greatest probability of occurrence. This would not be a very appropriate approach if the probabilities of the events were, say, $P(E_1) = .34$, $P(E_2) = .33$, and $P(E_3) = .33$. In this case event 1 is the most probable again, but it would seem inadvisable to concentrate all attention on E_1 and ignore the other two possible events.

There are two situations when the maximum likelihood criterion would seem appropriate. The first is the case in which one event is so much more probable than the others that it seems foolish to consider anything but that event. Thus, if $P(E_1) = .98$, $P(E_2) = .01$, and $P(E_3) = .01$, the maximum likelihood criterion seems appropriate. We all use this decision rule when we cross a street. Although there is a small chance that we will be hit by a car while crossing, the probability of our making it across safely is so high that we usually ignore the consequences involved with an accident's occurring even though these consequences might be very serious. A second situation where the maximum likelihood criterion is appropriate is the one in which choosing a course of action involves a great deal of preparation for carrying out the action even before the event is known. If the decision maker has time or resources to prepare for only one of the possible events, which one should he prepare for? The one that has the highest probability of occurring is the logical choice.

The **equal likelihood criterion** is used when the decision maker considers each of the events in the payoff table to be equally likely. This may be the case when he has absolutely no idea about the relative probabilities of the events and thus has to assume that each is as likely as any other. When this happens, the decision maker assigns equal probabilities to each of the events and proceeds as though he were calculating the expected monetary value of the actions. For Table 16.15, the equal likelihood decision maker would assume $P(E_1) = 1/3$, $P(E_2) = 1/3$, and $P(E_3) = 1/3$. Then he would calculate the EMV of each action using these probabilities:

$$\text{EMV}_1 = 8(1/3) + 10(1/3) + 6(1/3) = 24/3 = 8.00$$

$$\text{EMV}_2 = -6(1/3) + 12(1/3) + 18(1/3) = 24/3 = 8.00$$

$$\text{EMV}_3 = 12(1/3) + 15(1/3) - 10(1/3) = 17/3 = 5.67$$

$$\text{EMV}_4 = -4(1/3) - 8(1/3) + 20(1/3) = 8/3 = 2.67$$

Thus, the equal likelihood decision maker would be indifferent between actions 1 and 2 since they both have EMV's of 8.00 when the events are considered equally likely.

The preceding calculation illustrates that there is a somewhat faster way of determining the EMV of an action when all the probabilities are equal. This involves summing up all the payoffs in one column of the payoff table and then dividing the sum by the number of possible events. For A_1, the sum of the payoffs in Table 16.15 is 24, and EMV_1 is 24/3, or 8.00.

The equal likelihood criterion gives equal weight to every event and thus equal weight to every payoff in the table; it takes into account all the possible payoffs but does not allow for the fact that events might have unequal probabilities of occurring. There are likely to be few situations in which a decision maker cannot distinguish any difference between the relative probabilities of the events in his payoff table.

The **expected monetary value criterion** is sometimes called Bayes' criterion after the Reverend Thomas Bayes, an eighteenth-century English minister and mathematician. This criterion was fully explained in Section 16.3. It has the advantage of combining both the probabilities of the events and the payoffs of the actions in a logical way. As we have learned, the Bayesian decision maker makes good decisions in the long run, or on the average. It is only when some of the situations discussed previously arise and the monetary values in the payoff table are no longer good measures of the utilities of the decision maker that he should revert to some of the other decision-making criteria.

The Bayesian decision for Table 16.15 is found using the usual EMV calculations:

$$EMV_1 = 8(.45) + 10(.30) + 6(.25) = 3.6 + 3.0 + 1.5 = 8.1$$

$$EMV_2 = -6(.45) + 12(.30) + 18(.25) = -2.7 + 3.6 + 4.5 = 5.4$$

$$EMV_3 = 12(.45) + 15(.30) - 10(.25) = 5.4 + 4.5 - 2.5 = 7.4$$

$$EMV_4 = -4(.45) - 8(.30) + 20(.25) = -1.8 - 2.4 + 5.0 = 0.8$$

The action with the highest EMV (and also the lowest EOL) is action 1: $EMV_1 = EMV^* = 8.1$.

Most decision makers would be well off if they used only the criterion of expected monetary value in making their decisions, valuing the other criteria chiefly as tie-breakers. If two or more actions have the same expected monetary values, the decision maker can employ one of the other criteria to help him choose between the tied actions. If he desires to insure himself against large losses, he may revert to the pessimist's criterion to break the tie, for instance.

16.8 DECISION TREES

It was noted in Section 16.1 that the payoff table is a device for sharpening the statement of a decision-making problem so that the decision maker can better see the situation. Another graphical way of laying out a problem uses decision-tree diagrams. Any payoff table can be shown as a decision tree.

EXAMPLE 1

The decision tree in Figure 16.1 represents the problem faced by the business owner trying to decide whether or not to sell her business (Example 2 of Section 16.1). A decision tree like this one is read from left to right. First, it says, the owner must select between selling now or holding her business for one year. Then she finds out about the award of her contract. The decision tree is evaluated in the same way a payoff table is evaluated. The probabilities on the event branches are multiplied by the monetary figures at the ends of the branches. The weighted sum is then recorded in the circles from which the event branches emanate. The decision maker then chooses between the actions emanating from one action choice block and selects that action leading to the highest expected monetary value and crosses out the other actions. Thus, the tree is always evaluated from *right to left*. The evaluated decision tree for Example 1 is shown in Figure 16.2.

FIGURE 16.1 Decision tree for business sale

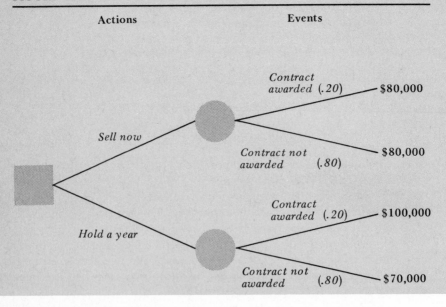

FIGURE 16.2   Evaluated decision tree for business sale

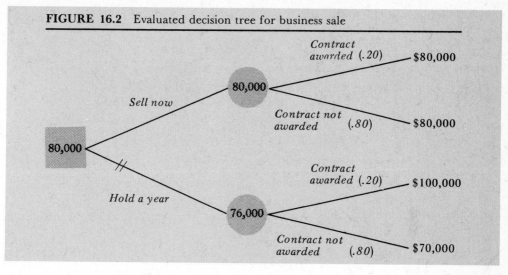

There are a few conventions which are usually followed to add clarity to decision-tree diagrams. Points in the tree where decisions or choices must be made are usually represented by rectangles and are called *decision nodes*. Points in the tree where the branches represent events that occur are drawn as circles. The events have probabilities attached to them, and the probabilities on the branches starting at a circle, or *event node*, must all sum to unity. In complex trees, monetary flows are often indicated on the branches of the tree inside triangles. See Figure 16.3 for an example of this.

We do not usually represent a payoff-table problem in the form of a decision tree for the simple reason that the payoff table is more compact. A payoff table containing ten events and ten actions could easily be put on this page, but a readable decision tree describing the same problem would certainly not fit the page. Still, despite their elementary and somewhat awkward appearance, decision trees can be used graphically to represent and solve quite complex problems where payoff tables will not work. A payoff table is appropriate when the decision maker must select one action from several alternative actions. When he faces a problem calling for a *sequence* of decisions to be made, or when there are *sequences* of events that determine the payoffs, the problem can be adequately represented only with a decision tree. Consider the following example.

EXAMPLE 2

Problem: An equipment manufacturer operates two plants, one quite large and the other small. The smaller plant will run out of work next month and

FIGURE 16.3 Decision tree for Example 2

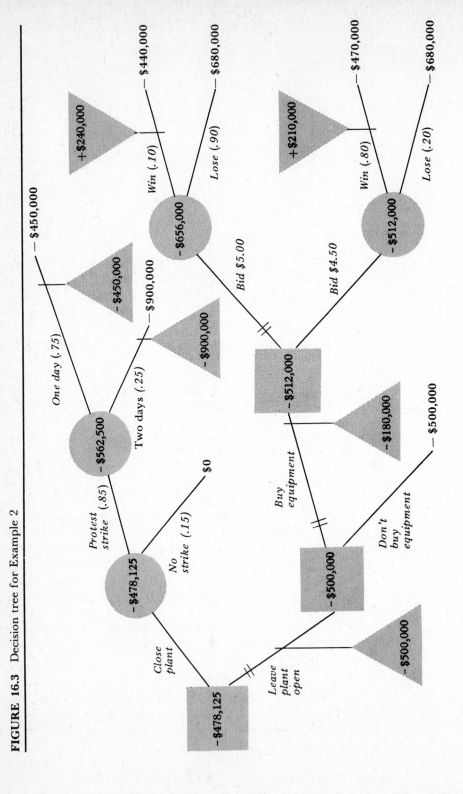

will have a one-week period when none of the currently scheduled projects can be worked on. Management is considering closing the small plant, and, according to labor contracts, it must make any closure announcement now. However, it fears that if an announcement is made, employees at the larger plant will stage a protest sympathy strike. Management feels there is about an .85 chance that this would happen. Information from the large plant indicates that there is about a .75 chance the strike would last one day and a .25 chance it would last two days. A strike at the large plant would cost the company about $450,000 per day.

If the smaller plant is left open during the slack week (five working days), there will no longer be any question of a strike but it will cost the company $100,000 per day more than if the plant were shut down. In order to cover part of these costs, the company might submit a bid to a local customer for manufacturing 60,000 units of a special product the company has never made before. In order to be eligible for this bid, the company would have to buy and install equipment costing $180,000 to show that it has the capability of doing the job, prior to entering a bid. The entire cost of the equipment would be charged against the project since there seems little likelihood the equipment would be used again in the near future. The management of the small plant cannot decide if a bid of $5.00 per unit or $4.50 per unit would be best (bids are being accepted only in increments of $.50). They feel there is a .10 chance they would win with a $5.00 bid and an .80 chance they would win with a $4.50 bid. They will not consider lower bids. Materials cost about $1.00 per unit. Should management announce a plant closure? If not, should it bid on the additional work? If so, what should be the bid?

Solution: The decision tree for this problem is presented in Figure 16.3. The figures at the ends of the branches are the net losses to the company if the sequence of decisions and events leads to that point. The figures in the triangles are losses and gains shown at the point where they occur; they are netted out to the end and can be ignored in the evaluation of the decision tree. The gains shown if the bid were won are calculated as contribution to fixed cost and profit, i.e., (revenue − materials costs) × 60,000 units. The tree has been evaluated from right to left, beginning at the net dollar losses.

Figure 16.3 shows that the company should go ahead and close the small plant since that action has the lowest expected cost, $478,125. Actually, the real costs will be either $450,000, $900,000, or nothing; but the $478,125 figure is a weighted average of these costs. It was found by weighting the $562,500 figure in the upper circle by .85 and the zero cost of no strike by .15. The $562,500 figure, in turn, was obtained by finding the expected cost of the one- and two-day strikes.

This example illustrates two important points. First, in evaluating decision trees we can use the expected value on one branch in the calculation of expected values further down the tree. And secondly, the use of a decision tree or payoff table cannot be merely a mechanical exercise. For instance, the difference between the expected monetary value of closing the plant and leaving it open is slightly under $22,000— not a very large amount relative to the rest of the figures in the problem. Thus, the company might want to take a more careful look at their decision. If they do, they will find that it is quite sensitive to the probabilities on which they base their calculations. If the probability of a strike after the plant closure were much higher than .85, the best decision would shift to leaving the plant open. In fact, if the probability of a strike rises to .89, then leaving the plant open becomes the best decision: The new EMV,

$$-\$562,500(.89) + \$0(.11) = -\$500,625$$

higher by $625 than the cost of leaving the plant open for five days. Similar analysis of the effects of changing other probabilities can be made. Since the probabilities used in this problem were all subjective in nature and thus subject to error, it would be wise to make such a sensitivity analysis. It might also pay the company to investigate further to see if it can get more reliable probability estimates. Without more reliable estimates, the alternatives of closing the plant and keeping it open seem to have very close to the same desirability. If company officials find it difficult to refine their probability estimates, they might wish to employ one of the alternative decision criteria mentioned in Section 16.7 to help decide between keeping the plant open and closing it. The company also might wish to consider its social responsibility in this problem and could decide in favor of keeping the plant open since the economic factors in favor of closure do not appear to be compelling.

When the time period covered by a decision tree is a year or more, the monetary values in the problem should be discounted using an appropriate discount rate. The decisions in some decision trees will result in cash flows over a period of years that amount to annuities. In these cases the present values of the annuities should be listed on the branches of the decision tree. Realistic decision-tree problems can yield rather complex trees. The evaluation of each branch of these trees can be especially difficult when the time value of money is considered and the values are discounted. However, there are computer programs available to help evaluate complex decision trees. These programs are simple to run, and the user does not need extensive computer programming experience to use them. Most of these programs require the user only to specify the beginning and endpoint of each branch on the tree, the probabilities associated with event branches, the cash flows associated with each branch, the time period over which the flows take place, and the desired discount rate. All these pieces of information are used as input to a preprogrammed set of computer instructions which cause the computer to construct a picture of the decision tree and evaluate the various branches. These programs even discount the monetary flows.

Such programs do not relieve the user of the responsibility of constructing a decision tree. This is often a very difficult part of the problem, since with the tree's structure depending completely on the problem at hand there can be few general rules governing decision-tree drawing. But these programs will do the tedious work of evaluating the tree and discounting the monetary values and ultimately listing the most desired courses of action. The use of such programs is especially valuable in doing sensitivity analysis on complex decision trees—that is, changing the probabilities of various events to see if the changes will alter the expected values of the actions enough to cause the decision maker to select a new set of actions. These probability adjustments can usually be made by changing one or two data entries and rerunning the program. Thus, the effects of several changes in probabilities on the branches of a decision tree can be evaluated in a matter of minutes.

PROBLEMS

1. Consider the following payoff table, where the figures are profits in thousands of dollars. Find the EMV of each action and select the best one.

EVENTS	ACTIONS PROBABILITIES	A_1	A_2	A_3
E_1	.20	7	28	−9
E_2	.35	10	−20	1
E_3	.45	−8	13	15

2. For the table in Problem 1:
   a. Find the opportunity loss table.
   b. Find the expected opportunity loss for each action.
   c. Find the EVPI for this problem.
   d. Find the EPUC for this problem.
3. Consider the following payoff table, which presents the costs of a project (in thousands of dollars).

EVENTS	ACTIONS PROBABILITIES	A_1	A_2	A_3	A_4
E_1	.20	20	20	26	30
E_2	.30	25	30	35	30
E_3	.50	20	18	10	22

   a. Eliminate any dominated actions in this problem.
   b. Find the EMV of the remaining actions and select the best one.
4. For the payoff table in Problem 3,
   a. Find the opportunity loss table.
   b. Find the EOL for each action.
   c. What is the maximum amount a decision maker faced with this table would be willing to pay for information about which event is going to occur?
5. The accompanying payoff table represents the cost to a certain county of building a section of new road under three alternative proposals. The cost depends not only on the nature of the proposal but also on the weather conditions that might exist during the winter.

(Figures in thousands of dollars)

EVENTS ╲ ACTIONS	PROBA-BILITIES	A_1 LET A CONTRACT	A_2 HAVE COUNTY DO IT	A_3 TWO-STAGE CONTRACT
E_1 Light winter	.25	120	70	100
E_2 Moderate winter	.50	120	105	120
E_3 Bad winter	.25	120	150	130

   The county has received a bid on the work indicating that the job could be done by a contractor for a flat $120,000. Alternatively, the contractor would be willing to begin the work in the fall as the first stage of a two-stage contract. Payment for the second portion of the work would depend on the severity of the winter. The costs of the two-stage contract are presented in the payoff table. Finally, the county could do the work itself, and past experience and cost estimates indicate that the costs depend to a great extent on the weather. The cost estimates for this proposal are also presented in the table.
   a. Find the action with the best EMV.
   b. If the county could pay for a special meteorological study of the likely winter conditions in this area, what is the maximum amount they should be willing to pay for it?
   c. What would the optimist's decision be?
   d. What would the pessimist's decision be?
   e. Find the decision assuming equally likely events.
   f. What is the maximum likelihood decision?
6. Several people looked at the same payoff table and made the comments listed here. Examine their comments and determine which of the following decision making criteria the different speakers would likely use:

> Optimist's
> Pessimist's
> Maximum Likelihood
> Equal Likelihood
> Expected Monetary Value

a. "No matter what I do, the worst always happens. I'm going to protect myself against the inevitable disaster."
b. "I play the averages. That way I come out best in the long run."
c. "There won't be any 'long run' for me if I don't strike it rich on this deal. I've got to have that big profit or I'm sunk."
d. "None of those probabilities look right to me. I just don't think you can tell what's going to happen."
e. "All of the probabilities in that table are too small to consider except for event 2."
f. "No matter what I do, I come out smelling like a rose. I might as well go for broke."
g. "I'm going to make my choice in such a way that I give weight to both the payoffs and the probabilities."
h. "I've only got time to prepare for one of those events. I just can't consider more than one at a time."

7. Refer to the payoff table in Problem **1**.
a. Find the optimist's decision.
b. Find the pessimist's decision.
c. What is the maximum likelihood decision?
d. Find the best decision assuming equal likelihood for all three events.

8. Consider the following payoff table. The figures are profits.

EVENTS ╲ ACTIONS	PROBABILITIES	A_1	A_2	A_3
E_1	.70	100	60	50
E_2	.30	70	200	400

a. Find the action with the highest EMV.
b. How much can the probabilities change, and in what direction, before the preferred action will change?
c. Under what probability combinations for E_1 and E_2 will action A_2 become the action with the highest EMV?
d. Assuming the original probabilities hold, how much can the profit $X_{23} = 400$ fall before A_3 is no longer a desirable action?

9. For the payoff table in Problem **8** find:
a. The optimist's decision.
b. The pessimist's decision.
c. The maximum likelihood decision.
d. The equal likelihood decision.
e. The EPUC.

10. The DAN-DEE Tool Company is planning its first (and probably last) production run of a new special-purpose machine. It already has five orders for this machine. The machine sells for $8500 and has variable costs of $2600 per machine. Fixed costs total $30,000 and are made up of new equipment costs and a manufacturing right the company has purchased. The entire fixed cost is to be charged to this production run.

The company is uncertain as to how many of these machines will eventually be ordered and thus how many it should produce. The company must make the production run this month in order to meet the delivery date on the first few machines. Uncertainty about the eventual total sales of the machine is expressed in the following probability distribution.

SALES	5	6	7	8	
P (SALES)	.25	.35	.30	.10	1.00

    a. What are the actions DAN-DEE could take?

    b. What are the events that might occur?

    c. Construct DAN-DEE's payoff table.

11. **a.** For the payoff table constructed in Problem **10**, find the action with the highest EMV.

    b. The company described in Problem **10** might perform a survey of potential customers for this machine. Such a survey would have a fixed cost of $500 and would cost about $50 per customer interview. What is the maximum number of customers that the company would be willing to interview?

12. The Peery-Nelson Company is trying to decide if it should market a new product. The company's profit (or loss) on the product depends on the percentage of the market it can capture. The break-even market share is 10%. The management feels that it will make $40,000 for every percentage point above 10% it is able to capture. But if the company fails to achieve a 10% market share, management has decided that losses will amount to

$$\$20,000 + \$30,000(10 - p)$$

where p is the market share expressed as a percentage (i.e., not as a decimal, but as a number between 0 and 100). The $20,000 cost is not incurred once the company achieves a 10% market share; it applies only if the company fails to meet the break-even point. Peery-Nelson feel that its chances at various possible market shares are

MARKET SHARE	PROBABILITY
7%	.05
8%	.10
9%	.12
10%	.18
11%	.25
12%	.20
13%	.10
	1.00

    a. What are the actions open to Peery-Nelson?

    b. Construct a payoff table for the company.

    c. Which action has the highest EMV?

    d. A market research firm will do a study for Peery-Nelson to determine just what market share it can capture. The survey will cost $72,000 and is guaranteed accurate. Should Peery-Nelson have the survey done?

13. Mary Hoffman, a financial analyst for International Electrical Controls, has just completed a month-long study to determine the potential profits from a three-part contract which is to be awarded by a computer manufacturer six months from now. The profit potentials depend on how many parts of the contract International gets (it has bid on all three). Ms. Hoffman has considered the fact that her company will enjoy the economies of lengthy production runs, but only up to a certain point in volume. She has also considered the special equipment whose costs will have to be charged to the contract.

The potential profits of the various-sized contracts depend on whether the company decides to do all the work itself, subcontract half of it to J. J. Electronics, or subcontract all of it to J. J. Electronics, the only subcontractor International would trust with work of this nature. Ms. Hoffman has worked with J. J.'s financial vice-president in determining her figures associated with the subcontracting. The vice-president has told her that there is no way that J. J. would consider taking any of the work unless International is willing to make a commitment right now as to how much of the work J. J. will get, assuming Hoffman's company wins some work. This is because of J. J.'s other possible opportunities and their need to prepare for such a large volume of work. In addition, Ms. Hoffman is informed, J. J. will charge International a flat $3000 "spadework" fee in the event that, having been promised all or part of the work now, they end up getting none due to International's winning none of the contract parts.

Ms. Hoffman's payoff table is as follows. She feels there is a .35 probability that International will get one or more parts of the contract. The probability that they get only one is estimated to be .20; that they get two is .10.

(All figures are in thousands of dollars)

EVENTS \ ACTIONS	PROBA-BILITIES	DO IT ALONE	SUBCONTRACT EVERYTHING	GO 50-50 WITH J.J.
Zero parts				
One part		20	5	10
Two parts		25	15	20
Three parts		−14	18	16

a. Fill in the first row of the table and the probability column.
b. Find the action with the highest EMV.
c. Construct an opportunity loss table and find the action with the lowest EOL.
d. What is the EVPI for this problem?
e. Show that $EMV_j + EOL_j = EPUC$ for all actions A_j, where $j = 1, 2, 3$.

14. Two professional football teams are playing in the Super Bowl. Assume there are only 48 seconds left in the game and team A is leading team B by the score of 27 to 21.
    a. Team B has fourth down and six on their own 33 yard line. The quarterback decides to go for a touchdown and throws a long, deep pass. What decision criterion is he using?
    b. Assume the pass in Part a was incomplete and team A takes over with 43 seconds left in the game. Team A then runs out the clock by running four straight plays in which the quarterback merely falls on the ball after taking it from the center. What decision criterion is team A using here?

15. Mr. Earnest Money speculates in real estate. Currently, he has an option to buy a piece of property. The value of the property depends on whether the City Planning Commission will rezone it to the classification Mr. Money desires. The commission will not meet until late next week, at which time it will decide to take one of three steps: rezone the property as Mr. Money desires, deny the rezoning application outright, or table the request until the next meeting some six weeks away. However, Mr. Money is in tight financial circumstances. He has to complete the deal on this piece of property within three weeks to free up his resources for another transaction at that time. The problem is compounded by the fact that his option to buy this property expires tomorrow, and the property owner wants him to renew it or buy the property outright today. The chances associated with the requested rezoning change seem to be about .15 for approval and about twice that for denial. The monetary values involved are the following:

Cost of the new option: $10,000   Cost of land purchased now: $150,000
Sales price if rezoning denied: $135,000   Sales price if rezoned: $250,000
Cost of land under new option: $155,000   Sales price if tabled: $170,000

    a. Construct the payoff table facing Mr. Money. Assume that if it is not profitable to purchase the land under a set of circumstances, he will not do so. (*Hint:* Make sure that one of the possible actions you consider is "Do nothing.")
    b. Are any of the actions dominated by the others? If so, which one(s)?
    c. Find the EMV of each action.
    d. What is the maximum amount Mr. Money would be willing to spend in finding out what the Commission is likely to do?

16. Solve Problem 1 using a decision tree instead of a payoff table.

17. The Pemform Company has 35 small retail stores located in the western United States. The San Jose store is across the street from a very large vacant tract of land. Two parties are currently seeking to buy the land. One would build a large shopping mall on the land; the other would use it to put up an amusement park. The Pemform Company retained a consultant who specializes in evaluating the effects which changing neighborhoods and traffic patterns have on businesses' revenues.

   According to the consultant, a shopping mall on the vacant land will mean an added $50,000/year in revenues for Pemform. However, the amusement park would change the surrounding area in such a way that the company could expect a $20,000/year drop in revenues.

The Pemform management is friendly with the real estate agent handling the negotiations for the land. At this time the real estate agent says there is a 40% chance the shopping mall purchasers will obtain the land, a 30% chance the amusement park people will win out, and a 30% chance that everything will fall through and the land will remain vacant.

The person who owns the building where Pemform's San Jose store is located has heard that the land across the street is "going to be developed." On this basis the owner is demanding a long-term lease at a rent increase amounting to an additional $7,000/year beginning this month.

The Pemform store could be moved to another nearby location where the rent and operating costs would be approximately the same as are currently being paid. The location consultant estimates that revenues in the new location would be approximately $10,000/year lower than at present, and would be unaffected by the development of the vacant land. The incremental costs of the move would be nearly negligible in the eyes of the management.

a. Draw a decision tree showing Pemform's problem.
b. Calculate the EMV values for each decision. (Ignore discounting of annuities).
c. What would be the optimist's decision for Pemform?
d. What would be the maximum likelihood decision for Pemform?
e. Could this problem be presented in a payoff table?

18. A manufacturing company faces a serious problem. Its chief raw material is copper. However, there is a national copper strike in progress. The company's supplier of copper can no longer ship copper to the company. However, the company could buy what it needs for a large contract from a reputable foreign supplier. Thus, the company's major problem is deciding if it will wait out the copper strike or if it will buy from the foreign supplier in order to satisfy the contract it is currently working on.

If the company decides to wait out the strike, then the President of the United States may invoke the Taft-Hartley Act forcing the strikers back to work. There seems to be about a 60% chance he will do so. If he does, then labor officials indicate there is about a 50–50 chance that the workers will obey the order. If they obey the order, the contract will be completed since copper will become available, and the company will make $400,000 on the contract. However, if the contract is not completed (due to the workers' not going back to work or for any other reason), the company will lose $200,000. If the President does not invoke the Taft-Hartley Act, there is still about a 10% chance that the strike will be settled in time for the company to finish the contract. But, of course, there has to be a 90% chance that the contract will not be finished under these circumstances and the company will lose the $200,000.

If the company decides to purchase copper from the foreign supplier, there is some question as to whether the supplier will be able to make delivery in time for the contract to be completed. The purchasing manager has stated, "I think there's an 80% chance that they can get the stuff here on time." Even if they deliver on time, however, there is only a .5 chance the company's own employees would not be willing to use foreign materials bought in order to help break another domestic union's strike. The ramifications of buying the foreign copper are also financial. Since the copper is a special type that cannot be used on other orders in the foreseeable future, the extra money spent on the foreign material would be lost if the contract is not finished. The extra cost (due to shipping and high prices brought on by the copper shortage) is $100,000. Thus, if the company uses foreign copper on the contract, its profit will be reduced to $300,000. The loss on the contract also goes to $300,000 if the foreign material is purchased but the contract can't be completed through late delivery or the company's own worker boycott of the material.

Construct the decision tree outlining the problem facing the company. Determine the best course of action using the EMV criterion.

19. The Martvig Construction Company is trying to decide which of two construction jobs it should bid on. The first is a job for constructing a bridge over a canyon. The second is the construction of a small highway interchange. Mr. Martvig estimates that there is a 50–50 chance he could win the bridge contract and a 60% chance he could win the interchange contract. However, he has determined that he will bid on only one of the two jobs. On the highway interchange job, he plans that his bid (if he makes it) will be for $2.0 million. He estimates his costs will run about $1.8 million. His bid for the bridge (if he makes it) will be $1.5 million. However, he is not sure what his costs will be. The costs depend on how much drilling will have to be done at the site where the bases to the bridge will be anchored on either side of the canyon. If the amount of drilling required is moderate, then the cost of meeting the contract will be about $1.0 million. However, the amount of drilling required may be extensive. Then the costs depend on who will do the drilling. Mr. Martvig has three drilling companies that he subcontracts from time to time. The companies are Miller Drilling, Bentley Properties, and Gahin Probers. If Miller does the job, they will charge $800 thousand more than if the moderate drilling is required. Bentley will charge $900 thousand more than if moderate drilling is required, and Gahin will charge $1 million more than if moderate drilling is required. One would think that Mr. Martvig would naturally want to work with Miller Drilling. However, due to the work schedules of the three companies, Mr. Martvig says there is only a .3 chance he can get Miller Drilling to do the extensive drilling—if it is required. Mr. Martvig thinks there is a .4 chance he will have to go with Bentley and a .3 chance he will end up with the high-priced Gahin doing the work. If the amount of drilling required is moderate, Mr. Martvig says that his own drilling people can handle the drilling job, and their cost has been built into the $1.0 million cost of meeting the contract just mentioned. Since the cost picture on the bridge contract is so heavily influenced by the type of drilling that might be required, Mr. Martvig asked a geologist to look at the situation. The geologist's evaluation of the situation was summarized in the last statement

of the report she submitted: "So many factors will determine what type of drilling will be required that we can only estimate the probabilities of what you will find once construction begins. We estimate there is a 30% chance that extensive drilling will be required. Thus, there is a 70% chance that moderate drilling will suffice."

Draw the decision tree facing Martvig Construction Company. At the ends of the branches, indicate the "profit" that will result if the company ends up at that point. Then indicate which bid the company should enter, on the basis of EMV.

20. The Boardman Aluminum Plating Company has been troubled recently by a process in which they do aluminum plating on an airplane engine part. The problem is due to irregular flows of electricity during the plating process. Based on historical data, the figures show that about 80% of the time they get a uniform flow of electricity and each batch of parts (there are 1000 in a batch) comes out 90% good. The other 20% of the time they seem to get an irregular flow of electricity during the plating process and end up with only 75% good parts.

Boardman engineers are currently examining three alternatives for solving this problem. The first is to do what they do now. They send the batches on to assembly where "touch-up" work on the defectives is done. Each such touch-up costs an added $12 per part. The second alternative is to set up a special new rework area where defectives could be reworked immediately after the plating process but before they go to assembly. All items coming out of such a rework area would be good ones. However, the added fixed and variable costs of the rework area figure out to be roughly $10 per part for the first 150 parts that had to be reworked and $15 per part for every part over 150. The extra cost is due to overtime pay required after 150 parts. The third alternative would be to install a voltage regulator on the plating process. The cost accountants have spread the cost of such a device and its maintenance over the useful life of the regulator and the plating process and have determined that it costs about $2200 per batch. Even with the regulator, however, only 90% of the parts would come out as good, regardless of the uniformity of the electricity flow.

Set up Boardman's payoff table and determine the EMVs for each alternative.

21. Mr. J. B. Highroller owns several metal fabricating firms in the Midwest. He has been quite successful in recent years and is now considering a major expansion. He plans to expand by purchasing 100% of the stock in either Company A or Company B.

Company A has 1 million shares of stock outstanding, and they could be acquired for $6 million. In calculating the desirability of the purchase Mr. Highroller would charge the purchase price of the stock against the stock's first three years' earnings. The annual earnings per share of Company A are currently $2.50 per share. However, the future earnings depend on antipollution legislation of the state where Company A is located. If Bill No. 1 is passed, J. B.'s accountants figure that Company A's earnings per share will drop to $2.25 for the next three years. If

Bill No. 2 is passed, the per share earnings are figured at $2.00 for three years. Mr. Highroller's executive vice-president has investigated the chances of passage for these bills and says they are:

> No legislation passed: 20%
> Bill No. 1: 30% chance of passage
> Bill No. 2: 50% chance of passage

Company B can be purchased for $5 million. Its earnings per share are currently $3, and there are 800,000 shares outstanding. The future earnings of the company depend on the outcome of a suit which has been filed against it by two competing firms. If Company B loses the court fight, it will have to pay damages and fines which are expected to reduce earnings per share to $1 for the next few years. Since Mr. Highroller calculates the desirability of his investments based on results in the first three years, the amount of time it will take to settle the suit affects his evaluation. J. B.'s lawyer estimates that there is a .4 probability that the suit will be ruled on one year from the purchase date, and a .6 probability that it will be ruled on two years after the purchase. She also estimates that there is a .5 probability of Company B's winning or losing the suit regardless of the date of its settlement.

a. Construct Mr. Highroller's decision tree. At the ends of the branches, indicate the annual cash flows over the three-year period.

b. Determine the EMVs of each potential purchase ignoring discounting.

c. Determine the EMVs of each potential purchase using discounted cash flows and a discount rate of 10%.

SAMPLE DATA SET QUESTIONS: *Refer to the 113 applicants for credit listed in the Sample Data Set Appendix of this book.*

a. Consider the following payoff table:

	PROBA-BILITIES	GRANT CREDIT	DENY CREDIT
Good Credit Risk			
Poor Credit Risk			

1. Discuss the "payoffs" that should be entered in the four cells of this table. What cost and profit items should be considered in arriving at these "payoffs"?

2. Discuss the probabilities that should be entered in this table. Do they differ from one applicant to the next? Could they be estimated for "the average applicant," based on data likely to be available in the department store's credit office?

b. The credit manager of the department store has said: "We use the expected monetary value criterion in determining whether to grant credit to individuals, where possible. We feel this is more appropriate than any of the other commonly used criteria." Discuss the validity of this statement.

REFERENCES

16.1 Baird, Bruce F. *Introduction to Decision Analysis.* Duxbury Press, North Scituate, Mass., 1978. Chapters 5–8.

16.2 Bierman, Harold, Jr., Charles P. Bonini, and Warren H. Hausman. *Quantitative Analysis for Business Decisions,* 5th edition. Richard D. Irwin, Homewood, Illinois, 1977. Chapters 3 and 4.

16.3 Hammond, John S. "Better Decision with Preference Theory," *Harvard Business Review,* November–December, 1967.

16.4 Harris, Roy D., and Michael J. Maggard. *Computer Models in Operations Management.* Harper & Row, New York, 1972. Exercise 4.

16.5 Miller, David W., and Martin K. Starr. *Executive Decisions and Operations Research,* 2nd edition. Prentice-Hall, Englewood Cliffs, 1969. Part II.

16.6 Schlaifer, Robert. *Analysis of Decisions Under Uncertainty.* McGraw-Hill, New York, 1969.

17
DECISIONS
AND
EXPERIMENTS

The decision problems of Chapter 16 all involved at least three elements: (1) they had alternatives from which the decision maker had to choose, (2) they had uncertain states of nature or events that might take place, and (3) they had variable payoffs which were conditional upon what action was taken and what event occurred. Without each of these elements the decision maker would have no problem. If he had no alternatives, he would face no decision; if he knew what event were going to occur, his decision would be trivial; and if all the payoffs were the same rather than being conditional upon the events and actions, then it would not matter what the decision maker decided to do. This chapter concentrates on the second element of these problems—the uncertainty of the events and how this uncertainty can be measured and expressed.

In certain types of betting games it is possible to calculate the exact probabilities associated with each of the game's events. But managers of public and private enterprises do not deal with betting games; they deal with the real world. Thus, the problem of assigning probabilities to the events facing them is not a matter that can be resolved with exactness. Sometimes, managers can estimate the chances that certain events will occur by looking at historical data and determining how often each of the events has occurred in the past. In a situation where no historical data are available or where the problem is one never faced before (like building a particular experimental rocket), a manager may have only subjective estimates of the probabilities involved. The problem with probabilities based on past experience is that in using them we automatically assume that the future will be like the past. As for subjective probabilities, they are at best only educated guesses and can be very inaccurate (see Section 16.2).

Nevertheless, a fair degree of reliability in probability forecasts is not completely out of the decision maker's reach. This is because one of the actions often open to a decision maker is to delay the decision and obtain information allowing him to specify more accurately the event probabilities of the problem. He usually does this by undertaking an objective survey, sample, or experiment. The sections of this chapter deal with methods that can be used to make sure that subjectively determined probabilities are to some extent reliable; methods of combining subjective or historical probabilities with the results of current, objective information to obtain better event probabilities; and guidelines for determining just when objective, experimental information should be sought.

17.1 SUBJECTIVE PROBABILITIES

When the decision maker faces events that affect the payoffs he may receive in a particular situation and when he has no historical information to guide his assessment of the event probabilities, he has no alternative but to use his best guesses as to what the probabilities of the events might be. When there are only a few events that might occur, this may not be too difficult a task.

Consider the case of a company—call it the HC&B Company— expecting a shipment of parts to arrive either this week or next week. Shipment receivers feel that there is a .4 chance the shipment will arrive this week and a .6 chance it will arrive next week. Also, they feel that there is about a .25 chance that the shipment may arrive damaged. The set of events faced by HC&B is shown in Figure 17.1 (which might be a set of event forks on a large decision tree). Factors

FIGURE 17.1   HC&B Company's shipment problem

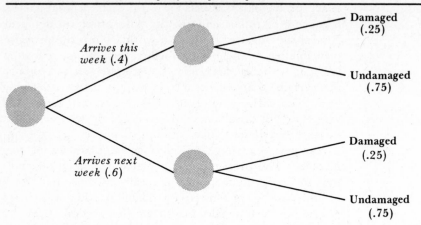

that might influence HC&B's estimate of the arrival probabilities would include their experience with the supplier, the last information they had concerning the location of the shipment, the delivery records of the railroads and trucking companies that might handle the shipment, and the season of the year. The factors influencing their subjective estimates of damage probabilities would be their record of past experience with damaged goods, their knowledge of shipping facilities at the supplier's location, the reputation for damaged goods of the railroads and trucking companies handling the shipment, and their own record for damaging goods as they are received.

The more knowledge the company has about factors such as these, the more likely their subjective probabilities are to be accurate. But even then, the probabilities are only guesses. The person making the subjective probability estimates does have one way of checking himself for realistic estimates. He can use probabilities like the values of .6, .4, .75, and .25 arrived at by the HC&B Company to compute the probabilities of *compound,* or joint, events and see if these probabilities seem realistic. For instance, if the arrival time and condition of the shipment upon its arrival are considered independent, then the joint probabilities of arrival time and shipment conditions can be found by multiplying probabilities together:

P(arrives this week damaged) = (.4)(.25) = .10

P(arrives this week undamaged) = (.4)(.75) = .30

P(arrives next week damaged) = (.6)(.25) = .15

P(arrives next week undamaged) = (.6)(.75) = .45

If the probabilities of the joint events seem unrealistic to the person making the estimates, he knows he must change the estimates of the individual events. If on examining the joint probabilities above he realizes that there is actually about a two-thirds chance that the shipment will arrive next week undamaged, then he knows that he must either change that two-thirds estimate or change the probabilities on arrivals and/or damage. If he holds to his estimate that P(arrives next week undamaged) = 2/3, then he must make an upward adjustment in either the P(arrives next week) or P(undamaged) or both.

Checking the consistency of subjective probabilities of simple events by calculating the joint probabilities of compound events is a fairly easy task when there are only a few events involved. However, many decision situations involve the possibility of numerous events. The task of finding subjective probabilities for all these events is not so straightforward.

Consider the problem faced by a manufacturing plant using a certain chemical in its operations. Assume for a moment that production planners feel usage of this chemical is most likely to fall in the range of 50,000 to 80,000 gallons. Perhaps they feel that there is quite a large chance—say as much as a 50 percent chance—that usage will lie outside this range and a very small chance that usage will be very high—as much as 120,000 gallons. Also assume they feel sure the usage will definitely not fall below 35,000 gallons. In this case there are 120,000 − 34,999, or 85,001, possible events if usage is to be measured to the nearest gallon. There are even more if we were to measure usage more finely.

It would be unrealistic to try to assign probabilities to every possible usage value measured to the nearest gallon. Our planners would do far better to treat the usage figure like a continuous variable and try to put some of their subjective feelings about usage to work to obtain a smooth usage probability curve like the one in Figure 17.2. This curve fits the description of the usage probabilities given previously. About 50 percent of the area under the curve is between 50,000 and 80,000 gallons. The curve terminates at 35,000 and 120,000 gallons and is slightly skewed to the upper side. The probability that usage will be between any two figures could be taken from this curve by measuring the area under the curve between those two points and expressing that area as a proportion of the total area under the curve. The areas could be found using the methods of integral calculus or rectangular approximation, that is, by using a series of rectangles (such as those in the figure) having a total area similar to the area under the curve. It is a simple matter to calculate the areas in the rectangles, and these areas are good approximations to the areas under the curve.

Obtaining a curve like the one in Figure 17.2 can be quite difficult.

FIGURE 17.2   Probability of chemical usage, showing rectangular approximation of areas

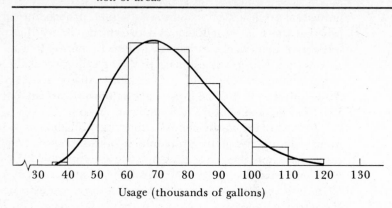

It is usually easier to construct a cumulative probability curve. The cumulative curve corresponding to the curve in Figure 17.2 is shown in Figure 17.3. This cumulative probability curve can be obtained in the following way: First, the person constructing the curve determines the usage figure that he believes divides all possible usage values into two equally likely halves. This value determines where the .5 probability

FIGURE 17.3   Cumulative probability curve for use of a chemical by a manufacturing plant

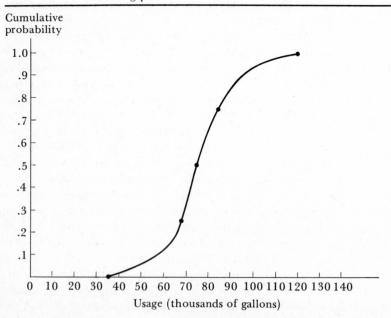

point will go on the cumulative curve. For this problem, assume the company determines that usages above or below 75,000 gallons are equally likely. Thus, the first point on the cumulative curve is (75, .5). Next, the two extreme points of the curve are plotted. Since the company is sure that usage will lie between 35,000 and 120,000 gallons, the cumulative probability that usage is 120,000 or less is 1.0 and the probability that it is 35,000 or less is .0. Thus, the next two points on the curve are (35, .0) and (120, 1.0). This process can be continued by determining the usage value that the company feels will be exceeded with only 25 percent probability. This value may be 85,000 gallons and thus gives the cumulative curve point (85, .75). Once again, the company would determine a usage value. This figure should be chosen so that there is only a 25 percent chance that usage would fall below it. Assume that this value is 68,000 gallons. This gives the cumulative probability curve the point (68, .25). The cumulative curve in Figure 17.3 was obtained by fitting a smooth curve through the five points developed previously. It would be possible to continue the process outlined and develop many more points for the curve, but the increased work may not be justified by the small amount of added accuracy.

The cumulative curve can be used to find the probabilities that usage will be between any two values, and rectangular approximation of areas is not needed. Assume, for instance, that the company's planners desire to know the probability that usage of the chemical will lie between 80,000 and 90,000 gallons. The cumulative probability figure for 80,000 gallons is read from Figure 17.3 as .66, and the figure for 90,000 gallons is .84. Thus, the probability that usage will lie in that 10,000-gallon range is .84 − .66, or .18.

Suppose that the company has a policy of ordering the chemical in question in 10,000-gallon lots in order to obtain the best possible bulk price. The chemical is known to deteriorate or spoil if it is not used over a specified period, so the planners want to order the correct amount. They could construct a payoff table showing the alternatives of ordering 40,000 gallons, 50,000 gallons, etc., up to 120,000 gallons. The events in this table should be the various usage levels expressed in 10,000-gallon ranges:

E_1 = 39,999 gallons or less E_6 = 80,000 to 89,999 gallons

E_2 = 40,000 to 49,999 gallons E_7 = 90,000 to 99,999 gallons

E_3 = 50,000 to 59,999 gallons E_8 = 100,000 to 109,999 gallons

E_4 = 60,000 to 69,999 gallons E_9 = 110,000 to 120,000 gallons

E_5 = 70,000 to 79,999 gallons

The probabilities associated with these events can be calculated from the cumulative probability curve in Figure 17.3. A very precise reading of the curve would give the following calculations.

$$P(E_1) \qquad\qquad = .02 \qquad P(E_6) = .84 - .66 = .18$$
$$P(E_2) = .05 - .02 = .03 \qquad P(E_7) = .94 - .84 = .10$$
$$P(E_3) = .12 - .05 = .07 \qquad P(E_8) = .98 - .94 = .04$$
$$P(E_4) = .32 - .12 = .20 \qquad P(E_9) = 1.00 - .98 = \underline{.02}$$
$$P(E_5) = .66 - .32 = .34 \qquad\qquad\qquad \text{Total} = 1.00$$

In each case the probability listed first is the cumulative probability that chemical usage will be *equal to or less than* the value at the upper end of the range specified by the event, and the probability that is subtracted is the probability that chemical usage will be *less than* the value at the lower end of the range specified by the event. The difference is then the probability that usage will lie *in* the range specified by the event. These nine probabilities could now be entered into the probability column of a payoff table. The payoff table would be nine rows (events, or actual amount of the chemical used) by nine columns (actions, or amount ordered) and would contain 81 entries of costs X_{ij} associated with the chemical order sizes and potential usage value over the next ordering period.

This table is not constructed here since we are concentrating only on the probabilities of the events. What we should note is that it has been possible to obtain nine rather precise and reasonable subjective probabilities for chemical usage from the cumulative probability curve in Figure 17.3. It would have been much more difficult to estimate these probabilities using direct guesses or rectangular approximations from the curve in Figure 17.2. The problem remains, however, that these are subjective probabilities and are not based on historical data or objective experimentation. The following section discusses methods by which subjective probabilities can be modified using objective experimental information to yield a revised distribution of event probabilities representing a combination of the subjective and the experimental information.

17.2   REVISION OF PROBABILITIES

Suppose a construction firm has been told that it will be awarded a contract to build a large flood-control system, contingent upon the state legislature's voting funds for the project. The construction company feels there is a .6 chance that the bill approving funds for the project will pass and a .4 chance it will not. These are subjective probabilities, based on the company president's "feel" for the mood of the legislature and some informal discussions with two golf partners who serve in the legislature.

Since the subjective probabilities are not at all lopsided in favor of the bill's passage, the company's management are not sure if they should begin substantive planning for the project, hire new personnel, and purchase new equipment. They realize that they might supplement their own subjective probabilities about the bill's passage with objective information published in a local newspaper. The paper regularly predicts the fate of pending legislation based on its political editor's interviews with key legislators. The paper lists its record of accuracy in the following way: For bills that eventually passed, the paper has correctly predicted passage 80 percent of the time and incorrectly predicted failure 20 percent of the time. For bills that eventually failed, the paper has correctly predicted failure 95 percent of the time and incorrectly predicted passage 5 percent of the time. If we let P denote that a bill passes, F denote that it does not pass, PP denote that passage was predicted, and FP denote that failure was predicted, then the paper's record of accuracy can be presented as *conditional* probabilities:

$$P(PP|P) = .80 \qquad P(FP|P) = .20$$
$$P(PP|F) = .05 \qquad P(FP|F) = .95$$

The first probability, for example, is the probability that the paper will predict passage, given that the bill will pass. Let us assume for a moment that the paper predicts passage for the bill funding the flood-control project. The company will be interested in the probabilities in the reverse order of those given previously, i.e., the probability that the bill will pass, given that it is predicted to pass, $P(P|PP)$, and the probability that it will fail, given that it is predicted to pass, $P(F|PP)$. They have the published accuracy record and the **prior probabilities** of the company president, $P(P) = .6$ and $P(F) = .4$, with which to work. These probabilities are called prior probabilities because they are the probabilities of passage for the bill *prior* to performing any experiment or obtaining any objective information.

A formula for combining the known prior and conditional probabilities in order to obtain $P(P|PP)$ and $P(F|PP)$ was presented as Expression (4.20) in Section 4.7. This formula is called *Bayes' rule* after the same Thomas Bayes for whom the Bayes decision criterion is named. That formula is presented here again using the notation developed in this problem.

$$P(P|PP) = \frac{P(P) \cdot P(PP|P)}{P(PP)}$$

(continued on next page)

$$= \frac{P(\text{P}) \cdot P(\text{PP}|\text{P})}{P(\text{P}) \cdot P(\text{PP}|\text{P}) + P(\text{F}) \cdot P(\text{PP}|\text{F})} \qquad \textbf{(17.1)}$$

and

$$P(\text{F}|\text{PP}) = \frac{P(\text{F}) \cdot P(\text{PP}|\text{F})}{P(\text{PP})}$$

$$= \frac{P(\text{F}) \cdot P(\text{PP}|\text{F})}{P(\text{P}) \cdot P(\text{PP}|\text{P}) + P(\text{F}) \cdot P(\text{PP}|\text{F})} \qquad \textbf{(17.2)}$$

In Section 4.7 we noted that expressions like (17.1) and (17.2) are somewhat formidable in appearance. However, probability diagrams, which are similar to decision trees, can be used to demonstrate the meaning of the formulas. The example of determining if the bill authorizing funds for construction of the flood control project will pass is diagrammed in Figure 17.4. Note that the events "Bill will pass" and "Bill will not pass" are assigned the company president's subjective probabilities of .6 and .4, respectively. The second set of forks in the tree present the newspaper's record for accuracy. The problem we face is how to revise the .6 and .4 probabilities to reflect the newspaper's record for accuracy.

FIGURE 17.4 Probability tree for passage or nonpassage of bill

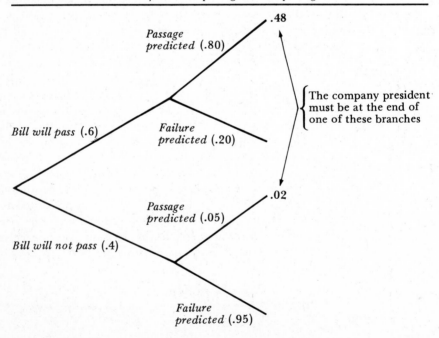

We assumed previously that the paper had predicted passage for the bill. That means the company president must be standing at the end of the top branch in the tree or at the end of the third branch down. This is because the president *knows* that the paper has predicted passage of the bill and he is no longer concerned with the event that failure might be predicted. The president just doesn't know whether it has predicted passage for a bill that will eventually pass or not pass. The joint probability that the bill will pass and that the paper would predict passage is $(.6)(.80) = .48$, which is listed at the end of the top branch. The joint probability that the bill will fail but that the paper would predict passage is $(.4)(.05) = .02$, which is listed at the end of the third branch. The chance of the company president's being at the end of either the first or the third branch is the sum of the two probabilities $.48 + .02 = .50$. The fact remains, however, that the president *is* at the end of one of those two branches since the paper predicted the bill would pass. So in determining the probabilities that the bill will pass or not pass, it seems reasonable that the president should allocate probabilities to passage and nonpassage that are proportional to the contribution that the "Bill will pass" and "Bill will not pass" branches make to the total probability of .50. That is,

$$P(\text{P}|\text{PP}) = \frac{.48}{.48 + .02} = .96 \qquad P(\text{F}|\text{PP}) = \frac{.02}{.48 + .02} = .04$$

In this process we have divided the .50 chance that the paper predicts passage into its component parts and have assigned relative weights of .96 and .04 to the chances that the prediction was made on a bill that will pass or a bill that will not pass. Again, this is possible since the president is no longer concerned with the event that the paper predicts failure.

This is exactly what Expression (17.1) accomplishes. If we examine the numerator we find $P(\text{P}) \cdot P(\text{PP}|\text{P})$, which is $(.6)(.80) = .48$. In the denominator we find the sum of two expressions: $P(\text{P}) \cdot P(\text{PP}|\text{P}) + P(\text{F}) \cdot P(\text{PP}|\text{F})$. These values are $(.6)(.80) + (.4)(.05) = .48 + .02$, which is the denominator used previously in our more intuitive approach. By a similar analysis we can show that Expression (17.2) also gives $P(\text{F}|\text{PP}) = .04$. The .96 and .04 figures are the *revised probabilities* of passage and failure of the bill based on both the prior probabilities of the company president and on the newspaper's prediction of passage. We can see that the chance of passage now seems very high. This is due to the fact that the newspaper predicts passage, and its record for accuracy is quite good.

Some people find that the computation of revised probabilities is more easily accomplished using a table. Table 17.1 shows the calculation of the revised probabilities in tabular form. The first column of the

TABLE 17.1   Revision of Probabilities for Bill Passage Given a Prediction of Passage

COLUMN 1	COLUMN 2	COLUMN 3	COLUMN 4	COLUMN 5
			JOINTS =	
POSSIBLE			*PRIORS ×*	
EVENTS	*PRIORS*	*CONDITIONALS*	*CONDITIONALS*	*REVISED*
Passage	.6	.80	.48	.96 = $P(P|PP)$
Failure	.4	.05	.02	.04 = $P(F|PP)$
	1.0		.50 (Marginal)	1.00

table contains the events that might occur—passage or failure of the project's funding bill. The second column contains the prior probabilities concerning these events—$P(P)$ and $P(F)$. The third column is the critical column. It contains the conditional probabilities. The conditional probabilities are always *the probabilities of observing the objective information obtained, given that the event specified in the row is true.* Thus, the .80 value in the first row of the third column is the conditional probability that the bill's passage would be predicted (the observed objective information), given that the bill will eventually pass (the possible event in the first row). The fourth column of Table 17.1 is the joint probability column. It is found by multiplying the prior column, term for term, by the conditional column. The .48 figure is the joint probability that the bill will pass *and* that success would be predicted for it, .6 × .80 = .48. This is also the figure that corresponds to the numerator in Expression (17.1). The .02 figure is the numerator of Expression (17.2) and is the joint probability that the bill will fail *and* that passage would be predicted for it, .4 × .05 = .02. The sum of the probabilities in the fourth column is the marginal probability. The fifth column contains the revised probabilities of passage and failure of the bill, $P(P|PP)$ and $P(F|PP)$, and these are found by dividing the marginal probability into each of its joint components: .48/.50 = .96 and .02/.50 = .04.

The revised probability that the bill will pass is now .96 as opposed to the prior value of .6. This is consistent with what one would expect since the objective information, the newspaper's prediction, favors the passage of the bill. Since the bill's passage now seems quite sure, the company can proceed with more confidence in making its plans for the future. It could use the revised probabilities in any payoff table where the events listed in the table were passage or failure of this particular bill.

To further illustrate the method of revising probabilities, Table 17.2 shows how the prior probabilities would have been revised *if the newspaper had predicted the bill would fail.* The conditional column in Table 17.2 shows the probabilities that the prediction of failure would

TABLE 17.2 Revision of Probabilities for Bill Passage Given a Prediction of Failure

POSSIBLE EVENTS	PRIORS	CONDITIONALS	JOINTS = PRIORS × CONDITIONALS	REVISED
Passage	.6	.20	.12	.24 = $P(P\|FP)$
Failure	.4	.95	.38	.76 = $P(F\|FP)$
	1.0		.50 (Marginal)	1.00

be made for a bill that eventually will pass and fail, respectively. That is, the conditionals are $P(FP|P)$ and $P(FP|F)$ and were given in the newspaper's record of accuracy. The joints are found, once again, by multiplying the prior column by the conditional column. The revised probabilities are then merely the individual joint probabilities divided by the marginal: $P(P|FP) = .12/.50 = .24$ and $P(F|FP) = .38/.50 = .76$.

Thus, if the newspaper had predicted failure for the bill, the company would have ended up feeling quite sure (probability .76) that the bill would, in fact, fail to pass the legislature. This is, of course, contrary to the subjective feeling of the company president, but the newspaper's record for accuracy for predicting bill failures in the past has been so outstanding that were they to predict failure, the revised probabilities heavily favor failure.

The method for revising probabilities demonstrated previously can be used in numerous situations for revision of priors. The only variations are that there may be more possible events and the method of finding the conditional probabilities differs from problem to problem.

EXAMPLE 1

Problem: A company is preparing to market a new product. The marketing vice-president has had a great deal of experience with new products and feels in a position to predict that the new product may capture 10, 20, or 30% of the market during its first year. His prior probabilities about the product's ability to capture those market shares are .25, .55, and .20, respectively. (Actually, the marketing vice-president probably realizes that the product might capture some intervening value like 18% but has used 10% to mean "between 5 and 15" and 20% to mean "between 15 and 25," and so forth.)

EVENTS	PRIORS	
.10	P(Market share = .10) =	.25
.20	P(Market share = .20) =	.55
.30	P(Market share = .30) =	.20
	Total	1.00

It is important to distinguish, in this case, between the events and the prior probabilities about the events. The events are listed as .10, .20, and .30, and these represent the market shares that might be captured. The priors are the probabilities of the events. Note that the priors add up to 1.00 but that the sum of the values in the event column is meaningless and does not have to add up to anything in particular.

The head of market research wishes to take a survey and use its results to check and, if necessary, modify these priors. That is, he would like to obtain some objective sample information which can be combined with the subjective priors to yield a set of revised probabilities concerning the various market shares which might be captured. In a market survey of 25 randomly selected customers, he finds only three, or 12%, who say they would buy the product. Intuitively he suspects that this poor showing of the product's saleability will change the priors about the market shares in the direction of placing heavier weight on the chance that there is only a 10% market awaiting the new product.

Solution: Table 17.3 shows the procedure for revising the priors. The table has the same structure as Tables 17.1 and 17.2. The only difference is in the way the conditional probabilities were obtained. In the construction company examples the conditionals were obtained from the newspaper's record for accuracy in predicting passage and failure of pending legislation. But when the events are *proportions*, as they are in this problem, then the conditionals can be obtained using the binomial formula or the binomial table (Table III in the Appendix). The conditionals, remember, are the probabilities of observing the objective information obtained, assuming each of the events is true.

Thus, the conditionals in Table 17.3 are the binomial probabilities of obtaining 3 successes in 25 trials from populations which contain 10, 20, and 30% successes, respectively. In the notation of Chapter 5, these probabilities are:

$$P(3) = \binom{25}{3}(.10)^3(.90)^{22} = .2265 \qquad \begin{cases} \text{assuming the true} \\ \text{market share is 10\%} \end{cases}$$

$$P(3) = \binom{25}{3}(.20)^3(.80)^{22} = .1358 \qquad \begin{cases} \text{assuming the true} \\ \text{market share is 20\%} \end{cases}$$

TABLE 17.3 Revision of Probabilities for New Product Market Share

POSSIBLE EVENTS	PRIORS	CONDITIONALS	JOINTS = PRIORS × CONDITIONALS	REVISED
.10	.25	.2265	.05663	.41
.20	.55	.1358	.07469	.55
.30	.20	.0242	.00484	.04
	1.00		.13616 (marginal)	1.00

$$P(3) = \binom{25}{3}(.30)^3(.70)^{22} = .0242 \qquad \left\{ \begin{array}{l} \text{assuming the true} \\ \text{market share is } 30\% \end{array} \right.$$

The joint probabilities are, once again, merely the product of the priors and the conditionals for each row of the table. And the revised probabilities are the joints divided by the marginal. Just as the head of market research suspected, the revised probabilities differ from the priors in that the chances of achieving a 10% market share now appear to be much greater than before and the chances of achieving a 30% market share now appear negligible.

It should be noted that when experimental, or objective, information is used to revise a set of prior probabilities, all the prior values usually change. The fact that both the prior and the revised probabilities of a 20 percent market share were .55 is only a coincidence of our example.

One may ask why we even bother to use the subjective prior probabilities in any of these problems where we have experimental, objective information upon which to base probability statements about the events. Why didn't we, for instance, simply use the methods discussed in Chapter 8 to construct a confidence interval for p, the potential market share for the new product, based on the value of f, the proportion of people in the sample saying they would buy the product? Some statisticians, those who have very little faith in the accuracy of the *subjective* probabilities that are revised using the methods described in this chapter, would argue that such a confidence interval based on only *objective* information would be far more trustworthy than a probability forecast incorporating guess-type subjective information. But the people who feel that the revision of probabilities using experimental objective information is a useful process support their view by saying that we all have hunches, insights, and intuitive feelings about the probabilities that certain events are going to take place. They suggest that we use these subjective feelings in our decision making anyhow and that their approach to the revision of event probabilities provides a framework within which to combine subjective feelings with objective information. Also, they claim, a person who is familiar with a situation can often come up with subjective event probabilities that are very accurate, and we would be foolish to throw out this valuable information and proceed only on the basis of the objective data.

Both positions, of course, are correct to a degree. When there is no basis upon which to arrive at prior judgments about the event that is going to take place, then we have no information with which to construct priors. In this case, it would probably be better to use only the objective information (perhaps in the form of a confidence interval) to estimate event probabilities rather than to assume that all possible events are equally likely and then revise this uniform prior distribution. On the other hand, if there is a great deal of historical information about which event might take place, or if the decision maker feels quite sure about subjective priors, it would be best to include this information in the decision-making process by constructing a prior probability distribution and then revising it using objective information.

17.3   WHEN TO EXPERIMENT

One final question has to be answered before we leave the topic of experimentation and revision of probabilities: How does one decide when it is appropriate to use prior probabilities in the decision tables presented in Chapter 16 and when it is best to revise these priors by performing some sort of experiment in the form of a sample or a survey or a test? There is no simple rule that can answer this question. Some mathematical methods can be employed in certain instances, but they are beyond the scope of this book (see Reference 17.1). However, there are a few general guidelines that can be adopted.

When the expected value of perfect information calculated from a payoff table is very high and the cost of the experiment is very low, then it will usually pay to perform the experiment and revise the payoff table's probability column. This may be the case in many problems. For instance, gathering more information often involves no more than making a few telephone calls. In the construction company example of Section 17.2, the company desired to know if a certain bill were going to pass the state legislature. It was able to revise its probabilities concerning the bill's passage by merely obtaining the local newspaper's forecast of the bill's fate and the newspaper's record for accurate predictions. This type of information gathering can usually be done very cheaply. On the other hand, a survey to determine consumer buying habits may cost several thousands of dollars and would not be a worthwhile experiment if it were undertaken in order to revise the probability distribution in a problem where the EVPI is $600.

More than the cost of the experiment is involved. Only experiments which promise to provide accurate information should be undertaken. For instance, it would not have been wise to incorporate the newspaper's prediction about the passage of the flood-control project's funding bill if the newspaper had been successful in its predictions only 35 or 40 percent of the time. But it may pay to perform a survey even though its cost is high and close to the EVPI of a problem if the information obtained is expected to be highly accurate. That is, we might be willing to spend almost as much as the expected value of perfect information if our information is nearly perfect in resolving the uncertainty we face.

PROBLEMS

1. Determine if the following probability statements are consistent with one another.

   a. There is a .4 chance the stock will drop. There is a .2 chance that Sue Bradley will sell the stock if it drops. There is about an 8% chance that the stock will drop *and* Sue Bradley will sell it.

   b. There is a .5 chance that the economy will be up next year. There is a .8 chance that our sales will go up when the economy is up. Thus, there is a .4 chance that our sales will be up next year.

   c. There is a .5 chance we will get the contract and a .1 chance that Smith will quit if we do. There is a .6 chance Smith will quit if we lose the contract. There is a .25 chance Smith will quit.

   d. There is a .7 chance that management will decide to market the QB-7. The chances of high, moderate, or low sales for the QB-7 are .4, .3, and .3, respectively. The probability that the product will be marketed *and* will achieve moderate sales or worse is .56.

2. A salesperson is trying to construct a probability curve which will indicate the likelihood of his selling various amounts of the company product next month. He feels there is a fifty-fifty chance of his selling more than or fewer than 2000 cases. He feels there is only a 25% chance that he can sell more than 2800 cases and only a 25% chance he will sell less than 1600. He has never sold more than 3200 nor fewer than 1300 cases in one month.

   a. Construct a cumulative probability curve for this salesperson.

   b. Use the curve to find the probability that next month's sales will be above 1800 cases.

   c. What is the probability that this salesperson will sell between 2200 and 3000 cases?

3. A certain large nonprofit organization is self-insured. A benefits officer is trying to establish probabilities associated with various levels of claims on the organization's health insurance next year. She feels there is a fifty-fifty chance that claims will be more than or less than $225,000. She feels that the chance of having claims of $350,000 or more is .25 and the chance that they will be $200,000 or less is about .4. There is virtually no chance that claims will be less than $150,000 or more than half a million dollars.

   a. Construct a cumulative probability curve for the benefits officer.

   b. Find the probability that health insurance claims will exceed one-quarter of a million dollars next year.

   c. Find that probability that claims will be between $250,000 and $400,000.

4. A student feels there is a .5 chance he will get a score of 80 or better on the next 100-point midterm exam. He feels there is only a .25 chance that he will get 90 or better, and he feels there is a .25 chance that he will receive a score of 55 or worse.

   a. Construct a cumulative probability curve for the student's exam score.

   b. Grades on the next midterm are to be given as follows:

TEST SCORE RANGE	GRADE
100–90	A
75–89	B
65–74	C
50–64	D
0–49	F

What are the subjective probabilities the student should assign to getting each of the five grades?

5. Use the method outlined in Problem **4** to determine a cumulative probability distribution for your scores on your next 100-point midterm exam. Then use the grading schedule in Problem **4** to assign probabilities to your possible grades.

6. Rework the example in Table 17.1, assuming the priors for passage and failure are .20 and .80, respectively.

7. A newspaper reporter does not know if she should go to the hotel where union and management contract negotiations are taking place. She feels that there is a .2 chance a contract agreement will be reached today and a .8 chance that there will be continued negotiations. Before she makes up her mind, she calls the hotel's night manager and asks how late negotiations lasted last evening. The night manager informs her that the negotiations lasted all night long. From her years of experience, the reporter knows that all-night negotiations have preceded 80% of the contract agreement announcements. But she also knows that 30% of the nonproductive bargaining sessions go all night and produce no agreement. Construct a table like Table 17.1 and find the prior and revised probabilities for a contract agreement and continued negotiations.

8. A company buys several shipments of the R28 part each year. It buys from two manufacturers, A and B, and the parts from the two manufacturers are identical in external appearance. A shipment of parts is taken from inventory and found to be 90% defective. The company is not sure to which manufacturer the shipment should be returned. Manufacturer A usually supplies one-third of the shipments and manufacturer B supplies two-thirds. Before taking the chance of shipping the defective R28 parts back to manufacturer B, the most probable supplier, the production manager finds that this particular shipment arrived in June of last year. Company records show that last year 40% of A's shipments arrived in June while 10% of B's arrived in that month.

   a. What are the prior probabilities for this problem?

   b. What are the conditional probabilities?

   c. Find the revised probabilities that the bad shipment came from manufacturers A and B.

9. An Illinois construction firm has bid on a state highway job in a neighboring state. The company felt that it had a .35 chance of getting the job. However, it needs to revise the probability of getting the job on the basis of some new information. This new information is that the highway department in the state where the bid was made has never awarded a highway construction contract to a company outside its own state.

a. What are the events that might take place?
b. What are the prior probabilities for these events?
c. What are the revised probabilities for the events?

10. Historical records for the production output of a machine show that the proportion of defectives produced by the machine is as follows:

PROPORTION p OF DEFECTIVES	PERCENT OF BATCHES WITH PROPORTION p DEFECTIVES
.01	.7
.05	.2
.10	.1

A sample of 20 pieces has just been taken from a new batch of the machine's output, and one was found to be defective. Construct a revised distribution for the proportion of defectives.

11. The Matthews Tennis Ball Company has been having trouble controlling the quality of its output. It manufactures tennis balls in lots containing 1000 cans, and there are three balls per can. The last 50 lots have had the following proportions of defects in the form of broken pressure seals on the cans:

FRACTION DEFECTIVE	NUMBER OF LOTS
.01	10
.05	25
.10	10
.20	5

A new lot is produced, and a sample of 25 randomly selected cans is inspected. Five of these are found to have broken pressure seals. Find a revised probability distribution for the proportion defective in the new lot.

12. The EVPI for a decision problem is $9700. The probabilities for the events in this problem's payoff table can be revised using an experiment which involves a sample. The setup cost of the experiment is $5200, and individual items in the sample cost $2 each for the first 1000 items. Then the cost drops to $1.50 on the remainder, but the first 1000 still cost $2 each. What is the maximum size sample that should be taken to revise the payoff table probabilities?

13. How high would the setup costs for the sample in Problem **12** have to be for you to be absolutely sure that the sample should not be taken?

14. In Problems **12** and **13**, what other piece of information would you want before you decided to take the sample?

15. Suppose that an experiment has been used many times in the past to predict whether event 1 or event 2 is going to take place. Past experience has shown that the experiment has been wrong in its prediction every time. Should this experiment be discontinued?

16. A production process at the Alpha-Omega Corporation is set so that its usual output is about 5% defective. On about 12% of the production runs, poor material is used, and this causes the defective rate to jump to 10%. The process can be adjusted to produce 5% defectives even with the poor material, but this adjustment costs $400. Defective items cost $2 each to replace. If the adjusted process is run using good material, the defective rate is still 5%.

   a. Construct a payoff table showing the events and alternative actions for this problem. Assume a run of 15,000 items is about to be made.

   b. What is the EVPI for this problem?

17. Assume that the process in Problem 16 can be run for a very short time to produce a sample that will indicate the approximate defective rate being turned out. Obtaining a sample of 15 items costs $30 in setup costs and $2.50 per item sampled. If one or more items are found defective in the sample of 15, the adjustment process will be performed for $400.

   a. What is the probability that the sample will cause the adjustment to be made when it does not need to be made?

   b. What is the probability that the adjustment will not be made after the sample when it really should be made?

   c. What is the cost of an unneeded adjustment?

   d. What is the cost of not adjusting when an adjustment is needed?

18. Draw a decision tree showing the problem faced by the Alpha-Omega Company in Problems 16 and 17. They must choose between A_1 = adjustment now, A_2 = no adjustment, and A_3 = sample and adjust if one or more defectives are found. Include all probabilities and all costs or losses from making wrong decisions at the ends of the branches on the tree. Then determine which action has the lowest expected cost.

19. Kelso Kemical Kompany (KKK) has just completed production of a batch of RB-7, a chemical product used by one of Kelso's customers, Benson Builders (BB). Just as the batch is about to be shipped to BB, KKK's production manager learns that one of the constituent chemicals of the RB-7 may have contained impurities that might prevent the RB-7 from performing satisfactorily for BB.

   If the product is shipped to BB and does not perform satisfactorily, BB's production run incorporating the RB-7 will be ruined and KKK will have to replace BB's RB-7 at a cost of $2000 and pay BB a penalty fee for their ruined production run. According to KKK's contract with BB, the penalty is $1000 if the ruined run is a "type I" run and $400 if it is a "type II" run. The production manager knows that BB's runs are approximately 50% of each type.

   If the questionable batch of RB-7 is scrapped, a new, guaranteed-good batch will cost an additional $2000. Rather than ship or scrap the present batch immediately, the production manager can subject it to a test to determine its quality. The cost of the test, which must be done on an overtime basis, is about $500. If the test shows the batch to be bad, the production manager will scrap the batch and run a new one despite his feeling that the test is only 90% accurate. That is, 10% of the batches the test identifies as good (or bad) are really bad (or good). He feels that the probability of the test's showing the batch to be good is .7. From past experience he feels there are really about two chances in three that the batch is good.

a.   Draw a decision tree for the production manager's problem. Label all branches clearly and enter the monetary consequences. Assign probabilities to the event forks.

b.   Use the decision tree as an aid in calculating the expected cost of each choice facing KKK; i.e., fill in the remainder of your decision tree. What should be done, and what is the expected cost of that action?

20.   A number of bankers are worried about actions the Federal Reserve may take next week. After the meeting of the Federal Reserve, they feel that monetary policies will be either tightened up, making credit harder to grant, or loosened up, making it easier to loan money. Mr. Timothy Campbell, president of Union County Bank, feels there is a 40% chance that credit policies will be tightened. He says, "There is no chance that credit will remain the same after next week's meeting. It's going to be tighter or looser, I'm sure of that." In order to get more information concerning the chances that credit will be tighter or looser after next week, Mr. Campbell contacted a respected economist, Dr. Jane Rock. Dr. Rock predicted that the Federal Reserve would loosen credit, but she gave this caution to Mr. Campbell: "Remember, Tim, I've only predicted 80% of the cases where credit was loosened. And in 15% of the cases where credit was tightened, I was caught predicting a loosening." Revise Mr. Campbell's prior probabilities on the basis of Dr. Rock's prediction.

21.   Mr. T. Robson does business with Union County Bank, discussed in Problem **20**. Mr. Robson needs to borrow $100,000 which he will use for one year. At the end of the year he will repay the lender the loan plus interest in a single payment. That is, there are no monthly or quarterly payments due on this loan. However, he has to decide if he will borrow the money from the bank or from a family trust which is willing to loan him the money. The loan from the trust will be at a 10% rate of interest, but the trust also would require some expenses for setting up the loan. They would be $800. The conditions of the loan at the bank are up in the air, and Mr. Robson has to apply for the loan right now. If credit is loosened by the Federal Reserve, as discussed in Problem **20**, the interest rate on Robson's loan will be 8%—if the loan is granted. There is about a 90% chance that it will be granted under the looser rules. If the loan is granted, it will be granted for the full amount. If the loan is denied, Robson will have to go to the family trust for the money. However, if credit is tightened by the Federal Reserve, the cost on Robson's bank loan would be 10% per year. Also, under the tighter rules there is only a 75% chance that the Robson loan will be approved by the bank. Even if the loan is approved, it may only be for half the $100,000 requested. There seems to be about a 70% chance that Robson will get the full amount requested and a 30% chance that he will get only $50,000 under these tighter conditions. The remaining $50,000 he would then borrow from the family trust at the 10% interest rate and with the same $800 initial fee. The initial loan processing fee at the bank is $300 plus one-fourth of 1% of the amount loaned. The processing fee is not charged if the loan is denied.

   a. Draw the decision tree facing Mr. Robson. At the ends of the branches enter the total cost of the loan—interest and fees.
   b. Determine the expected monetary value of each action using the prior probabilities of Mr. Campbell in Problem **20**.
   c. Repeat Part **b** using the revised probabilities found in Problem **20**.
   d. Does the revision of the probabilities change the preferred action?

SAMPLE DATA SET QUESTIONS: *Refer to the 113 applicants for credit listed in the Sample Data Set Appendix of this book.*

   a. The credit manager of the department store received an application from a woman who was 21 years of age. This woman had no job, listed no additional sources of income, owed nothing to creditors, and was single. Based on the data you see in the appendix, give a subjective probability that the credit manager would grant this woman credit.
   b. Assume that just prior to his decision on the woman mentioned in the previous question, the credit manager noted that the woman's address was in a very exclusive part of town. Upon further investigation, the credit manager found that this woman was the daughter of the president of the city's second largest bank. Discuss the factors that would cause you to revise the probability you listed in the previous question. What is your subjective probability now?

REFERENCES

17.1 Baird, Bruce F. *Introduction to Decision Analysis.* Duxbury Press, North Scituate, Mass., 1978. Chapters 9 and 10.

17.2 Bierman, Harold, Jr., Charles P. Bonini, and Warren H. Hausman. *Quantitative Analysis for Business Decisions,* 5th edition. Richard D. Irwin, Homewood, Illinois, 1977. Chapters 5 and 10.

17.3 Schlaifer, Robert. *Analysis of Decisions Under Uncertainty.* McGraw-Hill, New York, 1969. Chapter 8.

NONPARAMETRIC
METHODS

Many of the methods we have considered, such as those involving the *t* test, apply only to normal populations. The need for techniques that apply more broadly has led to the development of *nonparametric* methods. These do not require that the underlying populations be normal—or indeed that they have any single mathematical form—and some even apply to nonnumerical data. In place of parameters such as means and variances and their estimators, these methods use ranks and other measures of relative magnitude; hence the term **nonparametric.**

18.1 THE WILCOXON TEST

The *median* of a continuous population is a number ξ such that an observation X from the population has probability .5 of being less than ξ and probability .5 of being greater. In symbols,

$$P(X < \xi) = P(X > \xi) = .5 \tag{18.1}$$

The sample quantity corresponding to ξ was considered in Section 3.5. The population is *symmetric* about its median ξ if, for each positive x, the observation X has the same probability of being less than $\xi - x$ as of being greater than $\xi + x$. In symbols,

$$P(X < \xi - x) = P(X > \xi + x) \tag{18.2}$$

Geometrically, this means that the shaded areas in Figure 18.1 are equal for each x, so that the frequency curve looks the same if viewed through a mirror. A normal curve with mean ξ has median ξ and is symmetric about ξ, but a nonnormal curve like the one in the figure can also have this property.

Consider the null hypothesis that the population median is ξ_0 and the population is symmetric about ξ_0, together with the alternative hypothesis that the population median exceeds ξ_0 and the population is symmetric:

$$\begin{cases} H_0 : \xi = \xi_0 \text{ symmetric} \\ H_a : \xi > \xi_0 \text{ symmetric} \end{cases} \tag{18.3}$$

If the population is normal, the t test applies to this problem. The goodness-of-fit test described in Chapter 11 can be used to determine if a population is normal. If the population is nonnormal, the t test cannot be used in the hypothesis test, and another test is needed.

Let $X_1, X_2, \ldots, X_n$ be an independent sample from the nonnormal population. The **Wilcoxon statistic** W is computed in four steps.

STEP 1. Subtract the hypothesized median, ξ_0, from each observation to get $Y_i = X_i - \xi_0$.

FIGURE 18.1 A frequency curve symmetric about its median ξ

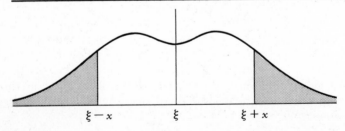

STEP 2.   Arrange the observations $Y_1, Y_2, \ldots, Y_n$ in order of increasing *absolute value*.

STEP 3.   Write down the numbers $1, 2, \ldots, n$ in order, each with the algebraic sign of the observation in the corresponding position in the arrangement of Step 1.

STEP 4.   Compute the sum W of these numbers with their appropriate signs.

For example, suppose the hypothesis is

$$H_0 : \xi = 3$$

$$H_a : \xi > 3$$

Assume a sample of $n = 4$ observations gives the following numbers: $+8.4, -4.3, -.8,$ and $+12.5$. The first step of the test gives Y_i values of $+5.4, -7.3, -3.8,$ and $+9.5$. Arranged in order of increasing absolute value, the Y_i values are

$$-3.8 \quad +5.4 \quad -7.3 \quad +9.5 \tag{18.4}$$

Here $+5.4$ precedes -7.3 because $|+5.4| < |-7.3|$ (even though $+5.4 > -7.3$). The numbers $1, 2, 3,$ and 4 with the corresponding signs attached are

$$-1 \quad +2 \quad -3 \quad +4 \tag{18.5}$$

The sum W is $-1 + 2 - 3 + 4$, or $+2$. The numbers $1, 2, 3,$ and 4 are the *ranks* of the quantities $3.8, 5.4, 7.3,$ and 9.5. In the testing Hypothesis (18.3), we reject H_0 in favor of H_a if W is excessively large.

For fixed n the distribution of W is the same for every population satisfying the null hypothesis H_0. There are published tables of this distribution for small values of n (see Reference 18.2). In the examples here we shall use the normal approximation, which is accurate for an n of 10 or more. The mean and variance of W are

$$\mu_W = 0 \tag{18.6}$$

and

$$\sigma_W^2 = \frac{n(n + 1)(2n + 1)}{6} \tag{18.7}$$

Thus, the standardized variable is W/σ_W. To test at the level α, we reject H_0 if W/σ_W exceeds the corresponding percentage cutoff point Z_α on the normal curve. This is the Wilcoxon test.

$$\begin{cases} H_0: \xi = \xi_0, \text{ symmetric} \\ H_a: \xi > \xi_0, \text{ symmetric} \\ R: W/\sigma_W > Z_\alpha \end{cases}$$

This testing problem can arise, for example, from paired comparisons, as in Section 9.9. We have n pairs of experimental units and two competing treatments (or a treatment and a nontreatment). Treatment 1 is given to a randomly selected element of pair i, which results in an observation, and Treatment 2 is given to the other element of pair i, which results in a second observation. Under the null hypothesis that the two treatments have the same effect, the difference, which we denote X_i, has median zero and is symmetric about zero. Without the assumption of normality required for the t test, we can use the Wilcoxon test.

EXAMPLE 1

Problem: In Example 1 of Section 9.9, we used the t test on the null hypothesis that a particular heat treatment has no effect on the number of bacteria in skim milk against the alternative that the treatment tends to reduce this number. Here $H_0: \xi = 0$ and $H_a: \xi > 0$. The X_i is the difference in the count (log DMC) for sample i before treatment and after treatment. The twelve differences are:

.03   .14   1.17   .15   −.02   −.04   .44   .22   −.01   .13   .19   .70

Solution: Since the hypothesized median is zero, we can omit Step 1 of the test and arrange the values in order of increasing absolute value; they are:

−.01   −.02   .03   −.04   .13   .14   .15   .19   .22   .44   .70   1.17

The numbers 1, 2, . . . , 12 with algebraic signs in the same pattern are:

−1   −2   +3   −4   +5   +6   +7   +8   +9   +10   +11   +12

The sum W of these comes to +64. By Formula (18.7),

$$\sigma_W = \sqrt{\frac{12 \cdot 13 \cdot 25}{6}} = \sqrt{650} = 25.5$$

so the standardized W value, W/σ_W, is $64/25.5$, or 2.51. The upper 5 percentage cutoff point $Z_{.05}$ on the normal curve being 1.645, we reject the null hypothesis of no effect.

The same method works for the opposite alternative

$$\begin{cases} H_0: \xi = \xi_0, \text{ symmetric} \\ H_a: \xi < \xi_0, \text{ symmetric} \\ R: W/\sigma_W < -Z_\alpha \end{cases}$$

and for the two-sided alternative

$$\begin{cases} H_0 : \xi = \xi_0, \text{symmetric} \\ H_a : \xi \neq \xi_0, \text{symmetric} \\ R : |W/\sigma_W| > Z_{\alpha/2} \end{cases}$$

In using the Wilcoxon test and other tests described in this chapter, we sometimes have to deal with tied observations or ranks. When two or more ranks are tied, we merely average the ranks and assign all the tied observations the average rank. For instance, in the five observations 3, 5, 7, 7, and 9 we see that the third and fourth observations are tied. The ranks to assign to the five observations are 1, 2, 3.5, 3.5, and 5. The method of using the average rank for tied observations will be used throughout this chapter. The two-sided Wilcoxon test and the use of tied ranks are demonstrated in the next example.

EXAMPLE 2

Problem: A market researcher constructed a questionnaire seeking consumer responses to a particular product. She hypothesized that people were equally split on whether they liked the product's package design. To test this hypothesis she asked the following question: "What is your feeling about the product's package design?" The respondents could then check a response on a seven-point scale that ranged from 1 = strongly dislike to 7 = strongly favor. The response in the middle was 4 = indifferent. In a pilot test of her questionnaire, she obtained the following responses to this question from twelve people:

$$7 \quad 3 \quad 5 \quad 4 \quad 7 \quad 1 \quad 2 \quad 2 \quad 5 \quad 7 \quad 6 \quad 5$$

Use the Wilcoxon W to test the hypothesis that the median response to the package design is 4. Let $\alpha = .05$.

Solution: The hypothesis we will test is $H_0 : \xi = 4$, and the alternative will be $H_a : \xi \neq 4$. If we subtract the hypothesized median of 4 from each response we obtain

$$3 \quad -1 \quad 1 \quad 0 \quad 3 \quad -3 \quad -2 \quad -2 \quad 1 \quad 3 \quad 2 \quad 1$$

Next, we arrange these values in increasing order of absolute value to give

$$0 \quad -1 \quad 1 \quad 1 \quad 1 \quad -2 \quad -2 \quad 2 \quad 3 \quad 3 \quad -3 \quad 3$$

The ranks can now be determined, but we must note there are several ties. Thus, we find the ranks are

$$1 \quad -3.5 \quad 3.5 \quad 3.5 \quad 3.5 \quad -7 \quad -7 \quad 7 \quad 10.5 \quad 10.5 \quad -10.5 \quad 10.5$$

The value of W is found by summing these ranks, and we find that $W = 22$. By Formula (18.7) we find

$$\sigma_W = \sqrt{\frac{12 \cdot 13 \cdot 25}{6}} = 25.5$$

Thus, the standardized value of W is $W/\sigma_W = 22/25.5 = .86$, which is not beyond $Z_{.025} = 1.96$, and we cannot conclude that the question responses have a population median different from 4, "indifferent."

18.2   THE SIGN TEST

The Wilcoxon test cannot be used unless the population is symmetric about its median. In the absence of symmetry, we can use the *sign test* instead. The sign test, being less sensitive than the Wilcoxon test, should be used only if symmetry cannot be assumed.

On the basis of an independent sample $X_1, X_2, \ldots, X_n$, we are to test the null hypothesis that the population median ξ is ξ_0 against the alternative that it exceeds ξ_0:

$$\begin{cases} H_0 : \xi = \xi_0 \\ H_a : \xi > \xi_0 \end{cases} \tag{18.8}$$

The sign test is simple. We count the number, y, of observations that exceed ξ_0 among $X_1, X_2, \ldots, X_n$. Because of Definition (18.1) of the median, the distribution of y under H_0 is binomial with $p = .5$, while under H_a it is binomial with $p > .5$. Thus, we can use the methods of Section 9.6.

EXAMPLE 1

Twenty secretaries using a new type of electric typewriter increased their typing speeds by the following amounts (in words per minute):

$$+7 \quad -6 \quad +3 \quad +1 \quad +6 \quad +4 \quad +9 \quad -5 \quad +9 \quad -7$$
$$-3 \quad +7 \quad -9 \quad +8 \quad +6 \quad -4 \quad +4 \quad +9 \quad -6 \quad +1$$

We are to test at the 5% level the null hypothesis that the median gain in typing speed is zero against the alternative that it is positive. There are 13 positive observations among the 20: $y = 13$. The mean and standard deviation of y are

$$n \times .5 = 10$$

and

$$\sqrt{n \times .5 \times .5} = \sqrt{5} = 2.24$$

so the standardized variable is $(y - 10)/2.24$, which has the value 1.34. The upper 5 percentage point $Z_{.05}$ is 1.645, so the data are consistent with the null hypothesis.

By an adaptation of the methods of Section 9.6, it is possible to test against a two-sided alternative ($\xi \neq 0$) as well.

The following example is presented to illustrate three points from the preceding discussion. First, it presents a situation where the hypothesized median is nonzero. Second, it shows that when one or more of the values is zero (which is neither positive or negative), the zero value is dropped from consideration and the sample size is reduced by one. Third, it demonstrates how to perform the sign test for small sample sizes.

EXAMPLE 2

Problem: Use the data in Example 2 of the previous section to test H_0: $\xi = 4$ against the alternative $H_a : \xi \neq 4$ using the sign test. Use $\alpha = .05$.

Solution: If we examine Example 2 of the previous section, we find that the list of responses, after 4 has been subtracted from each and they have been arranged by magnitude, is

$$-3 \quad -2 \quad -2 \quad -1 \quad 0 \quad 1 \quad 1 \quad 1 \quad 2 \quad 3 \quad 3 \quad 3$$

Four of these values are negative, seven are positive, and one is zero. Since zero is neither positive nor negative, we eliminate it from consideration and proceed as though we had only eleven observations. Then $y = 7$ and $n(.5) = 11(.5) = 5.5$. The standard deviation of the number of positive responses is

$$\sqrt{n \times .5 \times .5} = \sqrt{11 \times .5 \times .5} = 1.66$$

The standardized value is $(y - 5.5)/1.66 = .90$. There is seldom a situation in which a standardized variable of .90 would cause rejection of the null hypothesis. However, since our sample size of $n = 11$ (after one observation was dropped) is small, the test statistic cannot be compared to the normal distribution's critical values. Thus, we must use the exact binomial probability distribution to determine the probability of obtaining seven or more positive responses when we expect 5.5. If this probability is less than $\alpha = .05$, we will reject the null hypothesis. Thus, we seek:

$$P(r \geq 7 | n = 11, p = .5) = \sum_{r=7}^{11} \frac{11!}{r!(11 - r)!} (.5)^r (.5)^{11-r} = .27$$

Thus, if we reject $H_0 : \xi = 4$, we would have a 27% chance of being wrong. In the statement of the problem, we indicated that we would risk only a 5% chance of this type error, since $\alpha = .05$. Thus, we would not reject the null hypothesis.

In some cases with the Wilcoxon test, we are able to reject a null hypothesis that the less sensitive sign test will not reject. This raises the question of which result is correct. If we have reason to believe that the responses in the population are symmetric about their median, then the results of the Wilcoxon test are valid. However, if we feel that the responses from the population are likely to be skewed in one direction or the other, then the second test, the sign test, is more appropriate.

18.3   THE RANK SUM TEST

The rank sum test stands to the Wilcoxon test as the two-sample t test of Section 9.8 stands to the one-sample t test of Section 9.5. The rank sum test can be used to check whether two populations have the same median, and it does not require the assumption of normality.

Let $X_{11}, X_{12}, \ldots, X_{1n_1}$ be a sample of size n_1 from population 1, and let $X_{21}, X_{22}, \ldots, X_{2n_2}$ be a sample of size n_2 from population 2. Let H_0 be the null hypothesis that the two populations are the same, and let H_a be the hypothesis that the frequency curve for population 2 has the same shape as the curve for population 1 but is shifted to the right as shown in Figure 18.2.

The statistic for this problem, the *rank sum,* is computed in four steps.

STEP 1.   Combine the two samples into one, underlining the observations in the second sample to keep track of them.

STEP 2.   Arrange the observations in the combined sample in order of increasing size, carrying along the lines under the observations in the second sample.

STEP 3.   Write down the numbers $1, 2, \ldots, n_1 + n_2$ in order, underlining those for which the corresponding position in the arrangement of Step 2 is occupied by an element of the second sample (that is, by an underlined number).

STEP 4.   From the average of the underlined numbers subtract the average of the others. Call the difference V.

For example, suppose n_1 is 3, the observations being 3.1, -2.4, and 0.6; and suppose n_2 is 2, the observations being 7.7 and 2.8. The

FIGURE 18.2   The alternative hypothesis H_a

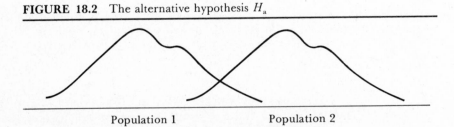

Population 1   Population 2

combined sample is:

$$3.1 \quad -2.4 \quad 0.6 \quad \underline{7.7} \quad \underline{2.8}$$

Arranged in increasing order, that is:

$$-2.4 \quad 0.6 \quad \underline{2.8} \quad 3.1 \quad \underline{7.7} \tag{18.9}$$

The numbers 1, 2, 3, 4, and 5 $(n_1 + n_2 = 5)$ with underlineations in the same pattern are:

$$1 \quad 2 \quad \underline{3} \quad 4 \quad \underline{5} \tag{18.10}$$

The underlined numbers average to $(3 + 5)/2$, or 4, the others to $(1 + 2 + 4)/3$, or 2.33. The difference V of these averages is $4 - 2.33$, or 1.66.

If we were to compute the two-sample t statistic (see Formula (9.17) in Section 9.8), we would first form the average $\bar{X}_2$ of the underlined numbers in Array (18.9), then the average $\bar{X}_1$ of the others, and then the difference $\bar{X}_2 - \bar{X}_1$ (which we would then normalize). To compute V, we apply the same procedure to Array (18.10). Although the numbers in (18.10) are different from those in (18.9), they come in the same order. Therefore, in the same way as does a large value of $\bar{X}_2 - \bar{X}_1$, a large value of V indicates that observations from population 2 tend to exceed those of population 1—indicates, that is, that the data in the pair of samples favor H_a over H_0. The statistic V can be thought of as a "nonparametric difference of means."

For each n_1 and n_2, the distribution of V under H_0 has the desirable property that it does not depend on the shape of the frequency curve common to the two populations. Tables are available for small values of n_1 and n_2 (see Reference 18.2). In the examples here, the normal approximation to this distribution can be used. The mean and variance of V are

$$\mu_V = 0 \tag{18.11}$$

and

$$\sigma_V^2 = \frac{(n_1 + n_2)^2 (n_1 + n_2 + 1)}{12 n_1 n_2} \tag{18.12}$$

We reject H_0 in favor of H_a if V/σ_V exceeds Z_α.

EXAMPLE 1

Problem: The cost of land per acre (in thousands of dollars) is presented below for 18 pieces of land in two different exclusive housing subdivisions.

We want to test at the 5% level the null hypothesis of no difference in land prices between the two subdivisions. The alternative hypothesis would be that the second subdivision's land prices are higher. Here n_1 is 10 and n_2 is 8.

	COST								
Subdivision 1	40.1	37.5	46.3	44.2	35.1	48.2	41.7	37.3	50.3	43.9
Subdivision 2	49.4	41.3	45.8	39.7	44.4	43.1	48.8	47.5

Solution: The two samples merged into one and ordered are:

35.1 37.3 37.5 39.7 40.1 41.3 41.7 43.1 43.9

44.2 44.4 45.8 46.3 47.5 48.2 48.8 49.4 50.3

The numbers 1, 2, ..., 18 ($n_1 + n_2 = 18$) underlined in the same pattern are:

1 2 3 4 5 6 7 8 9 10 11 12 13 14 15 16 17 18

The ranks for the second sample (the underlined numbers) average 88/8, or 11.0, and the ranks for the first sample (the remaining numbers) average to 83/10, or 8.3. The difference V is $11.0 - 8.3$, or 2.7. The standard deviation according to Formula (18.12) is

$$\sigma_V = \sqrt{\frac{18^2 \times 19}{12 \times 10 \times 8}} = 2.53$$

Thus, the standardized statistic V/σ_V has the value 2.7/2.53, or 1.07. Since this is less than 1.645, the upper 5% point for the normal curve, the null hypothesis is not rejected. Note that in this example we did not have to assume that the land prices in each of the subdivisions were normally distributed.

As in Section 18.1, ties are dealt with by averaging adjacent ranks. If in Example 1 the observation 37.5 were 39.7 instead, creating a tie for third from the bottom, the ranks 3 and 4 would each be replaced by 3.5.

EXAMPLE 2

Problem: A group of salespeople were given some sales training designed to boost their sales on a certain product. Another, independent group of salespeople were not given the training. The following data show the sales (in thousands of dollars) for the two groups during the month after the training.

SALESPERSON	UNTRAINED	TRAINED
1	2.6	4.3
2	1.5	1.2
3	4.2	4.6
4	2.5	2.8
5	3.0	4.6
6	1.8	2.0
7	2.6	2.9

Test the hypothesis that the training provides no improvement using the rank sum test. Let $\alpha = .05$.

Solution: If we were solving this problem using the techniques of Chapter 9, we would test the hypothesis $H_0: \mu_1 = \mu_2$ using the two-sample techniques of Section 9.8. However, that technique requires us to make the assumption that the populations of sales for the untrained and trained are normally distributed. This assumption is a rather strong one (but can be tested using a larger pair of samples than ours and the goodness-of-fit techniques of Chapter 11). Thus, we will proceed using the nonparametric test suggested in the statement of the problem.

The rank sum test: Our first step is to merge the fourteen sales figures, underlining those figures that are for salespeople in the training program.

   1.2 1.5 1.8 <u>2.0</u> 2.5 2.6 2.6 <u>2.8</u> <u>2.9</u> 3.0 4.2 <u>4.3</u> <u>4.6</u> <u>4.6</u>

Next, the numbers 1, 2, ..., 14 are placed in the same order, with appropriate values underlined to note the positions of the data values in the second group of numbers.

   <u>1</u> 2 3 <u>4</u> 5 6.5 6.5 <u>8</u> <u>9</u> 10 11 <u>12</u> <u>13.5</u> <u>13.5</u>

Note that the sixth and seventh values were tied for rank as were the thirteenth and fourteenth. The mean of the underlined numbers is $61/7 = 8.71$. The others average $44/7 = 6.29$. The difference, V, is $(8.71 - 6.29) = 2.42$. From Formula (18.12) we find that

$$\sigma_V = \sqrt{\frac{14^2 \times 15}{12 \times 7 \times 7}} = \sqrt{5} = 2.24$$

The standardized value is $V/\sigma_V = 2.42/2.24 = 1.08$, which is less than $Z_{.05} = 1.645$. Thus, we cannot reject the hypothesis.

18.4 SPEARMAN'S RANK CORRELATION MEASURE

Suppose four students, A, B, C, D, are ranked in two subjects, mathematics and history, with these results:

	STUDENT:			
	A	B	C	D
Mathematics	4	2	3	1
History	3	1	4	2

That is to say, in mathematics, student D is best (rank 1), B is next best (rank 2), C next (rank 3), and A is worst (rank 4); while in history the ranking (best to worst) is B, D, A, C. To check whether there is a connection between performance in the two subjects, we want a measure of the extent to which these two rankings agree.

This problem is similar to the problem in Section 12.5 of finding the correlation coefficient, r, between observations of two random variables. However, the significance of the r value in a correlation problem can be tested only if we assume that both variables follow the normal distribution. The rankings of students in mathematics and history do not follow the normal curve; they are uniformly distributed. Thus, we need a new method of finding correlations between two sets of *rankings* as opposed to between two sets of normally distributed variables.

Spearman's rank correlation measure, r_s, provides such a method. Spearman's measure is defined as

$$r_s = 1.0 - \frac{6\Sigma D_i^2}{n(n^2 - 1)} \tag{18.13}$$

where D_i is the difference between an item's ranking in one list and its ranking in the other. Tied ranks are handled, as usual, by averaging adjacent ranks. The n value is the number of items being ranked.

When the value of n is 10 or more, the following statistic is approximately t distributed with $n - 2$ degrees of freedom:

$$t = r_s \sqrt{\frac{n - 2}{1 - r_s^2}} \tag{18.14}$$

Note that r_s is a value between -1.0 and 1.0. If the two sets of rankings agree completely, then all the D values are zero, and $r_s = 1.0$. This, in turn, produces a t value in Formula (18.14) that is infinitely large— and therefore significant. If the two sets of rankings are in perfectly reversed order, then the sum of the D_i^2 figures divided by $n(n^2 - 1)$ and multiplied by 6 always equals 2.0. Thus, from Formula (18.13), $r_s = -1.0$, and again produces a significant t value that is infinitely negative. An r_s value of zero indicates that the rankings seem to have no positive or negative correlation.

We can find r_s for the four students mentioned at the first of this section in the following way. The differences in the rankings for mathematics and history are:

$$1 \quad 1 \quad -1 \quad -1$$

Then

$$\sum_{i=1}^{n} D_i^2 = (1 + 1 + 1 + 1) = 4.0$$

and

$$r_s = 1.0 - \frac{(6)(4)}{4(4^2 - 1)} = .6$$

Since $r_s = .6$, there seems to be a moderately high positive correlation in the rankings. But we cannot test the significance of this value using the t value of 18.14 since the sample size is not between ten and thirty items. For values of n outside this ten-to-thirty range, critical values of r_s can be found in special tables (see Reference 18.4). The following example shows how r_s is calculated and tested for significance using Relationship (18.14).

EXAMPLE 1

Problem: A securities analyst for a brokerage house wanted to investigate the degree of association between a company's profitability and its liquidity. He ranked ten firms (which he labelled A through J) in one industry according to their profitability and liquidity:

	A	*B*	*C*	*D*	*E*	*F*	*G*	*H*	*I*	*J*
					FIRM:					
Profitability	8	9	4	1	5	10	2	3	7	6
Liquidity	9	1	8	3	4	2	10	5	6	7

Use Spearman's rank correlation measure to test the null hypothesis H_0: *There is no rank correlation between profitability and liquidity* against the alternative H_a: *There is some rank correlation (positive or negative) between profitability and liquidity.* Let $\alpha = .05$.

Solution: The ten D_i^2 values are listed and summed next:

$$\sum_{i=1}^{n} D_i^2 = (1 + 64 + 16 + 4 + 1 + 64 + 64 + 4 + 1 + 1) = 220$$

Spearman's r_s value can be calculated from Formula (18.13) as

$$r_s = 1.0 - \frac{6(220)}{10(10^2 - 1)} = -.33$$

The significance of this negative value can be tested using Formula (18.14).

$$t = -.33 \sqrt{\frac{10 - 2}{1 - (.33)^2}} = -.99$$

This t value is greater than the $t_{.025,8} = -2.306$, so these data do not cause us to reject the hypothesis that profitability and liquidity rankings are uncorrelated.

It should be noted that Formula (18.13) correctly computes r_s only if none of the ranks in either set are tied. If there are only one or two tied ranks in the data, then Formula (18.13) is a good approximation to r_s. But if there are many tied ranks, a much more involved formula for r_s must be employed. This formula will not be presented here.

A second popular measure of rank correlation is Kendall's τ. The computation of Kendall's statistic is rather tedious and will not be presented here. A complete discussion of this measure can be found in references at the end of this chapter.

18.5   A SUMMARY NOTE ON NONPARAMETRIC METHODS

This chapter has presented the following nonparametric tests:

1. The Wilcoxon test. It is used to test the hypothesis that a population has a median ξ_0. It requires that we be able to assume the population has a symmetric shape. This test is parallel to the one-population t test of Chapter 9.
2. The sign test. This test is also used to test the hypothesis that a population has a median, ξ_0. It does not require the assumption of symmetry, but it is a somewhat weaker test. That is, it may not identify a population with a median different from ξ_0 as quickly as the Wilcoxon test, if the population is symmetric. This test is also parallel to the one-population t test of Chapter 9.
3. The rank sum test. This test is used to test the hypothesis that two populations have the same median. The only assumption required is that the shape of the two populations is the same (not necessarily symmetric). This assumption is a rather weak one since we are usually comparing populations of similar measures which are likely to be quite similar in shape. This test is parallel to the two-population t test of Chapter 9.
4. Spearman's rank correlation test. Spearman's r_s is used to test for correlation (positive or negative) between two sets of rankings of

TABLE 18.1 Summary of Four Nonparametric Tests

TEST NAME	NULL HYPOTHESIS	TEST STATISTICS	COMMENTS
Wilcoxon	$H_0: \xi = \xi_0$, symmetric	W $\sigma_w = \sqrt{\dfrac{n(n+1)(2n+1)}{6}}$	**a.** W is the sum of signed ranks. **b.** The population values must be symmetric around the median ξ. **c.** W is approximately normally distributed for n of 10 or more. **d.** For smaller n, tables of critical values for W are available.
Sign Test	$H_0: \xi = \xi_0$	y	**a.** y is the number of values greater than (or less than) ξ_0. **b.** No symmetry assumption is required. **c.** For small n, y is binomially distributed with $p = .5$. **d.** For large n, y is normal with mean $.5n$ and standard deviation $.5\sqrt{n}$.
Rank Sum	$H_0: \xi_1 = \xi_2$	V $\sigma_V = \sqrt{\dfrac{(n_1 + n_2)^2(n_1 + n_2 + 1)}{12 n_1 n_2}}$	**a.** V is the difference in the two samples' mean ranks in the combined data set. **b.** For small n_1 and n_2, tables of critical V values are available. **c.** For large n_1 and n_2, V is normally distributed.
Spearman's Rank Correlation	$H_0:$ rankings are not correlated	$r_s = 1 - \dfrac{6\Sigma D_i^2}{n(n^2 - 1)}$	**a.** D_i is the difference between an item's ranking in the two sets of rankings. **b.** For $n < 10$, special tables of critical r_s values are available. **c.** For $n \geq 10$, Expression (18.14) is t-distributed with $n - 2$ degrees of freedom.

the same items. No assumptions are involved. This measure is parallel to the correlation coefficient, r, of Chapter 12.

A summary of these tests is presented in Table 18.1.

These are only four of numerous nonparametric techniques. Also, there are variations on these techniques which are not presented here. References 18.2, 18.3, and 18.5 at the end of this chapter are entire books devoted to the wide field of nonparametric statistics. In short, this chapter only covers four basic tests that have parallel parametric tests. The parametric tests can be employed in cases where we are working with data that are normally distributed.

It is a general rule in statistical testing that the more powerful tests, the ones that more efficiently identify departures from the null hypothesis, require more restrictive assumptions about the populations being tested. Most of the powerful techniques described in Chapters 9 through 11 of this text require the somewhat restrictive assumption that the populations being dealt with be normally distributed. That assumption can be tested using the goodness-of-fit test of Chapter 11. If the assumption that the populations are not normally distributed must be rejected, then the techniques of this chapter can be used since they do not require the normality of the populations.

The great value of these nonparametric tests lies in their applicability to a broad class of populations. The only requirement for the use of nonparametric methods is that the sample data values can be ranked or ordered in some manner. The price we pay, however, for the broad applicability of these methods is a loss in the power of the tests to identify departures from the null hypothesis when they actually exist.

Because of this trade-off between the power and general applicability of techniques, statisticians generally follow this rule. If the population(s) being tested can be assumed to follow the normal distribution, the more powerful techniques of Chapters 9 through 11 should be used. If the population(s) exhibit(s) only a modest departure from normality, the more powerful parametric techniques are still usually applied, but with caution. Careful interpretation and qualification must be supplied with the results of the tests. Finally, if the population(s) cannot be assumed to have the normal distribution (perhaps due to the rejection of the normality hypothesis by the goodness-of-fit test), then nonparametric techniques should be applied.

PROBLEMS

1. Assume that the following measurements come from a population symmetric about its median ξ.

$$-12.6 \quad -20.5 \quad 8.1 \quad -27.3 \quad -21.2 \quad -29.4 \quad 15.5 \quad -25.3$$

$$15.9 \quad 20.3 \quad -11.7 \quad -9.3 \quad -28.8 \quad 32.7 \quad 19.2 \quad -27.1$$

Test $H_0: \xi = 0$ against $H_a: \xi < 0$ at the 2.5% level.

2. Use the sign test to test at the 5% level whether the following measurements come from a population with median 10. Use $H_a: \xi \neq 10$.

$$13.6 \quad 8.0 \quad 15.8 \quad 12.2 \quad 3.5 \quad 17.3 \quad 9.5 \quad 11.4 \quad 12.8 \quad 10.7$$

$$17.1 \quad 12.7 \quad 13.9 \quad 11.4 \quad 7.5 \quad 16.3 \quad 13.2 \quad 17.0 \quad 12.1 \quad 26.3$$

3. Do Problem 1 without the assumption of symmetry.

4. Use the rank sum test to check whether the following two samples come from the same population. Use $\alpha = .10$.

SAMPLE 1	120	136	107	109	129	117	125	110	124
SAMPLE 2	131	144	116	111	103	122	141	139	130
	133	132	135	148					

5. Do Problem **35,** Chapter 9, by the rank sum test.

6. Do Problem **36,** Chapter 9, by the rank sum test.

7. Several stock market analysts were asked to assess the probability that the Dow Jones Industrial Average will rise 10 or more points during the next month. Test the hypothesis $H_0: \xi = .5$ against the alternative $H_a: \xi > .5$ using $\alpha = .05$. Which test seems appropriate?

ANALYST	1	2	3	4	5	6	7
PROBABILITY	.6	.7	.4	.9	.7	.8	.5

8. The instructions attached to a standard examination administered to employees by a personnel manager read: "The median score for people across the country on this examination is 75." The personnel manager just finished giving this examination to eleven people in one department. Test the hypothesis that the median score of all people hired into this department is 75. Use $\alpha = .10$ and an alternative hypothesis $H_a: \xi \neq 75$. Use the sign test.

$$80 \quad 45 \quad 50 \quad 34 \quad 78 \quad 92 \quad 98 \quad 23 \quad 34 \quad 54 \quad 55$$

9. Rework Problem **8** using the Wilcoxon test.

10. A garment manufacturer has a shop containing several hundred commercial sewing machines. The employees who operate the machines attended a program on the safe and efficient use of their machines. The times to machine breakdown for a sample of $n_1 = 6$ machines before the training follow. The similar times to breakdown are listed for an independent sample of $n_2 = 6$ machines after the training. Use the rank sum test to determine if the median time to breakdown differs before and after training. Use $\alpha = .10$; the alternative hypothesis is $H_a: \xi_2 > \xi_1$. Conduct the test *without* pairing the data.

MACHINE	TIME UNTIL BREAKDOWN BEFORE	TIME UNTIL BREAKDOWN AFTER
1	23 weeks	42 weeks
2	32 weeks	39 weeks
3	44 weeks	24 weeks
4	12 weeks	26 weeks
5	55 weeks	48 weeks
6	6 weeks	15 weeks

11. Assume for the moment that a previous sample and test found that the hypothesis of normality for the two populations of times in Problem **10** could not be rejected. Use the methods of Section 9.8 to test the hypothesis of equal means for the data in Problem **10**. Use $H_a : \mu_1 < \mu_2$ and $\alpha = .10$. Under the circumstances described here, which of the two approaches would be preferred?

12. Work Problem **39** in Chapter 9 using
    a. The rank sum test.
    b. The Wilcoxon test.

13. Use the rank sum test to determine if there is a difference between the shelf lives of the same food product from two different firms. The shelf life figures (in days) follow and were determined by an independent testing agency. Use $H_0 : \xi_1 = \xi_2$ and $H_a : \xi_1 \neq \xi_2$ with $\alpha = .05$.

FIRM X	12	14	15	15	13	16	13	17	19	17	17
FIRM Y	20	10	19	21	15	16	18	14	15		

14. List two reasons why a paired comparison using the Wilcoxon test would not be appropriate for the data in Problem **13**.

15. The job performance of ten employees was ranked by two supervisors familiar with the work of all ten. Compute Spearman's rank correlation measure of the two sets of ratings. Is it significantly greater than 0 at the 2.5% level?

	EMPLOYEE:									
	A	B	C	D	E	F	G	H	I	J
Supervisor 1	5	6	3	9	4	8	1	7	10	2
Supervisor 2	3	4	1	8	5	10	6	7	9	2

16. Two interviewers at the same company rated several job applicants. Use Spearman's rank correlation measure, r_s, to show if the interviewers disagree on how they rank the applicants. Test the significance of r_s using the alternative hypothesis H_a: *There is negative rank correlation*. Let $\alpha = .01$.

	APPLICANT:										
	1	2	3	4	5	6	7	8	9	10	11
Interviewer 1	7	6	4	1	2	8	3	5	10	11	9
Interviewer 2	2	3	5	8	7	1	6	4	10	11	9

17. The second interviewer in Problem **16** complained that she thought the first interviewer was putting too much weight on an applicant's college GPA. To prove her point, she compared the applicants' rankings by GPA with the first interviewer's rankings. Calculate Spearman's rank correlation measure, r_s, and test to see if it is significantly different than zero at the $\alpha = .01$ level of significance. (*Hint:* Rank the GPA's first.)

| | **APPLICANT:** | | | | | | | | | | |
	1	*2*	*3*	*4*	*5*	*6*	*7*	*8*	*9*	*10*	*11*
GPA	2.1	2.4	3.2	3.9	3.7	2.1	3.5	3.5	2.0	2.0	1.9

18. At the beginning of the basketball season, local sports writers met to rank the teams in the Out Back Conference. The preseason rankings and postseason standings for the teams follow. Use Spearman's r_s to determine if the preseason rankings have any meaning. That is, test the hypothesis that there is no rank correlation against the alternative that the rank correlation between preseason polls and postseason standings is positive. Let $\alpha = .05$. (*Hint:* For $n = 5$, r_s must exceed .9 to reject the null hypothesis.)

TEAM	*PRESEASON STANDING*	*CONFERENCE FINISH*
Cup Cake Junction	1	3
Tumbleweed Tech	2	4
Hoopla College	3	1
Frisbee State	4	2
Washout United	5	5

19. Two loan officers examined eleven loan applications and ranked them according to their desirability to the bank. Compute the Spearman rank correlation r_s and determine if it is significant. Use the alternative that there is positive rank correlation and let $\alpha = .05$.

| | **LOAN:** | | | | | | | | | | |
	A	*B*	*C*	*D*	*E*	*F*	*G*	*H*	*I*	*J*	*K*
Officer 1	1	7	4	2	3	6	5	9	10	8	11
Officer 2	1	6	5	2	3	4	7	11	8	10	9

20. Two capital budgeting consultants were retained by Baird Enterprises to help the firm set its capital budgeting priorities for the coming year. The consultants ranked the projects in the following way:

| | **PROJECT:** | | | | | | | | | | | |
	A	*B*	*C*	*D*	*E*	*F*	*G*	*H*	*I*	*J*	*K*	*L*
Consultant 1	7	9	12	10	1	8	5	11	2	4	3	6
Consultant 2	9	11	10	8	4	6	5	12	1	2	7	3

Do the rankings of the consultants differ significantly from one another? Use the alternative that there is positive rank correlation and let $\alpha = .10$.

SAMPLE DATA SET QUESTIONS: *Refer to the 113 applicants for credit listed in the Sample Data Set Appendix of this book.*

    a.   Use the rank sum test to test the hypothesis that the medians of JOBINC are equal for the groups who were granted and denied credit. Let $\alpha =$ .025.

    b.   Rank the first twenty applicants in the "credit granted" group according to JOBINC and TOTBAL. Use only the first twenty who recorded a JOBINC figure. Test the hypothesis that there is no rank correlation between income and debt balance. Let $\alpha = .05$.

REFERENCES

18.1   Hogg, Robert V., and Allen T. Craig. *Introduction to Mathematical Statistics,* 3rd edition. Macmillan, New York, 1970. Chapter 11.

18.2   Kraft, Charles H., and Constance van Eeden. *A Nonparametric Introduction to Statistics.* Macmillan, New York, 1968.

18.3   Noether, Gottfried E. *Elements of Nonparametric Statistics.* John Wiley & Sons, New York, 1967.

18.4   Olds, E. G. "Distribution of Sums of Squares of Rank Differences for Small Samples," *Annals of Mathematical Statistics,* IX, 1938.

18.5   Siegel, S. *Nonparametric Statistics for the Social Sciences.* McGraw-Hill, New York, 1956.

CASES

The cases presented in this section are taken, for the most part, from actual situations. In some instances the data have been altered to simplify the solution.

The reader is encouraged to look for more than a numerical solution to each of these cases. The numerical solution is meaningful only in the context of the case. Its value to the decision maker is dependent upon the accuracy of the data from which it is computed, the validity of the assumptions underlying its computation, and the use to which it will be put. The reader should consider these issues, together with the computational aspects of each case.

THE NEW ENGLAND TIRE DEALERS ASSOCIATION

Several states have laws prohibiting the use of studded snow tires. Recently the legislature of one New England state was considering passage of such a law. Many automobile tire dealers traveled to the state capital to protest passage of the bill.

The supporters of legislation banning studded snow tires maintained that they are no safer than regular snow tires and that they damage highway surfaces so badly that the costs of repairs far outweigh any modest advantage the studs may provide. The tire dealers and their supporters maintained that studded tires produce significantly better automobile control on snow and ice and are well worth the cost of the highway damage they may do. Some of the tire dealers were undoubtedly motivated to oppose the studded tire ban by the extra charges they receive for inserting studs in tires. Most of them, however, sincerely believed that studded snow tires are safer.

E. W. Snell, president of the New England Tire Dealers Association, testified before a committee of the state legislature. In his testimony Mr. Snell stated the following: "In one of the Canadian provinces, 50% of the automobiles have studded snow tires. However, last winter 1408 automobile accidents were reported in the province. Only 168, or 12%, of these accidents involved cars with studded snow tires. What more proof do you need that studded snow tires are safer?"

THE STATE TAX COMMISSION VERSUS FLAGLE FAMILY ENTERPRISES

Mr. Patrick J. Flagle founded Flagle Family Enterprises several years ago using money he had borrowed from his father and four of his six brothers. Despite the fact that Mr. Flagle paid back the initial loans from the family after only three years of operation, he continued to carry "family" as part of the company's name since he felt he owed much of the business' current success to these initial loans.

The Flagle Family Enterprises owned several distributorships in the eastern states. Each distributorship was incorporated in the state in which it operated.

Mr. Flagle always believed in using to his advantage any information which might come his way. He was particularly interested last week in how he might use the information he saw in a newspaper article which appeared in the local paper of the city where his largest distributorship is located.

The article was written by a reporter who had interviewed several past and current employees of the state tax commission in an effort to determine how tax returns were selected for audit. The reporter gave several hints on how individuals might avoid audits of their personal returns.

However, Mr. Flagle was more interested in what the article said about the procedures used to audit businesses like his. Mr. Flagle was interested in determining the likelihood of his company's recently submitted tax return's being audited and how much such an audit might cost him in added taxes. Some excerpts from the article follow:

Not all business tax returns are audited. The Tax Commission's big computer has been programmed to look for "interesting" items. The exact nature of those interesting items is a closely held secret, but this reporter has been able to determine some of the rules that are used. Several "audit pools" are created as the computer scans the tax returns. Then only a percentage of the returns in each of the pools gets audited. For instance, if a company's wages and salaries exceed 9% of sales, the company's return goes into an audit pool. However, only 20% of the returns in the wage and salary pool ever get audited. Don't think you're safe from audit if you don't get audited for wages and salaries, however. You might get audited for being out of line on such things as travel, sales promotions, and commissions. These are all areas that the auditors have found business executives sometimes use to pad their expenses with items that should be disallowed. The following table shows what causes your tax return to be put into the various audit pools by the computer. The table also shows the percentage of each audit pool that eventually reaches the auditors' desks.

AUDIT POOL	PERCENT OF SALES YOU NEED TO EXCEED TO GET INTO POOL	PERCENTAGE OF POOL FINALLY AUDITED
Wages and salaries	9.0	20.0
Travel and entertainment	.5	35.0
Sales promotions	2.0	30.0
Depreciation	5.0	15.0
Interest	2.0	25.0
Maintenance	1.0	10.5
Commissions	1.5	20.5

After the computer has finished its first pass at the returns, a firm's name may appear in four or five audit pools. The computer then randomly selects the names of firms from the audit pools. After this pass at the returns in the pools, a firm's name may appear on the list to be audited zero, one, two, or up to seven times. Those firms listed one or more times are audited only once, of course.

Mr. Flagle felt sure he could use this information to determine if his firm's tax return were likely to be audited, but he was also interested in another table which appeared in the second article the reporter wrote on this subject. The second article dealt with the return to the state received for the auditors' efforts.

The reporter wrote:

It was not difficult to get the Tax Commission to show me data on the taxes they recover through their audits. They are very proud of the fact that this part of their agency is a "money maker." That is, they bring in more in additional tax assessments than the costs of the audits. For instance, their records show additional taxes collected through audits last year ran as follows:

ADDITIONAL COLLECTIONS	INDIVIDUAL RETURNS	CORPORATE RETURNS
−$10,000.00 to −$ 5,000.00	2	18
− 4,999.99 to − 2,000.00	5	41
− 1,999.99 to − 1,000.00	7	63
− 999.99 to Zero	120	68
.01 to 1,000.00	1250	421
1,000.01 to 2,000.00	921	250
2,000.01 to 5,000.00	32	116
5,000.01 to 10,000.00	6	82
10,000.01 to 15,000.00	0	21
15,000.01 to 20,000.00	0	5

The negative figures in the table indicate that some lucky people and companies who got audited received refunds. There were a few individuals and only one corporation that were required to pay more than $20,000 in added taxes.

FLAGLE FAMILY ENTERPRISES
Income Statement
January 1 through December 31

Net Sales		$5,968,344
Cost of Goods Sold		4,854,187
Gross Margin on Sales		1,114,157
Operating Costs:		
Wages and Salaries	$389,441	
Interest	125,784	
Advertising	62,560	
Office Expenses	20,438	
Insurance	16,859	
Payroll Taxes	19,011	
Sales Promotions	8,627	
Utilities	59,268	
Depreciation	11,590	
Maintenance	65,934	
Other Taxes	6,608	
Commissions Paid	93,040	
Freight	12,344	891,504
Earnings Before Taxes		$222,653

WESTERN PETROLEUM (A)

Western Petroleum is a small firm located in the southwestern United States. Its only business function is to create and market "oil drilling lease packages." An oil lease package consists of what amounts to "chances" to obtain oil leases on government-owned land in several states in the Southwest.

Each month the federal government conducts a lottery to determine who can obtain oil-drilling leases on various parcels of land the government has opened to oil drilling activities. The Bureau of Land Management (BLM) releases the numbers and locations of the parcels open for bid on the third Monday of each month. An application card and a $10 filing fee are due at the BLM one week later. Each applicant may submit only one card for each parcel of land. This restriction means that a private individual stands as much chance of winning a lease as does the largest oil company (which can submit only one application for the company, any of its employees, agents, or their immediate families). On the day of the lottery, a BLM employee literally draws three application cards out of a barrel. One drawing is made for each land parcel. The applicant who submitted the first card drawn is then given the opportunity to rent the drilling rights on the parcel for $1 per acre. If the winning applicant were disinterested in making the purchase, or unreachable for ten days after the drawing, then the lease would go to the person who submitted the second card drawn. If he or she were not interested, the third person would be contacted.

Almost all leases are rented by the winners, however, since the major oil companies stand ready to purchase any drilling rights at a moderate profit to the holder. The lease holder can seldom realize more than a $2500 profit on a lease since oil company officials realize that the typical winner does not have the resources to exploit the lease and actually conduct drilling operations.

Western Petroleum's president is Malcolm Maxfield. Mr. Maxfield describes his company's services as follows:

> We provide a convenient way for the average citizen to interface with the federal government in this oil leasing business. We take the hassle and some of the risk out of these transactions for our clients. We screen the properties on which bids are entered and bid on only realistic parcels. Since we are authorized to act as their agents, our clients do not have to endure that rather complicated registration process. The applications we submit are for the individual clients, and the BLM people accept them as though they came straight from the private firms or persons we represent.
>
> We also take some of the risk out of applying for these oil leases. We do this by offering packages that include both high- and low-profit potential lease parcels and low and high probabilities, respectively, of winning the lease. You see, this is a rather speculative business. For a $10 investment, someone can obtain a lease they can resell immediately

for anywhere from a $200 to $2500 profit. They don't even have to put up the money for the lease purchase. They just sign a form we give them, and their lease rights are turned over to an oil company. The oil company pays the BLM the rental fee and sends us the "profit" for our client. The client interacts only with us. We handle all arrangements and negotiations with the BLM and the oil companies.

We take a 20% fee on everything. Our clients pay $12 for each application they submit through us, and we get 20% of any profit they make on assigning the lease rights they win.

Mr. Maxfield has been concerned lately since some of his clients have been asking questions he can't answer very well. One of these clients is Stewart and Wagstaff Investment Advisors, Inc. Stewart and Wagstaff (S&W), in turn, have clients that entrust them with investment funds. Some of S&W's clients are currently supplying them with speculation funds. Thus, S&W approached Western Petroleum about six weeks ago and expressed some interest in buying several oil lease packages for their clients.

Due to the potential for a great deal of business from S&W, Mr. Maxfield is eager to accommodate them. First, he agreed to cut Western's commission from 20 to 14% on all business with S&W. Second, he agreed to remove from Western's mailing list all S&W clients. Third, he agreed to supply certain information to Mr. Saul Stewart by the first of next month. Meeting this third request is what has been occupying Mr. Maxfield's time for the past several days.

Mr. Stewart has expressed interest in buying several oil lease packages which Western calls its "Plan B—Moderate Return and Moderate Risk."* A table in Western's sales brochure describes Plan B as follows:

TABLE II: Description of Plan B

---

25 applications for Class I parcel leases
50 applications for Class II parcel leases
25 applications for Class III parcel leases

---

100 applications at $12 each = $1200 purchase price

Class I Parcels:	Proven oil or gas reserves found just outside the parcel's immediate area
Class II Parcels:	Geologically promising areas but no proven reserves
Class III Parcels:	Speculative areas where no geological studies have been conducted and no proven reserves are known

---

*Plan A is high return and high risk while Plan C is low return and low risk.

A letter from Mr. Stewart stated his information needs. In part, the letter said:

Since the selection of the winning application for the drilling rights on any individual parcel of land is a totally random event (a lottery) and since the various lotteries are independent of one another (winning one lease does not change your chance of winning another), it seems to me you could compute some probabilities for winning leases and some expected values for the various plans you offer. Based on your experience I would like you to estimate for me:

1. The probability that a Plan B will result in no winning lease (I would prefer a range of probabilities, assuming few to many applicants on each parcel).
2. The expected payoff to S&W for the purchase of a Plan B.

This letter caused Mr. Maxfield to conclude that these were legitimate questions which ought to be answered. He also felt the answers could be used in the marketing of all three of Western's plans. However, Maxfield felt uneasy about how he should proceed. He felt that Saul Stewart was more sophisticated than he in this probability area. Thus, he gathered the information he thought was relevant and put it in a letter to Mr. Stewart.

Dear Saul:

Thank you for your excellent letter of last week. You raised some interesting questions we had not considered previously—especially those about the probability of winning nothing and the expected value of our Plan B. In order to help answer your questions, I have put together this information:

PARCEL TYPE	PROFIT TO WINNER (BEFORE COMMISSION)	AVERAGE NO. OF APPLICANTS PER PARCEL	RANGE IN NUMBER OF APPLICANTS OVER LAST TWO YEARS
I	$2500	380	110 to 650
II	1200	80	25 to 125
III	200	15	10 to 40

These profit figures do not include the initial application fee. This table shows, as you would expect, that the real action is in the Class I parcels. This information would cause me to estimate the chance of winning a lease on any Class I parcel to be 1/380, but that chance may go as high as 1/110 or as low as 1/650 (on some of the real "hot" parcels that have high-yield wells near by). Similarly, the probabilities of winning the leases on Class II and Class III parcels appear to be:

CLASS	PROBABILITY	PROBABILITY RANGE
II	1/80	1/25 to 1/125
III	1/15	1/10 to 1/40

Since my background in probability is rather weak, I suggest that you and I meet for lunch next Thursday to discuss how this information can be used to answer your questions concerning Plan B and other questions that may come to mind. My secretary will be in touch with you to make an appointment.

Sincerely,

Malcolm Maxfield

NORTHEASTERN GAS COMPANY

Northeastern Gas Company is a natural gas utility which serves customers in a small area covering parts of two states on the East Coast. The company has limited natural gas sources and has had to put its business and industrial customers on an allocation during the past two winters. The company has worked for the most part on a voluntary basis with these customers and has had good success in reducing usage during times when the demand for gas would exceed the company's supply. The allocation periods have been limited to January and February, the coldest months of the year in Northeastern's service area.

In order to implement its allocation policies, the company has, in the past, established target "sendout" figures for its business and industrial customers. The sendout is the volume of natural gas sent out from the Company in a single day and is measured in units called MCFs (thousands of cubic feet).

The company has a sendout capacity of 310,000 MCF. On severely cold winter days, it has been the official policy of the company to reserve 180,000 MCF for its nonbusiness customers (homes, schools, hospitals, etc.). This reserve has proven adequate over the last few years but is not expected to prove adequate in the coming winter. That is, in the past a target sendout of 130,000 MCF for business and industrial customers has been adequate.

However, experience shows that if the target sendout is set at 130,000 MCF for business and industrial customers, then actual sendout might be slightly above or below that figure. In fact, on those days during the current winter when the target sendout was 130,000 MCF, the actual sendout averaged 130,500 MCF, with a standard deviation of 1230 MCF.

Mr. Ralph Maybe, Director of Operations for Northeastern, noted these figures and stated, "It looks as if we've got to set our business target sendout considerably lower than 130,000 MCF if we don't want to cut into the 180,000 MCF reserves for nonbusiness customers." In addition, he suggested that the 180,000 MCF figure was no longer a realistic reserve figure. "The growth of home building in our service area and the last few cold winters have shown us that 180,000 MCF is an adequate reserve under normal conditions, but we should be more realistic and establish variable reserve figures. Reserves for residential users should be higher on days that are predicted to be severely cold than on days that are only mildly cold."

On the basis of this reasoning, Mr. Maybe had one of his staff assistants prepare a list of recommended reserve levels, which follows.

PREDICTED LOW TEMPERATURE	SUGGESTED NONBUSINESS RESERVES
25°F or higher	180,000 MCF
20°F up to 25°F	185,000 MCF
15°F up to 20°F	192,000 MCF
10°F up to 15°F	199,000 MCF
5°F up to 10°F	207,000 MCF
0°F up to 5°F	216,000 MCF
−5°F up to 0°F	225,000 MCF
−10°F up to −5°F	238,000 MCF
−15°F up to −10°F	253,000 MCF
−20°F up to −15°F	271,000 MCF
−25°F up to −20°F	290,000 MCF

The assistant's report also noted that during the past few winters the Operations Department had been experimenting with the demand for gas when different target sendout figures were established. Another table in the report showed the following.

EXPERIMENTAL TARGET SENDOUT	NUMBER OF DAYS AT THIS LEVEL	MEAN SENDOUT ACHIEVED	STANDARD DEVIATION	COEFFICIENT OF VARIATION
100,000 MCF	7	99,600 MCF	1,050 MCF	1.05%
110,000 MCF	12	111,200 MCF	1,100 MCF	.99%
115,000 MCF	10	114,800 MCF	1,090 MCF	.95%
120,000 MCF	8	118,900 MCF	1,280 MCF	1.08%
125,000 MCF	20	125,400 MCF	1,225 MCF	.98%
130,000 MCF	35	130,500 MCF	1,230 MCF	.94%

The report concluded that "The coefficient of variation for the sendout figures seems to be rather stable over a fairly wide range of target sendout levels. It appears that this information could be used in conjunction with our knowledge that the sendout figures tend to follow a normal distribution about their mean to help us set daily target sendout levels for business and industrial users based on the forecasted low temperature for the following day."

Mr. Maybe was pleased with these results and gave his assistant permission to begin development of target sendout figures to correspond to the ranges of forecasted low temperatures identified in his first table. The only restriction he established was, "Make sure your target levels are set low enough so that we run only a 10% chance of having the business and industrial customers eating into the suggested reserves."

GOLD SEAL LABORATORIES

There are only a few laboratories across the country that test consumer products. Underwriters Laboratories of Chicago is the best known. Gold Seal Laboratories is another of these testing organizations, and it is located in Dayton, Ohio.

Ms. Patricia Seybolt is the chief testing engineer for the company and supervises the work of seven other engineers. The engineers have between four and six testing technicians reporting to each of them. If a company wishes to obtain the Gold Seal of approval from the laboratory to display on its products, it must allow a laboratory technician to come to its plant, randomly select items from the production line, and bring them back to Dayton for testing. The tests involve examining the items to see that they meet their design specifications and federal or local safety standards.

For instance, a clock-radio might be tested to make sure that the clock keeps reasonably accurate time, the alarm goes off when it is set, the case is shockproof, and that the circuitry has been connected just as the electrical schematic diagram shows it should be. Many items are tested for parts reliability. That is, an electrical appliance might be operated continuously for several hundred hours or until it breaks down in order to determine the operating life of the item.

Ms. Seybolt has just received some data on United Lighting, Inc.'s new ornamental outdoor arc lamp. The lamp is used on a post in the front yards of homeowners living on poorly lighted streets. United's lamp sells for a substantially higher price than the conventional lamps used for this purpose. However, it lights a much wider area, and United believes that it lasts much longer. Unfortunately, United has no hard data to prove this second claim. Sales of the lamp have been slow since the competition has been successful in making dealers believe that United's lamp is too expensive and has a rather short life.

Nearly two years ago, Gold Seal was asked to bid on the cost of testing some of United's arc lamps. Since United wanted to advertise the length of life of their lamp to dealers, they felt it necessary to use an independent testing agent.

Gold Seal won the contract to perform the tests. A Gold Seal technician visited the St. Louis plant of United Lighting and randomly selected twenty lamps for testing. The lamps were wired to a special control board which regulated the amount of time the lamps were lighted. This automatic board turned on all the lamps simultaneously and left them on for a randomly determined period of time, but the time was controlled so that it never exceeded twelve hours of continuous operation. When the lights were turned off, the control board monitored their temperature. Once the lights cooled to room temperature, they were turned on again for another randomly determined period.

When a lamp burned out, a small alarm sounded, and a technician recorded from a timer mounted on the control board the total amount of time the lamp had been operated. The test data showing the times until failure for all twenty lamps are shown in the following table. All that remains is for Ms. Seybolt to write her report to United Lighting concerning the length of life that can be expected from all their arc lamps. In her report, Ms. Seybolt intends to note that four of the lamps appeared to pop, crack, or actually explode when they failed. Since such events might involve a hazard to lamp users, she intends to make an estimate of the extent to which this characteristic can be expected among all United's new arc lamps.

ITEM NUMBER	HOURS TO FAILURE
1	13,140
2	17,555
3	9,490
4	8,090
5	12,234
6	13,876
7	14,570
8	18,442
9	11,109
10	14,007
11	14,990
12	7,901
13	13,893
14	15,737
15	12,459
16	13,952
17	9,333
18	10,873
19	11,537
20	10,755

HOLT, HARMAN, AND HANKS

Jay D. Wiest received his M.B.A. degree in 1973 and passed the C.P.A. examination in early 1975. He enjoyed his work with the firm of Holt, Harman, and Hanks. Mr. Hanks, one of the senior partners in the firm, had hired Wiest right out of State University where Jay was aided in his studies by a scholarship from the firm.

During his first year with Holt, Harman, and Hanks, Jay spent most of his time doing routine work for several of the firm's smaller clients. However, recently Jay had been assigned to the Lerner Equipment account, the firm's newest, biggest, and most exciting client.

Lerner Equipment holds the Southern Region dealership for Gunther Machinery, a German manufacturer of mills, lathes, and other metal fabricating equipment. Holt, Harman, and Hanks obtained the Lerner account recently after Lerner dismissed their former C.P.A. firm for failing to uncover fraudulent practices and embezzlement by Lerner's own accountant.

The fraud had been a rather clever one. Since Gunther is a German firm, Lerner is required to pay import taxes on all Gunther machines it brings into the country. However, Lerner's chief accountant and a Gunther employee had conspired to avoid part of the tax payments and channel the funds into their own pockets.

The Gunther employee arranged the Lerner account so that periodically Lerner would be sent invoices billing equipment at artificially low prices. For instance, a $17,000 machine might be invoiced at $13,000. Then Gunther would later send Lerner a bill for $4000 in fictitious consulting or advertising "fees." In this way Gunther received $17,000 for their machine, and Lerner paid the correct total price. The loser in this scheme, however, was the federal government. The import taxes were paid only on the invoice amount of $13,000 instead of the full price of $17,000. In this manner Lerner avoided paying import taxes on the $4000. Naturally, this is an illegal practice.

The money saved through this fraud did not stay in the company. Lerner's accountant knew that the amount being saved was substantial, so he periodically paid what he estimated had been saved to the B&R Import Service. When he was questioned about these payments, the accountant explained that B&R was a service that expedited certain foreign shipments and handled the payments of some import taxes. B&R were tax experts, the accountant claimed, and thus saved enough in taxes alone to pay for their expediting services. In reality, B&R Import Service was simply a post office box rented to the accountant!

During the year or so when this fraud was in operation, Lerner's costs as a percentage of sales changed very little. The illegal scheme saved import taxes which were then paid to the phony B&R Import Service.

The fraud was detected when the Gunther employee inadvertently mentioned the "special discounts" to a Lerner vice-president who he thought knew about the arrangement. The Lerner accountant was subsequently fired, and the C.P.A. firm was dismissed.

Jay Wiest's job, according to Mr. Hanks, was to estimate the proportion of Lerner's invoiced items from Gunther that contained the discounts. Jay was also to estimate the average size per item of the illegal discounts. These estimates were to be used to determine, first, if the government was going to require a 100% audit of the transactions between Lerner and Gunther and to determine, second, if and what type of prosecution was appropriate. Jay was told by Mr. Hanks that he was to work with Gunther's price list and the 25,825 items which Gunther had billed on 4321 invoices during the period when the fraud was in operation.

When Mr. Hanks discussed the assignment with Jay, he said, "Jay, I need a cost estimate on this job to give Mr. Lerner. We need to know the percentage of items with doctored prices to within ±5%. The Gunther people have started an investigation on their end and think the figure will run between 20 and 30%. We need to know the average size of the discounts to within about a hundred dollars. Remember, we'll be billing your time at $30 an hour."

Jay got to work on the job immediately. After about three hours of getting set up and acquainted with the Lerner bookkeeping system, he launched into the actual audit of the invoices. After another two hours he had sampled thirty-two items from the invoices, using a random number table to select first the invoice numbers to be examined and then the item on the invoice to be checked against the price list. (Joan Buckman, a clerk at Lerner's, brought out in batches the invoices Jay selected from his table. Jay then checked items on the invoices while Joan gathered a new batch based on the random number tables.)

Jay had found eight discounted items in the first two hours. The sizes of the discounts were 1.5, 5.0, 1.5, 7.0, 3.5, 5.0, 4.0, and 4.5 thousand dollars, respectively. He estimated he could average about twenty items per hour; however, he still had not submitted his cost estimate to Mr. Hanks.

HATFIELD CITY HOSPITAL

Hatfield City is located in the central region of a midwestern state. The city has a population of twenty thousand people, but it is the commercial center for an area of approximately fifty thousand people. The major medical facility in Hatfield City is the Hatfield City Hospital. The hospital has just over two hundred beds and a highly competent medical staff, considering the relatively small size of the city.

Hatfield City Hospital is owned by a nonprofit corporation called Midwest Health Care (MHC), which took over several unprofitable hospitals from churches and municipalities in the mid-1970s. The MHC Corporation was set up by a regional health planning group under the assumption that the group of hospitals could pool some of their resources and afford to hire competent central management. It was hoped that the new central management would be capable of recommending changes that would improve the profitability picture of the group to the point where, on the whole, they were breaking even. By standardizing management policies and procedures, sharing some expensive personnel and equipment, and pooling purchasing and personnel functions, the losses were eliminated at most of the hospitals in the MHC group. The entire MHC Corporation reached a break-even point in June of 1979.

A small central staff for MHC is located in Des Moines, Iowa. The staff conducts special studies and supervises the audits of the hospitals in the group. A recent study conducted at Hatfield City Hospital has caused some concern and debate. The study dealt with "lost charges."

A lost charge occurs when a physician orders a procedure, test, or medication for a patient but the patient's account is not billed for the cost of the service rendered. While the medical personnel attending the patients are quite careful about entering any action concerning the patient on the patient's medical record, there may be times when the patient's financial record is not charged. Any time this happens, the charge for service that should have been made to the patient's bill is "lost."

An audit can uncover these lost charges when an auditor compares the actions described on the medical record to the charges made on the financial record. Audits of the records at all the hospitals show that on a system-wide basis the MHC hospitals lose 9% of the charges, and the mean value of those lost charges is $22.50.

A recent audit of the records at Hatfield City Hospital involved examination of 412 randomly selected procedures, tests, or services taken from the medical records of patients admitted to the hospital in the past year. The audit result showed that in the sample there were 58 charges, or 14% of the 412, that were lost and thus not billed. The lost charges averaged $36.87 and had a standard deviation of $15.24.

The hospital administrator at Hatfield was upset by the audit results.

They seemed to imply that his people were being more lax in keeping track of appropriate financial charges. Moreover, they implied that either his personnel were compounding the problem by losing larger charges than were lost at the other hospitals in the system or the average charge for services at Hatfield City Hospital were higher than those of other hospitals.

The Hatfield City Hospital administrator wrote a memo about this matter to the MHC central office. In part, that memo stated:

> Naturally, I am concerned about the auditor's recent report concerning lost charges here at Hatfield City Hospital. However, before I take any precipitous action to "correct" this situation, I would like to be convinced that the audit results were not just a fluke. Isn't it possible that we are really at the system-wide standards of 9% losses and $22.50 average lost charge? Maybe the auditors just got an unusually bad sample from our records.
>
> Besides, even if we do lose more charges and even if they are higher than average, I'm not sure that this is something we ought to worry about. Over the past year, all MHC hospitals have agreed that when we lose money we can make up the difference by raising our basic room charges. We are currently well below the national average on that charge. Until the room charge gets up to the national average, I'm not convinced that we really need to worry about correcting any deficiencies (if there are any) in our accounting system here.

This memo was turned over to Ben Hanni, the MHC accountant who performed the original audit. He was asked if he could answer the questions raised by the Hatfield City Hospital administrator.

MOUNTAIN STATES FEED EXCHANGE

The Mountain States Feed Exchange (MSFE) has developed a new feed which it thinks will result in high weight gains for cattle in feed lots. At the same time, MSFE believes the new feed will help the animals resist certain diseases. To test the new feed, MSFE obtained the cooperation of Wilma Maxwell, who operates a feed lot in southern Colorado. Wilma fed the new feed to a group of animals in her feed lot between February 1st and May 12th, inclusive. The information on these animals is presented in the following table.

MSFE also obtained information concerning a number of animals that Wilma had in her southern Utah feed lot between January 15th and May 5th, inclusive, of the previous year. This group of animals was fed the standard MSFE feed. The record for these animals follows the table in summary form.

MSFE would like you to examine these data and write them a report discussing what the data show. Mr. Jeremiah Kroft, owner of MSFE, will review your report. He has a fair knowledge of statistics.

Record of Animals Using New Feed

BEGINNING WEIGHT	ENDING WEIGHT	FREE OF DISEASE?
375	552	yes
312	498	yes
433	720	yes
343	451	no
329	487	yes
396	556	yes
430	701	yes
438	578	no
450	739	yes
303	430	no
412	645	yes
440	689	yes
357	599	yes
437	687	yes
450	738	no
349	654	yes
387	729	yes
401	613	no
378	492	no
444	721	yes
316	617	yes
359	596	yes
369	637	yes
355	640	yes
323	589	yes
341	612	yes
377	712	yes
301	Died	Died

Record of animals receiving standard feed:

1. Mean beginning weight of 43 animals = 392.49 pounds. Standard deviation of beginning weights = 54.80 pounds.
2. Mean ending weight of 41 animals = 552.65 pounds. Standard deviation of ending weights = 133.77 pounds.
3. Mean daily weight gain (41 animals only) = 2.01 pounds. Standard deviation of daily weight gains = .45 pounds.
4. Number of sick animals (including two that died) = 12.

STATISTICS 101

Midcontinent Polytechnical Institute requires all students to take Statistics 101 in their freshman or sophomore years. Recently an alumnus, Mr. William T. Kahill, donated a large sum of money which he wished to be used to provide teaching aids for students of statistics. Mr. Kahill had nearly flunked out of Midcontinent Poly in 1938 due to his difficulty in mastering this subject.

The faculty of the Statistics Department used the funds to convert two classrooms into a statistics laboratory. These rooms contained desks, lounge chairs, chalkboards, and a small library of statistics reference books. In addition, teaching assistants were on duty fourteen hours a day. The teaching assistants helped students with their homework problems and checked out library materials to the students.

The students could not only check out books from the statistics library but also obtain electronic calculators, programmed learning aids, and a series of taped lectures on elementary statistics. The series of taped lectures was obtained only last term.

Dr. Walter Gossett, chairman of the Department of Statistics, was interested in determining the results that were being obtained using these taped lectures. Since Dr. Gossett taught a small section of Statistics 101 last term, he decided to compare the examination results of students who had used the tapes with those who had not. Thus, at the end of the term, he collected the sheets showing who had checked out the taped lectures. The results he found follow.

NAME	FINAL EXAM SCORE	CHECKED OUT TAPES?
Michael B.	82	yes
Harry O.	99	yes
Lucy McB.	77	no
Ralph T.	45	no
Mary L.	99	no
Jim C.	66	no
Bill R.	79	yes
David H.	80	no
Roy S.	89	yes
Gary G.	90	no
William S.	39	no
Harry T.	78	yes
Max W.	66	yes
Patty M.	86	no
Shauna T.	72	yes
Terry C.	70	yes
John B.	55	no

(Table continues on next page.)

NAME	FINAL EXAM SCORE	CHECKED OUT TAPES?
Bob G.	Dropped class	no
Betty S.	81	yes
Debbie B.	75	no
Susan D.	88	yes
Bruce B.	100	yes
Earl S.	25	no
Reed R.	95	yes
Edward B.	87	yes
George T.	65	no
Juan A.	83	yes
Janice N.	77	yes
Kent G.	94	yes

OUTER ISLAND RESORT

Outer Island Resort is located on a large island in Marshall Lake.* The lake is one of the most beautiful in the northwest United States. It is known for excellent fishing, and the hunting in the area brings people from all over the world.

The resort hotel was built in 1950 but operates only during the three-month summer season. The resort seldom makes money, but the owner, a wealthy eastern banker, does not worry as long as operations come close to breaking even. The owner seems to view the 150-room resort hotel almost as his personal summer retreat.

One of the most severe problems facing the management at the hotel is the high turnover rate during the "season." Last year it was 40% in just three months. That is equivalent to 160% on an annual basis.

There are two major problems presented by this midseason turnover. First, the typical individual who quits gives very short notice. The resort receives thousands of inquiries regarding summer jobs, but hires only a few hundred people. When a hired employee quits in midseason, the best job applicants among those not hired are no longer available. What was, prior to the season, an extensive labor pool, by midseason dries up to a shallow puddle.

Second, without adequate staffing, either guest services suffer or a greater burden is placed on the remaining employees—a burden

---

*Portions of this case appeared in an actual study reported in Croft, D. James, and Sorenson, Perry, "Screening Seasonal Hotel Workers," *The Cornell Hotel & Restaurant Administration Quarterly*, November 1974. Used with permission.

which breeds more dissatisfaction and terminations. At the height of the season, training of midseason replacements also becomes a problem. At the beginning of the season there is time dedicated to employee training, but by midseason, the pace of business has increased to the point where proper training cannot always be accomplished. A newly hired replacement simply may be thrown into a job and told to sink or swim. Thus, while employee turnover is a serious problem for any business, it is extremely critical for the resort.

While there is some disagreement in the industry as to which costs should be included in the employee turnover, an American Hotel and Motel Association report estimated the cost averages about $250 per employee turnover. This amount includes such items as:

> Midseason advertising
> Employment agency fees
> Interviewing
> Reference checks
> Clerical costs
> Computer check-in costs
> Identification cards and badge
> Job training
> Unemployment costs
> Room cleaning for new employees

These are only the quantifiable costs and indicate nothing about the potential effect of a termination on the quantity and quality of service rendered while a replacement is found and trained.

In order to discourage employees from terminating early, most seasonal resorts require some sort of "contract" or agreement. This is generally not legally binding on either the employees or employer, but it does assure an applicant of a job prior to his or her arrival at the resort. It encourages the applicant, through a low level of moral suasion, to stay in the job an agreed-upon period. Most often an employee will lose a bonus or forfeit a deposit and/or not be rehired next season if he or she does not stay the entire contract period. Outer Island Resort does use a contract, but it doesn't use a bonus or deposit system.

The problems involved with opening a large hotel, even under the best of conditions, are tremendous. Hotel consultants estimate a four-month period is required before all departments are running smoothly for a 400-room property. This is assuming the staff consists largely of trained professionals. In seasonal resort hotels, "Grand Opening" is an annual problem, and in most cases, the seasonal closing comes well before the shakedown period is complete. Unlike ongoing, year-round businesses, seasonal resorts must hire an entire staff every year, few of whom are experienced people.

For other than supervisory level positions, the hotel and concession employment requires little skill or technical knowledge. Rather, it requires a willingness to learn a job, work hard, and remain on the job through a rather short season. The resort selects many of its employees from among university students and other young people in the 18- to 25-year-old range.

Outer Island Resort tends to hire a high percentage of its employees from this age group because young people are:

1. Usually mature enough to live away from home.
2. Willing to work a six-day week for the wages offered (usually close to the legal minimum).
3. Able to correlate work and school schedules to agree with seasonal employment.

For the resort, the problem of employee selection for jobs open to the student group is oriented more toward choosing applicants who will stay for the duration of the season than toward any other single measure of employee success. This is due to the relatively low skill level required of most employees.

Harry Sorenson, the personnel manager at Outer Island, lives at the resort throughout the entire year. During the "off" months he has time to conduct employment interviews for the next season and to do staff work. This past winter he studied the turnover problem at the resort. He was interested in determining if there is any information on employee application forms that can be used to distinguish between those people who are likely to stay the entire season and those who are likely to quit early.

Mr. Sorenson extracted all the quantifiable data he could obtain from the application forms for employees of the past season. The information included twenty-six variables. They were:

1. Applicant's age (in months) at contract start date.
2. Marital status.
3. Number of dependents listed (including self).
4. Was an automobile license plate number listed in the question which requested this information? (This information was an indirect measure of whether an applicant owns an automobile.)
5. Miles from the applicant's home to Outer Island.
6. Number of years of education completed.
7. Is the applicant currently attending a college or university?
8. Number of former jobs listed on the applicant's work history.

Questions 9 through 13 deal with the applicant's reasons for applying to the company for a job. If one of the following reasons was given by the applicant, a "yes" was recorded as the response to that question. Otherwise, a "no" was recorded.

9. Applicant had visited the Marshall Lake area and would like to work there.
10. Applicant had worked for the company in previous years.
11. The job was recommended to the applicant by a friend.
12. The applicant had seen National Park Service information concerning jobs in the area.
13. The applicant had seen information about the job at school.
14. Was the applicant interviewed by a company recruiter?
15. Were there tips on job to which the applicant was assigned?
16. Did the applicant indicate a preference for any particular position(s) and receive his or her first, second, or third choice?
17. Did the applicant indicate a preference for one location (the company has facilities at three different locations in the Marshall Lake area).
18. Did the applicant indicate he or she would accept a position other than that (those) specified in Question 16?
19. Did the applicant indicate he or she would accept a position at a location other than that (those) specified in Question 17?
20. Has the applicant been employed before, or will this be the applicant's first job?
21. Number of days the applicant would be available for employment figured as the difference between the last day of the season and applicant's earliest possible report date.
22. The percentage overlap between the dates in Question 21 and the company's "season."
23. The number of days lag between the applicant's tentative starting date and the date on which the contract is to be offered.
24. Did the applicant attach a picture to the application form? (This is optional for all applicants.)
25. The length of the contract (in weeks) to be offered to the applicant.
26. Was the applicant living at home at the time of application?

Other items of information were also considered as possible indicators of an applicant's likelihood for staying through the season. For instance, in some cases it was possible to calculate the difference between the applicant's likely hourly wage rate with the company and the wage rate received on the last job. However, this information was not available for enough of the applicants to be used in a practical way.

After Mr. Sorenson made some preliminary calculations, he selected the ten characteristics he felt were the most important in distinguishing between the people he called the "stayers" and the "quitters." Even though his data came from 400 application forms, he was not sure which of the characteristics are truly different for the two groups of employees and thus deserving of his attention in trying to hire people who will stay the season.

CHARAC- TERISTIC NUMBER	CHARACTERISTIC	MEAN VALUE FOR THOSE WHO STAYED	MEAN VALUE FOR THOSE WHO QUIT
4	Did the applicant list an automobile license number on the application?	50% yes	33% yes
7	Was the applicant attending a college or university?	74% yes	57% yes
10	Had the applicant worked for the company before?	30% yes	15% yes
13	Did the applicant receive information about the job from school?	14% yes	6% yes
16	Was applicant placed in a job that was applied for?	89% yes	74% yes
17	Did the applicant indicate a preference for a location?	75% yes	57% yes
18	Would the applicant accept a position other than that applied for?	88% yes	83% yes
22	Percentage overlap between applicant's availability and the season	103%	97%
24	Inclusion of picture on application	41%	19%
25	Length of contract (in weeks)	12.8	13.9

GARDNER LEARNING SYSTEMS, INC.

Walter W. Gardner, Ph.D., resigned his appointment as professor of communications at Ohio State University in 1967. Since that time he has been the president and only full-time employee of Gardner Learning

Systems, Inc., a management consulting firm specializing in organizational communications.

Gardner Learning Systems offers courses in executive communications skills, public speaking, conducting meetings, and relations with the media. The courses are often offered at seminar facilities Dr. Gardner rents in New York, Chicago, and San Francisco. Dr. Gardner also presents "in-house" seminars and training programs for which he meets with managers in their own offices or conference rooms.

Training programs are received most readily when the trainers are able to show some measurable behavior change that results from the training. Dr. Gardner has been thinking of purchasing the rights to use a copyrighted course entitled "Better Listening" from Shaw Educational Materials. The reason Dr. Gardner is interested in purchasing the course is because he obtained such dramatic results when he used it in a trial program.

The trial program involved U.S. Department of Agriculture administrators. The trainees were asked to listen to two tape recordings before the training started. After each recording, they were asked to write down the key ideas covered by the speakers in the recording. Dr. Gardner then scored the answers to the two test tapes using a key provided by Shaw Educational Materials. Each test was scored on the basis of 100 possible points.

This pretraining test was followed by three hours of programmed learning in listening. The programmed learning training kits were designed and sold by the Shaw group. The learning experience was followed by a lunch or dinner break, and the posttraining test was administered after the meal. This second test was like the first. The course participants were asked to write down the main ideas covered by speakers on two tapes. Dr. Gardner also scored these results and was pleased with the improvement.

PARTICIPANT NUMBER	PRETEST SCORES:		POSTTEST SCORES:	
	Tape 1	Tape 2	Tape 1	Tape 2
1	35	21	95	100
2	56	28	46	80
3	15	77	44	66
4	100	91	82	100
5	56	57	75	80
6	72	56	100	80
7	19	8	54	56
8	15	59	50	66
9	76	90	82	80
10	22	59	100	36
11	32	77	100	100

(Table continues on next page.)

PARTICIPANT NUMBER	PRETEST SCORES:		POSTTEST SCORES:	
	Tape 1	Tape 2	Tape 1	Tape 2
12	30	40	77	66
13	32	41	64	80
14	17	0	90	80
15	55	46	61	74
16	20	23	42	66
17	61	64	78	100
18	56	82	100	100
19	56	62	95	100
20	72	64	95	80

Dr. Gardner suspected the results showed a real improvement since any of the four testing tapes could be used in the pretest or the posttest, and he had randomly selected the order in which the tapes were to be given. Shaw advertised that the tapes were "graded equal" in difficulty, but Dr. Gardner wasn't sure how to test this claim.

What he would like to know now is how he can present the results of this trial test in his sales presentations to prospective clients. Obviously, not all groups can expect the same overall results. However, Dr. Gardner would like to be able to say to a client, "The mean listening level of your people will increase between X and $Y\%$ as a result of this course, and I will give you quantitative documentation of this improvement."

METROPOLITAN CHARITIES

The Metropolitan Charities United Association raises money for charitable causes in a large eastern urban area. The director of the association, Mr. William P. Dankful, and his staff are interested in determining what contribution soliciting methods produce the best results.

The association has decided that it will test three solicitation methods: direct mail, telephone calls, and personal visits. In the test, three solicitors were used to make the telephone calls and the personal visits. All three made fifteen calls and fifteen visits each. This was done in order to eliminate any bias that might be introduced by the differing sales abilities of the three solicitors. Of course, none of the solicitors was involved in the direct mail solicitations since these were nonpersonal contacts.

The results of the test are presented opposite. The figures represent the amount of money donated to the association on each contact.

SOLICITORS	TELEPHONE CALLS	PERSONAL VISITS
1	$ .00	$ 2.00
1	.00	.00
1	5.00	10.00
1	10.00	.00
1	.00	.00
1	8.00	2.00
1	.00	1.00
1	15.00	125.00
1	.00	5.00
1	.00	.00
1	.00	12.00
1	10.00	.00
1	.00	.00
1	.00	50.00
1	5.00	.00
2	3.00	7.00
2	.00	10.00
2	.00	8.00
2	20.00	10.00
2	5.00	.00
2	.00	.00
2	.00	2.00
2	4.00	5.50
2	5.00	.00
2	.00	1.50
2	.00	30.00
2	.00	20.00
2	15.00	.00
2	25.00	10.00
2	.00	.00
3	.00	.00
3	15.00	.00
3	5.00	.00
3	.00	6.00
3	.00	4.00
3	3.00	.00
3	2.50	3.50
3	.50	110.00
3	.00	.00
3	100.00	9.50
3	.00	20.00
3	.00	.00
3	5.00	40.00
3	3.00	.00
3	.00	5.00

The responses to 50 direct mail solicitations were as follows:

$ .00	$10.00	$ .00	$ .00	$ .50
.00	.00	.00	.00	.00
.00	2.00	.00	.00	.00
5.00	5.50	.00	.00	75.00
.00	.00	50.00	.00	.00
.00	.00	.00	.00	.00
.00	.00	.00	.00	7.50
.00	.00	.00	2.50	.00
.00	.00	.00	.00	.00
.00	.00	.00	.00	.00

Mr. Dankful is interested not only in which method of solicitation produces the best results, but also in two other issues: (1) which method has the highest return on investment, and (2) which method produces more responses. Mr. Dankful is especially interested in this latter question since if he produces a high rate of response, he can sell his list of contributors to other professional fund raisers who want lists of people likely to contribute to worthy causes.

The costs of each contact differ. A direct mail contact is estimated to cost approximately $.20 for printing and mailing. The telephone contacts cost only the labor of the caller, telephone rental, and office space costs. These are estimated to be about $.35 per contact. (Calls are made until the potential donor is reached.) The personal visits cost the most. They involve labor costs of the visitor and mileage that must be paid for the visitor's automobile. These costs run about $3.00 per contact.

MIDNATION MEDICAL SCHOOL

"We all are aware of the serious maldistribution of physicians in this country today," began Dr. William Samuelson as he opened the meeting of the Midnation Medical School Admissions Committee. "Physicians are badly needed in our rural areas, but they concentrate in the urban areas. We need thousands more general practitioners, but most of our students seem to specialize. We have twenty times more applicants for positions in each freshman class than we can accept. At least 80% of those applicants are more qualified than you or I were when we were accepted to medical school years ago."

Midnation Medical School faces a problem that confronts medical schools across the country today. It has more qualified applicants than it needs. Now it is looking for entrance criteria other than past scholastic achievement, national test scores, and letters of recommendation. That is, Midnation Medical School is now interested in producing the "right" kind of physicians—those who will help fill the gaps in medical manpower of this country.

Dr. Samuelson proceeded to show the members of the Admissions Committee the results of a study which he had one of his graduate assistants do for him. The study seemed to show that there is a relationship between the type of city in which a medical student grew up and the type of city in which he eventually established a practice.

Dr. Samuelson's study involved 550 physicians who were graduates of Midnation Medical School. The hometowns of the physicians were classified as being rural, small, or urban in the following way:

Rural: population under 5000
Small: population between 5000 and 100,000
Urban: population over 100,000

The cities in which the physicians set up practice were classified in the same way. The following table shows this classification.

PHYSICIAN'S PRACTICE TOWN	PHYSICIAN'S HOMETOWN:			
	Rural	*Small*	*Urban*	*Total*
Rural	62	28	37	127
Small	25	73	61	159
Urban	18	64	182	264
Total	105	165	280	550

Some of the Admissions Committee members told Dr. Samuelson that he wasn't telling them anything they didn't know or couldn't have guessed. "You're just saying that small town kids tend to go back to the small towns," interjected Dr. Bills, a committee member. "Hold on just a minute," responded Dr. Samuelson. "I have a few things you probably haven't seen before. This next table shows the relationship between a physician's practice location and the type of hometown his wife came from."

PHYSICIAN'S PRACTICE TOWN	WIFE'S HOMETOWN:			
	Rural	*Small*	*Urban*	*Total*
Rural	64	29	33	126
Small	20	70	60	150
Urban	5	53	184	242
Total	89	152	277	518

"Thus it appears, gentlemen," Dr. Samuelson continued, "that we ought to begin asking our applicants about the hometowns their wives came from. It appears that the wife's hometown may be just as significant in identifying would-be rural practitioners as the applicant's hometown."

"In terms of identifying those applicants who are likely to go into general practice," Dr. Bills asked, "how will this help us?"

"The tables I have shown you so far won't help in that regard, but our study did uncover some interesting relationships between the MCAT scores and the practice preference of physicians." The MCAT scores to which Dr. Samuelson referred are the Medical College Admissions Test scores. Most medical schools require applicants to take the MCAT as part of the admissions procedure. Dr. Samuelson then showed the committee the following table. Physicians were classified according to their primary vocational interests as general practitioners (GPs), specialists, or researcher-academics. They also were classified by the way in which they scored on the MCAT when they were admitted to medical school. The first quartile refers to those who scored in the top 25% of those taking the test that year and who were admitted to medical school.

MCAT SCORE	TYPE OF PRACTICE:			
	General	Specialized	Researcher Academic	Total
1st quartile	33	69	30	132
2nd quartile	33	57	22	112
3rd quartile	46	92	20	158
4th quartile	35	80	14	129
Total	147	298	86	531

"This table indicates that if we want to turn out researcher-academics, we ought to admit only students who score very high on the MCAT. On the other hand, if we want to produce GPs, we ought to begin admitting more students in the lower quartiles of the MCAT."

This suggestion brought vigorous discussion from the Admissions Committee members. "That's outlandish," Dr. Bills blurted out. "We don't want to turn this place into a mediocre medical school with policies like that!" Dr. Horne, chairman of Community and Family Medicine, was cool toward the idea of admitting students with lower MCAT scores, but he did soften his stand by saying, "It's not as though we'd be letting in a bunch of dummies. Even the kids in the lower quartiles these days are smarter than we were when we were in the upper quartiles. When you think about it, we've been admitting on the basis of grade point average for too long. We've been admitting scholars to medical school. Do scholars make good bedside docs?"

The strongest opposition came from Dr. Wilder. He dealt very little with the students, except on research projects. He was known for his ability to secure government and foundation grants for medical research work. Dr. Wilder suggested, "You've got to do more homework before you can convince me of the validity of any of this. I'm not for admitting a bunch of country bumpkins to this school unless you can prove to me that your tables about hometowns are statistically significant. And, by the way, how come you started out with 550 in your study and get only 518 and 531 in your next two tables? Are you sure you followed good research techniques? I want to turn out a few more GPs like the rest of you, but your table about MCAT scores scares me. In the first place, I'm not sure there's any real relationship there, and in the second place, if there is, I'd just as soon we admit plenty of those high scorers who might be able to help me in my work."

MC FARLAND REALTY

Bill McFarland is a real estate broker who specializes in selling farmland in a large western state. Since Bill advises many of his clients about pricing of their land, he is interested in developing a pricing formula of some type. He feels he could increase his business greatly if he could accurately determine the value of a farmer's land.

Bill's friend, Mr. Martin Alluvial, is a geologist and has told Bill that the soil and rock characteristics in most of the area where Bill sells do not vary much. Thus, according to Mr. Alluvial, the price of the land should depend solely on the acreage, and any variation from a fixed price per acre is due only to psychological factors in the buyer and seller and not to the value or worth of the land.

Mr. McFarland has sold thirty-five plots of land in the past year. Their selling prices and acreages are listed in the following table.

SELLING PRICE (000)	ACREAGE	SELLING PRICE (000)	ACREAGE
60	20.0	92	30.0
130	40.5	77	25.6
25	10.2	122	42.0
300	100.0	382	133.0
85	30.0	5	2.0
182	56.5	42	13.0
115	41.0	60	21.6
24	10.0	20	6.5
60	18.5	145	45.0

(*Table continues on next page.*)

SELLING PRICE (000)	ACREAGE	SELLING PRICE (000)	ACREAGE
61	19.2	50	16.0
20	6.0	25	10.0
15	3.1	290	100.0
485	210.1	118	41.0
892	305.2	17	6.0
46	14.0	10	3.0
69	22.0	41	14.0
220	81.5	200	70.0
235	78.0		

Mr. McFarland would like to convince Melvin Curtis to sell his land. Mr. Curtis owns 250 acres of land in the western portion of what Mr. McFarland considers his "territory." However, Mr. Curtis is unwilling to sell until he can get an accurate feeling for the worth of his land. Bill got him to say that he would be willing to sell for approximately the same price per acre that people in the area were selling for last year "plus about 10% for inflation this year." However, Mr. Curtis said that he'd have to see some firm data on what land was going for last year before he would sell.

In the meantime, Bill McFarland needs to give Orin West an estimate of what he might get from his thirty-five acre plot. Mr. West wants his estimate early next week.

WESTERN PETROLEUM (B)

Mr. Malcolm Maxfield and Saul Stewart worked together to develop a report on the probability of winning no leases and on the expected returns from a Western Petroleum Plan B. Mr. Maxfield did most of the computation and writing work, with some technical aid from Mr. Stewart, who was more current on the theory needed in these problems. Reviewing the necessary probability theory, making the computations, and writing up the results took Maxfield approximately four days.

During this time Mr. Maxfield was impressed with the fact that the probabilities and expected returns were rather sensitive to the number of applications he assumed would be made for each parcel. For instance, he found that the probability of obtaining one or more parcels in a Plan B ranged from .993 (if the fewest people bid) down to .658 (if many people bid on each parcel). Under the optimistic assumption that few people would bid on each parcel, the expected number of parcels won by the holder of a Plan B was found to be 4.727. But under the pessimistic assumption that many people bid on each parcel, that expected number of parcels went down to 1.063.

Since the results of his computations were so sensitive to the number of people bidding on the parcels, Mr. Maxfield determined that he would continue his study of the situation in an attempt to find a means of estimating or forecasting how many people would be bidding for a parcel. He commented, "I know that Class I parcels are attracting more bids, but there is even a wide variation in the number of bidders on those parcels. If I could get a list of the characteristics of these parcels and how many people bid, maybe I could develop some kind of a forecasting equation based on the characteristics of the property."

With this objective in mind, Maxfield began gathering data on the characteristics and number of bidders on various parcels of land in the BLM lotteries during the past year. Since so many parcels of land had been involved, Maxfield could not find information on all of them. Thus, he took a random sample of several parcels and arranged their data by class in the following table.

NUMBER BIDDING	CLASS	MILES TO NEAREST PRODUCER	OUTPUT OF NEAREST PRODUCER	GEOLOGIST'S LIKELIHOOD ESTIMATE	PARCEL SIZE
300	I	8.0	5000	.40	3.5
375	I	3.2	5000	.40	4.0
200	I	8.0	4500	.15	10.2
531	I	.8	6500	.50	1.3
115	I	12.0	3600	.08	5.6
629	I	1.6	7800	.65	2.9
152	I	5.9	4200	.05	6.5
131	I	11.8	3600	.05	9.8
192	I	4.0	2600	.20	6.3
427	I	2.6	7800	.35	10.1
93	II	25.0	7100	.20	3.8
120	II	36.0	1500	.25	4.2
46	II	17.8	3500	.15	2.1
38	II	49.6	3800	.15	1.6
29	II	48.3	5100	.15	.8
77	II	18.5	4200	.20	2.8
102	II	17.1	5300	.33	3.5
82	II	16.5	3100	.25	3.0
53	II	26.1	4100	.18	3.0
48	II	18.1	3400	.15	2.3
10	III	135.1	3600	.02	4.6
21	III	151.5	4500	.04	10.1
16	III	80.5	3200	.03	8.6

(*Table continues on next page.*)

NUMBER BIDDING	CLASS	MILES TO NEAREST PRODUCER	OUTPUT OF NEAREST PRODUCER	GEOLOGIST'S LIKELIHOOD ESTIMATE	PARCEL SIZE
35	III	52.1	3900	.05	16.1
40	III	51.6	4100	.04	21.7
12	III	124.1	3600	.01	2.2
15	III	103.1	4700	.01	3.2
11	III	176.1	4300	.02	2.0
19	III	142.6	3600	.02	3.6
37	III	81.5	3100	.04	4.2

Each row in the table represents a parcel of land that the BLM put up for bid in the lotteries last year. The first column shows how many people bid on that parcel. The second column shows the parcel's classification. These classifications (I through III) are determined by the BLM. The third column in the table shows the number of miles (as the crow flies) from the center of the parcel to the nearest producing oil well. The output of the nearest producing well is listed in the fourth column. This output is measured in terms of barrels per month. The fifth column contains the subjective probability estimate that oil could be found in geological formations like those known to be on the parcel in question. These estimates were made by a geologist who served as a consultant to Western Petroleum. The final column in the table shows the acreage on the parcels. Thus, the first parcel listed in the table was one on which 300 people bid. It was a Class I parcel as defined by the BLM, and it was 8 miles from a well which produced 5000 barrels of oil per month. The geologist estimated that 40% of the formations like those found on this parcel yielded producing wells, and the parcel had an area of 3.5 acres.

Mr. Maxfield showed his table to Saul Stewart and asked him for suggestions on how it might be used to develop an estimating procedure for the number of bids to expect on each parcel. Mr. Stewart replied that multiple linear regression seemed like the technique that would help Maxfield. Stewart wrote down an equation for Maxfield:

$$\hat{Y} = a + b_1 X_1 + b_2 X_2 + b_3 X_3 + b_4 X_4 + b_5 X_5$$

and stated that $\hat{Y}$ would be the estimated number of bidders for a parcel and that X_1 through X_5 would be the five characteristics of the parcels Maxfield had listed in his table. In addition, Stewart offered to have an assistant at S&W enter Maxfield's data into a standard computer program to produce the regression equation. Malcolm was very pleased with the offer and turned his data over to Stewart.

On Wednesday of the next week, Maxfield and Stewart met for lunch to discuss the computer output Mr. Stewart's assistant had produced. (Some of the relevant portions of that output are presented at the end of this case). Saul Stewart explained the meaning of various figures in the computer output, and Maxfield felt that he understood the rough idea of how the output could be interpreted.

However, for over a week Malcolm was unable to get back to the computer output and study it due to the press of other business. When he did, he found that he was a little confused on some of the figures. In particular, he was quite upset about the fact that the variable X_2, miles to nearest producer, seemed to have a positive coefficient in his prediction equation. This indicated to him that a parcel far from a producing well would have more bidders than a parcel near a producing well. To resolve the confusion in his mind, Maxfield called Stewart with the idea of asking him for additional explanations and about the possibility of obtaining new computer runs of the data. Mr. Stewart, however, was out of town for the next two weeks.

VARIABLE	MEAN	STANDARD DEV	CASES
BIDS	131,9000	161,9573	30
CLASS	2,0000	,8305	30
MILES	47,6367	52,0051	30
OUTPUT	4343,3333	1428,9334	30
GEOEST	,1707	,1620	30
SIZE	5,4533	4,6296	30

	BIDS	CLASS	MILES	OUTPUT	GEOEST	SIZE
BIDS	1,00000	-,72735	-,56591	,65890	,87979	-,08082
CLASS	-,72735	1,00000	,83061	-,34870	-,65364	,14440
MILES	-,56591	,83061	1,00000	-,25269	-,62723	-,01385
OUTPUT	,65890	-,34870	-,25269	1,00000	,59948	-,04613
GEOEST	,87979	-,65364	-,62723	,59948	1,00000	-,28517
SIZE	-,08082	,14440	-,01385	-,04613	-,28517	1,00000

MULTIPLE R	,95337
R SQUARE	,90891
ADJUSTED R SQUARE	,88994
STANDARD ERROR	53,73065

ANALYSIS OF VARIANCE	DF	SUM OF SQUARES	MEAN SQUARE	F
REGRESSION	5,	691387,11388	138277,42278	47,89686
RESIDUAL	24,	69287,58612	2886,98275	

---------- VARIABLES IN THE EQUATION ----------------

VARIABLE	B	BETA	STD ERROR B	F
GEOEST	767,81263	,76797	108,41568	50,156
CLASS	-110,11453	-,56463	23,33311	22,271
MILES	1,30100	,41776	,38528	11,403
SIZE	8,07906	,23094	2,44708	10,900
OUTPUT	,01336	,11735	,00924	2,089
(CONSTANT)	57,03936			

THE TRINITY ALLIANCE CONGREGATION

In 1948 two United States-based churches combined their memberships. The first was the U.S. Trinity Church of God, which had been founded in 1890. At the time of the merger, U.S. Trinity had a membership of around 500,000. The second church was the smaller Holy United Congregation. This church was founded in 1930 and had a membership at the time of the merger of only 140,000.

The two churches were very similar in doctrine, organization, and rites. The 1948 merger was undertaken for both financial and doctrinal reasons. The Holy United Congregation needed financial support, and U.S. Trinity was interested in establishing a series of church-owned educational institutions, similar to the three Holy United had under way. Also, both churches preached the doctrine of "encompassing love." This doctrine stated, briefly, that God's encompassing love for mankind was the most important single characteristic binding people together and that this love is more important than any of the artificial differences that separate people from one another. Thus, the doctrine continued, people of varied backgrounds and beliefs should be able to live and work together by emulating God's encompassing love and forgetting their differences. The merger of the two churches into The Trinity Alliance Congregation was accomplished, in part, as a practical test of this doctrine.

The Council of Elders, which governed the merged church, was happily surprised at the smoothness with which the merger was accomplished. Their faith in the doctrine seemed justified. By the mid-1970s the church had grown to nearly two million members in the United States, Canada, Mexico, and several South American countries. This growth was due to two factors: an active missionary program in which retired couples served eighteen-month proselyting missions, and a doctrinal stand against birth control.

The Council of Elders was somewhat disturbed, however, by recent financial reports they had received from their twenty regional headquarters. It appeared that the current year's contributions to the Operating Fund were below last year's. The Operating Fund was used to pay for the church's clergy, building operations and maintenance, and the travel costs in the missionary program. (The missionary couples paid their own living expenses during their eighteen-month assignments.)

Several of the Elders felt that the contributions were down due to the extensive building program the church had undertaken recently. Millions of dollars were being spent for new buildings and facilities in most of the twenty regions, and especially in the newer regions where proselyting had been so successful in the 1960s. The chairman of the Council's Financial Planning Committee was worried about the relationship between contributions to the Operating Fund and the costs of building programs in the regions. It was his job to advise the regional staff members on the handling of contributions. Most contributions were collected and invested until needed at the regional levels. Very little money was sent to the worldwide headquarters.

In order to demonstrate the problem, Mr. Durke Jackson, chairman of the Financial Planning Committee, presented the following data to the Council and asked for advice on how he might go about finding a relationship between the various factors influencing contributions.

REGION	NO. OF MEMBERS (000)	CONTRIBUTIONS TO THE OPERATING FUND (MILLIONS):		BUILDING BUDGET (MILLIONS)
		Last Year	*Current Year*	
1. Atlantic	201.5	$24.3	$25.6	$1.23
2. South	152.3	12.2	13.1	.52
3. New England	172.0	19.1	19.0	1.82
4. Midwest	133.4	12.0	11.2	2.60
5. Southwest	105.2	7.2	6.9	.95
6. Mountain	55.2	6.6	5.1	2.09
7. Northwest	82.3	8.2	8.2	.94
8. West Coast	136.0	15.2	15.3	1.27
9. North Mexico	89.2	2.4	1.6	1.02
10. Central Mexico	135.2	3.5	2.6	1.22
11. South Mexico	50.6	1.6	1.6	.14
12. East Canada	126.7	11.4	11.6	1.00
13. Central Canada	58.3	5.3	5.3	.65
14. West Canada	101.1	9.1	9.0	1.11
15. Central America	63.4	1.8	1.9	.12
16. Brazil-Peru-Bolivia	105.0	2.4	2.3	.46
17. Guianas	25.0	.7	.7	.21
18. Argentina-Chile	86.0	3.0	3.2	.62
19. East Islands	42.0	.9	.8	.11
20. West Islands	38.0	.8	.9	.10

Mr. Jackson would like to be able to make some forecasts of contributions to the Operating Fund for the next year. He feels that he needs these most urgently in the eight United States regions. In developing his forecast, he plans to use the fact that in Regions 1 to 8 the construction budgets for the next year are expected to be: 1.50, .75, 2.52, 2.35, 1.15, 1.67, .95, and 1.53 million dollars, respectively.

FARNSWORTH COUNTY BUILDING DEPARTMENT

Frances Phillips has worked for Farnsworth County Building Department since she was in high school. She worked for the department part time during the school year and each summer, right through her four years of study at the local university. After graduation she began working full time and eventually worked into her current job, which involves a combination of personnel functions and purchasing.

Frances is faced with a decision that has been bothering her for a week. She must authorize purchase of some new welding equipment for the department, and she has to determine which company will be given the order for the equipment. The problem is compounded by the fact that the companies from which she might buy are putting on pressure to make a decision right now—before the final specifications for the equipment have been drawn up by the Building Engineering Division at the county's headquarters. These specifications were originally due to be completed six weeks ago, but higher priority projects have been occupying Building Engineering. Frances feels she has three alternatives:

1. Order the Felt Company equipment now for a cost of $15,000.
2. Order the Norton Company equipment at the specially discounted price of $10,000.
3. Delay her decision until after the specifications have been completed by Building Engineering.

Frances was at first doubtful about her authority to purchase prior to the completion of the specifications. But she checked with the County Attorney's Office and was informed that she is fully empowered to authorize any purchases she deems "in the county's best interests" and that the specifications are meant to help her in making her purchasing decision, but not to prevent her from making a prudent purchase. For instance, if the specifications are slow in being completed and a good purchase seems possible, she can act now. However, the county attorney advised Frances that any items that are bought before specifications are completed and that are later found not meeting the specification should be returned to the vendor or modified to meet the specifications. On the basis of this opinion, Frances feels that it would be permissible for her to make an early purchase of the welding equipment if she can document her case for not waiting for completion of the specifications.

The only problem with delaying the decision is that the future price will likely be greater for both companies' equipment. The Felt people are planning a 10% price increase on the first of next month—well before the specifications are due to be completed. Also, the Norton salesman will give the $10,000 price now but says that the price will go back up to its normal $14,000 level after the first of the month. He doesn't think there is a very good chance that Phillips could talk the factory into the $10,000 price after that date. He said, "There's only about a 10% chance that you could get that price later, but you can take the chance if you want to. My commission would be larger on the higher price."

It would appear the Norton equipment is preferable due to its lower price tag, but there is some question as to whether it will perform up to the specifications which are still being written. There is no doubt that the Felt welding equipment will fill the bill, but Frances has indicated to her associates that she is only "80% sure" that the Norton equipment can meet specifications. If it doesn't, the county could buy the Felt equipment or stay with Norton since the Norton people say there is a fifty-fifty chance they could modify their equipment so it will do the job specified. Norton would attempt the modifications only if the county buys the machine first. These modifications would, of course, add to the cost of the machine. The range of this added cost is "from $1500 to $6800," according to the Norton Company salesman. This range of prices holds now or at any purchase date in the foreseeable future. The cost of modifications would not have to be paid by the county if the modification attempt is unsuccessful. Norton would then buy back their machine for $8000, regardless of the original price paid.

In view of these complications, Frances Phillips has decided that she will have to approach the decision from a rational point of view and document carefully the reasons for her decision.

BLUE STAR CALCULATOR

In the fall of 1975 Bill McGraw and Leonard Alvey decided to make their break with Smithson Electronics, Inc. Both men had worked for Smithson for approximately fifteen years and had been satisfied with their treatment by the company for most of that time.

However, in 1974 George Smithson, Jr., had taken over as vice-president of Research and Development. George, Jr., son of the company's president, George Q. Smithson, knew little about research and development. McGraw and Alvey were development engineers under George, Jr.'s direction. The two men soon began to grumble about young Smithson's conservative attitude toward the development of new products.

McGraw and Alvey reached their breaking point in the summer of 1975. Many electronics companies across the country were getting into the market for hand calculators. Smithson Electronics had the technology, manufacturing facilities, and marketing organization to produce and sell hand calculators at a low start-up cost. McGraw and Alvey could not convince George Smithson, Jr. that Smithson should get into the market. He felt that such a venture was "too risky."

It was this decision which brought Bill and Leonard to write their letters of resignation. They spent two months raising money from friends and relatives and securing a small loan from a local bank. They opened the doors of Blue Star Calculator Company in January, 1976. Blue Star was to assemble the Tecktron Calculator, a simple hand calculator designed by McGraw and Alvey. The company bought the components from several sources, one of which was their old employer, Smithson Electronics.

Sales of the Tecktron Calculator began slowly. It was priced to sell for $74.95, which was about in the middle of the price range for calculators with similar features. After one full year of operation, Blue Star was not even close to profitability. Bill McGraw had taken a second mortgage on his home to pay off two of the smaller loans, and Leonard had borrowed enough from his father to pay off the bank. Both men were worried, but they hung their hopes on the Postal Service contract.

The Postal Service's Region Seven offices were in the market for hand calculators to be used by postal employees when making change and balancing their accounts at the end of the day. The purchasing officer was impressed with the Tecktron because of its simplicity. "Your machine has just the operations we need—no more and no less. If we buy your calculator, we won't be paying for a lot of fancy stuff our employees know nothing about and couldn't use on their jobs even if they did know," he told Leonard Alvey. Leonard had quoted the purchasing officer a very reasonable price and had high hopes of landing the contract.

That's when the bottom fell out from under Leonard and Bill's hopes. Just prior to the Postal Service's announcement as to which calculator would be bought, two different organizations published their evaluations of various simple hand calculators. The first evaluation appeared in *Appliance Buyer's Guide,* a publication which offered many consumer suggestions on the comparability of various appliances. The second evaluation was published in the monthly newsletter of Consumer Advisory Service, an organization with purposes similar to those of the *Appliance Buyer's Guide*'s publisher. The Advisory Service, however, evaluated a much wider range of products. *Buyer's Guide* rated the Tecktron calculator quite low. The Advisory Service was not quite so negative.

It was not entirely clear what criteria the two organizations were using in rating the calculators they tested, but both placed heavy emphasis in their ratings on what they called "probable maintenance requirements." Both admitted that many of the calculators were so new to the market that they had not had time to develop firm statistics as to the maintenance records of the machines. However, both organizations developed their evaluations of probable maintenance records based on design and quality of the calculators' components.

Buyer's Guide rated the Tecktron eighth among the twelve machines they examined for maintenance reliability, and the Advisory Service rated the Tecktron sixth among the same twelve machines. The persons rating the calculators also showed their evaluations of the twelve machines in the areas of ease of use (how simple to follow the instructions were) and overall adequacy. The positions in which the evaluators ranked the various machines in three categories are presented in the following table. In the table, ABG refers to the rankings of *Appliance Buyer's Guide,* and CAS refers to those of the Consumer Advisory Service.

The Postal Service purchasing officer for Region Seven had read the results of the evaluations and called McGraw. "I still think your machine is the one we should buy. But I'm over a barrel now. I've got to justify my purchases to my superiors, and it would look as though I'm buying inferior stuff if they see these two reports."

CALCULATOR BRAND	RANKS FOR PROBABLE MAINTENANCE REQUIREMENTS:		RANKS FOR EASE OF USE:		RANKS FOR OVERALL ADEQUACY:	
	ABG	CAS	ABG	CAS	ABG	CAS
Alpha	12	7	11	4	12	7
Cummins	1	11	3	5	2	10
Detro	4	2	1	6	4	1
Electra	5	12	4	10	6	12
Felton	2	9	6	1	3	8
Frazer	11	3	12	11	11	2
Hulett	6	8	9	12	5	9
Sharpy	9	1	7	9	10	3
TECKTRON	8	6	10	3	9	6
Ventura	7	4	8	7	8	4
Westlake	3	5	2	8	1	5
Zermot	10	10	5	2	7	11

McGraw tried to point out that the two people who rated the machines reached quite different conclusions on many of the machines. "Why look at how the Cummins machine turned out. *Buyer's Guide* rated it second overall, but the Advisory Service rated it tenth! The reverse was true for the Sharpy. *Buyer's Guide* rated it tenth, while the Advisory Service rated it third. It seems to me that these two groups must be using subjective criteria and judgments. Otherwise, they would have come in with ratings more consistent with one another. It also appears that there is an extremely high correlation between the ratings the machines were given in the maintenance reliability area and the overall rankings. If the evaluators were overly influenced by the maintenance ratings, then that's not giving the newer machines a fair shake. They admitted themselves that their maintenance data for the new machines was simply made up."

"What you say may be true," responded the purchasing officer, "but can you document some of what you say? I need some concrete statements to give my boss if he shows me these articles and asks why I didn't follow their evaluations."

SAMPLE DATA SET APPENDIX

The data set presented on the following pages represents a random sample of 113 people who applied for charge account privileges at a well-known department store on the East Coast. Each line of data represents one applicant and gives ten pieces of information about that person. The nature and measurement of the ten variables are discussed in the following.

CLASS   indicates whether the department store granted credit to the individual. The value 1 indicates credit was granted, and the value 0 indicates it was not. The first 63 people in the list were granted credit, and the last 50 were not.

SEX   indicates whether the applicant was male (indicated by a 1) or female (indicated by a 0).

AGE   indicates the applicant's age listed in years.

JOBYRS   indicates the number of years the applicant had held his/her current job. The value of 99 indicates that the individual was not employed in an income-producing job at the time the application was made.

JOBINC   indicates the monthly income the applicant was receiving from his/her job at the time of application. A value of 9999 indicates that the applicant had no monthly income.

ADDINC   indicates the amount of additional income (over and above that received from a regular job) the applicant received each month. The figures listed here most often included income from commissions or rental property.

TOTBAL   indicates the total balance of debt owed by the applicant (exclusive of a home mortgage) at the time of application.

TOTPAY   indicates the total monthly payments the applicant was making on the debt balance listed above.

SPINC   indicates the applicant's spouse's monthly income. Some applicants listed the value of 0, but many applicants merely left this item blank. A blank value is indicated in the data set by 9999.

MSTATUS   indicates the marital status of the applicant. Married applicants are indicated by a 1 and unmarried applicants (single, divorced, widowed) are indicated by a 0.

Questions concerning this sample data set are listed at the end of appropriate chapters in the text. The questions at the end of a chapter can be answered by applying the concepts discussed in the chapter to these data. An electronic computer is required for the solution of some of the more complex problems.

CLASS	SEX	AGE	JOBYRS	JOBINC	ADDINC	TOTBAL	TOTPAY	SPINC	MSTATUS
1	1	29	4	1200	200	5645	80	0	1
1	0	21	0	450	0	0	0	9999	0
1	1	23	1	700	0	1798	34	430	1
1	1	53	27	2000	0	0	0	9999	1
1	1	30	5	1200	0	3500	110	9999	1
1	1	25	3	925	0	828	103	500	1
1	1	47	20	1520	1100	0	0	9999	1
1	1	23	0	782	0	1626	79	9999	0
1	1	57	99	9999	880	0	0	850	1
1	1	34	2	2500	110	6000	70	9999	1
1	1	22	1	600	0	568	91	9999	1
1	1	44	8	1250	0	896	49	9999	1
1	0	53	9	600	755	0	0	9999	1
1	1	37	1	1200	0	0	176	9999	0
1	0	33	5	520	210	1000	28	9999	0
1	1	27	0	634	100	0	0	100	1
1	1	27	6	630	0	0	0	400	1
1	1	39	0	740	0	880	40	750	1
1	0	66	19	550	0	0	0	9999	1
1	1	35	3	1000	0	0	0	9999	0
1	1	37	0	1875	0	0	0	9999	1
1	1	40	11	2000	300	0	0	9999	1
1	1	24	4	1350	175	0	0	9999	0
1	1	60	99	9999	806	1740	36	9999	1
1	1	42	3	700	300	500	40	9999	0
1	1	48	7	4000	1000	16000	461	9999	0
1	1	31	4	800	0	0	0	500	1
1	0	29	3	600	150	0	0	9999	0
1	1	30	2	1000	0	1050	120	9999	0
1	1	78	30	1000	1620	0	0	9999	0
1	1	28	1	520	350	0	0	9999	1
1	1	22	1	650	0	0	0	250	1
1	1	39	5	800	0	0	0	9999	0
1	1	27	2	1100	0	800	55	9999	0
1	1	28	6	650	0	287	20	9999	1
1	1	65	99	9999	0	0	0	9999	1
1	1	56	25	2000	2000	0	0	9999	1
1	0	22	3	640	85	0	0	9999	0
1	1	22	6	750	0	0	0	9999	0
1	1	48	15	850	300	10800	163	150	1
1	1	63	6	1916	0	0	0	9999	1
1	1	28	0	9999	0	0	0	9999	1
1	1	32	1	2000	0	2800	0	9999	1
1	1	24	0	650	1000	0	0	9999	1
1	1	24	3	900	0	0	0	300	1
1	1	32	0	1450	0	2700	115	9999	1
1	1	70	99	9999	280	0	0	9999	1
1	1	35	0	700	100	700	60	9999	1
1	1	29	0	1060	0	457	51	9999	1
1	1	21	3	900	0	215	88	9999	0
1	1	28	8	1000	0	0	0	9999	1
1	1	60	6	2000	0	0	0	9999	0
1	1	27	2	1025	0	0	0	575	1
1	1	29	0	1336	115	0	0	9999	1
1	1	21	2	900	0	200	43	9999	0
1	1	50	7	1500	0	0	0	9999	1

CLASS	SEX	AGE	JOBYRS	JOBINC	ADDINC	TOTBAL	TOTPAY	SPINC	MSTATUS
1	1	42	13	3000	0	0	0	9999	1
1	1	25	1	713	83	0	0	9999	0
1	0	44	99	9999	0	6132	130	1000	1
1	0	26	16	1000	0	0	0	9999	0
1	1	34	13	1374	0	0	0	9999	1
1	1	22	0	833	0	0	0	9999	1
1	1	36	2	2200	400	0	0	9999	1
0	0	34	4	400	0	0	0	1000	1
0	1	21	1	540	0	469	46	0	0
0	1	40	3	1500	0	360	37	300	1
0	1	25	3	865	0	1000	103	500	1
0	1	22	9	570	260	200	25	0	0
0	1	34	20	1000	0	2400	205	0	1
0	0	63	8	600	0	0	0	0	0
0	1	28	0	440	0	120	20	0	0
0	0	29	0	600	300	0	0	9999	0
0	0	22	2	350	85	820	57	9999	0
0	1	30	1	1000	0	5146	217	9999	1
0	1	30	9	600	0	5000	288	9999	0
0	1	45	11	2225	0	4000	78	0	1
0	1	26	2	950	0	500	95	9999	0
0	1	28	1	400	240	0	0	500	1
0	1	40	5	1300	0	9000	200	250	1
0	0	25	0	600	43	169	15	9999	0
0	1	21	0	400	0	0	0	9999	0
0	1	24	0	755	0	0	0	9999	1
0	1	21	1	645	0	0	0	300	1
0	1	39	1	1000	116	1356	117	9999	1
0	0	29	5	539	0	220	40	9999	0
0	1	24	1	400	0	0	0	9999	0
0	0	23	99	9999	0	0	0	9999	0
0	1	28	2	660	196	890	17	660	1
0	1	22	0	1265	0	250	57	9999	0
0	0	24	2	400	0	50	0	200	1
0	1	29	10	1200	0	150	30	9999	0
0	1	21	0	520	0	0	0	9999	0
0	1	28	1	300	0	0	0	9999	0
0	1	22	3	700	0	0	0	420	1
0	1	19	3	700	0	0	0	9999	0
0	1	52	28	755	0	0	0	9999	1
0	1	32	1	750	310	0	0	450	1
0	0	24	0	500	309	0	0	9999	0
0	1	20	1	900	0	0	0	9999	0
0	0	20	1	376	0	200	20	9999	0
0	1	22	1	450	0	3063	106	9999	1
0	1	23	3	800	0	0	0	9999	0
0	1	32	10	800	0	2800	104	9999	1
0	1	35	0	450	0	0	0	9999	0
0	1	20	2	750	0	0	0	9999	0
0	1	34	0	600	175	2709	52	9999	1
0	1	32	2	1800	400	90	460	9999	1
0	1	35	1	1600	0	3900	163	9999	1
0	1	36	5	1300	800	765	54	350	1
0	1	27	0	660	0	768	64	556	1
0	1	23	2	700	0	385	20	9999	0
0	1	28	2	700	0	297	28	9999	0
0	1	28	2	1200	0	0	0	500	1

APPENDIX

TABLE I   Combinations of *n* Things Taken *r* at a Time*

SELECTED VALUES OF r:

n	0	1	2	3	4	5	6	7	8	9	10
0	1										
1	1	1									
2	1	2	1								
3	1	3	3	1							
4	1	4	6	4	1						
5	1	5	10	10	5	1					
6	1	6	15	20	15	6	1				
7	1	7	21	35	35	21	7	1			
8	1	8	28	56	70	56	28	8	1		
9	1	9	36	84	126	126	84	36	9	1	
10	1	10	45	120	210	252	210	120	45	10	1
11	1	11	55	165	330	462	462	330	165	55	11
12	1	12	66	220	495	792	924	792	495	220	66
13	1	13	78	286	715	1287	1716	1716	1287	715	286
14	1	14	91	364	1001	2002	3003	3432	3003	2002	1001
15	1	15	105	455	1365	3003	5005	6435	6435	5005	3003
16	1	16	120	560	1820	4368	8008	11440	12870	11440	8008
17	1	17	136	680	2380	6188	12376	19448	24310	24310	19448
18	1	18	153	816	3060	8568	18564	31824	43758	48620	43758
19	1	19	171	969	3876	11628	27132	50388	75582	92378	92378
20	1	20	190	1140	4845	15504	38760	77520	125970	167960	184756

*These values are sometimes called the *binomial coefficients* since they are the coefficients in the terms of the expansion of $(a + b)^n$. Note the symmetry of each row. This is due to the fact that

$$_nC_r = \frac{n!}{r!\,(n-r)!} = \frac{n!}{(n-r)!\,r!} = {_nC_{(n-r)}}$$

Thus, we can find $_{15}C_{12}$ even though there is no $r = 12$ column. We simply note that $_{15}C_{12} = {_{15}C_3} = 455$, which is readily determined from Table I.

TABLE II Random Numbers[1]

LINE	1–5	6–10	11–15	COLUMN 16–20	21–25	26–30	31–35
1	39591	16834	74151	92027	24670	36665	00770
2	46304	00370	30420	03883	94648	89428	41583
3	99547	47887	81085	64933	66279	80432	65793
4	06743	50993	98603	38452	87890	94624	69721
5	69568	06483	28733	37867	07936	98710	98539
6	91240	18312	17441	01929	18163	69201	31211
7	97458	14229	12063	59611	32249	90466	33216
8	35249	38646	34475	72417	60514	69257	12489
9	38980	46600	11759	11900	46743	27860	77940
10	10750	52745	38749	87365	58959	53731	89295
11	36247	27850	73958	20673	37800	63835	71051
12	70994	66986	99744	72438	01174	42159	11392
13	99638	94702	11463	18148	81386	80431	90628
14	72055	15774	43857	99805	10419	76939	25993
15	24038	65541	85788	55835	38835	59399	13790
16	74976	14631	35908	28221	39470	91548	12854
17	35553	71628	70189	26436	63407	91178	90348
18	35676	12797	51434	82976	42010	26344	92920
19	74815	67523	72985	23183	02446	63594	98924
20	45246	88048	65173	50989	91060	89894	36036
21	76509	47069	86378	41797	11910	49672	88575
22	19689	90332	04315	21358	97248	11188	39062
23	42751	35318	97513	61537	54955	08159	00337
24	11946	22681	45045	13964	57517	59419	58045
25	96518	48688	20996	11090	48396	57177	83867
26	35726	58643	76869	84622	39098	36083	72505
27	39737	42750	48968	70536	84864	64952	38404
28	97025	66492	56177	04049	80312	48028	26408
29	62814	08075	09788	56350	76787	51591	54509
30	25578	22950	15227	83291	41737	59599	96191
31	68763	69576	88991	49662	46704	63362	56625
32	17900	00813	64361	60725	88974	61005	99709
33	71944	60227	63551	71109	05624	43836	58254
34	54684	93691	85132	64399	29182	44324	14491
35	25946	27623	11258	65204	52832	50880	22273
36	01353	39318	44961	44972	91766	90262	56073
37	99083	88191	27662	99113	57174	35571	99884
38	52021	45406	37945	75234	24327	86978	22644
39	78755	47744	43776	83098	03225	14281	83637
40	25282	69106	59180	16257	22810	43609	12224
41	11959	94202	02743	86847	79725	51811	12998
42	11644	13792	98190	01424	30078	28197	55583
43	06307	97912	68110	59812	95448	43244	31262
44	76285	75714	89585	99296	52640	46518	55486
45	55322	07598	39600	60866	63007	20007	66819
46	78017	90928	90220	92503	83375	26986	74399
47	44768	43342	20696	26331	43140	69744	82928
48	25100	19336	14605	86603	51680	97678	24261
49	83612	46623	62876	85197	07824	91392	58317
50	41347	81666	82961	60413	71020	83658	02415

[1]Table II is abridged from *Table of 105,000 Random Decimal Digits,* Interstate Commerce Commission, Bureau of Transport Economics and Statistics, May 1949.

TABLE III   Binomial Probabilities

$$P(X \leq r) = \sum_{k=0}^{r} \binom{n}{k} p^k (1-p)^{n-k}$$

where X is the number of successes in n trials

n = 5

							p							
r	.01	.05	.10	.20	.30	.40	.50	.60	.70	.80	.90	.95	.99	r
0	0.9510	0.7738	0.5905	0.3277	0.1681	0.0778	0.0313	0.0102	0.0024	0.0003	0.0000	0.0000	0.0000	0
1	0.9990	0.9774	0.9185	0.7373	0.5282	0.3370	0.1875	0.0870	0.0308	0.0067	0.0005	0.0000	0.0000	1
2	1.0000	0.9988	0.9914	0.9421	0.8369	0.6826	0.5000	0.3174	0.1631	0.0579	0.0086	0.0012	0.0000	2
3	1.0000	1.0000	0.9995	0.9933	0.9692	0.9130	0.8125	0.6630	0.4718	0.2627	0.0815	0.0226	0.0010	3
4	1.0000	1.0000	1.0000	0.9997	0.9976	0.9898	0.9688	0.9222	0.8319	0.6723	0.4095	0.2262	0.0490	4

n = 10

							p							
r	.01	.05	.10	.20	.30	.40	.50	.60	.70	.80	.90	.95	.99	r
0	0.9044	0.5987	0.3487	0.1074	0.0282	0.0060	0.0010	0.0001	0.0000	0.0000	0.0000	0.0000	0.0000	0
1	0.9957	0.9139	0.7361	0.3758	0.1493	0.0464	0.0107	0.0017	0.0001	0.0000	0.0000	0.0000	0.0000	1
2	0.9999	0.9885	0.9298	0.6778	0.3828	0.1673	0.0547	0.0123	0.0016	0.0001	0.0000	0.0000	0.0000	2
3	1.0000	0.9990	0.9872	0.8791	0.6496	0.3823	0.1719	0.0548	0.0106	0.0009	0.0000	0.0000	0.0000	3
4	1.0000	0.9999	0.9984	0.9672	0.8497	0.6331	0.3770	0.1662	0.0473	0.0064	0.0001	0.0000	0.0000	4
5	1.0000	1.0000	0.9999	0.9936	0.9527	0.8338	0.6230	0.3669	0.1503	0.0328	0.0016	0.0001	0.0000	5
6	1.0000	1.0000	1.0000	0.9991	0.9894	0.9452	0.8281	0.6177	0.3504	0.1209	0.0128	0.0010	0.0000	6
7	1.0000	1.0000	1.0000	0.9999	0.9984	0.9877	0.9453	0.8327	0.6172	0.3222	0.0702	0.0115	0.0001	7
8	1.0000	1.0000	1.0000	1.0000	0.9999	0.9983	0.9893	0.9536	0.8507	0.6242	0.2639	0.0861	0.0043	8
9	1.0000	1.0000	1.0000	1.0000	1.0000	0.9999	0.9990	0.9940	0.9718	0.8926	0.6513	0.4013	0.0956	9

n = 15

							p							
r	.01	.05	.10	.20	.30	.40	.50	.60	.70	.80	.90	.95	.99	r
0	0.8601	0.4633	0.2059	0.0352	0.0047	0.0005	0.0000	0.0000	0.0000	0.0000	0.0000	0.0000	0.0000	0
1	0.9904	0.8290	0.5490	0.1671	0.0353	0.0052	0.0005	0.0000	0.0000	0.0000	0.0000	0.0000	0.0000	1
2	0.9996	0.9638	0.8159	0.3980	0.1268	0.0271	0.0037	0.0003	0.0000	0.0000	0.0000	0.0000	0.0000	2
3	1.0000	0.9945	0.9444	0.6482	0.2969	0.0905	0.0176	0.0019	0.0001	0.0000	0.0000	0.0000	0.0000	3
4	1.0000	0.9994	0.9873	0.8358	0.5155	0.2173	0.0592	0.0093	0.0007	0.0000	0.0000	0.0000	0.0000	4
5	1.0000	0.9999	0.9978	0.9389	0.7216	0.4032	0.1509	0.0338	0.0037	0.0001	0.0000	0.0000	0.0000	5
6	1.0000	1.0000	0.9997	0.9819	0.8689	0.6098	0.3036	0.0950	0.0152	0.0008	0.0000	0.0000	0.0000	6
7	1.0000	1.0000	1.0000	0.9958	0.9500	0.7869	0.5000	0.2131	0.0500	0.0042	0.0000	0.0000	0.0000	7
8	1.0000	1.0000	1.0000	0.9992	0.9848	0.9050	0.6964	0.3902	0.1311	0.0181	0.0003	0.0000	0.0000	8
9	1.0000	1.0000	1.0000	0.9999	0.9963	0.9662	0.8491	0.5968	0.2784	0.0611	0.0022	0.0001	0.0000	9
10	1.0000	1.0000	1.0000	1.0000	0.9993	0.9907	0.9408	0.7827	0.4845	0.1642	0.0127	0.0006	0.0000	10
11	1.0000	1.0000	1.0000	1.0000	0.9999	0.9981	0.9824	0.9095	0.7031	0.3518	0.0556	0.0055	0.0000	11
12	1.0000	1.0000	1.0000	1.0000	1.0000	0.9997	0.9963	0.9729	0.8732	0.6020	0.1841	0.0362	0.0004	12
13	1.0000	1.0000	1.0000	1.0000	1.0000	1.0000	0.9995	0.9948	0.9647	0.8329	0.4510	0.1710	0.0096	13
14	1.0000	1.0000	1.0000	1.0000	1.0000	1.0000	1.0000	0.9995	0.9953	0.9648	0.7941	0.5367	0.1399	14

TABLE III (*Continued*)

$n = 20$

r	.01	.05	.10	.20	.30	.40	*p* .50	.60	.70	.80	.90	.95	.99	r
0	0.8179	0.3585	0.1216	0.0115	0.0008	0.0000	0.0000	0.0000	0.0000	0.0000	0.0000	0.0000	0.0000	0
1	0.9831	0.7358	0.3917	0.0692	0.0076	0.0005	0.0000	0.0000	0.0000	0.0000	0.0000	0.0000	0.0000	1
2	0.9990	0.9245	0.6769	0.2061	0.0355	0.0036	0.0002	0.0000	0.0000	0.0000	0.0000	0.0000	0.0000	2
3	1.0000	0.9841	0.8670	0.4114	0.1071	0.0160	0.0013	0.0000	0.0000	0.0000	0.0000	0.0000	0.0000	3
4	1.0000	0.9974	0.9568	0.6296	0.2375	0.0510	0.0059	0.0003	0.0000	0.0000	0.0000	0.0000	0.0000	4
5	1.0000	0.9997	0.9887	0.8042	0.4164	0.1256	0.0207	0.0016	0.0000	0.0000	0.0000	0.0000	0.0000	5
6	1.0000	1.0000	0.9976	0.9133	0.6080	0.2500	0.0577	0.0065	0.0003	0.0000	0.0000	0.0000	0.0000	6
7	1.0000	1.0000	0.9996	0.9679	0.7723	0.4159	0.1316	0.0210	0.0013	0.0000	0.0000	0.0000	0.0000	7
8	1.0000	1.0000	0.9999	0.9900	0.8867	0.5956	0.2517	0.0565	0.0051	0.0001	0.0000	0.0000	0.0000	8
9	1.0000	1.0000	1.0000	0.9974	0.9520	0.7553	0.4119	0.1275	0.0171	0.0006	0.0000	0.0000	0.0000	9
10	1.0000	1.0000	1.0000	0.9994	0.9829	0.8725	0.5881	0.2447	0.0480	0.0026	0.0000	0.0000	0.0000	10
11	1.0000	1.0000	1.0000	0.9999	0.9949	0.9435	0.7483	0.4044	0.1133	0.0100	0.0001	0.0000	0.0000	11
12	1.0000	1.0000	1.0000	1.0000	0.9987	0.9790	0.8684	0.5841	0.2277	0.0321	0.0004	0.0000	0.0000	12
13	1.0000	1.0000	1.0000	1.0000	0.9997	0.9935	0.9423	0.7500	0.3920	0.0867	0.0024	0.0000	0.0000	13
14	1.0000	1.0000	1.0000	1.0000	1.0000	0.9984	0.9793	0.8744	0.5836	0.1958	0.0113	0.0003	0.0000	14
15	1.0000	1.0000	1.0000	1.0000	1.0000	0.9997	0.9941	0.9490	0.7625	0.3704	0.0432	0.0026	0.0000	15
16	1.0000	1.0000	1.0000	1.0000	1.0000	1.0000	0.9987	0.9840	0.8929	0.5886	0.1330	0.0159	0.0000	16
17	1.0000	1.0000	1.0000	1.0000	1.0000	1.0000	0.9998	0.9964	0.9645	0.7939	0.3231	0.0755	0.0010	17
18	1.0000	1.0000	1.0000	1.0000	1.0000	1.0000	1.0000	0.9995	0.9924	0.9308	0.6083	0.2642	0.0169	18
19	1.0000	1.0000	1.0000	1.0000	1.0000	1.0000	1.0000	1.0000	0.9992	0.9885	0.8784	0.6415	0.1821	19

$n = 25$

r	.01	.05	.10	.20	.30	.40	*p* .50	.60	.70	.80	.90	.95	.99	r
0	0.7778	0.2774	0.0718	0.0038	0.0001	0.0000	0.0000	0.0000	0.0000	0.0000	0.0000	0.0000	0.0000	0
1	0.9742	0.6424	0.2712	0.0274	0.0016	0.0001	0.0000	0.0000	0.0000	0.0000	0.0000	0.0000	0.0000	1
2	0.9980	0.8729	0.5371	0.0982	0.0090	0.0004	0.0000	0.0000	0.0000	0.0000	0.0000	0.0000	0.0000	2
3	0.9999	0.9659	0.7636	0.2340	0.0332	0.0024	0.0001	0.0000	0.0000	0.0000	0.0000	0.0000	0.0000	3
4	1.0000	0.9928	0.9020	0.4207	0.0905	0.0095	0.0005	0.0000	0.0000	0.0000	0.0000	0.0000	0.0000	4
5	1.0000	0.9988	0.9666	0.6167	0.1935	0.0294	0.0020	0.0001	0.0000	0.0000	0.0000	0.0000	0.0000	5
6	1.0000	0.9998	0.9905	0.7800	0.3407	0.0736	0.0073	0.0003	0.0000	0.0000	0.0000	0.0000	0.0000	6
7	1.0000	1.0000	0.9977	0.8909	0.5118	0.1536	0.0216	0.0012	0.0000	0.0000	0.0000	0.0000	0.0000	7
8	1.0000	1.0000	0.9995	0.9532	0.6769	0.2735	0.0539	0.0043	0.0001	0.0000	0.0000	0.0000	0.0000	8
9	1.0000	1.0000	0.9999	0.9827	0.8106	0.4246	0.1148	0.0132	0.0005	0.0000	0.0000	0.0000	0.0000	9
10	1.0000	1.0000	1.0000	0.9944	0.9022	0.5858	0.2122	0.0344	0.0018	0.0000	0.0000	0.0000	0.0000	10
11	1.0000	1.0000	1.0000	0.9985	0.9558	0.7323	0.3450	0.0778	0.0060	0.0001	0.0000	0.0000	0.0000	11
12	1.0000	1.0000	1.0000	0.9996	0.9825	0.8462	0.5000	0.1538	0.0175	0.0004	0.0000	0.0000	0.0000	12
13	1.0000	1.0000	1.0000	0.9999	0.9940	0.9222	0.6550	0.2677	0.0442	0.0015	0.0000	0.0000	0.0000	13
14	1.0000	1.0000	1.0000	1.0000	0.9982	0.9656	0.7878	0.4142	0.0978	0.0056	0.0000	0.0000	0.0000	14
15	1.0000	1.0000	1.0000	1.0000	0.9995	0.9868	0.8852	0.5754	0.1894	0.0173	0.0001	0.0000	0.0000	15
16	1.0000	1.0000	1.0000	1.0000	0.9999	0.9957	0.9461	0.7265	0.3231	0.0468	0.0005	0.0000	0.0000	16
17	1.0000	1.0000	1.0000	1.0000	1.0000	0.9988	0.9784	0.8464	0.4882	0.1091	0.0023	0.0000	0.0000	17
18	1.0000	1.0000	1.0000	1.0000	1.0000	0.9997	0.9927	0.9264	0.6593	0.2200	0.0095	0.0002	0.0000	18
19	1.0000	1.0000	1.0000	1.0000	1.0000	0.9999	0.9980	0.9706	0.8065	0.3833	0.0334	0.0012	0.0000	19
20	1.0000	1.0000	1.0000	1.0000	1.0000	1.0000	0.9995	0.9905	0.9095	0.5793	0.0980	0.0072	0.0000	20
21	1.0000	1.0000	1.0000	1.0000	1.0000	1.0000	0.9999	0.9976	0.9668	0.7660	0.2364	0.0341	0.0001	21
22	1.0000	1.0000	1.0000	1.0000	1.0000	1.0000	1.0000	0.9996	0.9910	0.9018	0.4629	0.1271	0.0020	22
23	1.0000	1.0000	1.0000	1.0000	1.0000	1.0000	1.0000	0.9999	0.9984	0.9726	0.7288	0.3576	0.0258	23
24	1.0000	1.0000	1.0000	1.0000	1.0000	1.0000	1.0000	1.0000	0.9999	0.9962	0.9282	0.7226	0.2222	24

TABLE IV   Poisson Probabilities

This table is a listing of the probability of exactly r successes for selected values of θ, defined by the Poisson probability function

$$P(r) = \frac{e^{-\theta}\theta^r}{r!}$$

Examples:

1.   If $\theta = 3.5$ then $P(r = 5) = .1322$ and $P(r = 0) = .0302$
2.   If $\theta = 8.0$ then $P(r \le 2) = .0003 + .0027 + .0107 = .0137$

θ

r	.1	.2	.3	.4	.5	.6	.7	.8	.9	1.0
0	.9048	.8187	.7408	.6703	.6065	.5488	.4966	.4493	.4066	.3679
1	.0905	.1637	.2222	.2681	.3033	.3293	.3476	.3595	.3659	.3679
2	.0045	.0164	.0333	.0536	.0758	.0988	.1217	.1438	.1647	.1839
3	.0002	.0011	.0033	.0072	.0126	.0198	.0284	.0383	.0494	.0613
4	.0000	.0001	.0003	.0007	.0016	.0030	.0050	.0077	.0111	.0153
5	.0000	.0000	.0000	.0001	.0002	.0004	.0007	.0012	.0020	.0031
6	.0000	.0000	.0000	.0000	.0000	.0000	.0001	.0002	.0003	.0005
7	.0000	.0000	.0000	.0000	.0000	.0000	.0000	.0000	.0000	.0001
8	.0000	.0000	.0000	.0000	.0000	.0000	.0000	.0000	.0000	.0000

θ

r	1.1	1.2	1.3	1.4	1.5	1.6	1.7	1.8	1.9	2.0
0	.3329	.3012	.2725	.2466	.2231	.2019	.1827	.1653	.1496	.1353
1	.3662	.3614	.3543	.3452	.3347	.3230	.3106	.2975	.2842	.2707
2	.2014	.2169	.2303	.2417	.2510	.2584	.2640	.2678	.2700	.2707
3	.0738	.0867	.0998	.1128	.1255	.1378	.1496	.1607	.1710	.1804
4	.0203	.0260	.0324	.0395	.0471	.0551	.0636	.0723	.0812	.0902
5	.0045	.0062	.0084	.0111	.0141	.0176	.0216	.0260	.0309	.0361
6	.0008	.0012	.0018	.0026	.0035	.0047	.0061	.0078	.0098	.0120
7	.0001	.0002	.0003	.0005	.0008	.0011	.0015	.0020	.0027	.0034
8	.0000	.0000	.0001	.0001	.0001	.0002	.0003	.0005	.0006	.0009
9	.0000	.0000	.0000	.0000	.0000	.0000	.0001	.0001	.0001	.0002
10	.0000	.0000	.0000	.0000	.0000	.0000	.0000	.0000	.0000	.0000

θ

r	2.1	2.2	2.3	2.4	2.5	2.6	2.7	2.8	2.9	3.0
0	.1225	.1108	.1003	.0907	.0821	.0743	.0672	.0608	.0550	.0498
1	.2572	.2438	.2306	.2177	.2052	.1931	.1815	.1703	.1596	.1494
2	.2700	.2681	.2652	.2613	.2565	.2510	.2450	.2384	.2314	.2240
3	.1890	.1966	.2033	.2090	.2138	.2176	.2205	.2225	.2237	.2240
4	.0992	.1082	.1169	.1254	.1336	.1414	.1488	.1557	.1622	.1680
5	.0417	.0476	.0538	.0602	.0668	.0735	.0804	.0872	.0940	.1008
6	.0146	.0174	.0206	.0241	.0278	.0319	.0362	.0407	.0455	.0504
7	.0044	.0055	.0068	.0083	.0099	.0118	.0139	.0163	.0188	.0216
8	.0011	.0015	.0019	.0025	.0031	.0038	.0047	.0057	.0068	.0081
9	.0003	.0004	.0005	.0007	.0009	.0011	.0014	.0018	.0022	.0027
10	.0001	.0001	.0001	.0002	.0002	.0003	.0004	.0005	.0006	.0008
11	.0000	.0000	.0000	.0000	.0000	.0001	.0001	.0001	.0002	.0002
12	.0000	.0000	.0000	.0000	.0000	.0000	.0000	.0000	.0000	.0001
13	.0000	.0000	.0000	.0000	.0000	.0000	.0000	.0000	.0000	.0000

TABLE IV (*Continued*)

θ

r	3.1	3.2	3.3	3.4	3.5	3.6	3.7	3.8	3.9	4.0
0	.0450	.0408	.0369	.0334	.0302	.0273	.0247	.0224	.0202	.0183
1	.1397	.1304	.1217	.1135	.1057	.0984	.0915	.0850	.0789	.0733
2	.2165	.2087	.2008	.1929	.1850	.1771	.1692	.1615	.1539	.1465
3	.2237	.2226	.2209	.2186	.2158	.2125	.2087	.2046	.2001	.1954
4	.1733	.1781	.1823	.1858	.1888	.1912	.1931	.1944	.1951	.1954
5	.1075	.1140	.1203	.1264	.1322	.1377	.1429	.1477	.1522	.1563
6	.0555	.0608	.0662	.0716	.0771	.0826	.0881	.0936	.0989	.1042
7	.0246	.0278	.0312	.0348	.0385	.0425	.0466	.0508	.0551	.0595
8	.0095	.0111	.0129	.0148	.0169	.0191	.0215	.0241	.0269	.0298
9	.0033	.0040	.0047	.0056	.0066	.0076	.0089	.0102	.0116	.0132
10	.0010	.0013	.0016	.0019	.0023	.0028	.0033	.0039	.0045	.0053
11	.0003	.0004	.0005	.0006	.0007	.0009	.0011	.0013	.0016	.0019
12	.0001	.0001	.0001	.0002	.0002	.0003	.0003	.0004	.0005	.0006
13	.0000	.0000	.0000	.0000	.0001	.0001	.0001	.0001	.0002	.0002
14	.0000	.0000	.0000	.0000	.0000	.0000	.0000	.0000	.0000	.0001
15	.0000	.0000	.0000	.0000	.0000	.0000	.0000	.0000	.0000	.0000

θ

r	4.1	4.2	4.3	4.4	4.5	4.6	4.7	4.8	4.9	5.0
0	.0166	.0150	.0136	.0123	.0111	.0101	.0091	.0082	.0074	.0067
1	.0679	.0630	.0583	.0540	.0500	.0462	.0427	.0395	.0365	.0337
2	.1393	.1323	.1254	.1188	.1125	.1063	.1005	.0948	.0894	.0842
3	.1904	.1852	.1798	.1743	.1687	.1631	.1574	.1517	.1460	.1404
4	.1951	.1944	.1933	.1917	.1898	.1875	.1849	.1820	.1789	.1755
5	.1600	.1633	.1662	.1687	.1708	.1725	.1738	.1747	.1753	.1755
6	.1093	.1143	.1191	.1237	.1281	.1323	.1362	.1398	.1432	.1462
7	.0640	.0686	.0732	.0778	.0824	.0869	.0914	.0959	.1002	.1044
8	.0328	.0360	.0393	.0428	.0463	.0500	.0537	.0575	.0614	.0653
9	.0150	.0168	.0188	.0209	.0232	.0255	.0281	.0307	.0334	.0363
10	.0061	.0071	.0081	.0092	.0104	.0118	.0132	.0147	.0164	.0181
11	.0023	.0027	.0032	.0037	.0043	.0049	.0056	.0064	.0073	.0082
12	.0008	.0009	.0011	.0013	.0016	.0019	.0022	.0026	.0030	.0034
13	.0002	.0003	.0004	.0005	.0006	.0007	.0008	.0009	.0011	.0013
14	.0001	.0001	.0001	.0001	.0002	.0002	.0003	.0003	.0004	.0005
15	.0000	.0000	.0000	.0000	.0001	.0001	.0001	.0001	.0001	.0002
16	.0000	.0000	.0000	.0000	.0000	.0000	.0000	.0000	.0000	.0000
17	.0000	.0000	.0000	.0000	.0000	.0000	.0000	.0000	.0000	.0000

TABLE IV   (*Continued*)

θ

r	5,1	5,2	5,3	5,4	5,5	5,6	5,7	5,8	5,9	6,0
0	,0061	,0055	,0050	,0045	,0041	,0037	,0033	,0030	,0027	,0025
1	,0311	,0287	,0265	,0244	,0225	,0207	,0191	,0176	,0162	,0149
2	,0793	,0746	,0701	,0659	,0618	,0580	,0544	,0509	,0477	,0446
3	,1348	,1293	,1239	,1185	,1133	,1082	,1033	,0985	,0938	,0892
4	,1719	,1681	,1641	,1600	,1558	,1515	,1472	,1428	,1383	,1339
5	,1753	,1748	,1740	,1728	,1714	,1697	,1678	,1656	,1632	,1606
6	,1490	,1515	,1537	,1555	,1571	,1584	,1594	,1601	,1605	,1606
7	,1086	,1125	,1163	,1200	,1234	,1267	,1298	,1326	,1353	,1377
8	,0692	,0731	,0771	,0810	,0849	,0887	,0925	,0962	,0998	,1033
9	,0392	,0423	,0454	,0486	,0519	,0552	,0586	,0620	,0654	,0688
10	,0200	,0220	,0241	,0262	,0285	,0309	,0334	,0359	,0386	,0413
11	,0093	,0104	,0116	,0129	,0143	,0157	,0173	,0190	,0207	,0225
12	,0039	,0045	,0051	,0058	,0065	,0073	,0082	,0092	,0102	,0113
13	,0015	,0018	,0021	,0024	,0028	,0032	,0036	,0041	,0046	,0052
14	,0006	,0007	,0008	,0009	,0011	,0013	,0015	,0017	,0019	,0022
15	,0002	,0002	,0003	,0003	,0004	,0005	,0006	,0007	,0008	,0009
16	,0001	,0001	,0001	,0001	,0001	,0002	,0002	,0002	,0003	,0003
17	,0000	,0000	,0000	,0000	,0000	,0001	,0001	,0001	,0001	,0001
18	,0000	,0000	,0000	,0000	,0000	,0000	,0000	,0000	,0000	,0000
19	,0000	,0000	,0000	,0000	,0000	,0000	,0000	,0000	,0000	,0000

θ

r	6,1	6,2	6,3	6,4	6,5	6,6	6,7	6,8	6,9	7,0
0	,0022	,0020	,0018	,0017	,0015	,0014	,0012	,0011	,0010	,0009
1	,0137	,0126	,0116	,0106	,0098	,0090	,0082	,0076	,0070	,0064
2	,0417	,0390	,0364	,0340	,0318	,0296	,0276	,0258	,0240	,0223
3	,0848	,0806	,0765	,0726	,0688	,0652	,0617	,0584	,0552	,0521
4	,1294	,1249	,1205	,1162	,1118	,1076	,1034	,0992	,0952	,0912
5	,1579	,1549	,1519	,1487	,1454	,1420	,1385	,1349	,1314	,1277
6	,1605	,1601	,1595	,1586	,1575	,1562	,1546	,1529	,1511	,1490
7	,1399	,1418	,1435	,1450	,1462	,1472	,1480	,1486	,1489	,1490
8	,1066	,1099	,1130	,1160	,1188	,1215	,1240	,1263	,1284	,1304
9	,0723	,0757	,0791	,0825	,0858	,0891	,0923	,0954	,0985	,1014
10	,0441	,0469	,0498	,0528	,0558	,0588	,0618	,0649	,0679	,0710
11	,0244	,0265	,0285	,0307	,0330	,0353	,0377	,0401	,0426	,0452
12	,0124	,0137	,0150	,0164	,0179	,0194	,0210	,0227	,0245	,0263
13	,0058	,0065	,0073	,0081	,0089	,0099	,0108	,0119	,0130	,0142
14	,0025	,0029	,0033	,0037	,0041	,0046	,0052	,0058	,0064	,0071
15	,0010	,0012	,0014	,0016	,0018	,0020	,0023	,0026	,0029	,0033
16	,0004	,0005	,0005	,0006	,0007	,0008	,0010	,0011	,0013	,0014
17	,0001	,0002	,0002	,0002	,0003	,0003	,0004	,0004	,0005	,0006
18	,0000	,0001	,0001	,0001	,0001	,0001	,0001	,0002	,0002	,0002
19	,0000	,0000	,0000	,0000	,0000	,0000	,0001	,0001	,0001	,0001
20	,0000	,0000	,0000	,0000	,0000	,0000	,0000	,0000	,0000	,0000

TABLE IV (*Continued*)

θ

r	7.1	7.2	7.3	7.4	7.5	7.6	7.7	7.8	7.9	8.0
0	.0008	.0007	.0007	.0006	.0006	.0005	.0005	.0004	.0004	.0003
1	.0059	.0054	.0049	.0045	.0041	.0038	.0035	.0032	.0029	.0027
2	.0208	.0194	.0180	.0167	.0156	.0145	.0134	.0125	.0116	.0107
3	.0492	.0464	.0438	.0413	.0389	.0366	.0345	.0324	.0305	.0286
4	.0874	.0836	.0799	.0764	.0729	.0696	.0663	.0632	.0602	.0573
5	.1241	.1204	.1167	.1130	.1094	.1057	.1021	.0986	.0951	.0916
6	.1468	.1445	.1420	.1394	.1367	.1339	.1311	.1282	.1252	.1221
7	.1489	.1486	.1481	.1474	.1465	.1454	.1442	.1428	.1413	.1396
8	.1321	.1337	.1351	.1363	.1373	.1381	.1388	.1392	.1395	.1396
9	.1042	.1070	.1096	.1121	.1144	.1167	.1187	.1207	.1224	.1241
10	.0740	.0770	.0800	.0829	.0858	.0887	.0914	.0941	.0967	.0993
11	.0478	.0504	.0531	.0558	.0585	.0613	.0640	.0667	.0695	.0722
12	.0283	.0303	.0323	.0344	.0366	.0388	.0411	.0434	.0457	.0481
13	.0154	.0168	.0181	.0196	.0211	.0227	.0243	.0260	.0278	.0296
14	.0078	.0086	.0095	.0104	.0113	.0123	.0134	.0145	.0157	.0169
15	.0037	.0041	.0046	.0051	.0057	.0062	.0069	.0075	.0083	.0090
16	.0016	.0019	.0021	.0024	.0026	.0030	.0033	.0037	.0041	.0045
17	.0007	.0008	.0009	.0010	.0012	.0013	.0015	.0017	.0019	.0021
18	.0003	.0003	.0004	.0004	.0005	.0006	.0006	.0007	.0008	.0009
19	.0001	.0001	.0001	.0002	.0002	.0002	.0003	.0003	.0003	.0004
20	.0000	.0000	.0001	.0001	.0001	.0001	.0001	.0001	.0001	.0002
21	.0000	.0000	.0000	.0000	.0000	.0000	.0000	.0000	.0001	.0001
22	.0000	.0000	.0000	.0000	.0000	.0000	.0000	.0000	.0000	.0000

θ

r	8.1	8.2	8.3	8.4	8.5	8.6	8.7	8.8	8.9	9.0
0	.0003	.0003	.0002	.0002	.0002	.0002	.0002	.0002	.0001	.0001
1	.0025	.0023	.0021	.0019	.0017	.0016	.0014	.0013	.0012	.0011
2	.0100	.0092	.0086	.0079	.0074	.0068	.0063	.0058	.0054	.0050
3	.0269	.0252	.0237	.0222	.0208	.0195	.0183	.0171	.0160	.0150
4	.0544	.0517	.0491	.0466	.0443	.0420	.0398	.0377	.0357	.0337
5	.0882	.0849	.0816	.0764	.0752	.0722	.0692	.0663	.0635	.0607
6	.1191	.1160	.1128	.1097	.1066	.1034	.1003	.0972	.0941	.0911
7	.1378	.1358	.1338	.1317	.1294	.1271	.1247	.1222	.1197	.1171
8	.1395	.1392	.1388	.1382	.1375	.1366	.1356	.1344	.1332	.1318
9	.1255	.1269	.1280	.1290	.1299	.1306	.1311	.1315	.1317	.1318
10	.1017	.1040	.1063	.1084	.1104	.1123	.1140	.1157	.1172	.1186
11	.0749	.0775	.0802	.0828	.0853	.0878	.0902	.0925	.0948	.0970
12	.0505	.0530	.0555	.0579	.0604	.0629	.0654	.0679	.0703	.0728
13	.0315	.0334	.0354	.0374	.0395	.0416	.0438	.0459	.0481	.0504
14	.0182	.0196	.0210	.0225	.0240	.0256	.0272	.0289	.0306	.0324
15	.0098	.0107	.0116	.0126	.0136	.0147	.0158	.0169	.0182	.0194
16	.0050	.0055	.0060	.0066	.0072	.0079	.0086	.0093	.0101	.0109
17	.0024	.0026	.0029	.0033	.0036	.0040	.0044	.0048	.0053	.0058
18	.0011	.0012	.0014	.0015	.0017	.0019	.0021	.0024	.0026	.0029
19	.0005	.0005	.0006	.0007	.0008	.0009	.0010	.0011	.0012	.0014
20	.0002	.0002	.0002	.0003	.0003	.0004	.0004	.0005	.0005	.0006
21	.0001	.0001	.0001	.0001	.0001	.0002	.0002	.0002	.0002	.0003
22	.0000	.0000	.0000	.0000	.0001	.0001	.0001	.0001	.0001	.0001
23	.0000	.0000	.0000	.0000	.0000	.0000	.0000	.0000	.0000	.0000
24	.0000	.0000	.0000	.0000	.0000	.0000	.0000	.0000	.0000	.0000

TABLE IV (*Continued*)

						θ				
r	9,1	9,2	9,3	9,4	9,5	9,6	9,7	9,8	9,9	10,0
0	,0001	,0001	,0001	,0001	,0001	,0001	,0001	,0001	,0001	,0000
1	,0010	,0009	,0009	,0008	,0007	,0007	,0006	,0005	,0005	,0005
2	,0046	,0043	,0040	,0037	,0034	,0031	,0029	,0027	,0025	,0023
3	,0140	,0131	,0123	,0115	,0107	,0100	,0093	,0087	,0081	,0076
4	,0319	,0302	,0285	,0269	,0254	,0240	,0226	,0213	,0201	,0189
5	,0581	,0555	,0530	,0506	,0483	,0460	,0439	,0418	,0398	,0378
6	,0881	,0851	,0822	,0793	,0764	,0736	,0709	,0682	,0656	,0631
7	,1145	,1118	,1091	,1064	,1037	,1010	,0982	,0955	,0928	,0901
8	,1302	,1286	,1269	,1251	,1232	,1212	,1191	,1170	,1148	,1126
9	,1317	,1315	,1311	,1306	,1300	,1293	,1284	,1274	,1263	,1251
10	,1198	,1209	,1219	,1228	,1235	,1241	,1245	,1249	,1250	,1251
11	,0991	,1012	,1031	,1049	,1067	,1083	,1098	,1112	,1125	,1137
12	,0752	,0776	,0799	,0822	,0844	,0866	,0888	,0908	,0928	,0948
13	,0526	,0549	,0572	,0594	,0617	,0640	,0662	,0685	,0707	,0729
14	,0342	,0361	,0380	,0399	,0419	,0439	,0459	,0479	,0500	,0521
15	,0208	,0221	,0235	,0250	,0265	,0281	,0297	,0313	,0330	,0347
16	,0118	,0127	,0137	,0147	,0157	,0168	,0180	,0192	,0204	,0217
17	,0063	,0069	,0075	,0081	,0088	,0095	,0103	,0111	,0119	,0128
18	,0032	,0035	,0039	,0042	,0046	,0051	,0055	,0060	,0065	,0071
19	,0015	,0017	,0019	,0021	,0023	,0026	,0028	,0031	,0034	,0037
20	,0007	,0008	,0009	,0010	,0011	,0012	,0014	,0015	,0017	,0019
21	,0003	,0003	,0004	,0004	,0005	,0006	,0006	,0007	,0008	,0009
22	,0001	,0001	,0002	,0002	,0002	,0002	,0003	,0003	,0004	,0004
23	,0000	,0001	,0001	,0001	,0001	,0001	,0001	,0001	,0002	,0002
24	,0000	,0000	,0000	,0000	,0000	,0000	,0000	,0001	,0001	,0001
25	,0000	,0000	,0000	,0000	,0000	,0000	,0000	,0000	,0000	,0000

TABLE V   Normal Probabilities: Areas of the Standard Normal Distribution

The values in the body of the table are the areas between the mean and Z.

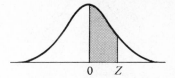

Z	,00	,01	,02	,03	,04	,05	,06	,07	,08	,09
,00	,0000	,0040	,0080	,0120	,0160	,0199	,0239	,0279	,0319	,0359
,10	,0398	,0438	,0478	,0517	,0557	,0596	,0636	,0675	,0714	,0753
,20	,0793	,0832	,0871	,0910	,0948	,0987	,1026	,1064	,1103	,1141
,30	,1179	,1217	,1255	,1293	,1331	,1368	,1406	,1443	,1480	,1517
,40	,1554	,1591	,1628	,1664	,1700	,1736	,1772	,1808	,1844	,1879
,50	,1915	,1950	,1985	,2019	,2054	,2088	,2123	,2157	,2190	,2224
,60	,2257	,2291	,2324	,2357	,2389	,2422	,2454	,2486	,2517	,2549
,70	,2580	,2611	,2642	,2673	,2703	,2734	,2764	,2793	,2823	,2852
,80	,2881	,2910	,2939	,2967	,2995	,3023	,3051	,3078	,3106	,3133
,90	,3159	,3186	,3212	,3238	,3264	,3289	,3315	,3340	,3365	,3389
1,00	,3413	,3438	,3461	,3485	,3508	,3531	,3554	,3577	,3599	,3621
1,10	,3643	,3665	,3686	,3708	,3729	,3749	,3770	,3790	,3810	,3830
1,20	,3849	,3869	,3883	,3907	,3925	,3944	,3962	,3980	,3997	,4015
1,30	,4032	,4049	,4066	,4082	,4099	,4115	,4131	,4147	,4162	,4177
1,40	,4192	,4207	,4222	,4236	,4251	,4265	,4279	,4292	,4306	,4319
1,50	,4332	,4345	,4357	,4370	,4382	,4394	,4406	,4418	,4429	,4441
1,60	,4452	,4463	,4474	,4484	,4495	,4505	,4515	,4525	,4535	,4545
1,70	,4554	,4564	,4573	,4582	,4591	,4599	,4608	,4616	,4625	,4633
1,80	,4641	,4649	,4656	,4664	,4671	,4678	,4686	,4693	,4699	,4706
1,90	,4713	,4719	,4726	,4732	,4738	,4744	,4750	,4756	,4761	,4767
2,00	,4772	,4778	,4783	,4788	,4793	,4798	,4803	,4808	,4812	,4817
2,10	,4821	,4826	,4830	,4834	,4838	,4842	,4846	,4850	,4854	,4857
2,20	,4861	,4864	,4868	,4871	,4875	,4878	,4881	,4884	,4887	,4890
2,30	,4893	,4896	,4898	,4901	,4904	,4906	,4909	,4911	,4913	,4916
2,40	,4918	,4920	,4922	,4925	,4927	,4929	,4931	,4932	,4934	,4936
2,50	,4938	,4940	,4941	,4943	,4945	,4946	,4948	,4949	,4951	,4952
2,60	,4953	,4955	,4956	,4957	,4959	,4960	,4961	,4962	,4963	,4964
2,70	,4965	,4966	,4967	,4968	,4969	,4970	,4971	,4972	,4973	,4974
2,80	,4974	,4975	,4976	,4977	,4977	,4978	,4979	,4979	,4980	,4981
2,90	,4981	,4982	,4982	,4983	,4984	,4984	,4985	,4985	,4986	,4986
3,00	,4987	,4987	,4987	,4988	,4988	,4989	,4989	,4989	,4990	,4990
3,10	,4990	,4991	,4991	,4991	,4992	,4992	,4992	,4992	,4993	,4993
3,20	,4993	,4993	,4994	,4994	,4994	,4994	,4994	,4995	,4995	,4995
3,30	,4995	,4995	,4995	,4996	,4996	,4996	,4996	,4996	,4996	,4997
3,40	,4997	,4997	,4997	,4997	,4997	,4997	,4997	,4997	,4997	,4998
3,50	,4998	,4998	,4998	,4998	,4998	,4998	,4998	,4998	,4998	,4998
3,60	,4998	,4998	,4999	,4999	,4999	,4999	,4999	,4999	,4999	,4999
3,70	,4999	,4999	,4999	,4999	,4999	,4999	,4999	,4999	,4999	,4999
3,80	,4999	,4999	,4999	,4999	,4999	,4999	,4999	,4999	,4999	,4999

TABLE VI   Values of *t* for Given Probability Levels[1]

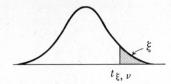

$$t_{\xi, \nu}$$

DEGREES OF FREEDOM, ν	PROBABILITY ξ OF A LARGER VALUE				
	.1	.05	.025	.01	.005
1	3.078	6.314	12.706	31.821	63.657
2	1.886	2.920	4.303	6.965	9.925
3	1.638	2.353	3.182	4.541	5.841
4	1.533	2.132	2.776	3.747	4.604
5	1.476	2.015	2.571	3.365	4.032
6	1.440	1.943	2.447	3.143	3.707
7	1.415	1.895	2.365	2.998	3.499
8	1.397	1.860	2.306	2.896	3.355
9	1.383	1.833	2.262	2.821	3.250
10	1.372	1.812	2.228	2.764	3.169
11	1.363	1.796	2.201	2.718	3.106
12	1.356	1.782	2.179	2.681	3.055
13	1.350	1.771	2.160	2.650	3.012
14	1.345	1.761	2.145	2.624	2.977
15	1.341	1.753	2.131	2.602	2.947
16	1.337	1.746	2.120	2.583	2.921
17	1.333	1.740	2.110	2.567	2.898
18	1.330	1.734	2.101	2.552	2.878
19	1.328	1.729	2.093	2.539	2.861
20	1.325	1.725	2.086	2.528	2.845
21	1.323	1.721	2.080	2.518	2.831
22	1.321	1.717	2.074	2.508	2.819
23	1.319	1.714	2.069	2.500	2.807
24	1.318	1.711	2.064	2.492	2.797
25	1.316	1.708	2.060	2.485	2.787
26	1.315	1.706	2.056	2.479	2.779
27	1.314	1.703	2.052	2.473	2.771
28	1.313	1.701	2.048	2.467	2.763
29	1.311	1.699	2.045	2.462	2.756
30	1.310	1.697	2.042	2.457	2.750
40	1.303	1.684	2.021	2.423	2.704
60	1.296	1.671	2.000	2.390	2.660
120	1.290	1.661	1.984	2.358	2.626
∞	1.282	1.645	1.960	2.326	2.576

[1]Table VI is abridged from Table III of Fisher and Yates: *Statistical Tables for Biological, Agricultural and Medical Research*, published by Longman Group Ltd., London (previously published by Oliver & Boyd, Edinburgh), by permission of the authors and publishers.

TABLE VII   95% Confidence Intervals (Percent) for Binomial Distributions[1]

NUMBER OBSERVED X	10		15		SIZE OF SAMPLE, n 20		30		50		100		FRACTION OBSERVED X/n	SIZE OF SAMPLE 250		1000	
0	0	31	0	22	0	17	0	12	0	07	0	4	.00	0	1	0	0
1	0	45	0	32	0	25	0	17	0	11	0	5	.01	0	4	0	2
2	3	56	2	40	1	31	1	22	0	14	0	7	.02	1	5	1	3
3	7	65	4	48	3	38	2	27	1	17	1	8	.03	1	6	2	4
4	12	74	8	55	6	44	4	31	2	19	1	10	.04	2	7	3	5
5	19	81	12	62	9	49	6	35	3	22	2	11	.05	3	9	4	7
6	26	88	16	68	12	54	8	39	5	24	2	12	.06	3	10	5	8
7	35	93	21	73	15	59	10	43	6	27	3	14	.07	4	11	6	9
8	44	97	27	79	19	64	12	46	7	29	4	15	.08	5	12	6	10
9	55	100	32	84	23	68	15	50	9	31	4	16	.09	6	13	7	11
10	69	100	38	88	27	73	17	53	10	34	5	18	.10	7	14	8	12
11			45	92	32	77	20	56	12	36	5	19	.11	7	16	9	13
12			52	96	36	81	23	60	13	38	6	20	.12	8	17	10	14
13			60	98	41	85	25	63	15	41	7	21	.13	9	18	11	15
14			68	100	46	88	28	66	16	43	8	22	.14	10	19	12	16
15			78	100	51	91	31	69	18	44	9	24	.15	10	20	13	17
16					56	94	34	72	20	46	9	25	.16	11	21	14	18
17					62	97	37	75	21	48	10	26	.17	12	22	15	19
18					69	99	40	77	23	50	11	27	.18	13	23	16	21
19					75	100	44	80	25	53	12	28	.19	14	24	17	22
20					83	100	47	83	27	55	13	29	.20	15	26	18	23
21							50	85	28	57	14	30	.21	16	27	19	24
22							54	88	30	59	14	31	.22	17	28	19	25
23							57	90	32	61	15	32	.23	18	29	20	26
24							61	92	34	63	16	33	.24	19	30	21	27
25							65	94	36	64	17	35	.25	20	31	22	28
26							69	96	37	66	18	36	.26	20	32	23	29
27							73	98	39	68	19	37	.27	21	33	24	30
28							78	99	41	70	19	38	.28	22	34	25	31
29							83	100	43	72	20	39	.29	23	35	26	32
30							88	100	45	73	21	40	.30	24	36	27	33
31									47	75	22	41	.31	25	37	28	34
32									50	77	23	42	.32	26	38	29	35
33									52	79	24	43	.33	27	39	30	36
34									54	80	25	44	.34	28	40	31	37
35									56	82	26	45	.35	29	41	32	38
36									57	84	27	46	.36	30	42	33	39
37									59	85	28	47	.37	31	43	34	40
38									62	87	28	48	.38	32	44	35	41
39									64	88	29	49	.39	33	45	36	42
40									66	90	30	50	.40	34	46	37	43
41									69	91	31	51	.41	35	47	38	44
42									71	93	32	52	.42	36	48	39	45
43									73	94	33	53	.43	37	49	40	46
44									76	95	34	54	.44	38	50	41	47
45									78	97	35	55	.45	39	51	42	48
46									81	98	36	56	.46	40	52	43	49
47									83	99	37	57	.47	41	53	44	50
48									86	100	38	58	.48	42	54	45	51
49									89	100	39	59	.49	43	55	46	52
50									93	100	40	60	.50	44	56	47	53
												*		†		†	

*If X exceeds 50, read $100 - X$ = number observed, and subtract each confidence limit from 100.

†If X/n exceeds .50, read $1.00 - X/n$ = fraction observed, and subtract each confidence limit from 100.

[1]Table VII is reproduced by permission from *Statistical Methods* by George W. Snedecor, fifth edition, © 1956 by Iowa State University Press, Ames, Iowa.

TABLE VIII   Percentage Points of the F Distribution[1]

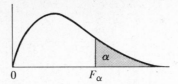

F DISTRIBUTION: .05 POINTS

ν_1 / ν_2	1	2	3	4	5	6	7	8	9
1	161.45	199.50	215.71	224.58	230.16	233.99	236.77	238.88	240.54
2	18.513	19.000	19.164	19.247	19.296	19.330	19.353	19.371	19.385
3	10.128	9.5521	9.2766	9.1172	9.0135	8.9406	8.8868	8.8452	8.8123
4	7.7086	6.9443	6.5914	6.3883	6.2560	6.1631	6.0942	6.0410	5.9988
5	6.6079	5.7861	5.4095	5.1922	5.0503	4.9503	4.8759	4.8183	4.7725
6	5.9874	5.1433	4.7571	4.5337	4.3874	4.2839	4.2066	4.1468	4.0990
7	5.5914	4.7374	4.3468	4.1203	3.9715	3.8660	3.7870	3.7257	3.6767
8	5.3177	4.4590	4.0662	3.8378	3.6875	3.5806	3.5005	3.4381	3.3881
9	5.1174	4.2565	3.8626	3.6331	3.4817	3.3738	3.2927	3.2296	3.1789
10	4.9646	4.1028	3.7083	3.4780	3.3258	3.2172	3.1355	3.0717	3.0204
11	4.8443	3.9823	3.5874	3.3567	3.2039	3.0946	3.0123	2.9480	2.8962
12	4.7472	3.8853	3.4903	3.2592	3.1059	2.9961	2.9134	2.8486	2.7964
13	4.6672	3.8056	3.4105	3.1791	3.0254	2.9153	2.8321	2.7669	2.7144
14	4.6001	3.7389	3.3439	3.1122	2.9582	2.8477	2.7642	2.6987	2.6458
15	4.5431	3.6823	3.2874	3.0556	2.9013	2.7905	2.7066	2.6408	2.5876
16	4.4940	3.6337	3.2389	3.0069	2.8524	2.7413	2.6572	2.5911	2.5377
17	4.4513	3.5915	3.1968	2.9647	2.8100	2.6987	2.6143	2.5480	2.4943
18	4.4139	3.5546	3.1599	2.9277	2.7729	2.6613	2.5767	2.5102	2.4563
19	4.3808	3.5219	3.1274	2.8951	2.7401	2.6283	2.5435	2.4768	2.4227
20	4.3513	3.4928	3.0984	2.8661	2.7109	2.5990	2.5140	2.4471	2.3928
21	4.3248	3.4668	3.0725	2.8401	2.6848	2.5757	2.4876	2.4205	2.3661
22	4.3009	3.4434	3.0491	2.8167	2.6613	2.5491	2.4638	2.3965	2.3419
23	4.2793	3.4221	3.0280	2.7955	2.6400	2.5277	2.4422	2.3748	2.3201
24	4.2597	3.4028	3.0088	2.7763	2.6207	2.5082	2.4226	2.3551	2.3002
25	4.2417	3.3852	2.9912	2.7587	2.6030	2.4904	2.4047	2.3371	2.2821
26	4.2252	3.3690	2.9751	2.7426	2.5868	2.4741	2.3883	2.3205	2.2655
27	4.2100	3.3541	2.9604	2.7278	2.5719	2.4591	2.3732	2.3053	2.2501
28	4.1960	3.3404	2.9467	2.7141	2.5581	2.4453	2.3593	2.2913	2.2360
29	4.1830	3.3277	2.9340	2.7014	2.5454	2.4324	2.3463	2.2782	2.2229
30	4.1709	3.3158	2.9223	2.6896	2.5336	2.4205	2.3343	2.2662	2.2107
40	4.0848	3.2317	2.8387	2.6060	2.4495	2.3359	2.2490	2.1802	2.1240
60	4.0012	3.1504	2.7581	2.5252	2.3683	2.2540	2.1665	2.0970	2.0401
120	3.9201	3.0718	2.6802	2.4472	2.2900	2.1750	2.0867	2.0164	1.9588
∞	3.8415	2.9957	2.6049	2.3719	2.2141	2.0986	2.0096	1.9384	1.8799

(Continued)

TABLE VIII (*Continued*)

F DISTRIBUTION: .05 POINTS

ν_2 \ ν_1	10	12	15	20	24	30	40	60	120	∞
1	241.88	243.91	245.95	248.01	249.05	250.09	251.14	252.20	253.25	254.32
2	19.396	19.413	19.429	19.446	19.454	19.462	19.471	19.479	19.487	19.496
3	8.7855	8.7446	8.7029	8.6602	8.6385	8.6166	8.5944	8.5720	8.5494	8.5265
4	5.9644	5.9117	5.8578	5.8025	5.7744	5.7459	5.7170	5.6878	5.6581	5.6281
5	4.7351	4.6777	4.6188	4.5581	4.5272	4.4957	4.4638	4.4314	4.3984	4.3650
6	4.0600	3.9999	3.9381	3.8742	3.8415	3.8082	3.7743	3.7398	3.7047	3.6688
7	3.6365	3.5747	3.5108	3.4445	3.4105	3.3758	3.3404	3.3043	3.2674	3.2298
8	3.3472	3.2840	3.2184	3.1503	3.1152	3.0794	3.0428	3.0053	2.9669	2.9276
9	3.1373	3.0729	3.0061	2.9365	2.9005	2.8637	2.8259	2.7872	2.7475	2.7067
10	2.9782	2.9130	2.8450	2.7740	2.7372	2.6996	2.6609	2.6211	2.5801	2.5379
11	2.8536	2.7876	2.7186	2.6464	2.6090	2.5705	2.5309	2.4901	2.4480	2.4045
12	2.7534	2.6866	2.6169	2.5436	2.5055	2.4663	2.4259	2.3842	2.3410	2.2962
13	2.6710	2.6037	2.5331	2.4589	2.4202	2.3803	2.3392	2.2966	2.2524	2.2064
14	2.6021	2.5342	2.4630	2.3879	2.3487	2.3082	2.2664	2.2230	2.1778	2.1307
15	2.5437	2.4753	2.4035	2.3275	2.2878	2.2468	2.2043	2.1601	2.1141	2.0658
16	2.4935	2.4247	2.3522	2.2756	2.2354	2.1938	2.1507	2.1058	2.0589	2.0096
17	2.4499	2.3807	2.3077	2.2304	2.1898	2.1477	2.1040	2.0584	2.0107	1.9604
18	2.4117	2.3421	2.2686	2.1906	2.1497	2.1071	2.0629	2.0166	1.9681	1.9168
19	2.3779	2.3080	2.2341	2.1555	2.1141	2.0712	2.0264	1.9796	1.9302	1.8780
20	2.3479	2.2776	2.2033	2.1242	2.0825	2.0391	1.9938	1.9464	1.8963	1.8432
21	2.3210	2.2504	2.1757	2.0960	2.0540	2.0102	1.9645	1.9165	1.8657	1.8117
22	2.2967	2.2258	2.1508	2.0707	2.0283	1.9842	1.9380	1.8895	1.8380	1.7831
23	2.2747	2.2036	2.1282	2.0476	2.0050	1.9605	1.9139	1.8649	1.8128	1.7570
24	2.2547	2.1834	2.1077	2.0267	1.9838	1.9390	1.8920	1.8424	1.7897	1.7331
25	2.2365	2.1649	2.0889	2.0075	1.9643	1.9192	1.8718	1.8217	1.7684	1.7110
26	2.2197	2.1479	2.0716	1.9898	1.9464	1.9010	1.8533	1.8027	1.7488	1.6906
27	2.2043	2.1323	2.0558	1.9736	1.9299	1.8842	1.8361	1.7851	1.7307	1.6717
28	2.1900	2.1179	2.0411	1.9586	1.9147	1.8687	1.8203	1.7689	1.7138	1.6541
29	2.1768	2.1045	2.0275	1.9446	1.9005	1.8543	1.8055	1.7537	1.6981	1.6377
30	2.1646	2.0921	2.0148	1.9317	1.8874	1.8409	1.7918	1.7396	1.6835	1.6223
40	2.0772	2.0035	1.9245	1.8389	1.7929	1.7444	1.6928	1.6373	1.5766	1.5089
60	1.9926	1.9174	1.8364	1.7480	1.7001	1.6491	1.5943	1.5343	1.4673	1.3893
120	1.9105	1.8337	1.7505	1.6587	1.6084	1.5543	1.4952	1.4290	1.3519	1.2539
∞	1.8307	1.7522	1.6664	1.5705	1.5173	1.4591	1.3940	1.3180	1.2214	1.0000

(*Continued*)

TABLE VIII   (*Continued*)

F DISTRIBUTION: .025 POINTS

ν_1 / ν_2	1	2	3	4	5	6	7	8	9
1	647.79	799.50	864.16	899.58	921.85	937.11	948.22	956.66	963.28
2	38.506	39.000	39.165	39.248	29.298	39.331	39.355	39.373	39.387
3	17.443	16.044	15.439	15.101	14.885	14.735	14.624	14.540	14.473
4	12.218	10.649	9.9792	9.6045	9.3645	9.1973	9.0741	8.9796	8.9047
5	10.007	8.4336	7.7636	7.3879	7.1464	6.9777	6.8531	6.7572	6.6810
6	8.8131	7.2598	6.5988	6.2272	5.9876	5.8197	5.6955	5.5996	5.5234
7	8.0727	6.5415	5.8898	5.5226	5.2852	5.1186	4.9949	4.8994	4.8232
8	7.5709	6.0595	5.4160	5.0526	4.8173	4.6517	4.5286	4.4332	4.3572
9	7.2093	5.7147	5.0781	4.7181	4.4844	4.3197	4.1971	4.1020	4.0260
10	6.9367	5.4564	4.8256	4.4683	4.2361	4.0721	3.9498	3.8549	3.7790
11	6.7241	5.2559	4.6300	4.2751	4.0440	3.8807	3.7586	3.6638	3.5879
12	6.5538	5.0959	4.4742	4.1212	3.8911	3.7283	3.6065	3.5118	3.4358
13	6.4143	4.9653	4.3472	3.9959	3.7667	3.6043	3.4827	3.3880	3.3120
14	6.2979	4.8567	4.2417	3.8919	3.6634	3.5014	3.3799	3.2853	3.2093
15	6.1995	4.7650	4.1528	3.8043	3.5764	3.4147	3.2934	3.1987	3.1227
16	6.1151	4.6867	4.0768	3.7294	3.5021	3.3406	3.2194	3.1248	3.0488
17	6.0420	4.6189	4.0112	3.6648	3.4379	3.2767	3.1556	3.0610	2.9849
18	5.9781	4.5597	3.9539	3.6083	3.3820	3.2209	3.0999	3.0053	2.9291
19	5.9216	4.5075	3.9034	3.5587	3.3327	3.1718	3.0509	2.9563	2.8800
20	5.8715	4.4613	3.8587	3.5147	3.2891	3.1283	3.0074	2.9128	2.8365
21	5.8266	4.4199	3.8188	3.4754	3.2501	3.0895	2.9686	2.8740	2.7977
22	5.7863	4.3828	3.7829	3.4401	3.2151	3.0546	2.9338	2.8392	2.7628
23	5.7498	4.3492	3.7505	3.4083	3.1835	3.0232	2.9024	2.8077	2.7313
24	5.7167	4.3187	3.7211	3.3794	3.1548	2.9946	2.8738	2.7791	2.7027
25	5.6864	4.2909	3.6943	3.3530	3.1287	2.9685	2.8478	2.7531	2.6766
26	5.6586	4.2655	3.6697	3.3289	3.1048	2.9447	2.8240	2.7293	2.6528
27	5.6331	4.2421	3.6472	3.3067	3.0828	2.9228	2.8021	2.7074	2.6309
28	5.6096	4.2205	3.6264	3.2863	3.0625	2.9027	2.7820	2.6872	2.6106
29	5.5878	4.2006	3.6072	3.2674	3.0438	2.8840	2.7633	2.6686	2.5919
30	5.5675	4.1821	3.5894	3.2499	3.0265	2.8667	2.7460	2.6513	2.5746
40	5.4239	4.0510	3.4633	3.1261	2.9037	2.7444	2.6238	2.5289	2.4519
60	5.2857	3.9253	3.3425	3.0077	2.7863	2.6274	2.5068	2.4117	2.3344
120	5.1524	3.8046	3.2270	2.8943	2.6740	2.5154	2.3948	2.2994	2.2217
∞	5.0239	3.6889	3.1161	2.7858	2.5665	2.4082	2.2875	2.1918	2.1136

(*Continued*)

TABLE VIII   (*Continued*)

F DISTRIBUTION: .025 POINTS

ν_1 ν_2	10	12	15	20	24	30	40	60	120	∞
1	968.63	976.71	984.87	993.10	997.25	1001.4	1005.6	1009.8	1014.0	1018.3
2	39.398	39.415	39.431	39.448	39.456	39.465	39.473	39.481	39.490	39.498
3	14.419	14.337	14.253	14.167	14.124	14.081	14.037	13.992	13.947	13.902
4	8.8439	8.7512	8.6565	8.5599	8.5109	8.4613	8.4111	8.3604	8.3092	8.2573
5	6.6192	6.5246	6.4277	6.3285	6.2780	6.2269	6.1751	6.1225	6.0693	6.0153
6	5.4613	5.3662	5.2687	5.1684	5.1172	5.0652	5.0125	4.9589	4.9045	4.8491
7	4.7611	4.6658	4.5678	4.4667	4.4150	4.3624	4.3089	4.2544	4.1989	4.1423
8	4.2951	4.1997	4.1012	3.9995	3.9472	3.8940	3.8398	3.7844	3.7279	3.6702
9	3.9639	3.8682	3.7694	3.6669	3.6142	3.5604	3.5055	3.4493	3.3918	3.3329
10	3.7168	3.6209	3.5217	3.4186	3.3654	3.3110	3.2554	3.1984	3.1399	3.0798
11	3.5257	3.4296	3.3299	3.2261	3.1725	3.1176	3.0613	3.0035	2.9441	2.8828
12	3.3736	3.2773	3.1772	3.0728	3.0187	2.9633	2.9063	2.8478	2.7874	2.7249
13	3.2497	3.1532	3.0527	2.9477	2.8932	2.8373	2.7797	2.7204	2.6590	2.5955
14	3.1469	3.0501	2.9493	2.8437	2.7888	2.7324	2.6742	2.6142	2.5519	2.4872
15	3.0602	2.9633	2.8621	2.7559	2.7006	2.6437	2.5850	2.5242	2.4611	2.3953
16	2.9862	2.8890	2.7875	2.6808	2.6252	2.5678	2.5085	2.4471	2.3831	2.3163
17	2.9222	2.8249	2.7230	2.6158	2.5598	2.5021	2.4422	2.3801	2.3153	2.2474
18	2.8664	2.7689	2.6667	2.5590	2.5027	2.4445	2.3842	2.3214	2.2558	2.1869
19	2.8173	2.7196	2.6171	2.5089	2.4523	2.3937	2.3329	2.2695	2.2032	2.1333
20	2.7737	2.6758	2.5731	2.4645	2.4076	2.3486	2.2873	2.2234	2.1562	2.0853
21	2.7348	2.6368	2.5338	2.4247	2.3675	2.3082	2.2465	2.1819	2.1141	2.0422
22	2.6998	2.6017	2.4984	2.3890	2.3315	2.2718	2.2097	2.1446	2.0760	2.0032
23	2.6682	2.5699	2.4665	2.3567	2.2989	2.2389	2.1763	2.1107	2.0415	1.9677
24	2.6396	2.5412	2.4374	2.3273	2.2693	2.2090	2.1460	2.0799	2.0099	1.9353
25	2.6135	2.5149	2.4110	2.3005	2.2422	2.1816	2.1183	2.0517	1.9811	1.9055
26	2.5895	2.4909	2.3867	2.2759	2.2174	2.1565	2.0928	2.0257	1.9545	1.8781
27	2.5676	2.4688	2.3644	2.2533	2.1946	2.1334	2.0693	2.0018	1.9299	1.8527
28	2.5473	2.4484	2.3438	2.2324	2.1735	2.1121	2.0477	1.9796	1.9072	1.8291
29	2.5286	2.4295	2.3248	2.2131	2.1540	2.0923	2.0276	1.9591	1.8861	1.8072
30	2.5112	2.4120	2.3072	2.1952	2.1359	2.0739	2.0089	1.9400	1.8664	1.7867
40	2.3882	2.2882	2.1819	2.0677	2.0069	1.9429	1.8752	1.8028	1.7242	1.6371
60	2.2702	2.1692	2.0613	1.9445	1.8817	1.8152	1.7440	1.6668	1.5810	1.4822
120	2.1570	2.0548	1.9450	1.8249	1.7597	1.6899	1.6141	1.5299	1.4327	1.3104
∞	2.0483	1.9447	1.8326	1.7085	1.6402	1.5660	1.4835	1.3883	1.2684	1.0000

(*Continued*)

TABLE VIII   (*Continued*)

F DISTRIBUTION: .01 POINTS

ν_2 \ ν_1	1	2	3	4	5	6	7	8	9
1	4052.2	4999.5	5403.3	5624.6	5763.7	5859.0	5928.3	5981.6	6022.5
2	98.503	99.000	99.166	99.249	99.299	99.332	99.356	99.374	99.388
3	34.116	30.817	29.457	28.710	28.237	27.911	27.672	27.489	27.345
4	21.198	18.000	16.694	15.977	15.522	15.207	14.976	14.799	14.659
5	16.258	13.274	12.060	11.392	10.967	10.672	10.456	10.289	10.158
6	13.745	10.925	9.7795	9.1483	8.7459	8.4661	8.2600	8.1016	7.9761
7	12.246	9.5466	8.4513	7.8467	7.4604	7.1914	6.9928	6.8401	6.7188
8	11.259	8.6491	7.5910	7.0060	6.6318	6.3707	6.1776	6.0289	5.9106
9	10.561	8.0215	6.9919	6.4221	6.0569	5.8018	5.6129	5.4671	5.3511
10	10.044	7.5594	6.5523	5.9943	5.6363	5.3858	5.2001	5.0567	4.9424
11	9.6460	7.2057	6.2167	5.6683	5.3160	5.0692	4.8861	4.7445	4.6315
12	9.3302	6.9266	5.9526	5.4119	5.0643	4.8206	4.6395	4.4994	4.3875
13	9.0738	6.7010	5.7394	5.2053	4.8616	4.6204	4.4410	4.3021	4.1911
14	8.8616	6.5149	5.5639	5.0354	4.6950	4.4558	4.2779	4.1399	4.0297
15	8.6831	6.3589	5.4170	4.8932	4.5556	4.3183	4.1415	4.0045	3.8948
16	8.5310	6.2262	5.2922	4.7726	4.4374	4.2016	4.0259	3.8896	3.7804
17	8.3997	6.1121	5.1850	4.6690	4.3359	4.1015	3.9267	3.7910	3.6822
18	8.2854	6.0129	5.0919	4.5790	4.2479	4.0146	3.8406	3.7054	3.5971
19	8.1850	5.9259	5.0103	4.5003	4.1708	3.9386	3.7653	3.6305	3.5225
20	8.0960	5.8489	4.9382	4.4307	4.1027	3.8714	3.6987	3.5644	3.4567
21	8.0166	5.7804	4.8740	4.3688	4.0421	3.8117	3.6396	3.5056	3.3981
22	7.9454	5.7190	4.8166	4.3134	3.9880	3.7583	3.5867	3.4530	3.3458
23	7.8811	5.6637	4.7649	4.2635	3.9392	3.7102	3.5390	3.4057	3.2986
24	7.8229	5.6136	4.7181	4.2184	3.8951	3.6667	3.4959	3.3629	3.2560
25	7.7698	5.5680	4.6755	4.1774	3.8550	3.6272	3.4568	3.3239	3.2172
26	7.7213	5.5263	4.6366	4.1400	3.8183	3.5911	3.4210	3.2884	3.1818
27	7.6767	5.4881	4.6009	4.1056	3.7848	3.5580	3.3882	3.2558	3.1494
28	7.6356	5.4529	4.5681	4.0740	3.7539	3.5276	3.3581	3.2259	3.1195
29	7.5976	5.4205	4.5378	4.0449	3.7254	3.4995	3.3302	3.1982	3.0920
30	7.5625	5.3904	4.5097	4.0179	3.6990	3.4735	3.3045	3.1726	3.0665
40	7.3141	5.1785	4.3126	3.8283	3.5138	3.2910	3.1238	2.9930	2.8876
60	7.0771	4.9774	4.1259	3.6491	3.3389	3.1187	2.9530	2.8233	2.7185
120	6.8510	4.7865	3.9493	3.4796	3.1735	2.9559	2.7918	2.6629	2.5586
∞	6.6349	4.6052	3.7816	3.3192	3.0173	2.8020	2.6393	2.5113	2.4073

(*Continued*)

TABLE VIII (*Continued*)

F DISTRIBUTION: .01 POINTS

ν_1 / ν_2	10	12	15	20	24	30	40	60	120	∞
1	6055.8	6106.3	6157.3	6208.7	6234.6	6260.7	6286.8	6313.0	6339.4	6366.0
2	99.399	99.416	99.432	99.449	99.458	99.466	99.474	99.483	99.491	99.501
3	27.229	27.052	26.872	26.690	26.598	26.505	26.411	26.316	26.221	26.125
4	14.546	14.374	14.198	14.020	13.929	13.838	13.745	13.652	13.558	13.463
5	10.051	9.8883	9.7222	9.5527	9.4665	9.3793	9.2912	9.2020	9.1118	9.0204
6	7.8741	7.7183	7.5590	7.3958	7.3127	7.2285	7.1432	7.0568	6.9690	6.8801
7	6.6201	6.4691	6.3143	6.1554	6.0743	5.9921	5.9084	5.8236	5.7372	5.6495
8	5.8143	5.6668	5.5151	5.3591	5.2793	5.1981	5.1156	5.0316	4.9460	4.8588
9	5.2565	5.1114	4.9621	4.8080	4.7290	4.6486	4.5667	4.4831	4.3978	4.3105
10	4.8492	4.7059	4.5582	4.4054	4.3269	4.2469	4.1653	4.0819	3.9965	3.9090
11	4.5393	4.3974	4.2509	4.0990	4.0209	3.9411	3.8596	3.7761	3.6904	3.6025
12	4.2961	4.1553	4.0096	3.8584	3.7805	3.7008	3.6192	3.5355	3.4494	3.3608
13	4.1003	3.9603	3.8154	3.6646	3.5868	3.5070	3.4253	3.3413	3.2548	3.1654
14	3.9394	3.8001	3.6557	3.5052	3.4274	3.3476	3.2656	3.1813	3.0942	3.0040
15	3.8049	3.6662	3.5222	3.3719	3.2940	3.2141	3.1319	3.0471	2.9595	2.8684
16	3.6909	3.5527	3.4089	3.2588	3.1808	3.1007	3.0182	2.9330	2.8447	2.7528
17	3.5931	3.4552	3.3117	3.1615	3.0835	3.0032	2.9205	2.8348	2.7459	2.6530
18	3.5082	3.3706	3.2273	3.0771	2.9990	2.9185	2.8354	2.7493	2.6597	2.5660
19	3.4338	3.2965	3.1533	3.0031	2.9249	2.8422	2.7608	2.6742	2.5839	2.4893
20	3.3682	3.2311	3.0880	2.9377	2.8594	2.7785	2.6947	2.6077	2.5168	2.4212
21	3.3098	3.1729	3.0299	2.8796	2.8011	2.7200	2.6359	2.5484	2.4568	2.3603
22	3.2576	3.1209	2.9780	2.8274	2.7488	2.6675	2.5831	2.4951	2.4029	2.3055
23	3.2106	3.0740	2.9311	2.7805	2.7017	2.6202	2.5355	2.4471	2.3542	2.2559
24	3.1681	3.0316	2.8887	2.7380	2.6591	2.5773	2.4923	2.4035	2.3099	2.2107
25	3.1294	2.9931	2.8502	2.6993	2.6203	2.5383	2.4530	2.3637	2.2695	2.1694
26	3.0941	2.9579	2.8150	2.6640	2.5848	2.5026	2.4170	2.3273	2.2325	2.1315
27	3.0618	2.9256	2.7827	2.6316	2.5522	2.4699	2.3840	2.2938	2.1984	2.0965
28	3.0320	2.8959	2.7530	2.6017	2.5223	2.4397	2.3535	2.2629	2.1670	2.0642
29	3.0045	2.8685	2.7256	2.5742	2.4946	2.4118	2.3253	2.2344	2.1378	2.0342
30	2.9791	2.8431	2.7002	2.5487	2.4689	2.3860	2.2992	2.2079	2.1107	2.0062
40	2.8005	2.6648	2.5216	2.3689	2.2880	2.2034	2.1142	2.0194	1.9172	1.8047
60	2.6318	2.4961	2.3523	2.1978	2.1154	2.0285	1.9360	1.8363	1.7263	1.6006
120	2.4721	2.3363	2.1915	2.0346	1.9500	1.8600	1.7628	1.6557	1.5330	1.3805
∞	2.3209	2.1848	2.0385	1.8783	1.7908	1.6964	1.5923	1.4730	1.3246	1.0000

(*Continued*)

TABLE VIII   (*Continued*)

F DISTRIBUTION: .005 POINTS

ν_1 / ν_2	1	2	3	4	5	6	7	8	9
1	16211	20000	21615	22500	23056	23437	23715	23925	24091
2	198.50	199.00	199.17	199.25	199.30	199.33	199.36	199.37	199.39
3	55.552	49.799	47.467	46.195	45.392	44.838	44.434	44.126	43.882
4	31.333	26.284	24.259	23.155	22.456	21.975	21.622	21.352	21.139
5	22.785	18.314	16.530	15.556	14.940	14.513	14.200	13.961	13.772
6	18.635	14.544	12.917	12.028	11.464	11.073	10.786	10.566	10.391
7	16.236	12.404	10.882	10.050	9.5221	9.1554	8.8854	8.6781	8.5138
8	14.688	11.042	9.5965	8.8051	8.3018	7.9520	7.6942	7.4960	7.3386
9	13.614	10.107	8.7171	7.9559	7.4711	7.1338	6.8849	6.6933	6.5411
10	12.826	9.4270	8.0807	7.3428	6.8723	6.5446	6.3025	6.1159	5.9676
11	12.226	8.9122	7.6004	6.8809	6.4217	6.1015	5.8648	5.6821	5.5368
12	11.754	8.5096	7.2258	6.5211	6.0711	5.7570	5.5245	5.3451	5.2021
13	11.374	8.1865	6.9257	6.2335	5.7910	5.4819	5.2529	5.0761	4.9351
14	11.060	7.9217	6.6803	5.9984	5.5623	5.2574	5.0313	4.8566	4.7173
15	10.798	7.7008	6.4760	5.8029	5.3721	5.0708	4.8473	4.6743	4.5364
16	10.575	7.5138	6.3034	5.6378	5.2117	4.9134	4.6920	4.5207	4.3838
17	10.384	7.3536	6.1556	5.4967	5.0746	4.7789	4.5594	4.3893	4.2535
18	10.218	7.2148	6.0277	5.3746	4.9560	4.6627	4.4448	4.2759	4.1410
19	10.073	7.0935	5.9161	5.2681	4.8526	4.5614	4.3448	4.1770	4.0428
20	9.9439	6.9865	5.8177	5.1743	4.7616	4.4721	4.2569	4.0900	3.9564
21	9.8295	6.8914	5.7304	5.0911	4.6808	4.3931	4.1789	4.0128	3.8799
22	9.7271	6.8064	5.6524	5.0168	4.6088	4.3225	4.1094	3.9440	3.8116
23	9.6348	6.7300	5.5823	4.9500	4.5441	4.2591	4.0469	3.8822	3.7502
24	9.5513	6.6610	5.5190	4.8898	4.4857	4.2019	3.9905	3.8264	3.6949
25	9.4753	6.5982	5.4615	4.8351	4.4327	4.1500	3.9394	3.7758	3.6447
26	9.4059	6.5409	5.4091	4.7852	4.3844	4.1027	3.8928	3.7297	3.5989
27	9.3423	6.4885	5.3611	4.7396	4.3402	4.0594	3.8501	3.6875	3.5571
28	9.2838	6.4403	5.3170	4.6977	4.2996	4.0197	3.8110	3.6487	3.5186
29	9.2297	6.3958	5.2764	4.6591	4.2622	3.9830	3.7749	3.6130	3.4832
30	9.1797	6.3547	5.2388	4.6233	4.2276	3.9492	3.7416	3.5801	3.4505
40	8.8278	6.0664	4.9759	4.3738	3.9860	3.7129	3.5088	3.3498	3.2220
60	8.4946	5.7950	4.7290	4.1399	3.7600	3.4918	3.2911	3.1344	3.0083
120	8.1790	5.5393	4.4973	3.9207	3.5482	3.2849	3.0874	2.9330	2.8083
∞	7.8794	5.2983	4.2794	3.7151	3.3499	3.0913	2.8968	2.7444	2.6210

(Continued)

TABLE VIII　(*Continued*)

F DISTRIBUTION: .005 POINTS

ν_1　ν_2	10	12	15	20	24	30	40	60	120	∞
1	24224	24426	24630	24836	24940	25044	25148	25253	25359	25465
2	199.40	199.42	199.43	199.45	199.46	199.47	199.47	199.48	199.49	199.51
3	43.686	43.387	43.085	42.778	42.622	42.466	42.308	42.149	41.989	41.829
4	20.967	20.705	20.438	20.167	20.030	19.892	19.752	19.611	19.468	19.325
5	13.618	13.384	13.146	12.903	12.780	12.656	12.530	12.402	12.274	12.144
6	10.250	10.034	9.8140	9.5888	9.4741	9.3583	9.2408	9.1219	9.0015	8.8793
7	8.3803	8.1764	7.9678	7.7540	7.6450	7.5345	7.4225	7.3088	7.1933	7.0760
8	7.2107	7.0149	6.8143	6.6082	6.5029	6.3961	6.2875	6.1772	6.0649	5.9505
9	6.4171	6.2274	6.0325	5.8318	5.7292	5.6248	5.5186	5.4104	5.3001	5.1875
10	5.8467	5.6613	5.4707	5.2740	5.1732	5.0705	4.9659	4.8592	4.7501	4.6385
11	5.4182	5.2363	5.0489	4.8552	4.7557	4.6543	4.5508	4.4450	4.3367	4.2256
12	5.0855	4.9063	4.7214	4.5299	4.4315	4.3309	4.2282	4.1229	4.0149	3.9039
13	4.8199	4.6429	4.4600	4.2703	4.1726	4.0727	3.9704	3.8655	3.7577	3.6465
14	4.6034	4.4281	4.2468	4.0585	3.9614	3.8619	3.7600	3.6553	3.5473	3.4359
15	4.4236	4.2498	4.0698	3.8826	3.7859	3.6867	3.5850	3.4803	3.3722	3.2602
16	4.2719	4.0994	3.9205	3.7342	3.6378	3.5388	3.4372	3.3324	3.2240	3.1115
17	4.1423	3.9709	3.7929	3.6073	3.5112	3.4124	3.3107	3.2058	3.0971	2.9839
18	4.0305	3.8599	3.6827	3.4977	3.4017	3.3030	3.2014	3.0962	2.9871	2.8732
19	3.9329	3.7631	3.5866	3.4020	3.3062	3.2075	3.1058	3.0004	2.8908	2.7762
20	3.8470	3.6779	3.5020	3.3178	3.2220	3.1234	3.0215	2.9159	2.8058	2.6904
21	3.7709	3.6024	3.4270	3.2431	3.1474	3.0488	2.9467	2.8408	2.7302	2.6140
22	3.7030	3.5350	3.3600	3.1764	3.0807	2.9821	2.8799	2.7736	2.6625	2.5455
23	3.6420	3.4745	3.2999	3.1165	3.0208	2.9221	2.8198	2.7132	2.6016	2.4837
24	3.5870	3.4199	3.2456	3.0624	2.9667	2.8679	2.7654	2.6585	2.5463	2.4276
25	3.5370	3.3704	3.1963	3.0133	2.9176	2.8187	2.7160	2.6088	2.4960	2.3765
26	3.4916	3.3252	3.1515	2.9685	2.8728	2.7738	2.6709	2.5633	2.4501	2.3297
27	3.4499	3.2839	3.1104	2.9275	2.8318	2.7327	2.6296	2.5217	2.4078	2.2867
28	3.4117	3.2460	3.0727	2.8899	2.7941	2.6949	2.5916	2.4834	2.3689	2.2469
29	3.3765	3.2111	3.0379	2.8551	2.7594	2.6601	2.5565	2.4479	2.3330	2.2102
30	3.3440	3.1787	3.0057	2.8230	2.7272	2.6278	2.5241	2.4151	2.2997	2.1760
40	3.1167	2.9531	2.7811	2.5984	2.5020	2.4015	2.2958	2.1838	2.0635	1.9318
60	2.9042	2.7419	2.5705	2.3872	2.2898	2.1874	2.0789	1.9622	1.8341	1.6885
120	2.7052	2.5439	2.3727	2.1881	2.0890	1.9839	1.8709	1.7469	1.6055	1.4311
∞	2.5188	2.3583	2.1868	1.9998	1.8983	1.7891	1.6691	1.5325	1.3637	1.0000

TABLE IX   Percentage Points of the χ^2 Distribution[1]

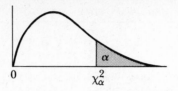

df	PROBABILITY OF A LARGER VALUE:				
	.100	.050	.025	.010	.005
1	2.71	3.84	5.02	6.63	7.88
2	4.61	5.99	7.38	9.21	10.60
3	6.25	7.81	9.35	11.34	12.84
4	7.78	9.49	11.14	13.28	14.86
5	9.24	11.07	12.83	15.09	16.75
6	10.64	12.59	14.45	16.81	18.55
7	12.02	14.07	16.01	18.48	20.28
8	13.36	15.51	17.53	20.09	21.96
9	14.68	16.92	19.02	21.67	23.59
10	15.99	18.31	20.48	23.21	25.19
11	17.28	19.68	21.92	24.72	26.76
12	18.55	21.03	23.34	26.22	28.30
13	19.81	22.36	24.74	27.69	29.82
14	21.06	23.68	26.12	29.14	31.32
15	22.31	25.00	27.49	30.58	32.80
16	23.54	26.30	28.85	32.00	34.27
17	24.77	27.59	30.19	33.41	35.72
18	25.99	28.87	31.53	34.81	37.16
19	27.20	30.14	32.85	36.19	38.58
20	28.41	31.41	34.17	37.57	40.00
21	29.62	32.67	35.48	38.93	41.40
22	30.81	33.92	36.78	40.29	42.80
23	32.01	35.17	38.08	41.64	44.18
24	33.20	36.42	39.36	42.98	45.56
25	34.38	37.65	40.65	44.31	46.93
26	35.56	38.89	41.92	45.64	48.29
27	36.74	40.11	43.19	46.96	49.64
28	37.92	41.34	44.46	48.28	50.99
29	39.09	42.56	45.72	49.59	52.34
30	40.26	43.77	46.98	50.89	53.67
40	51.81	55.76	59.34	63.69	66.77
50	63.17	67.50	71.42	76.15	79.49
60	74.40	79.08	83.30	88.38	91.95
70	85.53	90.53	95.02	100.43	104.22
80	96.58	101.88	106.63	112.33	116.32
90	107.60	113.14	118.14	124.12	128.30
100	118.50	124.34	129.56	135.81	140.17

[1]Table IX is abridged from Thompson, Catherine M.: "Table of Percentage Points of the χ^2 Distribution," *Biometrika*, Vol. 32 (1942), p. 187, by permission of *Biometrika* Trustees.

ANSWERS TO ODD-NUMBERED PROBLEMS

CHAPTER 1

1. Descriptive
3. Probability
5. **a.** Probability   **b.** Inference
9. Inflation and taxes
11. **a.** First island   **b.** Second island
    c. Many products for the old and fewer for the young.
13. **a.** Washington, D.C.   **b.** Other locations are states, not cities.
15. **a.** Induced bias
17. Self-selection is the single biggest problem.
19. Answer is in your current issue of the *Survey of Current Business*.
21. See the *Encyclopedia of Associations*.

CHAPTER 2

1. 7
3. **a.** 8   **b.** 9   **c.** .9   **d.** 8.4 to 17.3
5. **a.** 6   **b.** 23
   c. Boundaries: 55.5, 78.5, 101.5, 124.5, 147.5, 170.5, 193.5
7. **a.** 7   **b.** 25
   c. Boundaries: -.5, .5, 1.5, 2.5, 3.5, 4.5, 5.5, 6.5, 7.5
   d. Marks: 37, 62, 87, 112, 137
9. 6, 14, 18, 12
11. **a.** 3%   **b.** .67
    c. Boundaries: 2219.5, 2445.5, 2671.5, 2897.5, 3123.5, 3349.5
13. 7
15. **a.** Boundaries: 3.715, 4.715, 5.715, . . . , 11.715
    d. 7

CHAPTER 3

1. No, 9
3. **a.** $\sum_{i=1}^{4} X_i Y_i$ **b.** $\sum_{i=1}^{4} Y_i^2 f_i$

   c. $\left(\sum_{i=1}^{3} Y_i \right)^2$ **d.** $\sum_{i=1}^{3} (X_i - Y_i)$

   e. $\sum_{i=1}^{6} (-1)^{i+1} Y_i^i$ **f.** $\sum_{i=1}^{5} (-1)^{i+1} \frac{Y^i}{i}$

5. **a.** 82   **b.** 16   **c.** -13   **d.** 19   **e.** 58   **f.** 32
7. **a.** -1   **b.** 90   **c.** -9   **d.** -11   **e.** -23.5   **f.** 117
9. **a.** 943.23   **b.** 941.5   **c.** 1392   **d.** 127,477.22
   e. 357.04   **f.** 38%   **g.** 351.04
11. **a.** 30.0   **b.** 27.0   **c.** 22.4   **d.** 58.8   **e.** 73.5   **f.** 7.7, 8.6
13. $CV_1 = 12.80\%, CV_2 = 14.65\%$ 15. -1.25, 45,718.9, 213.82
17. 2.54   19. 45, 19.8, 4.45
21. 64.4, 65.56   23. 45.75, 63.83, 7.99, 17%
25. **a.** 595.65 and 650   **b.** 550.35, 42,744, 207
27. 41 to 59
29. **a.** At least 90%   **b.** About 67%
31. About 81%

CHAPTER 4

1. 8, 3, 4   3. 53
5. 720   7. 504
9. 120   11. Over 635 billion
13. 7560   15. **a.** 100   **b.** 210   **c.** 205   **d.** 200
17. 67,200   19. **a.** 120   **b.** 216   **c.** 60

23. **a.** .125   **b.** .250   **c.** .625   25. **a.** .5   **b.** .5   **c.** .25   **d.** .75
27. **a.** .48   **b.** .44   **c.** .08   29. **a.** .090   **b.** .067
31. .0025   33. 2/3
35. **a.** 272   **b.** .29   **c.** .50   37. .44
39. **a.** .48 **b.** .06 **c.** .82 **d.** .30 **e.** .40   41. .47, .53

CHAPTER 5

1. **a.** .50   **b.** .25   **c.** .60   **d.** .15
3. **a.** 5   **b.** .40   **c.** .40   **d.** .60   **e.** 3
5. **a.** 1, 1   **b.** Exceed by 1 second
7. **a.** 25% of the days saw 7 cars unloaded
   b. 7.75   **c.** .94   **d.** 0, not necessarily
9. **a.** .3669   **b.** .0355   **c.** .9827   **d.** .8491
11. **a.** .2500   **b.** .0244   **c.** .9011   **d.** 0
13. 2, 1.095
15. **a.** .0312   **b.** .9687   **c.** .3125
17. **a.** .1216   **b.** .3917
19. **a.** .8042   **b.** .2053   **c.** .5886   **d.** 1.000
21. **a.** .2916   **b.** .6561   **c.** .3439
23. **a.** .0205   **b.** .0000
25. **a.** .0547   **b.** .0116   **c.** Virtually zero
27. **a.** .1143   **b.** .5143
29. **a.** .3263   **b.** .3916   **c.** 0
31. .580
33. **a.** .1687   **b.** .0057   **c.** .3656   **d.** .0776   **e.** .9986   **f.** .1437
35. **a.** .0003   **b.** .4846   **c.** .9872
37. **a.** 2.25   **b.** .0507   **c.** .7038
39. **a.** .2240   **b.** .1606   **c.** .9502
41. .0680
43. **a.** .7734   **b.** .1747   **c.** .3721   **d.** .4013
45. **a.** .8907   **b.** .0170   **c.** .1210   **d.** .7324   **e.** .2017   **f.** .7348
47. **a.** 1.70   **b.** −.50   **c.** −3.55   **d.** 1.85   **e.** .60   **f.** 1.79
49. **a.** .8413   **b.** .3446   **c.** .5670   **d.** .2119
51. **a.** .5636   **b.** .7073
53. 8856, 4650
55. **a.** .3446   **b.** 0   **c.** .7874
57. **a.** .1587   **b.** .1894   **c.** .8490
59. **a.** .1296   **b.** .0614   **c.** .2995

CHAPTER 6

1. **a.** .0646   **b.** .1201   **c.** b
3. **a.** .1839   **b.** .0002   **c.** .1246   **d.** .8627
5. **a.** .1859   **b.** .0733   **c.** .0988
   d. Probability of success changes significantly from trial to trial

7. **a.** $\binom{10}{4}\binom{190}{2}\bigg/\binom{200}{6} = .000046$

   b. $P(r=4 \mid n=6, p=.05) = \binom{6}{4}(.05)^4 (.95)^2$

   c. .0003   **d.** Hypergeometric

9. **a.** $\sum_{r=8}^{12} \binom{1000}{r}\binom{4000}{25-r}\bigg/\binom{5000}{25}$

   b. .1087   **c.** .1056   **d.** .1313

   e. Normal, since p is not too extreme

11. **a.** 40   **b.** .2327   **c.** .9545

13. .0384

15. .4090

17. **a.** Hypergeometric   **b.** Normal   **c.** .0188

CHAPTER 7

1. 10, .16
3. The first is exactly normal. The second is approximately normal.
5. **a.** .0668   **b.** .0808   **c.** No
7. 9992 to 10008
9. .0179
11. **a.** .55   **b.** .0352
13. .0078
15. **a.** 32   **b.** 2.24   **c.** Exactly normal
17. .4176 (Due to rounding errors, the answer may go up to .43)
19. **a.** .08   **b.** .073   21. **a.** 0   **b.** .045   **c.** .3472

CHAPTER 8

1. **a.** 4, 36, 6, 2   **c.** 50, 4, 2, .67   **e.** 20, 16, 4, 1
3. **a.** .20, 4, 2, .0064   **c.** .36, 92.16, 9.6, .00058
5. **a.** 18.355 to 21.645   **b.** 47.348 to 56.652   **c.** 112.16 to 127.84
7. **a.** 2.763   **b.** 2.064   **c.** 1.725
9. **a.** 7.388 to 16.612   **c.** -1.208 to 17.208   **e.** -4.473 to 4.473
11. 82.62 to 117.38
13. 14.977 to 16.623 (assuming $t_{.025,99} \approx Z_{.025}$)
15. $-.869$ to 4.069   17. 810.0 to 1076.4
19. -58.26 to 55.76   21. 532.165 to 568.635
23. 13.349 to 16.651   25. 106.632 to 153.368
27. **a.** .52 to .96   **c.** .79 to .93   **e.** .10 to .14
29. **a.** .073 to .127   **c.** .206 to .794
31. .37 to .73, based on all 29 students and normal approximation to the binomial
33. .54 to .73   35. .061 to .139
37. .201 to .299   39. 67
41. 4595 more   43. 385
45. 9604   47. **a.** -30, 100   **b.** -50.86 to -9.14
49. 1507.2 to 2992.8   51. 887.2 to 8612.8, yes

53. 11.24 to 20.76

55. a. .28, .46, .18   **b.** .057   **c.** .047 to .313

57. −.025 to .125

CHAPTER 9

1. a. .33   **b.** Type I

3. a. Type I: Concluding uncontaminated when cases are contaminated
    Type II: Concluding cases are contaminated when they are not

   b. α should be small

5. a. $t = 57.7$, reject since $t_{.01,15} = 2.602$

   b. $t = -6.33$, reject since $-t_{.025,8} = -2.306$

   c. $t = -7.0$, reject since $-t_{.05,48} = -1.68$ (approximately)

7. $t = 8.0$, reject since $t_{.10,3} = 1.638$

9. $Z = -2.0$, reject since $-Z_{.025} = -1.96$

11. a. $t = 1.5$, do not reject since $t_{.05,8} = 1.860$

   b. $t = 1.5$, reject since $t_{.10,24} = 1.318$

13. $t = 10.33$, reject since $t_{.01,14} = 2.624$

15. $t = -3.59$, reject since $-t_{.05,60} = -1.671$

17. $t = 2.0$, reject since $t_{.10,35} = 1.307$ (approximately)

19. $t = 1.85$, reject since $t_{.05,11} = 1.796$

21. $Z = 2.08$, reject since $Z_{.05} = 1.645$

23. $Z = 2.00$, reject since $Z_{.10} = 1.282$

25. $Z = -1.86$, do not reject since $-Z_{.025} = -1.96$

27. a. $H_0: p \le .05$ and $H_a: p > .05$ **b.** 9

29. $t = -2.50$, reject since $-t_{.05,23} = -1.714$

31. $t = -2.28$, reject since $-t_{.05,22} = -1.717$

33. $t = 1.89$, reject since $t_{.05,14} = 1.761$

35. $t = -1.56$, do not reject since $-t_{.01,12} = -2.681$

37. a. $t = 2.92$, reject since $t_{.05,18} = 1.734$

   b. $t = 3.02$, reject since $t_{.05,9} = 1.833$

39. $t = 1.816$, reject since $t_{.10,13} = 1.350$

41. $Z = -1.37$, do not reject since $-Z_{.025} = -1.96$

43. a. $H_0: p_m \le p_w$, where p_w is proportion of women receiving promotions within two years.

   $H_a: p_m > p_w$

   b. $Z = 1.59$, do not reject since $Z_{.05} = 1.645$

45. a. .3483   **b.** .6966

47. $F = 4.41$, reject since $F_{.05,19,11} = 2.66$ (approximately)

49. $F = 2.25$, do not reject since $F_{.025,9,9} = 4.0260$

CHAPTER 10

1. $F = 4.21$, do not reject since $F_{.01,3,8} = 7.5910$

3. a.

SOURCE	df	SS	MS
Between	2	320	160
Within	27	972	36

b. $F = 4.444$, reject H_0 since $F_{.05,2,27} = 3.3541$

5. a.

SOURCE	df	SS	MS
Between	3	216	72
Within	11	146	13.27

b.

	GROUP:			
	1	*2*	*3*	*4*
$\overline{X}$	9	13	20	15
s	2.94	3.61	2.65	4.47

c. -19.6 to -2.4 *Note*: Use $t_{.005,11} = 3.106$ since the pooled variance estimate has 11 degrees of freedom.

d. $F = 5.4$, reject since $F_{.05,3,11} = 3.5874$

7. $F = 9.9$, reject since $F_{.01,2,18} = 6.0129$

9. 28

11. $k = 2$, $n_1 = 31$, and $n_2 = 41$

13. $t_{.025,70} = 1.997$, by interpolation

15. $t = -.92$, do not reject since $-t_{.025,70} = -1.997$

17. Calculated $t^2 = (-.92)^2 = .84 = $ Calculated F
Table $t^2 = (-1.997)^2 = 3.9880 \simeq 3.9877 = $ Table F

19. $F = 3.59$, do not reject since $F_{.05,1,14} = 4.6001$

21. a. $t = -2.49$, reject since $-t_{.025,18} = -2.101$

b. $F = 6.19$, reject since $F_{.05,1,18} = 4.4139$
Note: $(-2.49)^2 = 6.20$ which is very close to 6.19.

CHAPTER 11

1. Cannot reject assertion "curve" used.
$\chi^2 = 6.8$ and $\chi^2_{.05,4} = 9.49$

3. a. $f_1 = .267, f_2 = .300, f_3 = .433$

b. $\chi^2 = 14.0$, reject since $\chi^2_{.05,2} = 5.99$

5. $\chi^2 = 29.0$, reject since $\chi^2_{.01,6} = 16.81$

7. $\chi^2 = 260.4$, reject since $\chi^2_{.01,3} = 11.34$

9. $\chi^2 = 32.3$, reject since $\chi^2_{.05,2} = 5.99$

11. $\chi^2 = 7.2$, do not reject since $\chi^2_{.005,3} = 12.84$
Still cannot reject when $\alpha = .05$ since $\chi^2_{.05,3} = 7.81$

13. a. $\chi^2 = 0$

b. A chi-square value of zero indicates perfect agreement.

15. The last three classes had to be combined.
$E_0 = 27.3$, $E_1 = 35.4$, $E_2 = 23.0$, $E_{3+} = 14.3$
$\chi^2 = 4.12$, do not reject since $\chi^2_{.025,2} = 7.38$
Note that df $= 2$ since one degree of freedom was used in getting $\theta = 1.3$ from the data.

17. $\overline{X} = 83.75$, $s = 30.03$

$E_1 = 13.14$, $E_2 = 25.46$, $E_3 = 31.95$, $E_4 = 20.93$, $E_5 = 8.53$

$\chi^2 = .68$, do not reject since $\chi^2_{.05,2} = 5.99$

Note that df $= 2$ since two degrees of freedom were lost in obtaining $\bar{X}$ and s from the data.

19. $E_1 = 47.93$, $E_2 = 43.93$, $E_3 = 43.14$, $E_4 = 12.07$, $E_5 = 11.07$, $E_6 = 10.86$

    $\chi^2 = 1.45$, do not reject since $\chi^2_{.05,2} = 5.99$

21. **a.** $Z = -1.784$, do not reject since $-Z_{.025} = -1.96$

    b. $\chi^2 = 3.18$, do not reject since $\chi^2_{.05,1} = 3.84$

    c. $Z^2 = (-1.784)^2 = 3.18 = \chi^2$

23. $\chi^2 = 84.04$, reject since $\chi^2_{.01,4} = 13.28$

25. $\chi^2 = 1.14$, do not reject since $\chi^2_{.01,2} = 9.21$

CHAPTER 12

1. **a.** $\hat{Y} = .2 + .8X$ **b.** 4.2
3. **a.** $\hat{Y} = 2X$ **b.** 16
5. .966
7. **a.** .61   **b.** -1.23 to 1.23
9. **1.** Nonrandom X values
   2. Normally distributed Y values around the regression line
   3. Homoscedasticity
11. **a.** 2.68   **b.** .966
13. **a.** .51   **b.** $t = -11.8$, reject since $-t_{.05,5} = -2.015$
15. $\hat{Y} = -58 + 4X$
17. **a.** $\hat{Y} = 2.5 - 3X$ **b.** 2.5
    c. $t = -6.0$, reject since $-t_{.025,28} = -2.048$
19. **a.** $\hat{Y} = 2.9 - .3X$ **b.** 4.0, .25, .01
    c. $t = -3$, reject since $-t_{.005,23} = -2.807$ **d.** 1.87 to 3.93
21. **a.** 9.0   **b.** 8.3 to 9.7
23. a is in dollars.
    b is in dollars per pound.
    r^2 has no units.
25. **a.** a is in thousands of dollars, and b is in thousands of dollars per hour.
    b. \$139,200   **c.** \$120,000   **d.** \$0
    e. $X = 0$ is outside the range of observed X values.
27. **a.** $\hat{Y} = .0028 + .00657X$ **b.** .00627 to .00688
    c. $t = .26$, do not reject since $t_{.025,18} = 2.101$
29. **a.** $\hat{Y} = 5.9 + .155X$ **b.** .105 to .205
    c. $t = 8.2$, reject since $t_{.005,13} = 3.012$ **d.** 9.7 to 11.4
31. **a.** $\hat{Y} = 17.25 + 1.20X$ **b.** $\hat{Y} = 27.10 + 6.87X$
    c. $t = .94$, do not reject since $t_{.005,14} = 2.977$
    d. r^2 for salary on years is .54
    r^2 for salary on quota is .06
    Thus, "years" is the better predictor variable.
33. **a.** $r = .8$, $t = 4.0$, reject since $t_{.025,9} = 2.262$
    b. $r = .6$, $t = 3.0$, reject since $t_{.025,16} = 2.120$

35. a. $r = -.6, t = -3.0$, reject since $-t_{.025,16} = -2.120$
  b. $r = -.532, t = -3.01$, reject since $-t_{.025,23} = -2.069$
37. a. $t = 2.90$, reject since $t_{.025,6} = 2.447$
  b. $t = 2.88$, reject since $t_{.025,6} = 2.447$
  c. $F = 8.46$, reject since $F_{.05,1,6} = 5.9847$
  d. $2.90 \approx 2.88 \approx \sqrt{8.46} = 2.91$

CHAPTER 13

1. a, c, d
3. b, c
5. A stock of this type with no earnings and no growth is estimated to sell for $.72 per share. A $1 increase in earnings per share is associated with a $5.94 increase in price per share, while a 1% increase in growth is associated with $1.08 increase in price per share, all other things being held constant.
7. .37 to 11.51 The assumptions listed in Section 13.3 must be made.
9. $\beta_1 = 1.14$ and $\beta_2 = .44$ so estimates of Y are more sensitive to changes in X_1.
11. $F = 6.6$, do not reject since $F_{.025,2,3} = 16.044$
13. $S^2_{Y|12} = 271.1$, $R^2 = .88$
15. -14.7 to 39.3%
17. $F = 7.0$, do not reject since $F_{.01,2,2} = 99.000$
19. $\hat{Y} = \$365.70$
21. a. $t = .01$, do not reject since $t_{.10,2} = 1.886$
  b. $t = 1.27$, do not reject since $t_{.10,2} = 1.886$
  c. $F = 5.4$, do not reject since $F_{.05,2,2} = 19.000$
      The superintendent appears to be correct.
23. Note that $r^2_{12} = .84$, and this indicates multicollinearity between X_1 and X_2.
25. a. $F = 49.451$, reject since $F_{.05,5,14} = 2.9582$
  b. $t = -1.8$, do not reject since $-t_{.025,14} = -2.145$
  c. $t = 1.46$, do not reject since $t_{.025,14} = 2.145$
27. a. Age, which has a coefficient of variation of 72.1%
  b. Age, $r^2 = .6749$
  c. Age and frequency. $r^2 = (-.72264)^2 = .52221$
  d. Hours of operation and Frequency. $r^2 = (.00526)^2 = .000028$
29. a. $\hat{Y} = 353.12 + 5.14\text{Age} - 3.01\text{Frequency}$
  b. $S^2_{Y|23} = 14,583$
  c. $R^2_{Y|23} = .70292$
  d. $t = -1.19$, do not reject since $-t_{.025,15} = -2.131$
  e. 353.12 is the predicted repair cost of a pump that is new and has no monthly maintenance. Mean monthly repair costs seem to rise $5.14 per month and drop $3.01 for each added routine maintenance performed each year.
  f. $F = 17.74553$, reject since $F_{.05,2,15} = 3.6823$
  g. 1.874
31. $r_{23} = -.72264$ and indicates that s_b is inflated for the regression equation coefficients on age and frequency.

CHAPTER 14

1. **a.** 393   **b.** 131   **c.** 32   **3.** Many legitimate examples are possible.

5. Many legitimate examples are possible. Suggestions: stock prices of companies whose international operations are nationalized, the Dow-Jones industrial average as it reacts to economic news, the price of gold as it reacts to news, or any one of many others.

7. **b.** Curved

    c. No, seasonal variation takes place *within* one year; these data are annual.

9. **a.** July of the first year   **b.** June of the last year

11.

YEAR	MERGER	YEAR	MERGER
1962	26.13	1970	116.13
1963	28.38	1971	137.00
1964	30.88	1972	148.75
1965	37.38	1973	160.25
1966	44.38	1974	173.00
1967	54.25	1975	191.75
1968	72.63	1976	221.13
1969	92.50		

13. .8 since the closer α is to 1.0, the more responsive the smoothed average is to current data.

15. The smoothed series is identical to the actual series.

17. **b.** $\hat{Y} = 1.011 + .1047t$ **c.** 3.31

19. First Method: 109.0, 96.7, 87.2, 107.1
    Second Method: 121.2, 100.4, 80.3, 98.1

21. The year in question has higher values and a different pattern than the other years. The intercept (and thus the general level of the trend line) will be raised by these data.

23.

JANUARY	0.0	JULY	484.9
FEBRUARY	−59.9	AUGUST	457.4
MARCH	413.0	SEPTEMBER	356.8
APRIL	639.4	OCTOBER	370.0
MAY	587.6	NOVEMBER	182.1
JUNE	558.4	DECEMBER	33.3

    The coefficient on the trend value is 20.2.

25. **a.** Winter 137.6; spring 98.1; summer 68.3; fall 96.0

    b. Winter 0.0; spring −8.35; summer −14.90; fall −10.65

27. Irregular variation. Probably due to an unusually high number of winter snow and ice storms.

29. 117.5, 113.6, 105.6, 91.8, 76.8, 94.5, 102.1, 117.2, 88.3
    No marked cycle can show up in only three years of data.

31. $429.7 billion

33. 17,700; 11,250; 13,500; 17,550

35. **a.** $244,000; $219,444

    b. ($244,000 + $219,444)/2 = $231,722 multiplied by 12 gives $2,780,664

37. **a.** $152,950   **b.** $339,853   **c.** $342,998

    d. The one for 1988 since that date is closer to the present.

e.

JANUARY	$190,988	*JULY*	$346,750
FEBRUARY	$218,880	*AUGUST*	$382,470
MARCH	$212,629	*SEPTEMBER*	$477,651
APRIL	$340,461	*OCTOBER*	$583,832
MAY	$286,226	*NOVEMBER*	$669,551
JUNE	$231,686	*DECEMBER*	$224,960

39. 154,028
41. a. Winter 21,000; spring 59,500; summer 35,000; fall 24,500
  b. 33,333   **c.** $4 \times (33,333) = 133,332$

CHAPTER 15

1. a. 100.0, 101.3, 105.2, 113.0, 121.6
  b. 82.3, 83.3, 86.5, 92.9, 100.0
3. 146.0   **5.** 158.4
7. 160.0   **9. a.** 151.9   **b.** 953.8
11. a. 151.3   **13. a** and **c** only. The rest need current
  b. 281.8   period quantity values.
15. 135.3   **17.** 200.0
19. -5.5%
21. a. Yes, $575,803.80, in 1967 dollars.
  b. 5.5%
  c. $11,044,303.80
  d. $16,540,230
25. a. $54.90   **b.** $D = 3.7432$ **c.** $57.70
27. Consider such things as the items in the two indices' market baskets, the possible differences in the base periods, and the weights that might have been used.

CHAPTER 16

1. $EMV_1 = 1.3$, $EMV_2 = 4.45$, $EMV_3 = 5.30$, which is the highest.
3. a. A_4 is dominated by both A_1 and A_2.
  b. $EMV_1 = 21.5$, $EMV_2 = 22.0$, $EMV_3 = 20.7$, which is the lowest.
5. a. $EMV_2 = 107.5$ **b.** $7,500
  c. Have the county do it and expect a light winter.
  d. Let a contract and insure a maximum cost of 120.
  e. A_2 at 108.33   **f.** A_2 since this is the lowest value in row E_2.
7. a. A_2 at 28   **b.** A_1 at -8 **c.** A_3 at 15   **d.** A_2 at 7.0
9. a. A_3 **b.** A_1 **c.** A_1 **d.** A_3 **e.** 190
11. a. Make 7, EMV = $4075   **b.** 46
13. a. Payoffs are: 0, -3, -3
    Probabilities are: .65, .20, .10, .05
  b. Do it alone has EMV of $5800
  c. Do it alone has EOL of $1600
  d. $1600
  e. $5.8 + 1.6 = 7.4$; $1.45 + 5.95 = 7.4$; $2.85 + 4.55 = 7.4$

15. **a.**

EVENTS	PROBABILITIES	DO NOTHING	RENEW	BUY OUTRIGHT
Rezone	.15	0	85	100
Deny	.30	0	−10	−15
Table	.55	0	5	20

   b. No single action dominates any other action.
   c. $EMV_1 = \$0$; $EMV_2 = \$12,500$; $EMV_3 = \$21,500$
   d. EVPI = $4,500
17. **b.** EMV for remaining is $7,000
      EMV for moving is −$10,000
   c. Remain and expect the shopping mall to be built and net +$43,000.
   d. Remain since the shopping mall has the highest chance of occurrence.
   e. Yes. Events are: shopping mall, amusement park, or no change. Actions are: remain or move.
19. EMV for bid on interchange is $120,000.
      EMV for bid on bridge is $115,000.
21. **b.** EMV for Company A is $525,000.
      EMV for Company B is $1,080,000.
   c. Discounted EMV for Company A is −$591,097.
      Discounted EMV for Company B is $102,930.

CHAPTER 17

1. **a.** OK
   b. OK only if the company's sales cannot possibly rise when the economy is down, i.e., P(sales up|economy down) = 0.
   c. No, P(Smith quits) = .35
   d. No, P(marketing *and* moderate or worse sales) = .42
3. **b.** About .45   **c.** About .30
5. *Hint:* Be conservative!
7. P(agreement) = .40, P (no agreement) = .60
9. **a.** Win job and lose job
   b. P(Win) = .35; P(Lose) = .65
   c. P(Win) = .00; P(Lose) = 1.00
11. The revised probabilities are: .0001, .0838, .3639, .5522. If Appendix Table III, rather than Formula (5.13), is used to compute the conditionals, the answers look more like .0000, .0845, .3634, .5521.
13. $9700
15. No, do the opposite of what the experiment suggests.
17. **a.** .5367   **b.** .2059   **c.** $400   **d.** $1500
19. Expected costs are: rerun a new batch, $2000; Test and decide, $1289; and Ship current batch as is with no test, $900.
21. **b.** EMV for borrow from trust fund is $10,800.
      EMV for borrow from bank is $9570.75.
   c. With revised probabilities, EMV for borrow from bank is $8996.03.
   d. Applying to the bank is still preferred.

CHAPTER 18

1. $W/\sigma_W = -1.29$, do not reject since $-Z_{.025} = -1.96$
3. $y = 6$, $Z = -1$, do not reject since $-Z_{.025} = -1.96$
5. $V/\sigma_V = 1.91$, do not reject since $Z_{.01} = 2.326$
7. Sign test. We have no reason to assume symmetry. Since $n = 6$ (n is reduced by 1 for the seventh analyst), we can calculate α directly as $P(r \geq 5 | n=6, p=.5) = .11$.
9. $W/\sigma_W = -1.87$, reject since $-Z_{.05} = -1.645$
11. $t = -.40$, do not reject since $-t_{.1,10} = -1.372$
13. $V/\sigma_V = .99$, do not reject since $Z_{.025} = 1.96$
15. $\Sigma D^2 = 44$, $r_s = .73$, $t = 3.02$, reject since $t_{.025,8} = 2.306$
17. $\Sigma D^2 = 10.5$, $r_s = .95$, $t = 9.13$, reject since $t_{.01,9} = 2.821$
19. $\Sigma D^2 = 26$, $r_s = .88$, $t = 5.56$, reject since $t_{.05,9} = 1.833$

INDEX